Life-Cycle Modelling for Innovative Products and Processes

IFIP – The International Federation for Information Processing

IFIP was founded in 1960 under the auspices of UNESCO, following the First World Computer Congress held in Paris the previous year. An umbrella organization for societies working in information processing, IFIP's aim is two-fold: to support information processing within its member countries and to encourage technology transfer to developing nations. As its mission statement clearly states,

> IFIP's mission is to be the leading, truly international, apolitical organization which encourages and assists in the development, exploitation and application of information technology for the benefit of all people.

IFIP is a non-profitmaking organization, run almost solely by 2500 volunteers. It operates through a number of technical committees, which organize events and publications. IFIP's events range from an international congress to local seminars, but the most important are:

- the IFIP World Computer Congress, held every second year;
- open conferences;
- working conferences.

The flagship event is the IFIP World Computer Congress, at which both invited and contributed papers are presented. Contributed papers are rigorously refereed and the rejection rate is high.

As with the Congress, participation in the open conferences is open to all and papers may be invited or submitted. Again, submitted papers are stringently refereed.

The working conferences are structured differently. They are usually run by a working group and attendance is small and by invitation only. Their purpose is to create an atmosphere conducive to innovation and development. Refereeing is less rigorous and papers are subjected to extensive group discussion.

Publications arising from IFIP events vary. The papers presented at the IFIP World Computer Congress and at open conferences are published as conference proceedings, while the results of the working conferences are often published as collections of selected and edited papers.

Any national society whose primary activity is in information may apply to become a full member of IFIP, although full membership is restricted to one society per country. Full members are entitled to vote at the annual General Assembly, National societies preferring a less committed involvement may apply for associate or corresponding membership. Associate members enjoy the same benefits as full members, but without voting rights. Corresponding members are not represented in IFIP bodies. Affiliated membership is open to non-national societies, and individual and honorary membership schemes are also offered.

Life-Cycle Modelling for Innovative Products and Processes

Proceedings of the IFIP WG5.3 international conference on life-cycle modelling for innovative products and processes, Berlin, Germany, November/December 1995

Edited by

Helmut Jansen and Frank-Lothar Krause
IPK-Berlin
Berlin
Germany

Published by Chapman and Hall on behalf of the
International Federation for Information Processing (IFIP)

CHAPMAN & HALL
London · Glasgow · Weinheim · New York · Tokyo · Melbourne · Madras

Published by Chapman & Hall, 2–6 Boundary Row, London SE1 8HN, UK

Chapman & Hall, 2–6 Boundary Row, London SE1 8HN, UK

Blackie Academic & Professional, Wester Cleddens Road, Bishopbriggs, Glasgow G64 2NZ, UK

Chapman & Hall GmbH, Pappelallee 3, 69469 Weinheim, Germany

Chapman & Hall USA, 115 Fifth Avenue, New York, NY 10003, USA

Chapman & Hall Japan, ITP-Japan, Kyowa Building, 3F, 2-2-1 Hirakawacho, Chiyoda-ku, Tokyo 102, Japan

Chapman & Hall Australia, 102 Dodds Street, South Melbourne, Victoria 3205, Australia

Chapman & Hall India, R. Seshadri, 32 Second Main Road, CIT East, Madras 600 035, India

First edition 1996

© 1996 IFIP

Printed in Great Britain by Hartnolls Ltd, Bodmin, Cornwall

ISBN 0 412 75590 4

Apart from any fair dealing for the purposes of research or private study, or criticism or review, as permitted under the UK Copyright, Designs and Patents Act, 1988, this publication may not be reproduced, stored or transmitted, in any form or by any means, without the prior permission in writing of the publishers, or in the case of reprographic reproduction only in accordance with the terms of the licences issued by the Copyright Licensing Agency in the UK, or in accordance with the terms of licenses issued by the appropriate Reproduction Rights Organization outside the UK. Enquiries concerning reproduction outside the terms stated here should be sent to the publishers at the London address printed on this page.
  The publisher makes no representation, express or implied, with regard to the accuracy of the information contained in this book and cannot accept any legal responsibility or liability for any errors or omissions that may be made.

A catalogue record for this book is available from the British Library

∞ Printed on permanent acid-free text paper, manufactured in accordance with ANSI/NISO Z39.48-1992 and ANSI/NISO Z39.48-1984 (Permanence of Paper).

CONTENTS

Preface — xi
Conference Chair, International Program Committee
   and Local Organization Committee — xiii
Session Chairmen — xiv

Part One   Opening

1. Life cycle modelling as a management challenge — 3
   G. Spur

2. Potentials of information technology for life-cycle-oriented product and process development — 14
   F.-L Krause and Chr. Kind

Part Two   Advanced Strategies

3. Sustainable industrial production – environmental issues in product development — 31
   L. Alting

4. Towards a comprehensive life cycle modelling for innovative strategy, systems, processes and products services — 43
   V.A. Tipnis

5. Disassembly factories for recovery of resources in product and material cycles — 56
   G. Seliger, C. Hentschel and M. Wagner

Part Three   Life Cycle Modelling

6. Lifecycle modelling as an approach for design for X — 71
   W. Eversheim, U.H. Böhlke and W. Kölscheid

7. Product life cycle modelling for inverse manufacturing — 80
   F. Kimura and H. Suzuki

8	Some theoretical issues and methods of life-cycle engineering V. Tarassov	90
9	Global engineering network – applications for green design H. Schott and H. Birkhofer	93
10	The green browser: a proposal of green information sharing and a life cycle design tool Y. Umeda, T. Tomiyama, T. Kiriyama and Y. Baba	106
11	Towards a new specification method for an automated system L. Jacquet, Y. Sallez and R. Soenen	116

Part Four   Decision Support

12	An approach to planning of textile manufacturing operations: a scheduling method D. Mourtzis, N. Papakostas and G. Chryssolouris	131
13	A unified decision support tool for product mangement H.E. Cook	146
14	Scenario-management during the early stages of product development J. Gausemeier, A. Fink and O. Schalke	158
15	The FABERCOAT decision support system: a design tool for plasma sprayed coatings M. Foy, M. Marchese and G. Jacucci	170
16	Modeling high precision assembly processes using discrete event simulation – an alternative solution for decision making in manufacturing II. Astinov and P. Hadjijski	183
17	A market model of manufacturing control T. Kis, A. Márkus and J. Váncza	195

Part Five   Assembly and Disassembly

18	Integration of assembly considerations in product design M.A. Willemse and T. Storm	209
19	Modeling and planning of disassembly processes E. Zussman, B. Scholz-Reiter and H. Scharke	221
20	Recycling and disassembly of electronic devices K. Feldman and O. Meedt	233

Contents

21	Information management to support economical disassembly of technical products *D. Spath and C. Tritsch*	246
22	A proposal of CPR-Graph method for assembly sequences generation *J. Wang, K. Ohkura and K. Ueda*	256
23	A rule based system for design for manufacture and assembly *E. A. Warman*	268

Part Six Rapid and Virtual Prototyping

24	Choosing the right rapid prototyping technology for each phase of product development *H. Grabowski, J. Erb and K. Geiger*	281
25	Innovative product development and advanced processes by solid freeform manufacturing *D. Kochan*	293
26	CAD data preparation in a virtual prototyping environment *H. Kress, J. Rix and I. Kiolein*	301
27	Identification of process relevant form structures for stereolithography within solid models *F. Mandorli and U. Cugini*	313
28	Rapid prototyping – new manufacturing tools for improving design and prototype cycle *F. Klocke, S. Nöken and H. Wirtz*	325
29	Rapid prototyping of engineering methodologies *N. Wang and J. Cheng*	337

Part Seven Sustainable Manufacturing

30	Architecture consideration for sustainable manufacturing processes re-engineering *Z. Deng*	349
31	Manufacturing objects as uniform NC-interface for machining and measuring *A. Storr, D. Handel, C. Itterheim, B. Rommel and H. Ströhle*	356
32	The post-mass production paradigm, knowledge intensive engineering and soft machines *T. Tomiyama, T. Sakao, Y. Umeda and Y. Baba*	369

Part Eight Design for Environment

33 Multi-lifecycle design strategies: applications in plastics for durable goods 383
D.H. Sebastian, M. Xanthos, E. Ehrekrantz, M.C. Leu, K.K. Sikar, R. J. Candill and R. Magee

34 Environmentally sound computer aided design 395
R. Anderl and J. Katzenmaier

35 Model based approach to life cycle simulation of manufacturing facilities 408
H. Hiraoka, D. Saito, S. Takata and H. Asama

36 Methods for continual improvement of products and processes 420
R. Züst and G. Caduff

37 Development of environmentally friendly products – methods, material and instruments 432
H. Birkhofer and H. Schott

38 Life cycle assessment (LCA) – a supporting tool for vehicle design? 444
C. Kaniut and H. Kohler

Part Nine Specific Methods

39 Next generation product development 461
D. Haban, T. Haase and A. Strobel

40 Dynamically modified method of data model in the product development process 467
R. Ning and B. Li

41 Industrial methods for product and process development – a case study 475
J. Vallhagen

Part Ten Recycling

42 Design for ease of recycling 489
W. Beitz

43 Design for recyclability – an analysis tool in the "engineering workbench" 501
H. Meerkamm and J. Weber

Part Eleven Feature Technology

44 SESAME – simultaneous engineering system for applications
    in mechanical engineering   515
    H.K. Tönshoff, T. Baum and M. Ehrmann

45 Feature based modelling of conceptual requirements for styling   527
    F.-L. Krause, J. Lüddemann and E. Rieger

46 Flexible definition of form features   540
    *R. Geelink, O. W. Salomons, F. van Slooten, F.J.A.M. van Houten
    and H.J.J. Kals*

Part Twelve Distributed Product Development and Manufacturing

47 A conceptual system support framework for distributed
    product development and manufacturing   553
    B.E. Hirsch, T. Kuhlmann, C. Maßow, R. Oehlmann and K.-D. Thoben

48 Virtual product development for total customer solutions   565
    K. Preiss and L. Bagrit

49 Distributed and multicriteria management tools for
    integrated manufacturing   576
    D. Trentesaux, J.F.N. Tchako and C. Tahon

50 Integrated enterprise modelling for business process reengineering 589
    K. Mertins and R. Jochem

Index of contributors   601
Keyword index   603

Preface

The aim of the conference PROLAMAT'95 on „Life Cycle Modelling for Innovative Products and Processes" was to present, discuss and summarize requirements and solutions for sustainable product development and manufacturing processes. The employment of information technology to support the development of new strategies will be the main focus. This volume contains the papers presented at the conference which provide opportunities to identify the state-of-the-art and address future requirements.

A variety of branches with their specific products and manufacturing processes were the basis for intensive discussions. The relationships between aims for sustainability, costs, quality, and performance are of significant interest. Changes in organizational structures, outsourcing and globalization are important parameters for novel product development and manufacturing strategies. The link to standardization will be emphasised.

The papers in this volume are presented under the following headings:

- Advanced Strategies;
- Life Cycle Modelling;
- Decision Support;
- Assembly and Disassembly;
- Rapid and Virtual Prototyping;
- Sustainable Manufacturing;
- Design for Environment;
- Specific Methods;
- Recycling;
- Feature Technology;
- Distributed Product Development and Manufacturing.

A large number of papers was submitted for consideration. Members of the International Program Committee worked assiduously to select appropriate papers. Thanks are due to them.

Furthermore, we express our thanks to the local organization committee as well as to the officials of the „International Federation for Information Processing" (IFIP) for their efforts, which helped to make this conference possible.

Frank-Lothar Krause　　　　　　　　　　　　　　　　　　　　　　　　Helmut Jansen

Conference Chair

Frank-Lothar Krause

International Program Committee

M. Abramovici (D),
L. Alting (DK),
R. Anderl (D),
W. Beitz (D),
I. Bey (D),
O. Björke (N),
H. van Brussel (B),
U. Cugini (I),
L. Eberhard (D),
J. Encarnacao (D),
W. Eversheim (D),
K. Feldmann (D),
J. Gausemeier (D),
B. Girard (F),
H. Grabowski (D),
D. Haban (D),
G. Hermann (H),
B.E. Hirsch (D),
G. Jacucci (I),
H. Jansen (D),
F. Kimura (J),
T. Kjellberg (S),
D. Kochan (D),
F.-L. Krause (D), *Chair*
S.C.Y. Lu (USA),
M. Mäntylä (SF),
L. Nemes (AUS),
R. Ning (PRC),
G. Olling (USA),
K. Preiss (ISR),
G. Seliger (D),
M. Shpitalni (ISR),
R. Soenen (F),
D. Spath (D),
G. Spur (D),
A. Storr (D),
V.A. Tipnis (USA),
T. Tomiyama (J),
M. Veron (F),
E.A. Warman (UK),
M. Wozny (USA),
R. Züst (CH).

Local Organization Committee

Y. Ficiciyan,
W. Grottke,
H. Jansen, *Chair*
K. Mertins,
J. Reinecke,
K.-V. von Schöning,
W. Turowski,
A. Ulbrich.

Session Chairmen:

PART ONE:		OPENING
Session 0:		G. Olling, Chrysler Corporation, Auburn Hills MI, USA.

PART ONE: OPENING
Session 0:   G. Olling, Chrysler Corporation, Auburn Hills MI, USA.

PART TWO: ADVANCED STRATEGIES
Session 1:   R. Züst, BWI, ETH Zürich, Switzerland.

PART THREE: LIFE CYCLE MODELLING
Session 2:   K.-V. von Schöning, INPRO, Berlin, Germany.
Session 6:   M.J. Wozny, National Inst. of Standards & Technology, Gaithersburg MD, USA.

PART FOUR: DECISION SUPPORT
Session 3:   K. Mertins, FhG-IPK/IWF, Berlin, Germany.
Session 7:   M. Mäntylä, Helsinki Univ. of Technology, Espoo, Finland.

PART FIVE: ASSEMBLY AND DISASSEMBLY
Session 4:   W. Grottke, VW-GEDAS, Berlin, Germany.
Session 8:   M. Abramovici, Ruhr-Universität Bochum, Germany.

PART SIX: RAPID AND VIRTUAL PROTOTYPING
Session 5:   W. Turowski, AEG Schienenfahrzeuge, Hennigsdorf, Germany.
Session 12:   D. Kochan, SFM-GmbH, Dresden, Germany.

PART SEVEN: SUSTAINABLE MANUFACTURING
Session 9:   A. Ulbrich, FhG-IPK / IWF-TU Berlin, Germany.

PART EIGHT: DESIGN FOR ENVIRONMENT
Session 10:   L. Eberhard, Daimler-Benz, Stuttgart, Germany.
Session 13:   H. Jansen, FhG-IPK, Berlin, Germany.

PART NINE: SPECIFIC METHODS
Session 11:   G. Hermann, Technical University of Kecskemét, Hungary.

PART TEN: RECYCLING
Session 14:   T. Kjellberg, IVF/KTH, Stockholm, Sweden.

PART ELEVEN: FEATURE TECHNOLOGY
Session 15:   H. Hoffmann, Univ. Técnica Federico Santa María, Valparaíso, Chile.

PART TWELVE: DISTRIBUTED PRODUCT DEVELOPMENT AND MANUFACTURING
Session 16:   F.-L. Krause, FhG-IPK / IWF-TU Berlin, Germany.

PART ONE

Opening

1

Life Cycle Modeling as a Management Challenge

G. Spur
IWF/IPK Berlin
Pascalstr. 8-9, 10587 Berlin, Germany, 030 / 314 23 349

Abstract
The author deals with the effects a cyclic product definition has on the product design. Special attention is paid to the changes affecting the management and the internal organization of companies.

Keywords
Recycling, Product life cycle, Management duties

1 INTRODUCTION

The current economic system is defined by and dependent on economic growth in terms of quality and quantity. Since the beginning of the industrialization this growth has led to an increasing use of resources, which is only recently restricted by economic policy due to the shortages of raw material and energy sources. Until today, economic growth and resource consumption cannot be decoupled. To solve this problem, our industrial production processes have to be changed significantly (Spur, 1995).

First trends showing that modern industrial nations are able to reach this aim are already visible. It must be assumed that achieving this objective is a necessary condition in order to maintain a leading position among the world's industrial nations.

The production processes have developed according to the historically grown industrial structures. The change in industrial production is in terms of technology and economy increasingly interlaced with the social development of our society. Nowadays we are acting within a worldwide industrial production system, which allows for an increasing production of goods and services while employing a decreasing number of people.

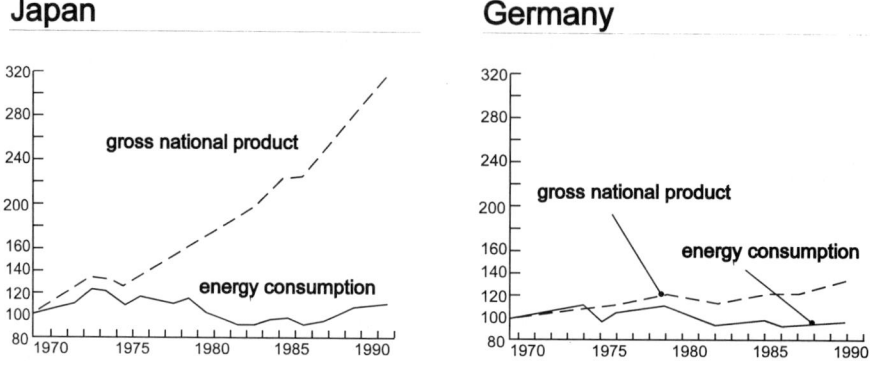

Figure 1   Gross national product and energy consumption (Index 1970 = 100) (Hennicke, 1995)

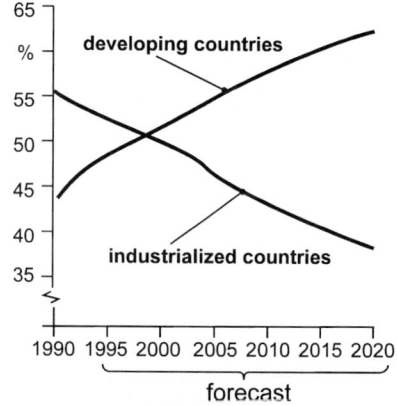

Figure 2   Share of world national product (Kunerth, 1995)

In the current social discussions the occurring changes in social circumstances, for example the loss of jobs, is rated negatively. The importance of work as the sense in life has developed historically. To overcome the problem of unemployment and to create new jobs has become a major problem and the most important subject of economic and social politics.

On a political level the necessity to increase the productivity derives from the competition for economic importance between differently developed countries. On the company level it is enforced by a growing global competition.

2 PROBLEMS

The further development of our industrial societies is characterized by a growing sensitivity towards environmental issues. The growth of our present social productivity is linked to increasingly negative effects, occurring especially during the second half of this century. None of these have been taken into account or even realized by those who started and promoted industrial progress.

Waste problems, technical accidents with severe consequences, environmental pollution and climatic changes are terms that have become an apocalyptic vision of modern society for some people. The linear structure of our consumer society that is composed of production, consumption and waste is increasingly criticized, starting about 30 years ago.

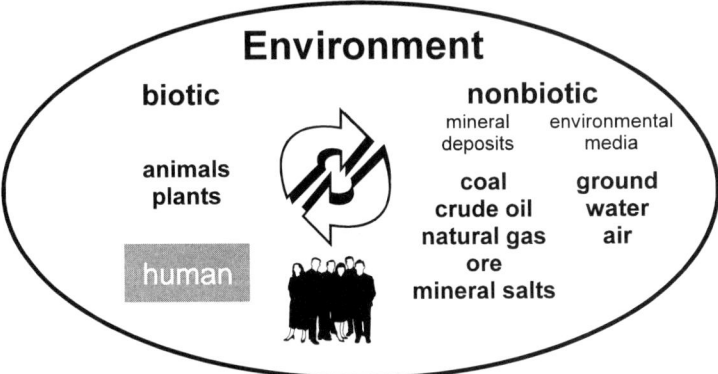

Figure 3   Relationship human - environment (Hüttl, 1995)

In opposition to this linearity some social groups came up with the idea of a cyclic structure of the economy. The global economic development and the increasing world population have a massive impact on the balance of nature and lead to a growing strain on the environment. The rising consciousness for environmental issues in societies has started innovative processes that will fundamentally change all fields of technological development. This will be a global strategy acting on new superior ideals.

3 PRODUCT LIFE CYCLE

The understanding of its fields of activity has fundamentally changed for the engineering business. Engineers are a part of society that deals with the development and improvement of production processes. The aim is to realize production processes and economic solutions that are better and more intelligent by integrating this life cycle philosophy into technology and economy. The idea of environmental conservation is not in opposition to technological and economic progress. A constructive environmental policy has to be linked to economic procedures, so that innovative impulses result from it.

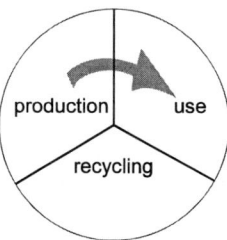

Figure 4   Product life cycle

The definition of the term "product life cycle" has to be seen in connection with this development. It is the conception of a circulation of produced goods between the fields of production and consumption. New in this understanding is that used products get back into the production process. The word recycling has been developed for this part of the life cycle.

For a long time the term recycling has been conceived as hostile to or at least obstructing progress and was used mostly with a moral and ideological purpose. The reasons for this conception are its origin as outlined above and the refusal of some social groups to deal with what was only addressed as "the waste problem". Producers, consumers and people dealing with the waste as products, whose usage and exploitation has ended, just wanted to get rid of the problem. This motive was so dominant that the term disposal was used for both recycling and waste. A very inappropriate term for two so different subjects that are so different.

Today we are in a phase of fundamental change. A successful field of economy is dealing with disposal yielding two-digit growth rates. Recycling is a market of new possibilities and challenges.

4   RECYCLING

In the producing industry recycling is seen as a strategy enabling a repeated use of products or components. The aims of recycling are the saving of the environment as a supplier of natural resources and the reduction of enviromental pollution (Dyckhoff, 1994).

The demand for recycling exists for the whole process line of production. This comprises the preproduction which refines the raw materials, the material processing and the manufacturing of the end-product as the last step. The production potential in a company contains all goods which participate in the manufacturing process. The main material is used for the product, process materials are supplied which support the manufacturing process.

Output operators of the production system are the main products which realize the expected benefit. Further products with or without market value and disturbing products which cause ecological damage are produced as well.

The question of undesirable by-products and disturbing products has been almost completely neglected in traditional business administration (Dyckhoff, 1994). On the other hand the existence of coupled products cannot be ignored in connection with recycling strategies. The following fields of recycling can be distinguished:

- main products as disused commodity goods,
- by-products as accumulated waste-products, and
- disturbing products like emission in form of energy and substances of different conditions.

To initiate recycling processes, it is necessary to develop procedures which cause a dissolution of the functional combination between used main products and by-products. These procedures are described as disassembling processes. An increasing requirement for a market for recycled products becomes evident. Furthermore a market for recycled parts and components can already be recognized, which has a functional use in sense of exchangeability a well as being a market for reusable materials.

A basic requirement for a successful recycling process is the fulfilment of the quality demands established by the market. From this fact the demand for a suitable information system can be derived (Dyckhoff, 1994).

Such an innovation causes the emergence of special technical systems, procedures, facilities and means for the environmental product development which will cope with resource and waste disposal requirements, especially in the early phase of product development.

A very important role in the developing field of recycling and waste plays the technology of disassembling. This is a rather new field of technology which becomes more and more important, especially for the exploitation of products from the automotive and machine building industry. Disassembly is the entirety of planned processes leading to the separation of multicomponent systems into singular components, parts and shapeless material.

For this new field of research and eventually economic area new processes and systems have to be found in close cooperation between research and development. The future-orientated technology has to be approached through new ways. In "think-tanks" or centers of innovation creative development processes can be stimulated and promoted without the usual burden of routine work. The developed solutions can be realized through prototypes. An independent new market of innovation could be established.

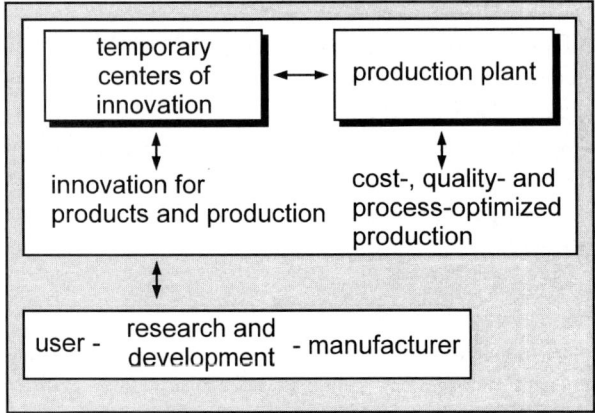

Figure 5 Center of innovation (Spur, 1995)

5 CONSEQUENCES OF THE PRODUCT LIFE CYCLE CONCEPTION

Due to the replacement of the idea of a **linear** product life by the conception of a product life **cycle**, nowadays the perspective towards produced goods assumes a new quality and evokes progress. The responsibility of the producer used to be at a minimum once the product was sold. Today there are laws of liability for products, and taking back used products becomes obligatory. The producer cannot get rid of his product anymore, recycling is a legal reality. There are various levels on which this has to take effect:

- product policy,
- product strategy,
- product development and design and
- product manufacturing.

The overall process of economy becomes more complex through the integration of this cyclic aspect. The process is like positive feed-back in the sense of control engineering. Increasing demands for development and design as well as necessary logistics are the results. But this also incorporates great possibilities and chances.

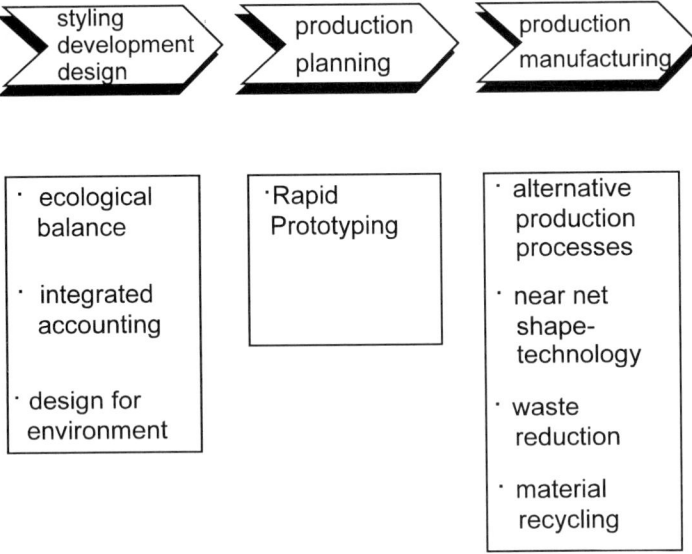

Figure 6   Sectors for integration of environmental aspects in production (Pollmann, 1995)

The constraints for identification and documentation of products and developments can also be useful for other requirements such as quality assurance, maintenance, marketing and development processes. Thus the time needed for development and manufacturing can be reduced and customer demands can be better taken into account. Not only design-to-cost but design-to-customer is the future policy. The contact to the customer and user must be intensified, and the customer must be regarded as a partner, who will keep the product for a certain time and who trusts the producer to guarantee optimum performance.

Thus the legal realities indirectly change themselves: the classic sales contract giving unlimited rights to the owner is of decreasing importance. Through the obligation to recycle products the producer is taken into the responsibility and cannot leave it to the user, when and how the usage of the product is terminated. One could imagine that in the future the producer decides at what time the product is taken out of usage, similar to what is now done in recalling faulty products. This leads to a rational and controlled recycling. Legal forms like the sale of usage or contracts similar to leasing contracts today could become common practice.

The company is no longer competing just with other market participants but also has to develop a „co-evolution" with all participants in the economic process. This thesis seems to be confirmed by an increasing level of qualification of employees, changing market demand and the exchange of new technological fields. Management has to keep its eye on the interaction of the social and technological-economic environment. Obviously there are limits put on a non-holistic and non-cyclic view at the production process.

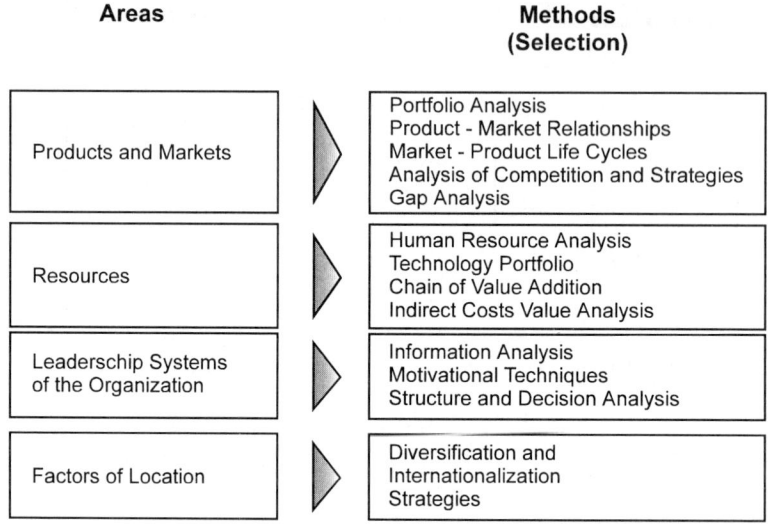

Figure 7   Methods of strategic planning (Spur, 1990)

6 SUSTAINABILITY AS A MOTTO FOR MANAGEMENT

A central point for the product to be recycled is the continuous environmental compatibility. This term is derived from forest economy, an economic field which has always calculated and acted in temporal terms of generations. It means that the economy has to act to preserve its existence in the long term. The aim is to look at the product life cycle in total and optimize the processes and decisions so that the damage to the environment is kept at a minimum. In this context the necessity to plan products including the anticipation of their impacts and qualities for the total time of usage becomes more and more important. New professions such as the environmental engineer will be needed. The production system has to be analyzed in all its complexity.

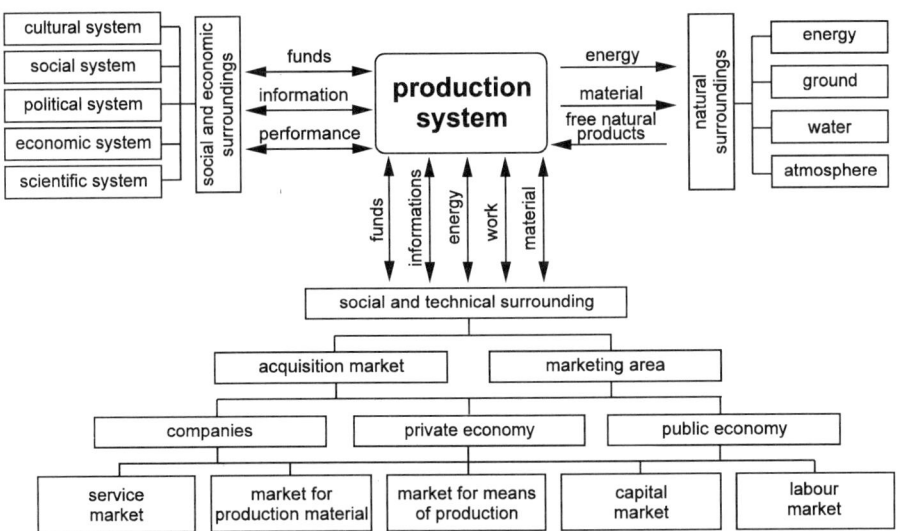

Figure 8 Surrounding of an industrial production system (Spur, 1995)

The demands originating from the changing competition of the producers and vendors on the global market are for the singular companies

- price consciousness,
- diversity,
- variability,
- speed,
- flexibility, and
- innovative ability.

The changes in which modern companies live necessarily lead to changes in means and methods which are implied to meet these requirements. The development of strategies and conceptions for recycling is one of the important challenges that the deciding managers have to face. Thus qualified knowledge is the ground on which they have to deal with new terms and new methods such as ecological balances. There have to be changed conceptions of aspects like (Tipnis, 1993)

- Economy,
- Environment,
- Ecology,
- Empowering,
- Education,
- Exellence, and
- Ethics.

Figure 9   Key-management qualifications (Spur, 1995)

7   CONCLUSIONS

The aspect of environmental preservation has become an integral part of modern economy. The definition of the term product life cycle and the connected change of focus in production must be accepted. The economic processes have to be transferred to a new, more intelligent level. The technological and structural adjustments have to be realized by a visionary company policy. Recycling has to be conceived as a chance for market leadership.

8 REFERENCES

Dyckhoff, H.; Dormstädter, A. and Soukal, R. (1994) Recycling *Handbuch Produktionsmanagement* (ed.. Corsten), Gabler, Wiesbaden.
Hennicke, P. (1995) Arbeit und Produktion, Proceedings PTK 95, Berlin.
Hüttl, R. F. (1995) Umweltingeneurwesen in Forschung und Wirtschaft, Proceedings PTK 95, Berlin.
Kunerth, W. (1995) Leistungspotentiale der deutschen Wirtschaft, Proceedings PTK 95, Berlin.
Pfeiffer, W.; Schultheiß, B.; Staudt, E. (1975) Recycling- Systemtechnischer Ansatz zur Berücksichtigung von Wiederverwendungskreisläufen in der langfristigen Unternehmensplanung, *Systemtechnik - Grundlagen und Anwendung*, (ed. G. Ropohl), Carl Hanser Verlag, Munich, Vienna.
Pollmann, W. (1995) Technologien umweltverträglicher Produktion, Proceedings PTK 95, Berlin.
Spur, G. (1995) Marktführerschaft als Managementaufgabe, Proceedings PTK 95, Berlin.
Spur, G. (1990). Corporate management in the industrial society of the future, Robotics & Computer-Integrated Manufacturing, Vol. 7, No. 1/2 pp. 3-14.
Tipnis, V. A. (1993) Evolving issues in product life cycle design, *CIRP*, Vol. 42, pp. 169-173.

9 BIOGRAPHY

Prof. Dr. h.c. mult. Dr.-Ing. Günter Spur, born in 1928, studied mechanical engineering and manufacturing technology at the TH Braunschweig, where he received his doctoral degree. From 1962 he was employed in leading positions at the machine tool factory Gildemeister & Comp., Inc., Bielefeld. In 1965 Prof. Spur was offered the professorship of Machine Tools and Manufacturing Technology at the Technical University Berlin (IWF-Berlin). Since 1976 Prof. Spur has also been the Director of the Fraunhofer-Institute for Production Systems and Design Technology (IPK-Berlin).

2
Potentials of information technology for life-cycle-oriented product and process development

F.-L. Krause, Chr. Kind
Fraunhofer-Institut für Produktionsanlagen und Konstruktionstechnik (IPK),
Institut für Werkzeugmaschinen und Fertigungstechnik (IWF-TUB)
Pascalstr. 8-9, 10587 Berlin, Germany
Tel: +49(30)39006 244   Fax: +49(30)393 02 46
E-mail: frank-l.krause@ipk.fhg.de

Abstract
Due to the increasing responsibility of companies and increasing demand on the market for high quality and environmentally conscious products, consideration of life-cycle requirements in product and process development has become more important in recent years. The modification of product and process development to take account of the complex interrelationships within product life-cycles requires organisational as well as administrative changes and an application of specific methods and tools to support manager and staff. In this article, potential aspects of information technology leading to the attainment of this goal are presented. The main emphasis is placed on environmental aspects.

Keywords
life-cycle, product and process development, information technology

1   INTRODUCTION

At the most recent UN climate conference in Berlin, it became evident that sustainability must become one of the key future aspects and the goal of development activities of industry, politics and research with a view to preserving the reserves of natural resources and decrease pollution to the greatest possible extent (Spur, 1995a). On the basis of increasing ecological awareness in modern societies, many international enterprises have realized the necessity for life-cycle concerns and environmentally conscious design and manufacturing.

Different activities of enterprises as well as of government make evident that the complex correlations which have to be considered in reaching the goal of life-cycle-oriented product and process development and environmentally conscious design and manufacturing cannot be controlled by organizational changes only. To use the resources which are available worldwide in an optimal way, specific methods and tools are required. To accomplish their tasks appropriately both managers and staff have to apply these methods and tools which are based on modern information and communication technology (Spur, 1995b). In this article potentials of information technology to support life-cycle-oriented product and process development will be presented with a main emphasis on environmental aspects.

2 ECOLOGICAL ACTIVITIES OF INDUSTRY AND GOVERNMENT

Since a positive image concerning environmental issues is highly recommended from customers and market today, in many international corporations environment becomes a management task. On this basis, big international companies initiated various activities. Some selected main points will be presented in the following.

At Daimler Benz and in the corporate units, „chief environmental officers" have been appointed who must report directly to the head of their respective company management board (Daimler Benz, 1995). Similar to Daimler-Benz at the United Technologies Corporation (UTC, 1995), an „environment, health and safety officer" reports directly to the presidents of the business units. Environmentalism was declared to have the highest ranking position across the breadth of Electrolux's process and operating philosophies (Dzierwa, 1995). Toyota initiated a „Toyota Earth Charter" which includes development of a totally clean system from product development to scrapping after usage. As a result, the Toyota Environmental Activities Plan defines practical goals (Iwai, 1995). Likewise, the BMW management has initiated various environmental activities (Milberg, 1995).

The environmental reports mentioned above not only point out some organizational changes towards more ecological oriented companies but also show that environmental burdens have been strongly reduced within the last 10 years. For instance, UTC has managed to cut hazardous waste generation in half since 1988, Daimler-Benz has increased the recycling of waste material by 36% since 1993, BMW has reached a rate of recycling up to 85% using a current model, and Electrolux has reduced the negative impact on the ozone layer caused by new refrigerators in recent years to zero. As a result of the 5R Campaign (refine, reduce, reuse, recycle, and retrieve energy) at Toyota, the amount of waste per volume of sales has been decreased by 90% since 1973.

In addition to the voluntary initiatives of many enterprises, additional activities of various governments pursue two different courses in general. On the one hand, laws and guidelines are to be enacted, on the other, research work specific to environmental is to be founded.

In Germany, two important examples of regulations are the new „Kreislaufwirtschaftsgesetz" (life-cycle management law) and the VDI guideline 2243. The „Kreislaufwirtschaftsgesetz", which will come into effect in October 1996, forms the legal basis in the area of waste disposal. According to the principle of origination, the responsibility for waste disposal as recognized by this law will be assigned on the basis of the place of generation of waste. The design for recycling of technical products is described in the VDI-guideline 2243 (VDI2243,

1993). This guideline represents the general correlation of recycling processes and design recommendations for the development of technical products.

The life-cycle issue constitutes the focus for a crucial point of research within the new program „production 2000" of the German BMBF (Ministry for Education, Science, Research and Technology), called „Life-Cycle-Oriented Resource Management" (BMBF, 1995). One of the subtasks of this crucial point of research is the development of tools for a life-cycle-oriented product and process development.

Another example of a life-cycle-oriented research project in Germany financed by the DFG (German Research Community) is the new Sfb 281 „Disassembly Factories" (Sfb281, 1995). In this project, the opportunities and potentials for recovery of resources in product and material life-cycle are to be studied.

3 LIFE-CYCLE-ORIENTED PRODUCT AND PROCESS DEVELOPMENT

According to the second draft of ISO 14040 (1995), life-cycle comprises the consecutive and interlinked stages of a product or service system, from the extraction of natural resources to final disposal. The extensive responsibility of a company for its products on the one hand and the necessity of a closer contact to customers on the other lead to a planning and controlling of all phases of the product life-cycle. The most important phase of this product life-cycle management is product and process development, during which most of the life-cycle specific features of a product are determined. This means that a life-cycle-oriented product and process development is necessary to produce a product which meets all life-cycle-oriented requirements and, as a major aspect of life-cycle, to minimize negative impacts of this product and its production on environment.

Because of the determination of future product behavior in the early phases of product life-cycle the entire life-cycle of a product has to be taken into account at the planning and design phase of a product. This means not only to develop specialized technologies or products to meet extrapositioned environmental requirements but to consider specific consequences on environment of the product in every phase of product development, production, usage including both transport and maintenance, and disposal. Besides the product itself the processes to develop, produce, use, and dispose the product has to be considered as well.

According to this distinction the impact of products on environment can be divided into three major categories:
- environmental impact caused directly by the product,
- environmental impact caused by processes related directly to the product, and
- environmental impact caused by the overhead.

From the point of view of the product a direct impact is evoked by the product itself during manufacturing, transport, use and disposal. This includes for example the consumption of raw material to manufacture the product, energy consumption of the product during usage, and the emission of toxic substances in the disposal phase.

An indirect impact in the first grade is caused by processes directly related to life-cycle phases of the product. These are the consumption of environment resources and emissions of processes to manufacture, transport and dispose a product. Examples are given by the consumption of energy and process materials during manufacturing and the energy consumption and pollution of transport systems.

In the third category, all products and processes are collected which have an impact on environment and which have relations to the product and its processes except those mentioned in the second category. These environmental impacts commonly are called overhead. For example the manufacturing overhead results from the operation of a company not directly related to the manufacture of products such as air conditioning or heating of buildings (Alting, 1995).

Figure 1 shows the four major stages of life-cycle and the environmental loads which are caused in these stages. Since the product does not yet exist materially in the product development phase, there are only little burdens on environment in this phase. This is based on the assumption that material related processes like the manufacturing of prototypes which is actually a subtask of product development will be considered as a part of the production phase.

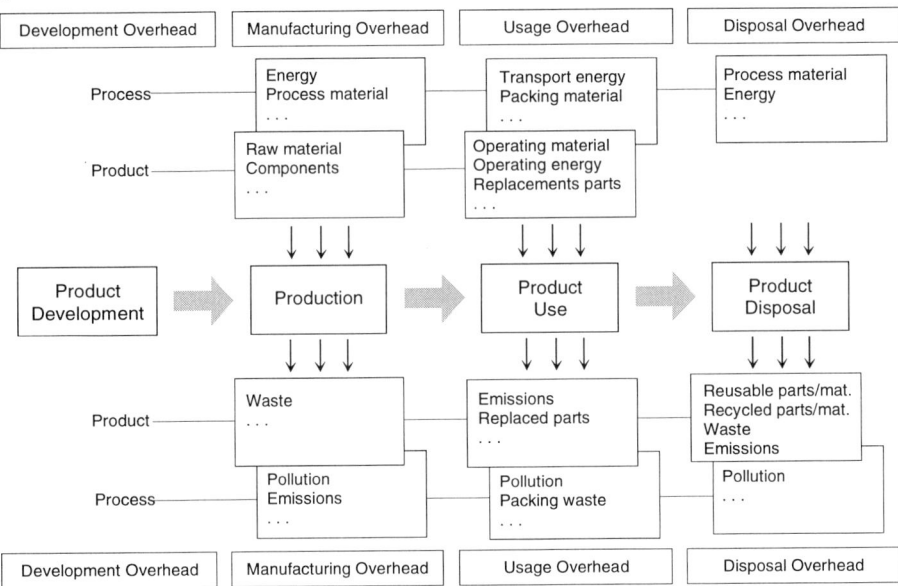

Figure 1 Life-cycle stages and relation to environment.

By way of contrast in the development phase most of the criteria which determine environmental burdens are defined. On this basis life-cycle-oriented product and process development becomes one of the most important phases in life-cycles. Since life-cycle-oriented data are not specified exactly because technologies, cost structures and regulations of today cannot be transferred into future conditions which are valid at product disposal, it is necessary to supply the designer with the most up-to-date information so that he is able to base his decisions on a precise prediction of technological development.

To minimize the impact on environment and reduce ecological burdens, Wenzel (1994) defined five improvement strategies which have an effect on one or more phases of product life-cycle:
• better material handling,

- chemical savings or substitutions,
- thermal energy economization,
- electricity economization, and
- overhead reduction.

The application of these strategies with regard to the environment impact categories listed above requires organizational changes as well as support by appropriate methods and tools. In this connection, modification of organization means adaptation of business processes to the demands of life-cycle orientation, corresponding modification of workflows and introduction of new methods of working. Since life-cycles of products have to be regarded as units, only an integrated approach will lead to a truly economical solution.

In all phases of product life-cycle, computer based tools can be used to improve the performance. Comparable to organizational aspects, tools have to support integrated aspects as well. There should be no gaps between different computer-based applications within product life-cycle. In addition to the integration of product life-cycle phases on the one side information technology supports the parallelization of work activities both within the phases and overlapping. On the other side information technology makes it feasible for persons who participate in product life-cycle to collaborate in distributed systems.

4 TOOLS AND METHODS FOR LIFE-CYCLE-ORIENTED PRODUCT AND PROCESS DEVELOPMENT

4.1 Tools and methods in product life-cycle

To reach the goal of environmentally conscious product and process development, a considerable number of measures in the areas of organization, mode of operation and information technology are necessary. These fields are closely connected to each other.

To support life-cycle-oriented product and process development, different requirements with respect to information technology can be derived from the major goals of life-cycle orientation stated in section 3. These requirements mainly originate in the huge need for information to fulfill the demands of environmentally conscious product life-cycles. The flood of information a designer must theoretically take into account must be reduced, be presented appropriately and be made processable by methods of simulation, by adequate application systems and knowledge based information and decision support systems. In addition, the product developer must be able to evaluate the environmental consequences of the product or its processes in subsequent phases. Therefore, for a capable design system which meets the requirements of a life-cycle-oriented design process, these additional modules are of significant value.

Figure 2 represents some aspects for reaching the goal of environmentally conscious product life-cycle using computer based systems. Two general areas can be distinguished. On the one side there are basic tools which can be employed in all phases of the product life-cycle. An information technology infrastructure connects these phases and enable a life-cycle oriented exchange of all kinds of data. On the other side application oriented tools support persons to fulfill their tasks in specific phases of product life-cycle. Other tools and methods such as economic or quality management tools and tools to realize technical solutions are not mentioned explicitly.

Since the support of ecological product development using information technology is the main subject of this article, in the following sections some tools and methods with a direct influence on environmental issues are specified with regard to their application within product development.

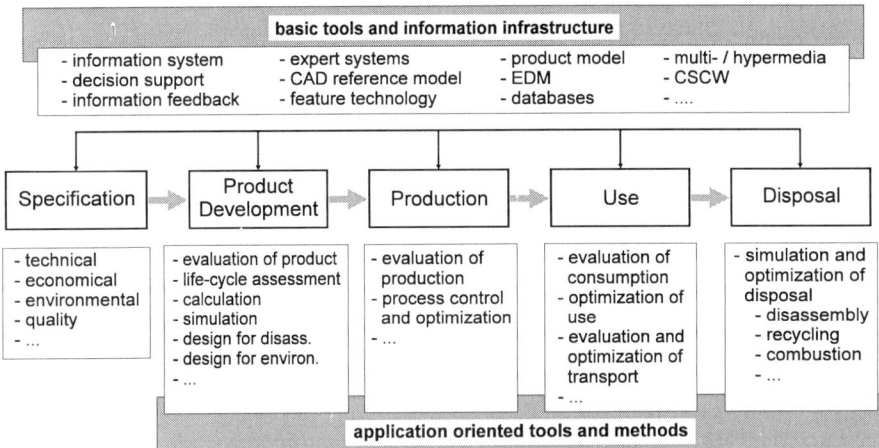

Figure 2 Support of life-cycle-oriented product and process development using information technology.

4.2 Basic tools for life-cycle-oriented product development

Information systems
The different requirements which have to be considered during product development lead to a high demand for information. The amount of information even increases with the additional goal of considering ecological requirements. For instance, there are today more than 8000 laws and regulations in Germany on environmental issues (Milberg, 1995). Having the right information at the right time is therefore one of the most important aspects within life-cycle-oriented product development.

Information should be supplied to the designer as early as possible within the product development process. This means that as early as the product planning phase, there must be computer-based information systems which support the product developer. But, according to the strategy of least commitment, decisions should be made as late as possible. This gives the designer the possibility to react to changing requirements or new environmentally oriented guidelines or research results with a minimum effort.

The designer needs appropriate information to consider all life-cycle-oriented demands. On a high level of generality, this environmentally oriented information consists mainly of:
- material and component descriptions (pollutant concentration, energy consumption, feasibility for recycling, etc.),
- process descriptions (energy consumption, pollution, generation of process waste, etc.),
- descriptions of ecological systems, and
- guidelines, laws, and standards.

The designer needs specific tools to acquire this information with a minimum of effort. He should not be required to read through a large quantity of material and invest a great deal of time finding the required information. Information systems are able to support the designer and supply him with the appropriate information. Expert systems can be employed to work up this information and make decisions easier and more clear. These systems also support multiple criteria decision making to consider not only environmental requirements but also demands concerning economy, functions, safety, etc.

Engineering data management (EDM) systems are also information management tools which process product data in life-cycles. The main tasks are:
- to enable the access on product data, documents and their relations for persons and application systems concerned and
- to control the workflow within an enterprise (Ploenzke, 1994).

One major, life-cycle-oriented function of EDM-tools is to keep a historical record of product changes from initial idea through all phases of product life-cycle up to final disposal, and make this information available as required.

New information technologies such as multimedia or hypermedia systems offer a good approach to the solution of the problem of information supply. Because of the well structured data within a hypermedia system and the ability to intuitively navigate through this structure, it becomes easier to find specific information. The employment of multimedia systems which integrate text and pictures with video and audio data and animation improve the intelligibility and clearness of information. These facts will also raise the motivation of the user to employ the information system in his daily work. To achieve this goal, it is necessary to integrate the information system into the work environment of the user. Add-on systems are required to the individual application systems used by the persons within the product development process. The user should not be forced to leave the system he knows, to look for the desired information in another system, and finally with an expenditure of much effort transfer the data to his own work environment.

Another possibility for the support of the information supply is provided by the application of modern communication technologies. Via video communication, colleagues can cooperate within a team or experts can be consulted in a video conference to ask for specific information or to get help in solving a given problem. As in the case of the standalone information system, the video connection is to be established without leaving the specific work environment. Tools for a computer-supported cooperative work (CSCW) offer a high potential for a systematic communication between all persons involved in the product life-cycle.

Comparable to the tools for early information supply, the designer also requires tools for modeling the results of his work as early as possible. Using these tools, he should, for example, be able to define a requirement model which contains ecological requirements in addition to functional and other requirements. To identify and evaluate environmental influences on the product and vice versa at this early stage of product development, specific tools are again required.

From the point of view of different design aspects, environment-relevant product information must be represented in specific partial product models as part of a segmented total product model (Krause, 1992). Within this model, all product-related data which are used or generated within the product life-cycle are stored. The most important partial model to support design for environment is the environment model which includes but is not limited to regulations, laws and standards effective in the marketing areas, description of environmental related product characteristics, and environmental conditions of ecological systems the product will be used in.

An approach of a life-cycle modeling system within an architecture for concurrent engineering enabling technologies has already been presented (Kimura, 1992). To meet the modeling requirements all phases of a product life have to be taken into account. Other demands on this modeling system arise for an open architecture in order to enable a product dependent configuration and to be sharable because modeling functions will be performed in a team simultaneously.

Frequently, the complexity of ecological systems cannot be modeled precisely due to unknown factors or unknown interrelations within these systems. This makes it necessary to use specific information processing systems to work up information in the way the designer needs it. For example, systems for life-cycle assessment or life-cycle analyses and systems for simulation of products and processes can be used by the designer without having a mathematically exact model of the object to be examined. Furthermore, knowledge-based information systems offer a high potential for the consideration of a large amount of heterogeneous knowledge and experiences of environment experts.

Life-cycle-oriented information technology infrastructure
As shown in the preceding section the transfer of information and communication between different partners within the product life-cycle is crucial. This leads to an information technology infrastructure which is specifically oriented towards product life-cycle and is characterised by two main aspects, Figure 3:
- The information is locally distributed.
- The information is chronologically distributed.

The local distribution of information can be explained by the large number of information sources on environmental issues which have to be considered during product and process development. There is information about materials or components which a designer wishes to use for his product, planned for use in a specific, potentially ecologically sensitive environment. In another case, the process planner has to decide which cooling lubricant is to be employed in order to minimize toxic effects on the worker.

In many cases the exact location of the information source is not known. And since increasing amounts of information world-wide (the total knowledge in the world doubles every three to four years) and the shortening of product development times lead to the necessity for more rapid provision of information, it becomes even more difficult to find the right information within am acceptable time-frame. Therefore, the importance of information broker functionality has increased in recent years and a specific information industry has been created. While today the designer mainly is required to contact other persons who operate information retrieval systems, in the future automated systems will be available which can be operated by the designer himself.

With respect to the chronological distribution of information another task of the information technology infrastructure is the storage of information. These information stores do not only support an easy input and output of data, but with respect to life-cycle orientation they must also store the data for long periods of time. Such long-term storage is necessary since one might need information generated in an early phase of product life in a final phase, and the time-frame between these two phases can be quite long. For instance, for the disassembly of a product a disassembly plan is needed. This plan has either already been generated during product development or it must be newly generated based on the product structure. In both cases, data originating in product development has to be available for product disassembly. Another example is supplied by the material used for specific parts. The designer lays down the material during product development and describes the exact compound. If this information is not directly connected to the part, as is done today with some plastic parts, it is necessary to extract this information from the product description generated in the product development phase.

To fulfill the requirements described above, databases have been employed in recent years. Nowadays, they are used to store and make available all kinds of environmentally relevant data. This includes standards and guidelines on environmental issues. To guarantee the topicality and validity of information, the databases have to be kept up-to-date and should be improved continuously. The users have to be trained to work with the database appropriately. Intelligent search mechanism are very important to find the relevant data as fast as possible. Some modern databases are also able to store multimedia material.

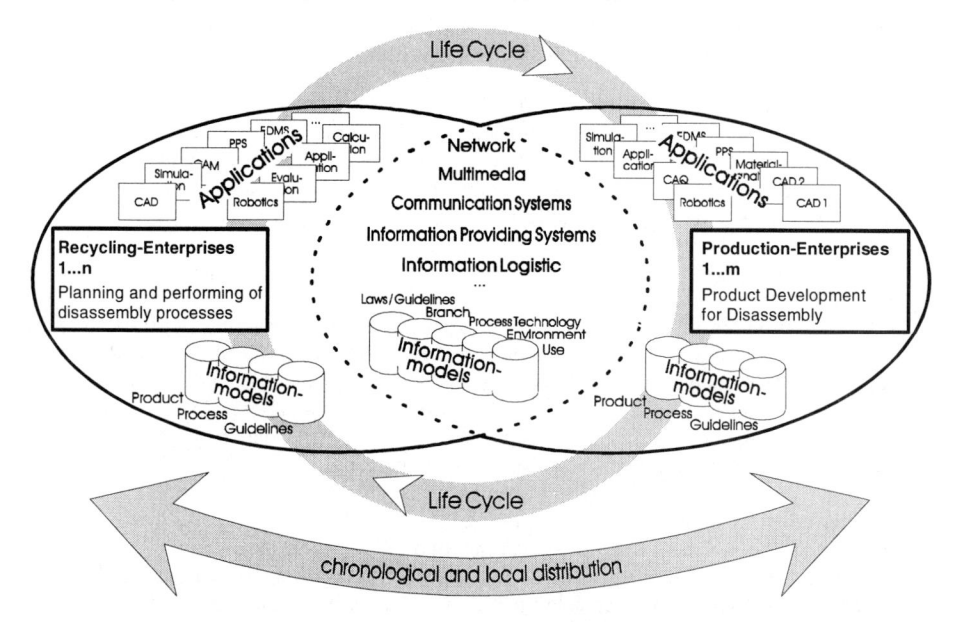

Figure 3 Local and global infrastructure of information technology for life-cycle-oriented product and process development.

An important element of the information technology infrastructure for product development are CAD systems. In the compound project „CAD reference model" the setting up requirements on the configuration of future CAD systems will be described in order to optimize integrated product development (Dietrich, 1994). This description has been formulated as an open reference architecture which already considers life-cycle aspects. A system part enable the connection of various application systems and a integrated product model represents the basis for all product data including life-cycle-oriented data.

In order to realize an information technology infrastructure which meets the requirements described above the definition of standards is necessary. This standards will make it possible, for instance, to link open systems for communication purposes and to read data after a long-term storage.

Information feedback

Feedback is the capture of information about the entire product life-cycle and its evaluation and use for the correction and verification of every preceding phase of life-cycle in order to optimize the results of these phases (Krause, 1993). The feedback of information in particular supplies information essential to the new or continuing development of products (Woll, 1994).

Information from all phases of the product life-cycle has to be fed back into the early phases of product development. This information can, for instance, originate in manufacturing and assembly, usage and service, disassembly and recycling. To reach this goal, all relevant information must first be captured over the course of the entire life-cycle in a feedback model. Subsequently and as required, information is worked up and distributed to all target systems.

A life-cycle-oriented information feedback defines specific requirements with respect to the information technology infrastructure described in the relevant section above.

Feature technology

Feature technology offers a high potential for the improvement of life-cycle-oriented product and process development (Krause, 1995). For processes in design and process planning which are integrated into a feature-based process chain, features can be used to reference specific environmentally oriented data. These features can be called environmental features. For instance, a manufacturing feature which describes a specific manufacturing process, becomes an environmental feature if aspects such as energy consumption or waste incurred are represented. Semantic modeling helps the designer to consider specific ecological requirements.

Since all ecological information for the product itself and the processes of the product life-cycle are available, environmental features can be used for life-cycle assessments. On this basis, it is possible to define a feature-based process chain which supports environment-oriented product and process development.

4.3 Application oriented tools and methods

There are various application oriented methods and tools to support life-cycle-oriented product development. Many of them such as life-cycle assessment (LCA) or simulation systems which are described in the following have the tasks of evaluating product behavior and environmental consequences in subsequent phases. Material evaluation is one of the key tasks in this area.

Other tools support designers with respect to specific design strategies such as design for environment (DfE) or design for disassembly (DfD). Design for environment focuses, for example, on energy economization. In order to support an optimal recycling tools or methods for design for disassembly are employed which are able to group parts into subassemblies with fixed disposal requirements and to determine the optimum disassembly depth (Alting, 1995).

Product and process simulation
One important method of information processing in the product development is the simulation of product behavior and of production processes. These simulation processes may lead to results which can be used to optimize products and processes with respect to environmental and life-cycle aspects.

Based on a product model which represents all life-cycle-oriented product characteristics, it is possible to predict product behavior, the product's impact on natural environment during use and disposal, and the reaction of the product to environmental effects in the usage phase. Additionally, simulation tools can be used to analyze the interacting behavior of different products in a specific environment and the opportunities for mutual increase or decrease of effects on environment.

All processes within life-cycle and their behavior and impact on natural environment can be simulated as well. Commonly, the analyzed processes are part of the production, usage and disposal phase. A critical subprocess in production might be, for example, the assembly of different components into complete products. Comparable to this is the disassembly subprocess in the disposal phase. Subprocesses to be analyzed within the usage phase might be, for example, the transportation processes before or after the use of the product.

A model that takes specific life-cycle issues into account by representation of relevant constraints must constitute the basis of such simulation tools. In the case of disassembly, for example, which is an important process for part recycling, the model must represent easily removable fasteners between two components. Methods and tools which focus on product structure in order to improve disassembly mainly have the tasks of grouping parts into subassemblies with fixed disposal requirements and of calculating the optimum disassembly depth. Since the design for disassembly is a multiple-objective task and the constraints valid for the future disposal cannot be predicted precisely, tools should be able to handle these imprecise data.

Generally, the simulation of the disassembly process is based on the product model which represent the very new state of the product. Since in reality the product will wear out during use, some product features may change due, for instance, to erosion or corrosion. A new approach will consider these changes by defining a usage model. On the basis of this model, the effects of usage caused by physical loads on the disassembly process will be simulated, Figure 4.

Even the product development process itself may become more ecological by simulation techniques. High performance computer tools make it possible to replace real prototypes with virtual prototypes. Energy and material consumed in the production of the real prototype are economized, which is especially useful in the case of destructive testing of many prototypes. This leads to a future aspect of simulation technology. By applying virtual reality tools, the behaviour of a virtual product in a virtual environment can be predicted.

Tools for life-cycle assessment
Life-cycle assessment (LCA) offers an important method for optimizing products and processes with respect to environmental aspects. By applying these methods, all products and processes can be evaluated with regard to their impact on the environment. LCA is a systematic set of procedures for compiling and examining the inputs and outputs of materials and energy and the associated environmental impacts directly attributable to the functioning of a product or service system throughout its life-cycle (ISO14040, 1995).

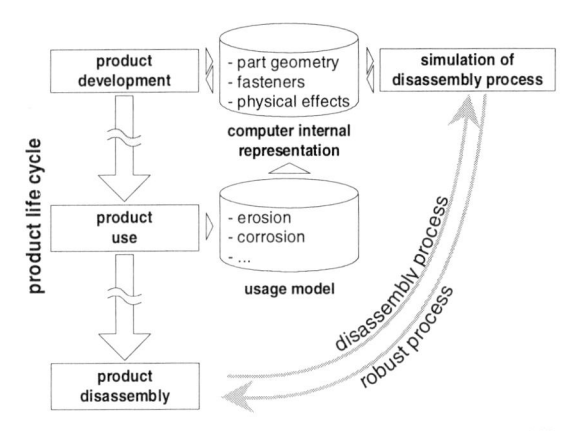

Figure 4 Simulation tool to support design for disassembly in product life-cycle.

There are two major problems of LCA's. On the one hand, it is often not possible to exactly differentiate the ecological system to be evaluated from the surrounding environment. The more one considers indirect aspects of the impacts of a product on environment, the more difficult the LCA's for specific products become. On the other hand, methods for the comparison of different burdens on environment are controversial. Because of the lack of objective values, it is hardly possible to predict whether the results of the five overall improvement strategies will have a corresponding impact on the environment.

Alting (1995) presents an extensive list of tools and methods used world-wide for LCA studies. These tools can in most cases be used with complex products. To support engineers with little or no environmental knowledge in relating to the output, approximately 2/3 of all tested tools feature an impact assessment function. Most of them have a database for storing basic ecological data. They are normally oriented around specific issues like energy and transportation, or processes from chemical industries.

LCA's should result in the identification of improvement potentials. A total ecological assessment is only useful for specific parts, products and processes. In this case, supplier-specific data can be used. Since this data might reflect economic aspects which do not become public, the data should be handled with care. Neutral and independent institutes offers an appropriate way to ensure this requirements.

5 CONCLUSION

To strengthening their market position enterprises pursue different strategies such as improving of product quality, shortening of development cycles, and decreasing of costs. As a new aspect environmental consciousness with respect to products and processes arose in recent years. On this basis, many enterprises realize the necessity for raising the environmental quality of their products and processes in order to improve their corporate image with respect to environmental aspects. Legislation and guidelines will also force companies to initiate measures aiming at products and processes which meet environmental requirements.

To attain this goal product and process development must be adopted to the demands of life-cycle-orientation. In addition to organizational changes the application of specific tools and methods is necessary. To this end an information technology infrastructure must be established which is able to supply, transfer and store chronologically and locally distributed data within the product life cycle. Existing tools either support life-cycle specific subtasks within the different phases of the product life-cycle or are basic tools that are deployed over all phases. A true integration of these computer based tools with respect to life-cycle product and process development has not yet been achieved.

6 REFERENCES

Alting, L. and Jorgenses, J. (1993) The Life Cycle Concept as a Basis for Sustainable Industrial Production. Annals of the CIRP, Vol. 42/1, 163-167.
Alting, L., (1995) Design for life-cycle. To be published in Annals of the CIRP, vol. 44/2.
BMBF (1995) (eds.) Rahmenkonzept „Produktion 2000" 1995 - 1999. Strategien für die industrielle Produktion im 21. Jahrhundert. Bonn.
Daimler-Benz (1995) Environmental Report 1995. Stuttgart.
Dietrich, U., Kehrer, B., Hayka, H., and Jansen, H. (1994) Systemarchitektur des CAD-Referenzmodells unter den Aspekten Kommunikation, Produktdatenmanagement und Integration. Proceedings GI-Fachtagung CAD 94, March 1994, Paderborn, 343-366.
Dzierwa, R. (1995) Environmentality. Appliance, February 1995, E-31-E-32.
ISO 14040 (1995) 2 Draft: Life Cycle Assessment - Principles and Guidelines
Iwai, T. (1995) Toyota's Activities for the Environment. International Journal of Environmentally Conscious Design & Manufacturing, Vol. 4, No. 1. Albuquerque, New Mexico, 29-41.
Kimura, F., Kjellberg, T., Krause, F.-L., Lu, S., and Wozny, M. J. (1992) The First CIRP International Workshop on Concurrent Engineering for Product Realization. Tokyo, Japan, 27.-28. June, Annals of the CIRP, Vol. 42/2, 743-746.
Krause, F.-L. (1992) Leistungssteigerung der Produktionsvorbereitung. Proceedings Produktionstechnisches Kolloquium Berlin, 21.-23. October 1992, 166-184.
Krause, F.-L., Ulbrich, A. and Woll, R. (1993) Methods for quality driven product development. Annals of the CIRP, Vol. 42/1, 151-154.
Krause, F.-L., Ciesla, M., Rieger, E., Stephan, M., Ulbrich, A. (1995) Features - Semantic Objects for the Integration of Tasks in the Product Development Process. Computers in Engineering (eds. Busnaina, A., Rangan, R.), Boston, Massachusetts.
Milberg, J. (1995) Die Automobilindustrie als Wirtschaftsfaktor. Proceedings Produktionstechnisches Kolloquium Berlin, 3.-4. October 1995, 92-103.
Ploenzke (1994) (ed.) Engineering Daten Management Systeme. Ein Technologiereport der Ploenzke AG, Kiedrich/Rheingau.
Sfb 281 (1995) SFB Demontagefabriken. ZwF 90, 3, 80.
Spur, G. (1995a) Lebenszyklusorientiertes Produktmanagement. ZwF 90, 9, 404 - 405.
Spur, G. (1995b) Marktführerschaft als Managementaufgabe. Proceedings Produktionstechnisches Kolloquium Berlin, 3.-4. October 1995, 7-19.
UTC (1995) Environment, Health and Safety Progress Report. Hartford, Connecticut.
VDI 2243 (1993) Konstruieren recyclinggerechter technischer Produkte. Grundlagen und Gestaltungsregeln, Blatt 1, Beuth, Berlin.
Woll, R. (1994) Informationsrückführung zur Optimierung der Produktentwicklung. Reihe Produktionstechnik Berlin, Band 134, Hanser, München, Wien.

7 BIOGRAPHY

Prof. Dr.-Ing. Frank-Lothar Krause, born 1942, studied Production Technology at the Technical University Berlin. In 1976, he became Senior Engineer for the CAD Group at the Institute for Machine Tools and Production Technology (IWF) of the TU Berlin and earned his doctorate under Prof. Spur. Since 1977, he has been Director of the Design Technology Department at the Fraunhofer Institute for Production Systems and Design Technology (IPK Berlin). He earned the qualification as a university lecturer in 1979 and has been University Professor for Industrial Information Technology at the IWF of the TU Berlin since 1990.

Dipl.-Ing. Christian Kind, born 1964, studied Production Technology at the Technical University of Berlin. After finishing his diploma in 1992, he joined the Institute for Machine Tools and Production Technology (IWF) of the TU Berlin, where he works as a research engineer in the department of Industrial Information Technology.

PART TWO

Advanced Strategies

3

SUSTAINABLE INDUSTRIAL PRODUCTION ENVIRONMENTAL ISSUES IN PRODUCT DEVELOPMENT

L. Alting
Institute of Manufacturing Engineering
Technical University of Denmark Building 425
2800 Lyngby, Denmark
Phone +45 45 882522, Fax +45 45 933435
E-mail pila@unidhp.uni-c.dk

Abstract
The term sustainable industrial production is discussed based on the challenges facing the industrialised world. The focus is especially on methods and tools to enable the product development function to consider the environmental impacts in a product's life cycle phases at the development stage. This is an important element of developing a new industrial culture.

Keywords
Sustainable industrial production, environmental assessment, product development.

1 INTRODUCTION

During the last few decades it has become clear that our present industrial production and consumption culture is facing dramatic changes in the future due to:
- pollution and waste problems;
- nonrenewable resource consumption;
- rapid growth in the world population (with large production/consumption demands).

The developing countries are fighting poverty and health problems and we have no choice but to support their struggle for economic growth, which is a prerequisite for a more stable world. Our best contribution is to develop a new, sustainable industrial culture, which can be scaled up by a factor of 5-6 compared to the present level without creating environmental and resource problems.

It is with this background that the term sustainability appears more and more frequently at international and national meetings, in journals, etc. Figure 1 shows this schematically with a few of the major events listed.

The coming of the eco-era

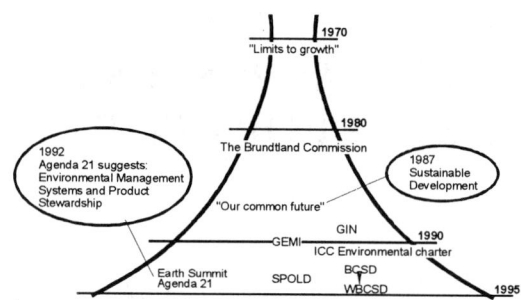

Figure 1 The coming of the eco-era.

The term sustainability is not very well defined. It is often used only to cover environmental issues, but basically it has a much broader content. In the Brundtland report, Our common future (1987), it is defined as:

"Sustainable development meets the needs of the present without

compromising the abilities of future generations to meet their own needs".

Seen from a company viewpoint Figure 2 illustrates the groups that influence/determine the long term/sustainable future of the company. These interest groups have different value criteria determined by societal culture, education, ethical values, etc. To ensure a long term survival the company must understand the value criteria of these interest groups and learn to communicate according hereto. If a company does not understand this new "language", it becomes very vulnerable in public discussions.

Interest/Groups	"Values"
Share holders	1. Economy - dividend; 2. Secure investment; 3. Positive image, good working place, ethical/responsible behaviour
Customers	1. Quality, right price, stability; 2. Environmental, health and resource responsible; 3. Positive image, ethical behaviour
Suppliers	1. Economic stability; 2. Products giving the suppliers good references (Environment, health, resources)
Employees	1. Good and stable work place; 2. Positive contribution to the local community/international; 3. Ethical behaviour (accepted values)
Local community	1. Economic contribution; 2. Environmental health and resource responsible; 3. Ethical behaviour, contribution to local pride
Interest groups/ organisations	1. Environmental, health, and resource responsible; 2. Ethical behaviour
National	1. Economic contribution; 2. Environmental, health, resource responsible; 3. Ethical behaviour, positive image, pride contribution
International	1. Economic strength; 2. Environmental, health, resource responsible; 3. Ethical behaviour; 4. Positive contribution to sustainable industrial culture

Figure 2 Groups that influence/determine the long term/sustainable future of a company.

Saemann (1995) concentrates the above conditions for a long term success into simultaneous and equivalent considerations of three responsibilities:
- economic;
- social/societal;
- environmental.

In this paper mainly the environmental responsibility will be discussed and to this Saemann says "comprehensive, acceptable environmental performance results only from a strategy, which involves all business functions and employees".

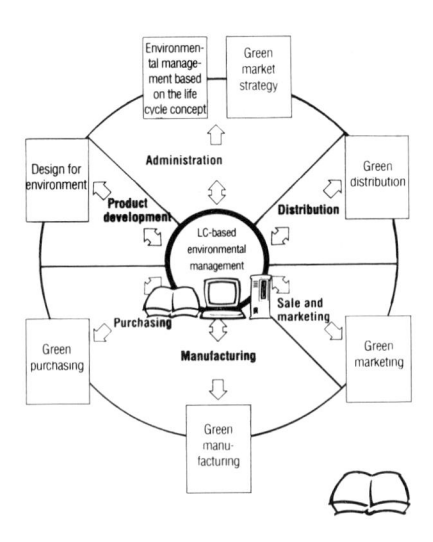

Figure 3 Environmental management system based on the life cycle concept.

Figure 4 lists some main elements of environmental responsible production (based on Alting, L. and Legarth J. B. (1995), Our common future (1987), Saemann, R (1995)). It is important to notice that in the last few years, the environmental focus in industry and in legislation has shifted from production processes to products in their life cycle. This is due to the fact that minimising environmental effects in a life cycle phase may increase pollution in another and the necessity to minimise the overall environmental consequences of our "product consumption", see also section 2.

Before turning to the product development issues a short discussion will be devoted to the enforcement of the transition towards a sustainable industrial culture.

On both national and international level the two major considered driving forces are:

- legislation (force compliance);
- market instruments (economic incentives).

Legislation is necessary, but it is not sufficient and enforcement problems are large. Furthermore, legislation always comes late and does not necessarily reflect an understanding of mechanisms in environmental protection. Market instruments are discussed heavily at the moment and used increasingly, but it is difficult always to find and instrument where the environmental incentive and the economic incentive work in the same direction. Otherwise, it is just a new type of tax, which does not change behaviour.

Function	Elements
Company board and executive management	1. Strategy/vision (product focused); 2. Corporate culture; 3. Steady and visible commitment and communication, planning, reporting system, management system; 4. Leadership style (empowerment", emphasis on ethics; 5. Product management - product life cycle assessments - long-term composition of product ranges - acquisitions/disinvestments
R & D Product development	Life cycle assessment of environmental consequences - in the conceptual stage of R&D projects, during product and process development, in the transition from research to products/processes, during development of design guidelines for environmentally friendly material selection, production, distribution, usage and disposal
Production	1. Investment proposals (plant design, siting, size, etc.); 2. Operation and maintenance instructions; 3. Cleaner technologies; 4. Material/waste-/energy/pollution minimisation; 5. Yearly documentation/reporting procedures
Purchasing/logistics	1. Supplier/subcontractor certification/inspection; 2. Mode of transportation; 3. Location, size, type of warehouses (distribution)
Marketing/sales	1. "Green products", documentation rules; 2. Support for solving of customers' environmental problems
Administration	1. Reduction of overall energy consumption (overhead); 2. Recycling of paper, machines, etc.

Figure 4 Main elements of environmental responsible production.

Besides legislation and market instruments a third and maybe the most important element in enforcement is personal and institutional ethics, Saemann, R (1995).

Here ethics is understood as ethics of responsibility (for actions and their consequences) and ethics of communication in all our actions and decisions. This has to do with the previously mentioned language of communication to interest groups.

For us in the technological area, Saemann, Saemann, R (1995), says that it is important to develop an "ethics in technology", if we are:

- To remain essential contributors to the future of mankind on our planet.
- To reenter the mainstream of public acceptance.

This "ethics of technology" will contain, Saemann, R (1995): personal responsibility, responsibility towards fellow men and society, protection of environment and contribution to economic success, freedom in technical research and its limitations, preservation of our basis of life, benefits versus risks/damages, innovation, technical competence, holistic, comprehensive view, two-way communication and truthfulness.

By applying such an "ethics of technology" the engineering community will create a "sustainable" basis for its contribution to a new world.

2 ENVIRONMENTAL ISSUES IN PRODUCT DEVELOPMENT

As seen from the discussions in the introduction (see Figure 4), the responsibility towards the environment is tied up with all functions in a company. The executive management has the responsibility for the overall policy/strategy/vision and for the frames established for each function in the company. Each function has some influence on the environmental impact caused by a product in all its life cycle phases and it is important to establish, how this influence is carried out within the overall frame. It is important here to realise that many product manufacturers within their own company only see a smaller or larger fraction of the environmental impacts from their products, as material suppliers, subcontracts, users and disposers will see the remaining fraction. This means that even if your own production is clean, the product may in its life cycle have a considerable environmental impact.

Therefore, it is important that environmental impacts are related to a product's whole life cycle and that the company's environmental responsibility is oriented towards this in all functions. This is also reflected in the change of focusing environmental protection from production processes to products, see Figure 5.

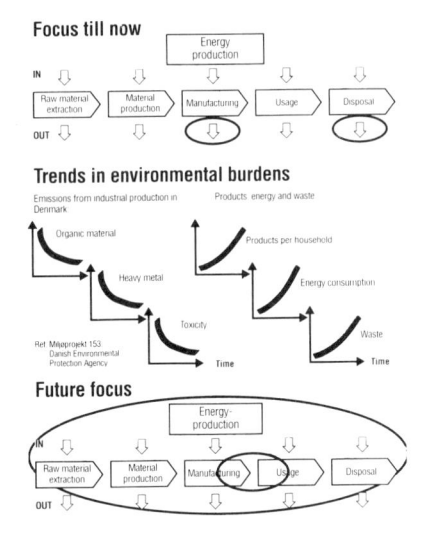

Figure 5 Environmental focus.

At the same time that pollution from production processes has decreased, our consumption of products, energy and production of waste has increased much more causing a totally much larger environmental impact. Therefore, the only alternative is to shift focus to products and to minimise the environmental impact in the life cycle. The result hereof is that future environmental control is mainly in the hands of the customer in interaction with the companies. In the light of the discussion in the introduction, legislation, market instruments and ethics of responsibility (also of the customer) will govern this interaction. To be able to cope with these issues the company must develop its products in a life cycle perspective and be able to document the environmental impacts in all life cycle phases.

The product development function is the major player in determining the environmental properties of the product, see Figure 6.

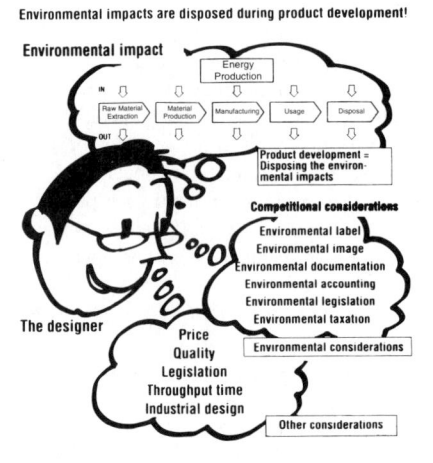

Figure 6 Environmental considerations in product development.

Here it is seen that by selecting product concepts, structures, materials, processes, etc., the environmental properties in all life cycle phases are determined. The product developer will have to consider the environmental performance along with all the other requirements to the product to come up with a competitive product.

The main issue here is which methods and tools are necessary and available to enable the product developer to handle the environmental aspects in an internationally accepted manner.

2.1 Methods and tools for the product developer

A major issue is here how to measure and assess the environmental consequences and how to do this before the product exists.

At the Technical University of Denmark a number of methods and tools to support the development function have been developed during the last couple of years and they will be published in the Spring of 1996.

Here the main contents of the tools are described. Figure 7 shows that the "yardsticks" to measure environmental impacts are:

- resources;
- environmental effects;
- health effects.

Sustainable industrial production 37

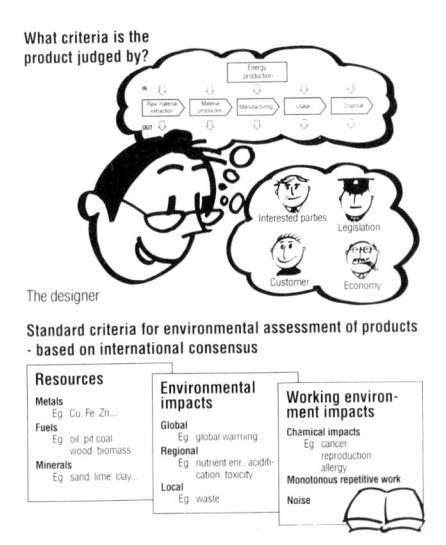

Figure 7 Environmental assessment of products, criteria for resources, environmental impacts, working environment impacts.

In this context mainly the environmental effects are discussed. The basis for calculating the environmental effects is the LCA (Life Cycle Assessment) methodology developed by SETAC, SETAC (1993). Figure 8 shows the necessary steps: modelling, inventory, standardisation, normalisation and valuation.

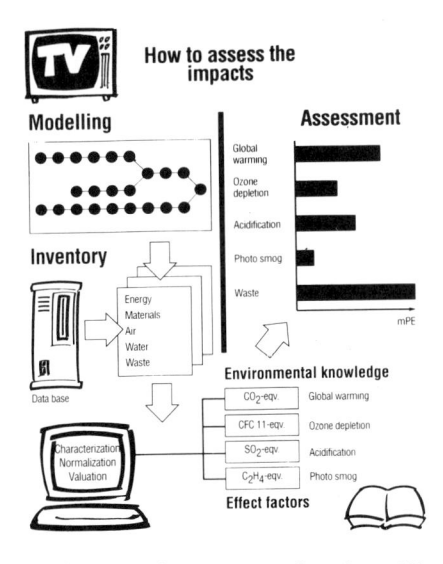

Figure 8 Environmental assessment of products, life cycle modelling.

Figure 9 shows in principle the large data collection and handling problem that LCA contains.

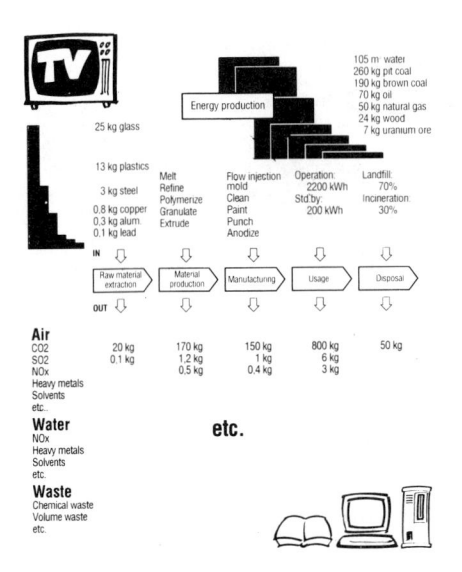

Figure 9 Environmental assessment of products, modelling/inventory.

After the standardisation step, the environmental impacts (divided into global, regional and local effects) are measured in substances emitted from a product in its life cycle for example kg CO_2 equivalents. To transform this number into a meaningful number it is normalised by the present yearly CO_2 emissions created per world citizen (approx. 8000 kg CO_2/year), i.e., transformed into PE (person equivalents). As to regional and local effects the present emissions per Danish citizen are used to transform the effects to PE or mPE (milli PE). Figure 10 shows environmental profiles based on these principles.

The described LCA can be used to diagnose environmental improvement potentials in an existing product, see the left profile for a refrigerator when changing the coolant from a CFC to a propane/butane. It is seen that ozone depletion disappears and CO_2 emission is reduced to 50% (R134a) and to 25% (propane/butane). But at the same time in the last case photosmog creation is increased by 18% - which is best, 25% reduction in the CO_2 impact or 18% increase in photosmog? An LCA can thus be used to focus environmental efforts, see the profile to the right. Here it is seen, from where environmental impacts result: materials, chemicals used in processes, thermal energy, electrical energy, and overhead. Overhead is impacts created by comfort energy in administration, production, etc., and other resources not associated directly with the product production, i.e., that do not change with changed products.

From this profile it is seen that by removing chemicals in production, eco-toxicity, human toxicity and ozone depletion can be reduced drastically.

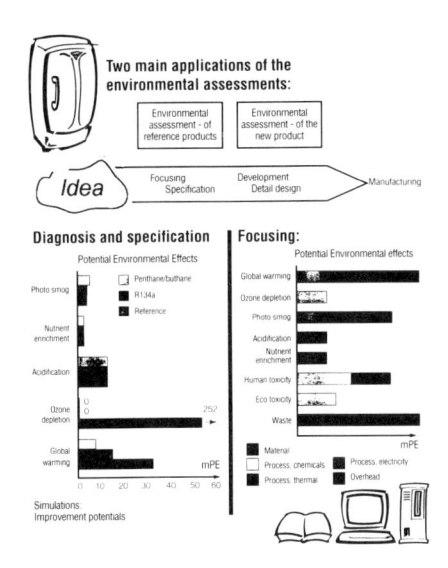

Figure 10 Environmental assessment of products, environmental diagnosis.

Another way of showing the origin of the environmental effects is illustrated in Figure 11, which is the profile for a good television. Again here it is easy to identify, where it is possible to improve the profile.

LCA can also be used on the conceptual level, but to get data it is necessary to use a reference product, for example an existing product or a synthetic product (known technology, competitors product, etc.).

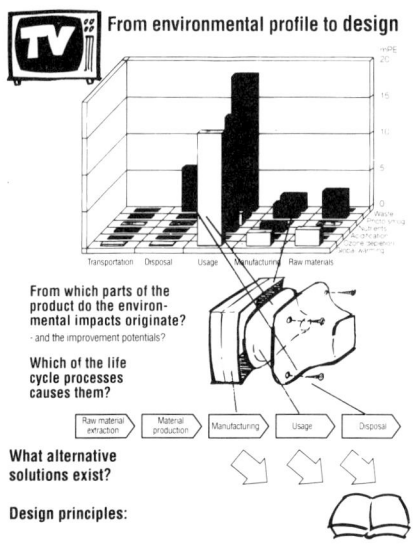

Figure 11 Design for environment.

To carry out a LCA requires much work and it is necessary to use computer based tools. These tools should also enable simulation of changes in materials, processes, etc. I.e., at the design stage the product developer can simulate the consequences of different solutions.
Some tools are available, Alting, L. and Legarth J. B. (1995), but these are mainly developed for and by environmental specialists. In the next years, new user-friendly computer based tools will emerge.

2.2 Design rules

If it is considered too complex to perform LCA's and the knowledge is not available in the company, simpler design rules may be developed, see Figure 12.

Life cycle phase	
Pre-manufacture	
Strategy	**Relevance**
Use of recycled materials	Resource depletion, environmental burdens
Use of less energy intensive materials	Environmental burdens
Environmentally conscious component selection	Supplier performance, environmental burdens
Use of renewable materials	Resource depletion
Manufacture	
Use high-through-put processes	Environmental burdens, working environment
Use material saving processes	Resource depletion, Environmental burdens
Overhead reduction	Environmental burdens
Transportation/distribution	
Improved logistics	Environmental burdens
Low volume / weight	Environmental burdens
Use recycled materials for packaging	Resource depletion, environmental burdens
Use	
Low energy consumption	Resource depletion, environmental burdens
Design for maintenance/long life	Resource depletion
Disposal	
Design for disassembly	Resource depletion
Material quality preservation	Resource depletion, environmental burdens

Figure 12 Prevailing design strategies.

These design rules must be based on a LCA for the specific product group and cover all life cycle phases for the product. Depending on the level of detailing very specific rules can be made for the specific phase. Figure 13 shows design for disassembly rules.

Benefits	Design rules
Less disassembly work	1. Combine elements; 2. Limit material variability; 3. Use compatible materials; 4. Group harmful materials into subassemblies; 5. Provide easy access to harmful, valuable or reusable parts
Predictable product configuration	1. Avoid ageing and corrosive material combination; 2. Protect subassemblies against soiling and corrosion
Easy disassembly	1. Accessible drainage points; 2. Use fasteners easy to remove or destroy; 3. Minimise number of fasteners; 4. Use the same fasteners for many parts; 5. Provide easy access to disjoining, fracture or cutting points; 6. Avoid multiple directions and complex movements for disassembly; 7. Set centre-elements on a base part; 8. Avoid metal inserts in plastic parts
Easy handling	1. Leave surface available for grasping; 2. Avoid non-rigid parts; 3. Enclose poisonous substances in sealed units
Easy separation	1. Avoid secondary finishing (painting, coating, plating, et.); 2. Provide marking or different colours for materials to separate; 4. Avoid parts and materials likely to damage machinery (shredder)
Variability reduction	1. Use standard subassemblies and parts; 2. Minimise number of fastener types

Figure 13 Generally accepted DfD design rules.

For many small and medium sized companies these types of design rules will be fundamental.

3 PERSPECTIVES

It is important that environmental issues are seen as a part of the sustainable industrial culture and as a part of the strategy/vision for the environmental conscientious company, see Figure 4.

In the next years, the international standardisation work in ISO (14 000 series) will facilitate the introduction of LCA in the companies, but many companies have already achieved significant results.

In a few years computerised LCA tools will be part of designers' CAD tools making environmental life cycle assessment a natural basis for all product development.

4 REFERENCES

Alting, L. and Legarth J. B. (1995) Life cycle engineering and design. CIRP keynote paper, August 1995, to be published in CIRP Annals, vol. 1995.

Our common future (1987) Oxford University Press, (The Brundtland report)

Saemann, R (1995) Environmental strategies, ethics and management. Swiss Academy of Engineering Sciences, CAETS-Convocation, Sweden 1995.

SETAC (1993) Guidelines for life cycle assessment: A code of practise. Edition 1, SETAC workshop, Portugal 1993.

5 BIOGRAPHY

Professor Leo Alting is head of The Institute of Manufacturing Engineering and Division Manager for the Division of Materials and Process Engineering, The Institute for Product Development, Technical University of Denmark. Mr. Alting is involved in many industrial projects also on a European basis as well as many research programmes. Mr. Alting has established a rapidly growing development and research group on Environmental Life Cycle Assessment of industrial products as a support function for industrial companies to develop environmentally friendly products. Prof. Alting is a member of The Danish Academy of Technical Sciences, The Royal Swedish Academy of Engineering, CIRP and several institutions and societies.

4

TOWARDS A COMPREHENSIVE LIFE CYCLE MODELING FOR INNOVATIVE STRATEGY, SYSTEMS, PROCESSES AND PRODUCTS/SERVICES

Vijay A. Tipnis
Synergy International, Inc., 285 Heards Ferry Road, Atlanta, GA, 30328 USA,
Tel: (404) 843 0694;   Fax: (404) 851 0134;   e-mail: vtipnis@aol.com

Abstract
Life cycle modeling of products, processes, and systems has become crucial for environmental stewardship as well as for corporate competitive strategy. Experience to date demonstrates that although life cycle modeling is conceptually obvious, it is not easy to perform. Among the difficulties faced are: (1) How to limit the scope and yet produce useful information? (2) How to model the entire life cycle when only limited information exists on the life cycle performance? (3) How to ensure that the life cycle model is comprehensive and yet manageable? (4) How to ensure data integrity and consistency? (5) How to establish competitive, yet environmentally responsive stewardship? The framework presented in this paper suggest a plausible comprehensive life cycle modeling methodology that addresses that above questions.

Key Words
Life Cycle Modeling, Environmental Stewardship, Competitive strategy

1 INTRODUCTION

Traditionally, manufacturer's product warranty extended to a limited period (from a few days to a few months) from the purchase of a product. The warranty covered product against certain specified material and manufacturing defects with caveats for proper use and maintenance, and against abuse. Only during this limited warranty period, manufacturers are held liable for failures in the stated performance of the product and certain harmful side effects during its proper use provided the product is properly maintained. Although this practice of limited product warranty continues, it is no secret that the growing environmental and safety regulations in many countries essentially hold the manufacturers responsible for the harmful and toxic emissions, recycling, and disposal of the product, packaging, and waste by-products generated during the entire life cycle of the product as well as for any resulting harm to the users, the society, and the environment. How to evaluate and design products against these life cycle liabilities has, therefore, become a formidable challenge to the manufacturers.

Initially, most manufacturers considered strict environmental regulations as unwelcome restrictions on their operations and products, putting them at a competitive disadvantage against the manufacturers from regions with less strict environmental regulations. Eventually, some leading manufacturers began to carve out a competitive advantage from improvements in products and processes spurred by the need to satisfy the strict environmental regulations. The competitive advantage came to these leading manufacturers from innovative products and processes, and also, from their customer preference to their products because of their environmental stewardship.

Environmental stewardship involves (1) putting protection of our fragile environment above parochial competitive gains, (2) curtailing urges to stimulate customer 'wants' of the 'throwaway' consumerism, and (3) environmentally-safe technologies for improvement and protection of the environment. The habits of predatory competitive behavior which exploits consumers, regions, and nations by environmentally harmful consumption of materials and energy and waste disposal should not be tolerated by environmental stewardship. Are we ready for this challenge? Some are not convinced of the urgency and gravity of the environmental challenge. Others fear that they may be at a disadvantage if their competitors do not embrace environmental stewardship as well. However, environmental stewardship is providing a powerful competitive advantage to some of the world's leading corporations. Thus, a new manufacturing paradigm, called the Paradigm 'E' emerged: *Compete on Ecology* (Tipnis, 1993). According to the Paradigm 'E', the life cycle design of products, processes and systems as well as environmental stewardship became a matter of competitive strategy in the global competitive market place. The potentially catastrophic global effects of the explosive growth in consumerism and consequently, pollution predicted from the systems dynamics models (Meadows and Meadows, 1972 and 1992), must be halted. The sooner the Paradigm 'E' is adopted by manufacturers and consumers, the closer we come to avoiding the global catastrophe.

One of the first tasks to Compete on Ecology is to integrate environmental considerations into the business and product strategies (Tipnis, 1994b). Most corporate environmental organizations are rarely involved in the mainstream corporate product or production decisions; they are considered 'add-ons' that deal with the environmental regulations and constraints. Without such an integration, the following important transformation tasks (Tipnis, 1993) cannot be done well: (1) implement abatement technologies to meet current environmental regulations in the short run and initiate projects for pollution prevention at the source through redesign of products, processes, systems, and practices, (2) ensure that containment is provided to existing processes exhibiting risks of harmful emissions so as to minimize potential hazard to workers and users, (3) initiate product recycling programs to ensure that products already in use are collected and recycled per regulations, and (4) redesign products to minimize recycling and harmful emissions to the work place, surrounding community, and the ecosystem. Life cycle modeling is a basic tool that applies directly or indirectly to all of the above challenges.

Life cycle, as applied to living creatures, is an easily understood concept: from womb-to-tomb. For man-made artifacts such as material products, when the same life cycle concept is applied the question arises as to what should be the starting point of the product life and what should be the end of the product's life. Indeed, the entire industrial ecosystem must be included if every material and energy consumed in the product's life cycle is to be accounted for, including the product's reuse, re-manufacturing, and recovery of materials and energy. Moreover, the harmful effects of emissions and waste dumped into nature from the manufacture and use and disposal of products are, in most cases, cumulative and time-delayed. How to account for the contributions made form individual product to the initial gradual, cumulative, and eventually exponential degradation of the ecosystem is also a part of the challenge of life cycle modeling. The overall challenges of life cycle modeling and the state of the art is found in (Inoue, 1992, Jovanne, 1994, Alting, 1995, Geiger, 1995, and Yoshikawa & Kimura, 1995, among others).

Clearly, it is not an easy task to make life cycle modeling a rigorous and hence a dependable tool. Such a tool must not only satisfy the environmental regulations on toxic materials inventory and environmental impact assessment, but also more importantly, be useful for evaluation of technological alternatives involving trade-offs between performance, quality, productivity and life cycle cost versus the environmental requirements. In this paper, I have attempted to briefly outline a framework for this task.

2. LIMITATIONS OF THE CURRENT LIFE CYCLE ASSESSMENT (LCA) METHODOLOGIES PRACTICED IN ENVIRONMENTAL ASSESSMENTS:

For the past 15-20 years or so, Life Cycle Assessment (LCA) methods have evolved in response to the need for determination of material flows during the entire life cycle of a product, a process or a system.

LIFE CYCLE INVENTORY ASSESSMENT (LCI): As applied to environmental and ecological issues, LCA was primarily restricted to inventory assessment (also, known as 'Cradle-to-Grave' and more recently, as 'Concept-to-Grave' analysis during the entire life cycle) of material flows within an ecosystem focused on the quantification of environmental releases and materials (prone to toxic or harmful emissions) used in the entire life cycle of a product, package, process, or activity. Similarly, efforts have been made to assess the impact of energy generation, transmission, transportation, and consumption on environment. Although, LCA primarily deals with the material flows, the realization that the energy and material flows are intertwined within any industrial ecosystem has led several efforts to include both the flows in environmental assessment. However, despite some significant advances, the early efforts suffered from the following limitations:

(1) Most investigations were a 'onetime snap shot' of a complex phenomena.

(2) Most investigations were reported as "take my word" reports. There was no practical method to verify the validity, completeness, or comprehensiveness of the assessment or the accuracy of the data.

(3) The lack of rigor presented a particularly vexing problem of credibility of the results and recommendations from these reports.

(4) No practical method existed to 'reuse' or 'update' the assessment as the parameters or data values changed.

Society of Environmental Toxicology and Chemistry (SETAC) undertook the mission of creating a technical frame work for LCI (SETAC, 1991) and a code of practice guidelines for LCA (SETAC, 1993). SETAC has set the following goals for bring greater credibility and usability of the data generated during LCAs: (a) reviewers must be able to gain good understanding: applicability, validity, and limitations of an investigation, (b) the purpose and the goals of the LCA must be clearly defined, (c) the system boundary, data requirements, assumptions, and limitations must be defined, (d) the spatial (global, regional, local) & temporal (time horizon) must be stated, (e) a measure of performance must be defined before hand, and (f) data-quality must be predefined.

Clearly, SETAC needs a comprehensive LCI model to accomplish its goals. Despite considerable work SETAC recommendations cannot be implemented until a rigorous modeling methodology, associated databases, and computational and simulation tools are developed and validated.

SETAC ENVIRONMENTAL IMPACT ASSESSMENT (EIA): While LCI includes material flows during the life cycle phases (Raw Material Acquisition Distribution & Transportation, Materials Processing, Use/Reuse/Maintenance, Recycle, and Waste Management), EIA (SETAC, 1992) deals with the impact of harmful emissions on the following impact categories:

Human health: Acute Effects (Accidents explosions, fires) and Chronic effects: (Diseases)

Ecological Health: Population, Community, Ecosystem

Resource depletion: Stock Resources (Energy & materials), Flow renewables (energy & materials)

Social Welfare: Air, Water, Land quality/quantity and Agricultural productivity

In EIA, the concept of stressor is introduced to assess the contribution of individual impact levels from a given category of emissions. For example, typical human health impact stressors are chemical toxicity, high energy radiation, and condition creating potential hazards for accidents. SETAC, realizing that it is extremely difficult to establish a clear-cut cause-effect relationship between the emissions and the specific stressors, suggests that a site specific cause and effect is not necessary for EIA. However, few corporations will be satisfied to base crucial decisions which may lead to litigations if site specific data were not used to support the claim. Emissions (BOD, COD, VOD) and waste release pathways (Airborne Emissions, Effluents, Solid Waste release, Fugitive release, Land release, Ground and Water Release) must be clearly and accurately mapped to establish their potential impact on the specific stressors. How to do this mapping still remains a challenge to life cycle modeling.

IMPROVEMENT ANALYSIS: SETAC is currently working on establishing guidelines for improvement analysis to make improvements based on the findings of the application of LCI and EIA to specific cases. How well business needs will be addressed in these guideline remains to be seen.

3   CORPORATE COMPETITIVE STRATEGY AND NEEDS FOR LIFE CYCLE ASSESSMENT IN PRODUCT, PROCESS, AND SYSTEM IMPROVEMENT

Corporate competitive strategy and needs for life cycle assessment of product, process, and system improvement have been always intertwined; one cannot be done without the other (Tipnis, 1994b). Until the emergence of the environmental regulations, corporate use of life cycle assessment was confined only to the warranty period. Also, little if any attention was paid to emissions, waste disposal or recycling. Furthermore, the products, processes, and systems were not designed with the life cycle impacts in mind. This has profoundly changed with the emergence of the Paradigm 'E': Compete On Ecology. Now there is a great deal of urgency in designing for environmental stewardship. Consequently, the need for comprehensive and dependable life cycle assessment has become acute. However, corporations are not just interested in using the life cycle assessment for satisfying the environmental regulations as SETAC is emphasizing, they are more interested in formulating alternative competitive strategies based on the projections from life cycle analysis of product, process, and system performance, quality, reliability, warranty, product liability, and harmful side effects that may adversely affect their environmental stewardship. This makes life cycle assessment the very backbone of the corporate competitive strategy. Therefore, a proper architecture and structure must be provided for life cycle assessment to incorporate the current product, process, and system assessment tools under the umbrella of the life cycle assessment.

One of the early uses of systematic product life cycle assessment can be found in aerospace industry where the fuel usage, parts replacement, and maintenance dominate the overall life cycle of an aircraft (Tipnis, 1988). Consequently, during the design phases of an aircraft emphasis is placed on the improvements having significant impact on the frequency and cost of repair, and also, on the weight and fuel savings. Life cycle models are used to identify improvements based on the alternatives of materials, processing, and design configurations. Indeed, these improvements continue with the aging of the aircraft fleet as new materials or technologies are evaluated and implemented with every design update. Automotive, and other industries have now become interested in life cycle costs and consequences to their customers as well. The situation in electronics products is different. The rapid technological progress is making the life cycle shorter and shorter and hence the electronic products become obsolete long before they start to wear-out. Here attention has to be paid to disassembly, recycling and recovery of precious raw materials. Therefore, life cycle models have become crucial for evaluation of design alternatives in materials, processing and assembly methods. Furthermore, environmental regulations have forced all industries to evaluate the consequences of emissions during manufacturing, usage, and disposal/recycling phases. This too must be addressed by the life cycle models.

4 PROPOSED FRAMEWORK FOR LIFE CYCLE MODELS

At the outset it should be recognized that there are a variety of product, process, and system models in use to serve a variety of needs during design, manufacturing, and use phases of the life cycle of products, processes and systems (see Krause et al, 1994, for example). From simple input-output models to complex dynamic, physical, and stochastic models are needed to evaluate, simulate, and estimate performance, quality. reliability, productivity, cost and other attributes. The proposed framework does not replace these models; nor does it hinders their further evolution. The proposed framework provides simply an umbrella architecture under which these models can be represented and interconnected. The emphasis of the framework is conceptual and computational simplicity without restricting these models to any specific model forms. This is accomplished through the umbrella of Activity Modeling methodology. As a prelude to the framework acknowledge the following:

PRODUCT & PROCESS MODELS: Every product is manufactured through a series of sequential steps, called processes, where the raw material is transformed into finished end product (Tipnis, 1988). During use of the end product, the product performance gradually degrades due to wear and tare and part failures. Although, repair and maintenance processes may restore the product's performance and use, it is discarded when it can no longer be repaired or maintained economically. However, the products life ends from it being replaced by a newer and often more capable product. This product obsolescence may happen to the entire product or process technologies themselves when manufacturing or use paradigm-shifts occur. Indeed, planned technological product obsolescence is a successful corporate competitive strategy. Thus, many discarded products may still have substantial residual life. How to account for waste of materials and energy in this residual life is also a part of the challenge of life cycle modeling.

Interestingly, software can also be represented by an analogous series of processes which transform input data into output results. Information models of products, processes, and systems can, therefore, be represented like those of the materials and energy transformation models.

SYSTEMS: Everything can be considered a system that is composed of two or more indivisible components depending on the level of abstraction and point of view of the observer . For simplicity of abstraction in modeling complexities, system boundaries are predefined. Everything outside the system is considered as environment which imposes certain influences not in control of the system. System and environment exchange certain material, energy, and information flows. Furthermore, it is possible to consider the system to be a part of a supersystem in which it serves a certain useful function, and every system can be decomposed into subsystems and components which are interlinked into a specific structure to produce the desired system performance. Humans as workers in the systems or users of the system can be represented as a part of the supersystem which the system serves. All of the above entities can be directly or indirectly exposed to the surrounding natural environment as shown in Figure 1. The task for life cycle modeling is to construct an appropriate framework in which the system architecture (hierarchy) and structure (connections) can be first represented and then evaluated consistently and rigorously.

PHYSICAL MODELS: Every processing unit that performs material and/or energy transformation has one or more physical (also, chemical, biological or nuclear) processes which can be represented by appropriate physical models. Recently, a unified methodology to define a variety of physical transformations in manufacturing systems is proposed (Bjorke, 1995). The connection between the processing unit models, which are generally empirical (Tipnis, 1988), and the underlying physical models determines how well the process can be modeled and, therefore, controlled.

THE PROPOSED FRAMEWORK: The overall framework of the proposed life cycle model shown in Figures 2 & 4 consists of successive decompositions of the environment-supersystem-system-subsystem architecture defined in Figure 1. As shown in Figure 4, the hierarchical nature of the life cycle model allows a top down decomposition of each phase of the life cycle. Activity modeling methodology described below can be used for this purpose. Granted, the activity modeling methodology provides essentially an input-output model with a static snapshot view of the life cycle. However, its capability to represent all sorts of activities: material, energy, data, manual, and automated transportations, storage, transformations, inspection, etc. make it a viable candidate for umbrella under which comparative, dynamic, simulation, and other physical models can be defined. Furthermore, the activity models can be used as a basis for Colored Petrinet and dynamic simulation models as well as for discrete event simulators of manufacturing systems, and also for project management, and cost estimation using Activity Based Costing. Ease of use and availability of graphics software makes these models readily integrable and upgraded. Connections to spreadsheet computations and to relational databases are also built-in in some of the commercially available activity modeling software. Therefore, a common architecture that can be shared by design, manufacturing, suppliers, users, recyclers is feasible without the need for additional software coding. In the following the methodology of activity modeling is defined with an application to life cycle modeling.

5   ACTIVITY MODELS

Activity Modeling is the process of defining each activity (function, operation, process, action, etc.) in a system which involves several inter-linked subactivities. Activities may be transformation, movement, generation, use, or disposal of material, energy, data, and information. A specific activity analysis of interest is the Structured Analysis and subsequently refined version known as IDEF (ICAM, 1975) developed by the U. S. Air Force for defining complex set of interacting activities involved in the life cycle of an aircraft. Although IDEF belongs to a group of activity modeling software known as Computer Assisted Software Engineering (CASE) tools, it is particularly suited to define complex systems involving material, energy, data and information transformation because of the strict grammar and convention within it. The following conventions are followed to bring consistency and rigor to the life cycle modeling

Purpose & Viewpoint Of Activity Models: Activity Models become more useful when the purpose (why & so-what) and viewpoint (for whom is the model constructed) are clearly stated at the beginning. It is most important to define **the boundaries of the system** to be modeled based on the purpose. Also, the viewpoint must match the purpose as well as **the structure of the model** (system, sequence, processing unit, operations, and processes). The model must be **resolved** (exploded or decomposed) only in those areas pertinent to the purpose per the viewpoint.

Activity Model IDEF Convention (Refer to Figure 2): **Activity:** Any action that is necessary to convert the inputs into outputs. Activity must be a **Verb**. **Input & Output:** Any sets of parameters which are transformed by an activity. These parameters assume specific values at instances depending on the states of the system and the activity. Inputs & outputs are **Nouns** (Specifically, **Objects**). Material, Energy & Data are typical inputs and outputs. All inputs and outputs are shown as sets and subsets; inputs entering from the left face of the activity box and outputs exiting from the right face of the activity box. When input or output from one activity box becomes a control to another activity box, it **qualifies** (limits, constraints, or guides) the activity that can be controlled in that box. Inversely, if a control from one activity box becomes an input to another activity box, it provides a constrained input to that box. **Controls** (also, called **Constraints**): These specify the limits, range, procedures, working regions permissible for the activity to take place. Any points laying outside the region are considered not permissible for the activity. These are **Qualifying Phrases**. **Mechanisms:** The entities, also known as **Resources**, that make the activity happen: Machines, Equipment, Computers, Robots, Organizations, and Individuals. These are **Nouns** (Specifically, **Subjects**).

Activity & Data Explosion Relationships: Activity can be subdivided into its subactivities as long as all the subactivities are members of the subset. The subsets should be exclusive members of the activity; no subactivity can be member of two activities. **Data (Inputs & Outputs):** can be subdivided into their appropriate subelements per the data set they belong to. Each subelement is an exclusive member of a single data set in a given activity. Also, **Join** and **Branch** of the data set divides the data set into (Branch) or combines the data elements into (Join) per the set operation stated in the assumptions of the model. This maintains data integrity for the purpose of exporting data to the data dictionary and data computations.

Activity Model Construction Process (See Figure 3): Activity Model construction should follow the following steps:

1. Define the purpose, the viewpoint and the scope.

2. Prepare an overall schematic diagram showing the ecosystem boundaries (spacial, functional, and temporal).

3. Depending on the purpose, the viewpoint and the scope, establish the boundaries of the system to be modeled.

4. Prepare A-0 diagram with the highest level activity necessary per the purpose, the viewpoint, and the system boundary. Identify all pertinent inputs, outputs, constraints, and mechanisms. Make sure that these are the supersets of all other elements of the system. For avoiding undue model complexity, limit inputs, outputs controls, and mechanisms to three or less supersets. It is often necessary to reach a consensus among participants on the specific definitions of the terms (Input, Constraints, Output, and Mechanisms). A common vocabulary is essential to achieving clarity in the model.

5. Resolve the system into subsystems per the viewpoint. Keep the system, sequence, processing unit, operations, processes sequence intact. For each explosion prepare a subset of the schematic so as to ensure that there is no overlap or missing portions form the overall schematic. (A schematic diagram of the explosions of different levels of details of the paint system is shown in Figure 5 for the case of automotive paint line).

6. For each subactivity, identify modes of operation at the lowest possible level. This ensures that the modes of operation are kept distinct from the model structure.

7. Only explode or resolve to the least extent necessary to answer the purpose of the model, and only those portions of the system that are directly involved in the viewpoint. This keeps the model building effort manageable in size and scope.

8. Prepare the tentative **final results** frames (For example, Excel Outputs and Worksheets) before resolving the model. This will help limit the explosion process to only those pertinent to the final answer. (The typical results are obtained by comparing As-Is vs. To-Be models as shown in Figure 3).

9. Activity Model is best built for a specific system; generic models can then be formed from a few specific cases, if necessary. For the model validation, specific data are needed. Only validated models should be used to draw recommendations.

Activity Modeling Software Tools: IDEF modeling tools are readily available from a number of software houses in the U. S. A. The one used in the Synergy Methodology® (Tipnis, 1993) allows connections to a relational database, spreadsheets, and also, allows Activity Based Costing/Management (ABC & ABM) to be performed. Furthermore, it has hyper-text and hyper-graphics capability for linking schematic diagrams, text, and imported scanned images with the IDEF logical activity diagrams. The software runs on Macintosh, DOS & Windows PC, and UNIX computers. As shown in the next section, the Product Life Cycle Economic Modeling has been linked to the Activity modeling through Excel spreadsheet worksheets. Furthermore it can serve as a front-end to Colored Pertinet software, and discrete event simulators. The links to specific product, process and system models can be constructed through the data dictionary provided. However, any integration with other models may need to be coded in. Furthermore, some models such as System Dynamics models (Meadows et al, 1972, 1992) of life cycle may best be constructed from scratch to serve the specific needs.

6 DISCUSSION OF QUESTIONS FOR ATTAINING COMPREHENSIVE LIFE CYCLE MODELING

The proposed framework for comprehensive life cycle modeling must provide satisfactory answers to key questions drawn from the environmental and business considerations stated in the introduction. SETAC documents (1991-93) provide guidelines form environmental view point. The business considerations, especially for competitive product strategy and product planning are some of the key steps because these early steps control the entire design and deployment of product, process, and system life cycle as pointed out in (Tipnis, 1993, 1994a). At this early stage of the proposed framework for life cycle modeling, I invite everyone concerned to participate in the discussion of the questions from their own perspective and experience, as I am attempted to do below.

QUESTION 1: How to limit the scope and yet produce useful information? This is the first challenge one faces in every life cycle modeling. As the material, energy and information flows become identified, it is tempting to include all pre- and post- activities that affect the life cycle under investigation; soon model grows to all involved material, energy and information flows for the product, processes, and systems in the entire industrial ecosystem, making it virtually impossible to get any useful information.

Clearly the scope of any specific life cycle modeling must be limited. To date the best method to do this is to identify the portion of the total physical system that best satisfies the intended purpose of the investigation. For example in the life cycle assessment of the automotive paint line (Tipnis, 1993), the purpose was to improve quality (reduce defects) of the painted surface at the same time meet projected stricter environmental regulations for effluent disposal into river. The scope was chosen to include all cleaning, degreasing, and painting processes, as well the waste water and solid waste treatment systems involved (see Figure 4). This limited the scope yet enabled useful information to be investigated for the trade-off between quality improvement in painted auto bodies and the restrictions on emissions of effluents from the paint line The limiting of the scope permitted the life cycle model to be manageable and useful.

QUESTION 2: How to model the entire life cycle when only limited information exists on the life cycle performance?

At the early design stages of a new product, little if any information exists on the projected performance of the product. The reliability of the components and subsystems in the product must be estimated from prior similar product's performance and warranty data and projected for the new product. It this stage, construction of a conceptual life cycle model with all the down stream product life cycle phases is a sound starting point as was found in the recent project at Xerox (Tipnis, 1994b). Although limited information existed about the downstream phases, the identification and construction of an activity model of the product life cycle allowed each aspect with potential influence on the product performance could be evaluated through a sensitivity model. Aspects with greatest projected influence were then investigated through subsequent Robust Design experiments.

QUESTION 3: How to ensure that the life cycle model is comprehensive and yet manageable?

How comprehensive a life cycle model should be depends on the purpose and scope. In activity modeling, the first step is to clearly define the purpose and the scope so as to limit the scope to the specific objectives of the purpose. It is necessary to identify the following:

(1) The environment, supersystem, system, subsystem, and components involved (see Figure 1 for guidance).

(2) The main function and the auxiliary function the system to be modeled must serve.

(3) The harmful side effects (environmental and others) resulting while the system performs the main and auxiliary useful functions.

(4) The specific objectives of the purpose must be stated in terms of the type of improvement desired in the main, and auxiliary function, and reduction in the harmful side effects. This must be stated in quality, reliability, productivity, life cycle cost, and environmental impact reductions.

(5) The scope must be limited to the material, energy, and information flows within the bounds of the specific objectives.

(6) Although the life cycle activity model may start form top down, it only needs to be decomposed in those life cycle phase concerned with the specific objectives.

(7) The model can be made as comprehensive as necessary by decomposing it through activities, processing units, material, energy, and information transformation processes and even to the physical transformations for materials or algorithmic transformations for data using appropriate modeling methodologies. The specific activity model decomposition provides the necessary inputs, output, constraints for these models.

Using this scheme, the paint line activity model was made comprehensive yet manageable to about 250 activities (Tipnis, 1993, 1994a, 1994b).

QUESTION 4: How to ensure data integrity and consistency?

It is important to start with a clear definition of the data elements and their set relationships. In almost all the product, process, and system life cycle modeling projects, the need for common and unambiguous definitions for each of the terms involved was found to be essential because different divisions of the same corporation use the same terms with quite different meanings. A dictionary of the terms and their meaning and interpretation was found to be very essential to ensure data consistency and integrity. Furthermore, the decomposition of the data through the activity model required set operation to be identified and predefined to avoid ambiguities and overlaps in the meaning of the data elements. This can best be illustrated by the need to identify each of the different types of waters as they pass through each processing step of the post-paint waste-water cleaning and recovery processes (Tipnis, 1993).

QUESTION 5: How to establish competitive, yet environmentally responsive stewardship?
This question poses a new challenge that can only be satisfactorily addressed by adopting the paradigm 'E', including the recommended seven steps to adopt the paradigm provide specific guidelines (Tipnis, 1993, 1994a). Implementation of these guidelines require construction of the appropriate product, process and system life cycle models right from the construction of product strategy and product planning stages because no other phase determines product's success or failure more than product strategy and product planning. No matter how well the subsequent phases of concurrent engineering, pre-production, production and others are executed, the competitive posture for the product is frozen in this phase. How to design a winning competitive product strategy and how plan and execute product introduction timing have become key challenges in the global competitive market place. Moreover, the environmental stewardship adopted in the competitive business strategy must form the basis for the design of the entire product life cycle; the materials, energy, processes, systems, and the technologies for assembly, disassembly, recycling, and disposal must be decided in this phase as well. Therefore, how to generate, evaluate, and select competitive product concepts that embody environmental stewardship is crucial. A more compete answer to this question is found elsewhere (Tipnis, 1994b).

In closing, it is important to point out the importance of economics to competitive strategy and environmental stewardship because there is a need to construct comprehensive life cyclc cost models (Tipnis, 1994a) that must accompany application of life cycle models to the evaluation of alternatives in competitive product strategies as well as environmental stewardship (Tipnis, 1993). Because no matter how much we desire to improve the environment, ultimately, we must make tough economic choices: who pays, how much, when, and so-what are the payoffs? Without exception, all environmental decisions require economic sacrifices: consumers must pay more, communities must tax more, corporations must spend more, and governments must allocate more budgets to environmental actions. And yet, economics of environmental improvements is one of the least well understood subjects.

7 REFERENCES

Alting, L. and Lengrath, J. B. (1995) Life Cycle Engineering and Design. CIRP Annals Vol 44/2/95
Altshuller, G. (1985) *Creativity As An Exact Science,* (Translation) Gordon Breach Publisher
Bjorke, O. (1995) *Manufacturing System Theory,* Tapir Publishers
Geiger, M. (1995) Towards Clean Forming Techniques. CIRP Annals Vol 44/2/95.
Inoue, H. (1992) *Ecofactory: Concepts and R&D Themes,* JETRO Special Issue, Tokyo.
Jovane, F. et al (1994) A Key Issue in Product Life Cycle: Disassembly. CIRP Annals Vol 43/2/
Krause, F.-L., Kiesewetter, T. and Kramer, S. Distributed Product Design. CIRP Annals, Vol. 43/1/1994 149-52.
Meadows, D. H., Meadows, D. L., Randers, J., and Beherns III, W. W., (1972) *Limits to Growth,*

Signet Books.
Meadows, D. H., Meadows, D. L., Randers, J. (1992) *Beyond The Limits,* Chelsea Green Publ.
SETAC (1991) *A Technical Framework for Life-Cycle Assessments,* The SETAC Foundation, USA.
SETAC (1992) *A Conceptual Framework for Life Cycle Impact Assessment,* SETAC Foundation.
SETAC (1993) *A Code of Practice Guidelines for Life Cycle Assessment,* SETAC Foundation.
Tipnis, V. A. (1988) Process and Economic Models for Manufacturing Operations, *Design and Manufacturing of Integrated Manufacturing Systems,* edited by W. Dale Compton, National Academy Press.
Tipnis, V. A. (1991) Product Life Cycle Economic Models. Annals of CIRP, Vol 40/1/1991, 463-66
Tipnis, V. A. (1993) Evolving Issues In Product Life Cycle Design. Annals of the CIRP, Vol 42/1/1993, 169-73.
Tipnis, V. A. (1994a) Towards A Comprehensive Methodology for Competing on Ecology. Proceedings of the Electronics and Environment Symposium, IEEE, 139-48.
Tipnis, V. A.(1994b) Challenges in Product Strategy, Product Planning and Technology Development for Life Cycle Design. Annals of the CIRP, Vol43/1/1994, 157-62.
Yoshikawa, H. and Kimura, F. (1995) Inverse Manufacturing. CIRP Discussions of STC Dn.

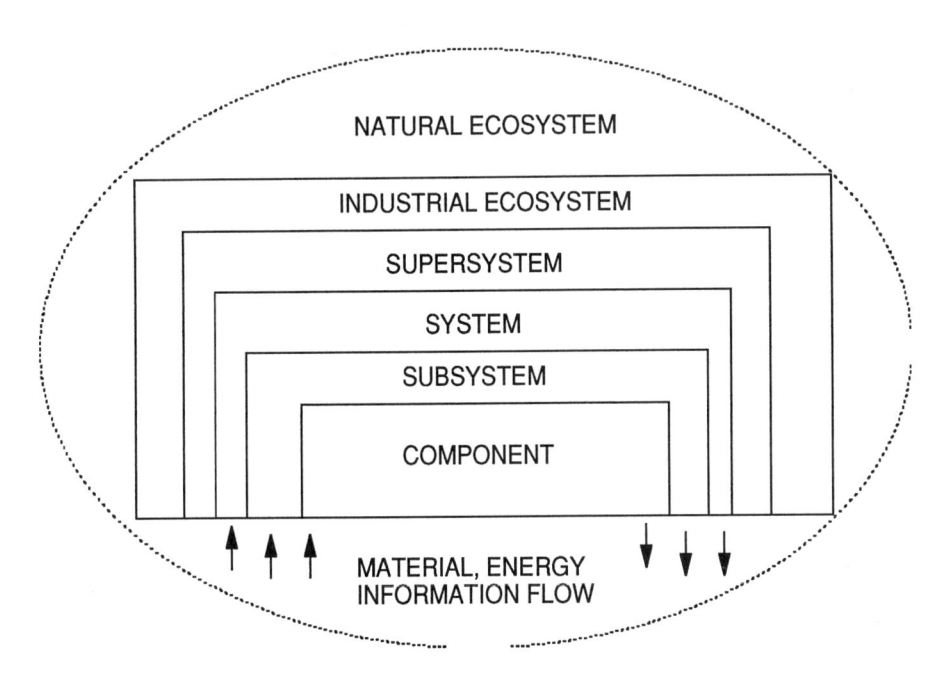

Figure 1 Environment, Supersystem, System, Subsuystem, Component Architecture Natural Ecosystem & Industrial Ecosystem

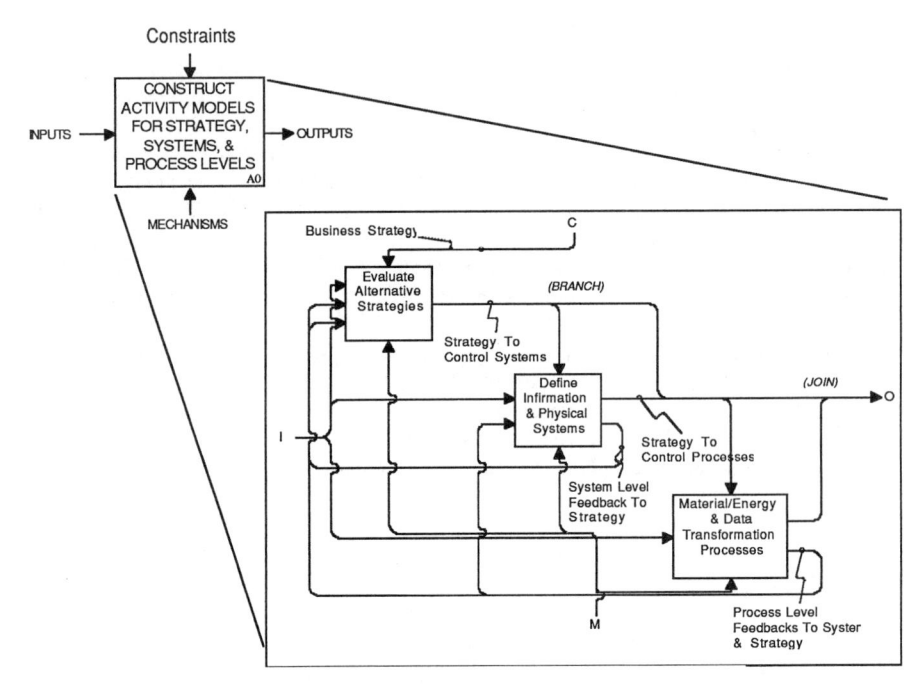

Figure 2 Decomposition & IDEF Convention in Activity Life Cycle Modeling

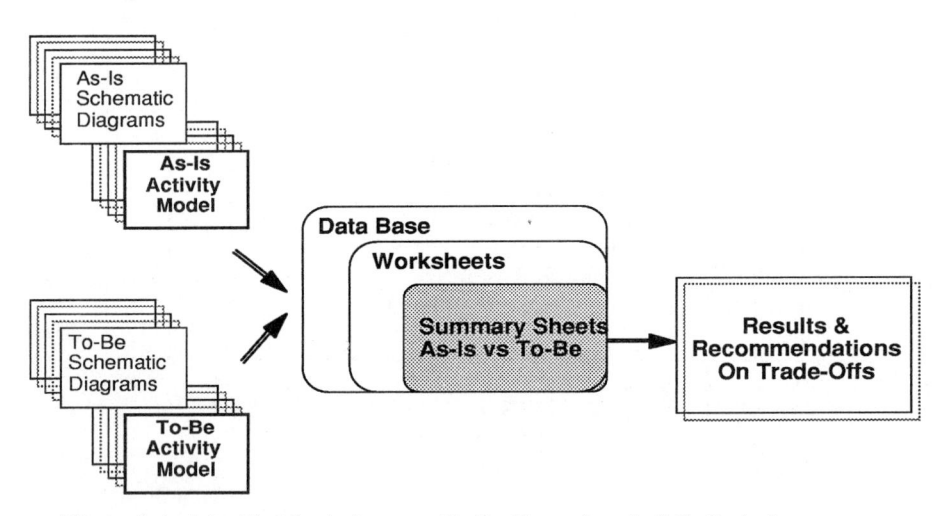

Figure 3 Activity Models: As-Is versus To-Be Alternatives for Life Cycle Assessment

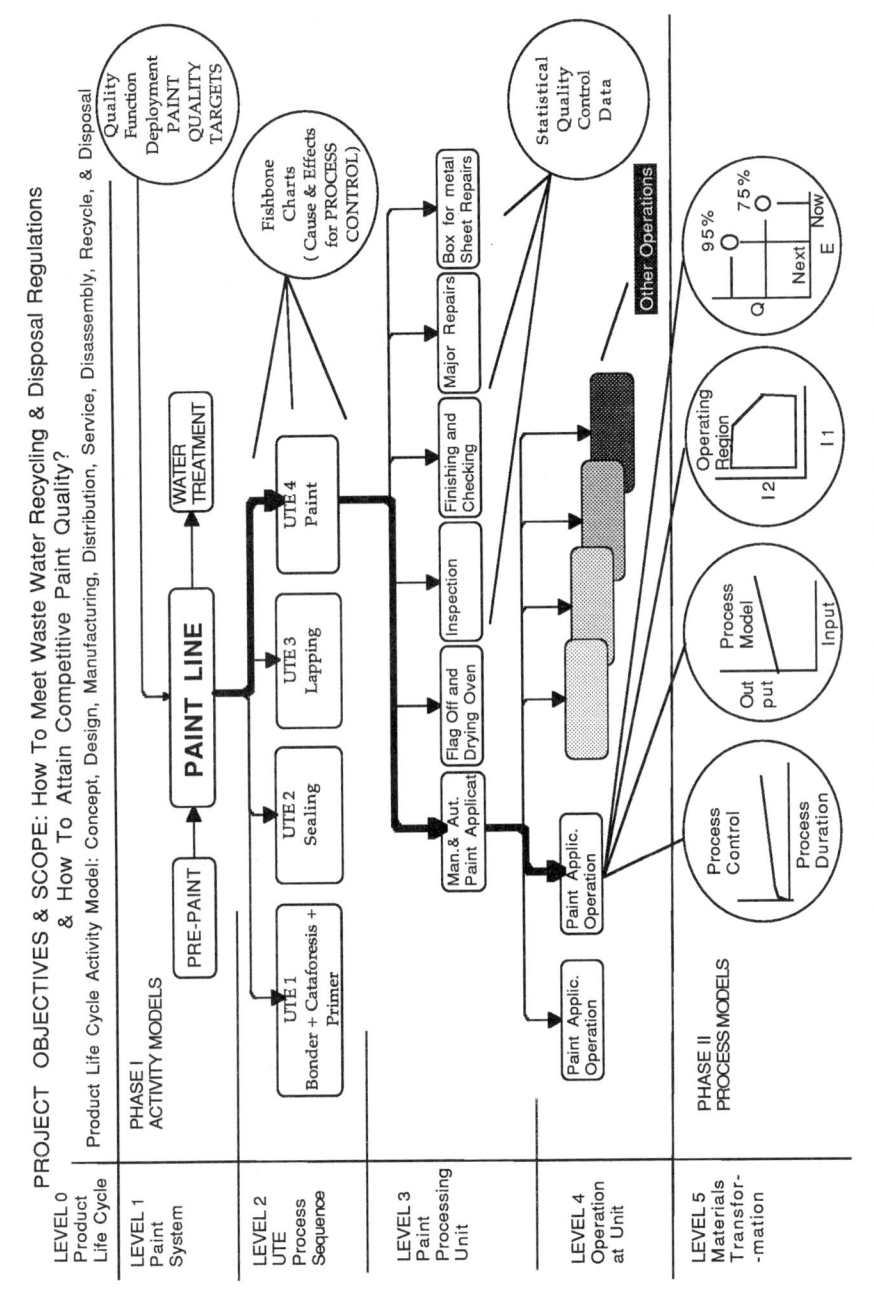

Life cycle modelling for innovative strategy, systems, processes   55

Figure 4 Schematic Life Cycle Model of An Automotive Paint Line: Decomposition

5

Disassembly Factories for Recovery of Resources in Product and Material Cycles

G. Seliger, C. Hentschel, M. Wagner
Technical University Berlin
Institute for Machine Tools and Manufacturing Technology
Department of Assembly Technology
Pascalstr. 8-9, 10587 Berlin, Germany
Phone: ++49-30-314-22014, Fax: ++49-30-314-22759
E-mail: sfb281@ipk.fhg.de

Abstract

The worldwide increase of waste and restricted resources require the recycling of worn-out products. Disassembly of complex consumer products offers advantages to other recycling technologies, like material recovery and reuse of components. Current disassembly processes, however, can not cope with the increasing amount and variety of products that have to be discarded each year.

The multifarious aspects of disassembly factories are addressed in an integrated project by an interdisciplinary research team at the Technical University Berlin. The closely correlated study fields are disassembly processes and tools, logistics and urban planning, product evaluation and disassembly planning as well as design for disassembly. An outline of the research program is given and the guiding theses are presented.

Two examples illustrate the range of the research activities. In a bottom-up approach fundamental knowledge about disassembly processes is acquired and used for the design of new tools. With the strategy of creating new acting surfaces disassembly tools become independent of shape variations and additional changes to the products during usage. The top-down approach provides methods for the planning of disassembly factories. Grouping of product variants is used to support the design of disassembly systems. To cope with the uncertainties due to usage influences on the product, fuzzy set theory is applied for planning purposes.

Keywords

Disassembly, Interdisciplinary Research, Tools, Planning

1 INTRODUCTION

1.1 Situation in Germany

The restricted natural resources in Germany have supported the early development of ecological awareness and have led to early recycling activities, especially in the former German Democratic Republic. But those activities were restricted to collection systems for materials like glass or paper. The speed of innovation together with customers´ demands for a wide product variety causes an expanding stream of technical consumer products that have to be discarded every year – for example, in 1994 more than 900 000 tons only in the Western part of Germany. Technical consumer products set high requirements to recycling processes because of their high grade of integration of different components, their highly complex composition of mechanical, electrical and electronic parts and their different materials that are being used. Up to now the majority of these products end their life cycle in a landfill, although valuable materials and still operational components are thrown away and the expenses for refuse dumps have been exploding throughout the recent years (figure 1).

Law making boards in Germany have reacted by issuing laws and regulations which aim at the reduction of waste itself and at keeping harmful substances from becoming a part of household trash. This kind of legislation will substantially change former disposal procedures.

The 'Decree on Product and Material Recycling Loops' (Gesetz zur Förderung einer rückstandsarmen Kreislaufwirtschaft und Sicherung einer umweltverträglichen Entsorgung von Abfällen), a law supporting a product and material loop with regard of an ecologically sound waste disposal policy, approved by the German Bundestag in April 1994, aims at:

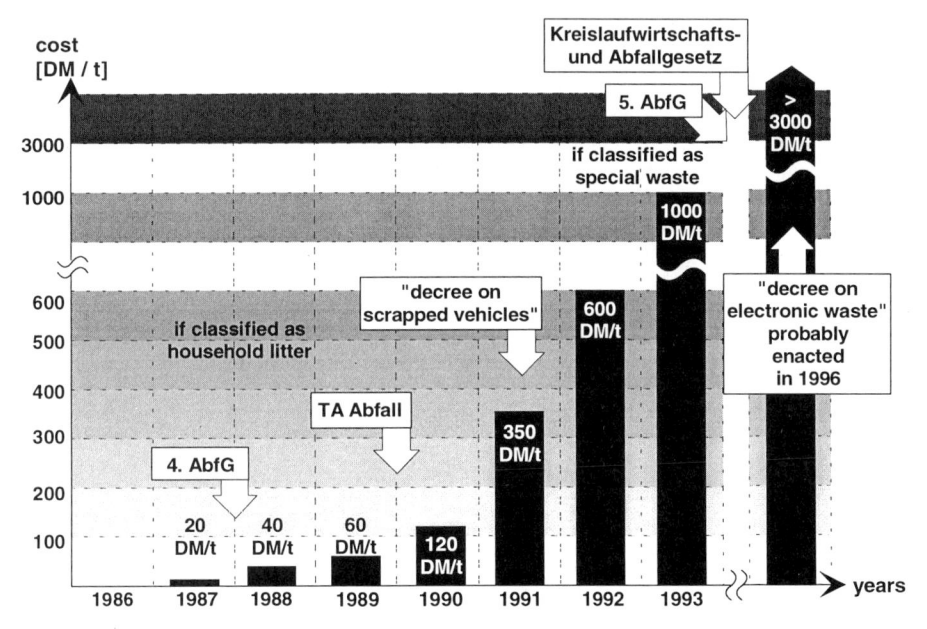

Figure 1 Development of disposal costs in Germany.

- the extension of the manufacturers and the distributors' responsibility for an end-of-life treatment of used products,
- the reutilization of residues,
- the reutilization of secondary resources,
- an ecologically harmless disposal of unavoidable waste.

The 'Decree on Electronic Waste' (Elektonikschrottverordnung) is supposed to dedicate manufacturers and distributors to take back used electronic products and to dispose them of properly. The 'Technical Regulation for Waste Disposal' (Technische Anleitung Abfall) establishes a nationwide standard planning procedure for disposal facilities. The 'Technical Regulation for Shredder Residues' (TA-Shredderrückstände) is a special regulation saying that all unavoidable waste must be processed, so that material or energy is reutilized, or that it must be disposed of in special landfills. According to this regulation, an expensive subterranean storage becomes mandatory if waste from a shredder or mill contains more than 10 mg/kg of PCB and/or if a maximum hydrocarbon content of 4% is exceeded.

In addition to those legal constraints manufacturers face a growing ecological consciousness of their customers. Thus, companies can turn the recycling regulations into a positive image factor for selling their new products.

1.2 Disassembly - gate to recycling

Currently, complex technical consumer products like cars or household appliances were mainly shredded and disposed of. Today's state-of-the-art shredding processes only allow recovery of metals, which may be soiled with fractions of other materials so that they do not live up to required quality standards. Non-metal materials are hardly separable from the shredder scrap and have often to be disposed of in special landfills at high expenses. They are thus completely taken out of any material loop.

Two approaches allow to change the situation: First the improvement of the shredding and sorting technology, and second further development in the field of disassembly. Disassembly is superior in comparison to shredding, when different materials should not get mixed in the process (figure 2). This is the case for harmful substances or valuable materials being concentrated in the product. In addition to this advantage, disassembly allows recovery of complete components and subassemblies to be put back into use.

A drawback of disassembly is its still lacking economic efficiency. Today, disassembly is mainly done manually, rarely supported by adequate tools, and requiring high labor costs. Additionally, manual labor is physically demanding, dirty, and, at times dangerous. Today, disassembly capacities are too restricted and too low in capacity with respect to future demands resulting from used goods returned from customers.

Processing Technologies (e.g. Shredding)

+ simple, cheap process
+ high mass flow rate
- no recycling of components
- restricted material recycling
- spreading of former local concentrated toxics
- high sorting effort
- remainings partially special waste

Material mix -> landfill !!

Disassembly

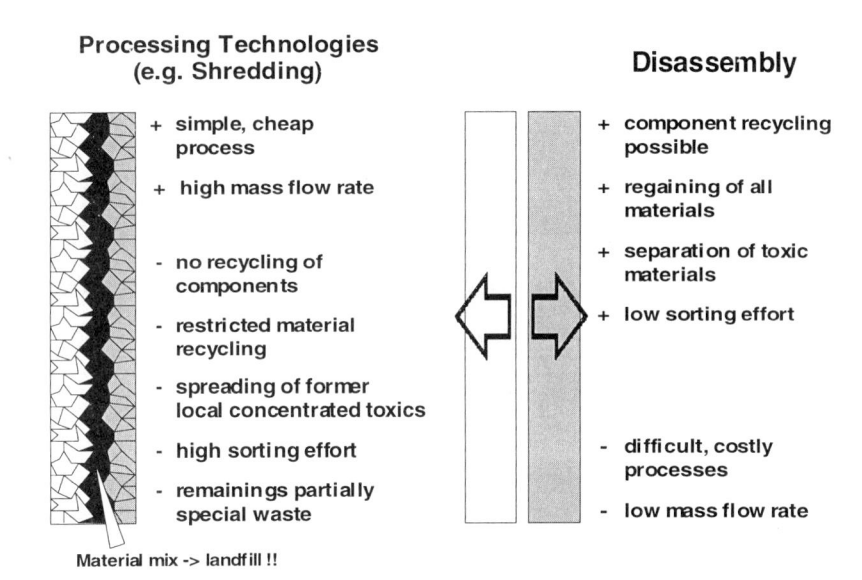

+ component recycling possible
+ regaining of all materials
+ separation of toxic materials
+ low sorting effort
- difficult, costly processes
- low mass flow rate

Figure 2 Comparison of Processing Technologies and Disassembly.

2 DESIGNING DISASSEMBLY FACTORIES

2.1 Integrated research project at the Technical University Berlin

The complex constraints of disassembly processes produce a challenge that can not be solved with traditional engineering strategies. The urgent situation due to the amount of worn out products and the tightening legal situation as well as the manifold correlations with non-technical sectors requires a holistic approach.

For this reason an interdisciplinary research team was formed at the Technical University Berlin with scientists from many different fields. Their background ranges from manufacturing technology, machine tools and assembly technology via engineering design and industrial information technology to material flow and logistics as well as architecture (figure 3). The objective is to create and realize scenarios of efficiently organized disassembly factories with highly mechanized, and partly automated processes. Therefore, the goals of this research program are as follows :

- design and realization of prototypical disassembly facilities by systematic research of disassembly processes and the deduction of tools,
- design and realization of logistics structures appropriate for in-house, as well as external flow of used products and the deduction of urban planning methods,
- basics for the evaluation of the recyclability of used products, as well as for planning and controlling of disassembly processes,

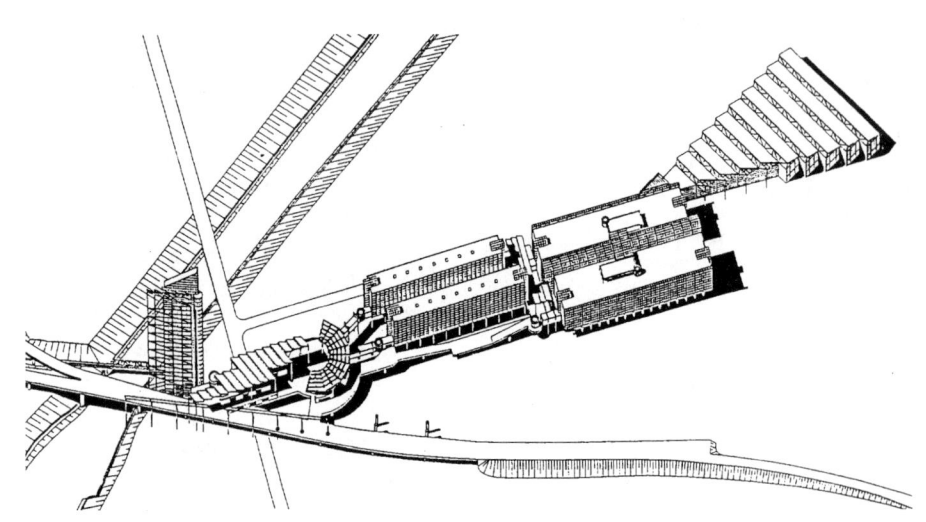

Figure 3 Architecture study of a drive-in disassembly factory for automobiles (L. Niewald / P. Bayerer).

- methods and information techniques to support the design stage of new products in order to increase their disassembly / recycling potential.

These four major objectives are being reflected in four different parts of the overall project.

Team A: Processes and tools
In order to technically control and economically optimize disassembly processes, a basic knowledge of these processes is required. In the field of separation processes, methods such as drilling and sawing as well as hydro and laser cutting need to be considered regarding their suitability for disassembly purposes. A further goal of this part is the analysis of detachable joints and the development of devices for the loosening of joints and the handling of objects. Furthermore it is planned to develop new clamping principles that are flexible enough for disassembly purposes. Another focus of the study is labor security and the development of protection means and regulations for it. The development of sensors and data processing methods for disassembly is the first step towards automation.

Team B: Logistics and urban planning
Disassembly factories with their widely scattered input of used products depend on locations close to densely populated areas and optimized logistic networks. Therefore, it is necessary to integrate logistics, disassembly factories and urban structures to closed product and material loops. New concepts for external logistics must be matched with the logistics of customers as well as with suppliers and must be suitable for different types of cities. Internal logistics also face new constraints, like maximized stocks becoming advantageous due to little or zero cost for used products.

Team C: Product evaluation and disassembly planning
The optimal combination of disassembly and other recycling processes may differ for each product. Means for effective evaluation of recyclability need to be provided to determine the best treatment for a given used product and to predict recycling options during the design of new products. These tools include accounting systems for a wide range of different items like emissions, resource consumption or social costs, so that different recycling scenarios can be compared. But even for given disassembly processes planning and control of facilities has to react upon uncertainties resulting from usage influences on products. Flexible planning and control systems will be developed that offer an on-line control potential reaching far further than system available today.

Team D: Design for disassembly
Disassembly efficiency will be increased if design for ease of disassembly such as reusable component or pure material use is considered. Great emphasis on product design for ease of disassembly must be put on joints and product structure. Deduction of design criteria provides a great deal of support for assessment during the design stage. New simulation tools allow to improve predictions about end-of-life product conditions. They will be integrated together with the results of the other research teams into an information network that provides and receives information for producers, users and recyclers.

These teams are organized in many different sub-projects that combine their results in continuous working meetings.

2.2 Research Guiding Hypotheses

The ongoing research work is guided by the following hypotheses (figure 4). They are constantly questioned and reevaluated based on the obtained results.
- Disassembly is a paramount part of recycling, because recovery of pure materials and valuable parts is made possible only through disassembly. To date, disassembly does not exist on a large scale. This situation is the result of the fact that disassembly is not yet economically efficient.
- A stiffening of federal laws has become foreseeable and economical efficiency is an entrepreneurial challenge. This research program is meant to allow these aims to be reached. For improved economic efficiency, the development of disassembly methods and tools, adapted logistics systems and factory buildings, used product evaluation methods and ease for disassembly and recycling product design is necessary.
- Disassembly factories are considerably different from assembly factories for a number of reasons. Disassembly must be performed for a much larger quantity of widely varying products, with missing product information, and under adverse circumstances.
- Research will alternate in a field, ranging from one centralized disassembly factory to various decentralized disassembly factories. A centralized factory provides the advantages of increasing economies of scale at the cost of expensive transportation. Decentralized facilities become advantageous if tools and processes represent efficient for small lot size disassembly. These two scenarios will compete against each other and the outcome will have an impact on the layout and design of future disassembly factories.
- In addition to this, there is another consideration between manual and automated disassembly. Even though automation will help rationalize disassembly processes, disassembly proc-

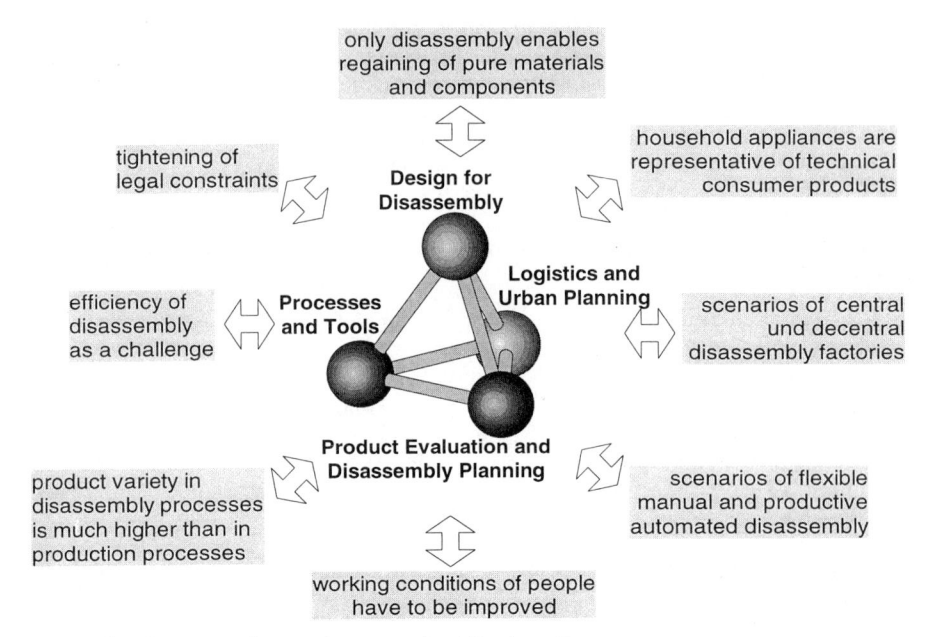

Figure 4 Structure of research team and guiding hypotheses.

esses cannot operate without the flexibility of human workers. Due to a large quantity of different products, missing product information and adverse circumstances of disassembly itself, hybrid disassembly processes are needed to combine human abilities and the advantages of automated devices.

- White goods products like washing machines, dishwashers or ovens are representative for many other goods that will have to be disassembled in the future. This follows from the fact that materials and joining techniques used in these products can be found in many other products sold today. In addition, white goods products' life cycles range from 10 to 15 years, during which the product's condition changes considerably. Experiences and knowledge gained from disassembly of white goods can easily be transferred to other groups of products such as cars.

The current research phase having started this year runs for three years and is funded by the German research foundation (Deutsche Forschungsgemeinschaft).

3 EXAMPLES OF RESEARCH ACTIVITIES

In addition to the interdisciplinary approach concurrent research techniques are applied using bottom-up as well as top-down strategies. Bottom-up means basic research to gain comprehensive knowledge about disassembly processes. Top-down provides the methods to manage the disassembly of used products in an industrial scale. In the following, one example is given for each of these two approaches.

3.1 Disassembly tools with in-process created acting surfaces

While destructive disassembly, by cutting of components for example, aims at material recycling, non-destructive disassembly dismantles the product according to its structure. This allows even the reuse of components, which provides advantages for products like electric tools (Dieterle, 1993). Partially destructive disassembly also preserves valuable components, while inferior parts like joining elements for example may be damaged in order to speed up the overall disassembly process. This strategy has already been applied to the disassembly of hose joints (Seliger, 1995).

The transmission of forces and torques to loose detachable joinings and to handle solids generally operates with frictional connection or form closure between endeffectors and objects. Therefore endeffectors highly depend on the object's geometry and state of acting surfaces. The wide variety of products, the diverse joining techniques and the unpredictable changes to the original shape during usage set very high flexibility requirements to disassembly tools. Screw joints with the diverse types and sizes of screw heads are one example (figure 5). Usually, the endeffector is adapted to the actual geometry of the object, leading to a corresponding number of tools to be provided or to flexible systems with a resulting high complexity (Hesse, 1989). Other approaches to reduce flexibility requirements for handling systems (Wagner, 1995) work only for new products. In addition shape detection and recognition as well as control units are needed to run such a closed-loop system. The resulting high expenses and time consuming tool changes prevent cost-effective disassembly processes.

A new approach has been suggested in the research proposal: The endeffector itself creates new acting surfaces before transmitting forces and torques. Thus the endeffector is independent of the acting surfaces that already exist and is robust against uncertain alterations like corrosion or others. One example is an endeffector that welds to the object, using material closure to create a new acting surface. An example for providing form closure is the imprinting of solids to stamp the object's surface. A tool with in-process creation of new acting surfaces

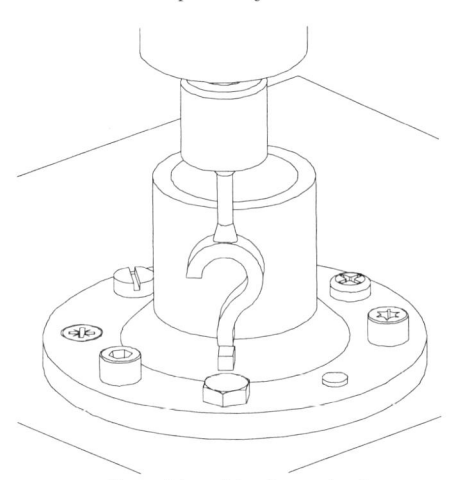

Geometric variety of screw heads
(with disassembly tool under development)

Possible effects of usage
(bottom part of washing machine)

Figure 5: Challenges to the disassembly of screw joints.

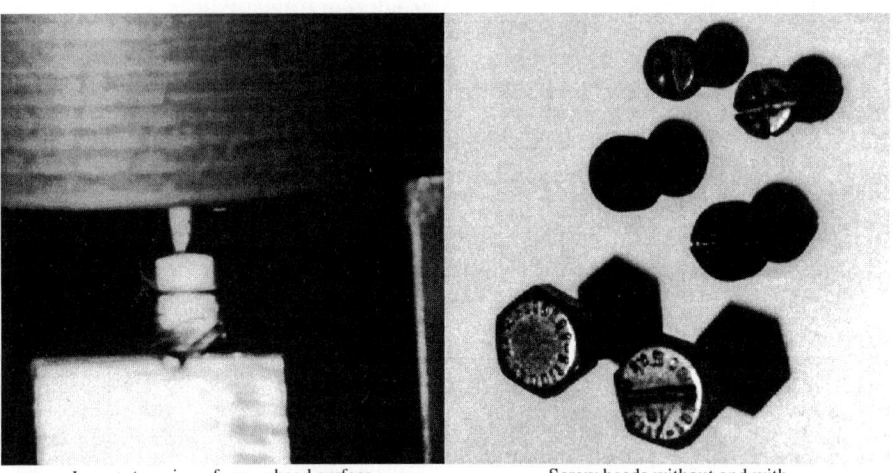

| Impact stamping of screw head surface with impulse forces | Screw heads without and with in-process generated acting surfaces |

Figure 6: Strategy for a new tool for the disassembly of screw joints.

works for a wide range of objects and needs only information about the position of the object to be dealt with. Thus the interaction of endeffector and object can run as an open-loop system. Those characteristics already have positive effects on manual disassembly, but they are especially advantageous for automated disassembly processes.

Since most joining elements and many parts in consumer products are made of metals, high forces and torques are required to stamp the surfaces. Experiments with endeffectors based on left-turning drills showed reaction forces that exceed by far the limits set for hand-held tools. But as high forces and torques are only required for a short time during the stamping process, they can be generated by linear and rotary impulses with minimized reaction forces to the tool housing. First experiments with high-speed impact stamping showed a good potential for the development of flexible disassembly tools (figure 6).

The design of tools based on this process should also allow to incorporate existing acting surfaces, if they are advantageous for the disassembly process. The integration of strategies for the case of failure of the original disassembly process enhances overall productivity. For the given example of screw joints, this could be a conventional drilling process, that removes the screw head.

3.2 Disassembly planning by fuzzy grouping of used products

Disassembly for recycling is confronted with a huge variety of products from various producers and production years, as it usually cannot restrict itself to a spectrum of products from only one manufacturer. One well known approach to cope with product variety is group technology, first introduced in the 60ies for classifying parts according to their similarities (Mitrofanow, 1960; Opitz, 1960) This idea has been developed in many directions, for example, to analyze process flows for organizing production according similar processes (Burbidge, 1975).

Disassembly factories for recovery of resources 65

Group technology is also promising for planning disassembly and recycling. Already from the products' design attributes, important disassembly and recycling information can be derived. One example is the country of origin, that is determining material parameters. TV sets, e.g., that were produced by European manufacturers mainly contain bariumoxide as x-ray absorber in their cathode ray tube (CRT) glass, while CRTs from Asia contain an important amount of strontiumoxide instead (Collentro, 1993). Other design attributes refer to the size and weight of the product, determining the amount of materials to be recovered, or to its year of production, determining prevailing joining techniques or product structure. Grouping based solely on design attributes would be insufficient for disassembly purposes. In disassembly, one has to deal with used products, showing various deviations from their original state and condition. Therefore, also usage attributes have to be defined. Usage attributes refer to damage, wear and hazardous potential due to missing parts, dirt, deformation or breakage.

The drawback of current grouping approaches is, that they require certain, definite information about the products. For those actually coming back from the usage market, however, often neither information about the product's design nor about their usage attributes exist. The disassembly process has to be started with a high level of uncertainty on the products' characteristics. Therefore, a new approach for grouping is suggested based on fuzzy set theory to take these uncertainties into consideration (Hentschel, 1995).

Fuzzy set theory (Zimmermann, 1991) allows unprecise information about the product characteristics by member ship functions. The latter determine the degree to which a product belongs to a group. This is again demonstrated with the example of CRTs from used TV sets and monitors. A CRT consists of three main parts: screen, cone and neck with electron gun (figure 7).

The screen glass is coated from the inside with a fluorescent layer. This layer contains cadmium sulfide used as green fluorescence in color tubes and orange fluorescence in some monochrome tubes. If shredded, this material would contaminate the entire shredder scrap. Isolation of the fluorescent layer allows recovery of nearly all other materials and reduces input

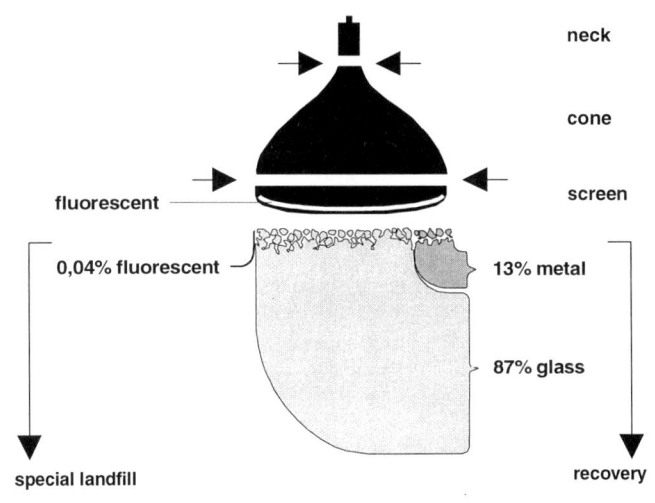

Figure 7 Recyclability of Cathode Ray Tubes

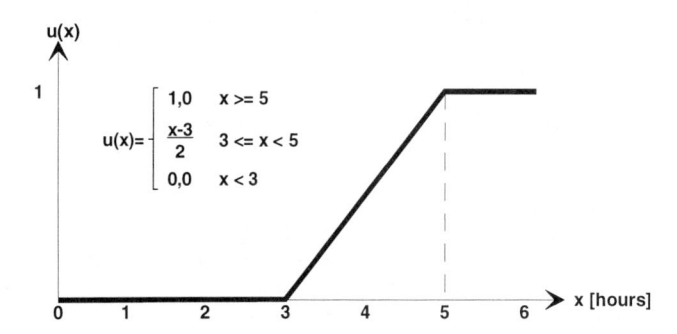

$$u(x) = \begin{cases} 1.0 & x \geq 5 \\ \dfrac{x-3}{2} & 3 \leq x < 5 \\ 0.0 & x < 3 \end{cases}$$

Figure 8 Membership function for encrustation of fluorescent layer due to hours passed since ventilation of color CRTs.

into special landfills. In their original condition, CRTs contain a vacuum inside, which allows easy remove of the fluorescent layer. Used CRTs, however, are often ventilated due to inadequate collection and transport to the recycler. The longer the CRT has been ventilated, the more encrustation of the fluorescent layer may occur due to moisture getting into the tube. This is to be avoided, as encrustation makes it difficult to separate the fluorescent layer from the screen glass or renders all the efforts to access this layer being done in vain. As the degree of encrustation cannot be described exactly, the definition of a membership function makes a fuzzy classification criteria (figure 8). Encrustation of the fluorescent layer caused by moisture absorption takes place within five hours. For these tubes either a special facility has to be provided or the parts would have to be shredded and deposited in special landfills. The optimum would be such that time between ventilation and separation of the fluorescence does not exceed three hours.

Currently, an algorithm is developed, that groups products based on a larger number of crisp and fuzzy design and usage attributes. Based on the membership functions, used products with fuzzy characteristics can be assigned to groups facilitating planning and control of disassembly systems.

4 SUMMARY

A tightening legal situation and exploding cost for landfills require the recycling of used products in addition to ecological reasons. The disassembly of worn-out products allows to recover components and subassemblies in product cycles and eases following recycling processes like material reprocessing. Highly productive disassembly factories are needed to deal with the increasing streams of used technical consumer goods. Due to the diverse correlation of disassembly with processes beyond the technical field researchers from engineering sciences to logistics and architecture joined a special research program at the Technical University Berlin to contribute their expertise to the design of disassembly factories. The concurrent approach in many different fields is outlined using two examples for the development of disassembly tools and planning methods. The results of the research work that has started this year will not only converted into prototypes of disassembly factories throughout the following phases, but will also provide a scientific background for further legislation in Germany.

5 REFERENCES

Burbidge, J. L. (1975) *The Introduction of Group Technology*. Wiley & Sons, New York.
Collentro, J. S. (1993) *CRT Disposal: Recycle or Pay the Cost*. In: Information Display, Nr. 1, 9-11.
Dieterle, A. (1993) *Flexible Demontage von Elektrokleingeräten*. Produktionsautomatisierung 6/93, 13-15.
Hentschel, C., Seliger, G. and Zussman, E. (1995) *Grouping of Used Products for Cellular Recycling Systems*. In: Annals of the CIRP Vol. 44/1/1995, 11-14.
Hesse, S., Mansch, I. (1989) *Auswahl flexibler Greifer*. Robotersysteme 5, 87-104.
Mitrofanow, S. P.(1960) *Wissenschaftliche Grundlagen der Gruppentechnologie*. VEB Verlag Technik, Berlin.
Opitz, H. (1971) *Werkstückbeschreibendes Klassifizierungssystem*. Verlag Girardet, Essen.
Seliger, G. and Wagner, M. (1995) Demontagewerkzeuge schaffen Zeit- und Kostenvorteile. *Der Betriebsleiter*, 10/95, 36-39.
Wagner, M. and Seliger, G. (1995) *Flexible and Safe Automation with Gripping Interfaces*. In: Proceedings of the 27th CIRP International Seminar on Manufacturing Systems, 491-498.
Zimmermann, H. J. (1991): *Fuzzy Set Theory and its Applications*. Kluwer Academic Publishers, Boston, 2nd Edition.

6 BIOGRAPHY

Prof. Dr.-Ing. Günther Seliger, born 1947, studied industrial engineering at the Technical University Berlin and received his doctor´s degree from Prof. Spur at the Institute for Machine Tools and Manufacturing Technology in 1983. After holding different positions at the Institute for Machine Tools and Manufacturing Technology and at the Institute for Production Systems and Design Technology of the Fraunhofer Society he became professor for assembly technology at the Technical University Berlin in 1988. He is spokesman of the special research program 281 "Disassembly Factories", which started in January 1995.

Dipl.-Ing. Claudia Hentschel, born 1964, studied industrial engineering with focus on mechanical engineering at the Technical University Berlin and the Ecole Nationale des Ponts et Chaussées, Paris. In 1991 she started working at the department for assembly technology as research assistant and since 1995 she is supporting the spokesman of the special research program 281 "Disassembly Factories".

Dipl.-Ing. Martin Wagner, born 1963, studied mechanical engineering with focus on controls and technical mechanics at the University of Stuttgart and at the San Diego State University. Since 1992 he is a research assistant at the departement for assembly technology, specializing in the development of disassembly tools within the special research program 281 "Disassembly Factories".

PART THREE

Life Cycle Modelling

6

Lifecycle Modelling as an Approach for Design for X

Walter Eversheim, Uwe H. Böhlke, Wilfried Kölscheid
Fraunhofer-Institut für Produktionstechnologie
Steinbachstraße 17, D-52074 Aachen, Germany
Phone: ++49/ (0)241/ 8904100
Fax: ++49/ (0)241/ 8904198
e-mail: boh@ipt.rwth-aachen.de

Abstract
This contribution deals with a new evaluation method concerning product design. This method is founded on the theory of lifecycle assessment and is transformed in the computer aided device CALA (Computer Aided Lifecycle Analysis). The method as well as the device were developed at the FhG-IPT (Fraunhofer-Institute for Production Technology).

Keywords
Evaluation method, lifecycle assessment, Computer Aided Lifecycle Analysis, CALA, enviroment orientated production.

1 Introduction

The modern customer requires products which not only function unobjectionably. These products should possibly also be rated as no environmental impact during their usage and at their sanitation. The enterprises of the production industry thus have to accept the challenge to develop products, which must comply with - partly contradictory - customer demands (Elbach 1992 and Eversheim 1993). In order to meet this challenge in an appropriate way devices and methods are needed which already in product development establish the basis for production of succesful high quality and environmentally acceptable products (Lehmann 1990 and Böhlke 1994).

2 Evaluation Methods and Tools for Product Design

Lifecycle Model

One condition for the development of the requested devices and methods is to add the phases "product usage" and "product sanitation" up to the present only considered product lifecycle phase "production". Two aspects of the lifecycle model are of special significance: A coordination of objectives and demands in product development have to include all product lifecycle phases.

A product developed according to recycling demands after the usage phase renders the reusing of some of its components possible in production or in product usage (Warnecke 1994). The transformation of demands out of different product lifecycle phases in product development is classed under the term "Design for X".

The term "Design for X" means, that in product development several differently oriented restrictions must be taken into account (Figure 1). "Design for Manufacture and Assembly" (DFM, DFA) propagated by Boothroyd (Boothroyd 1992) concerns in this connection the phase of production. In product development product demands concerning product usage and product sanitation are transformed with help of "Design for Disassembly" (Alting 1995 and Boothroyd 1995). An overriding "Design for Energy Saving" (DFES) aims at production as well as at product usage and product sanitation.

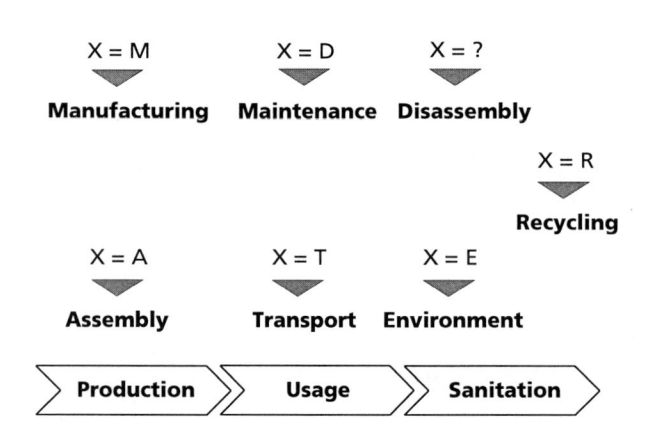

Figure 1 Restrictions and Demands on the Design.

Methods for Product Design

At present the criteria at product design are only transformed in chronological order. As a result of this sequential proceeding the possibilities of a comprehensive optimization of product are strongly reduced. In order to cope with the demands on a comprehensive product design it is necessary to realize isochronously product demands such as reduced employment of energy and material and reduced pollutant emissions (Figure 2). In this connection it is essential to develop evaluation methods and evaluation devices supporting a comprehensive product design.

Lifecycle modelling as an approach for design for X 73

Proceeding at Product Design

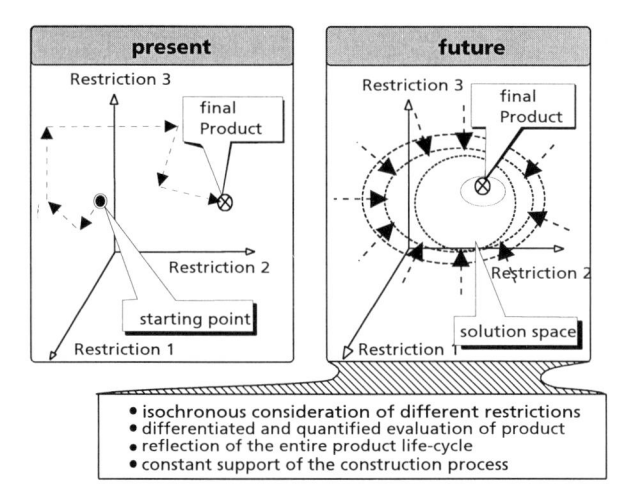

Figure 2 Product Design adjusted to Restrictions.

The existing methods and devices only deal with particular restrictions. The directive catalogues, for example, only offer help for manufacture. The catalogues for relative costs only support a cost-oriented product design, whereas standards and laws treat law conditions (VDI 1993). A methodical approach considering comparatively many restrictions is the product line analysis. The product line analysis is based on a comprehensive analysis of seperate products: i.e., starting from the revelation of raw material to the production phase and up to the usage and sanitation phase the product line analysis considers the entire energetic, ecological and social effects of a product realization. The same situation is given regarding the computer aided devices for product design. The predominant portion of these devices is made for the optimization of only one restriction. An example for this situation are the existing data bases offering informations about product assembly, further the FEM-systems and the 2D resp. 3D-CAD systems providing information about design optimization of products. The same applies to systems for cost calculation. Merely the devices leading to a comprehensive product design in the scope of ecological optimization offer an approach to a comprehensive product optimization. The Computer Aided Lifecycle Analysis, shortly CALA, is such a system which enables the user to optimize the product ecologically during the entire product lifecycle (Eversheim 1993 and Böhlke 1994).

In the following the basic method of product line analysis, respectively the structure of the resulting system CALA, is explained in order to show how to proceed with the conception of a mutidimensional optimization of a product.

Module Structure of the Evaluation Method

The product line analysis is an evaluation approach considering the resource demands caused by product realization. In concrete terms this means that energy and material streams are examined from the input as well as the output view. The corresponding limits of balance of a product cover the entire product lifecycle: i.e. production, usage and sanitation (Figure 3).

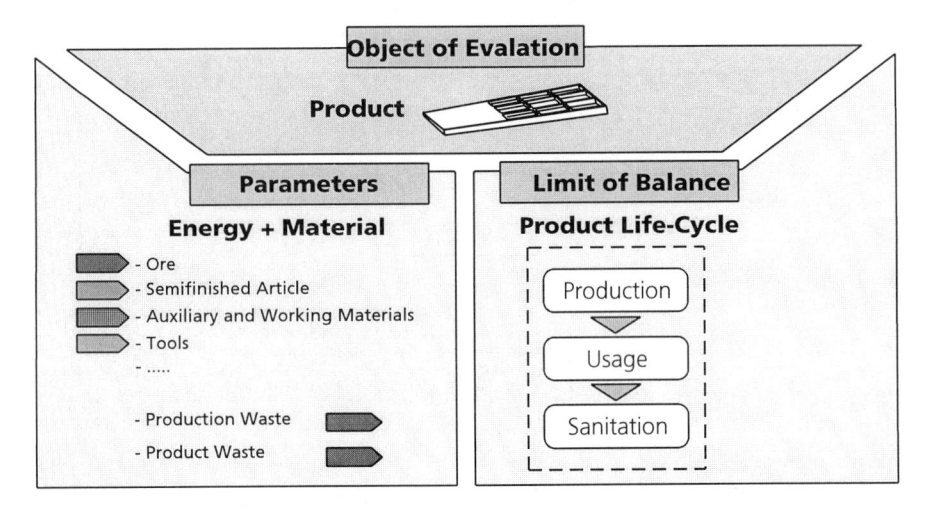

Figure 3 Entire Evaluation of employment of Energy and Material.

The IPT developed a method to achieve the objectives of product line analysis by a systematic proceeding. This proceeding is founded on four modules: data capture, balancing, evaluation and analysing. Structured by these four modules the method enables the user to solve ecology-oriented tasks by systematic analysing of data (Figure 4).

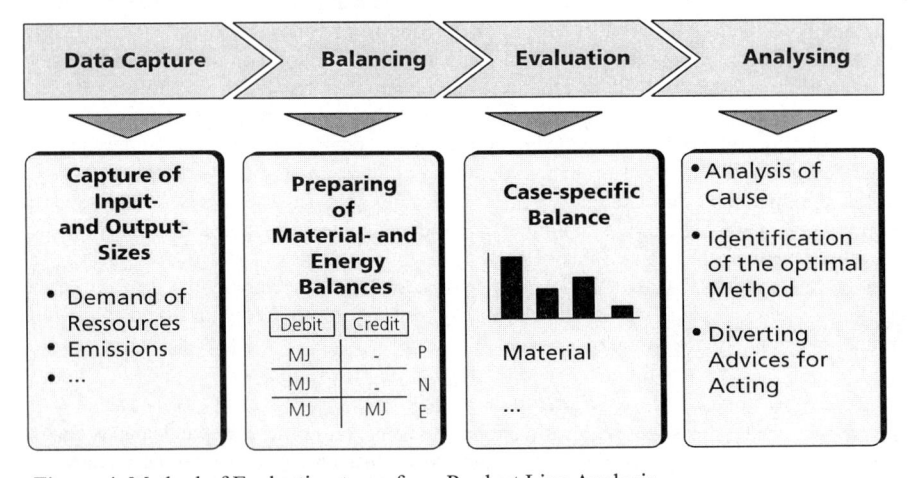

Figure 4 Method of Evaluation to perform Product Line Analysis.

Lifecycle modelling as an approach for design for X 75

With module "data capture" the different needs of resources of a product realisation as for instance the direct demand of energy and materials as well as the direct demand of process materials are considered. As mentioned above the entire product lifecycle is taken into account. This first step also includes the starting sizes product, waste and pollutant arisings and possible energy usage. The second module "balancing" also considers indirect demands of resources, based on the prepared data. The relevant applications of energy and material are lead back and balanced to the calculation size "primary energy" in the form of a primary energy balance. In this step also the energy arisen by recycling is credited (Figure 5). On top of that it is proved if the captured sizes have to be furtherly detailed. The third module "evaluation" accumulates the captured and balanced demands of resources and renders them interpretable. On this basis the user can already take environmental-oriented selection decisions.

Description of the Product Life-Cycle by Process Elements

Balance of Material and Energy for each Process Element

- Materials
- Auxiliary- and working Materials
- Energy

Conversion aided by process -specific Ciphers

- Waste
- Contaminants
- Residual Energy

Data Capture and Data Administration

Manufacturing

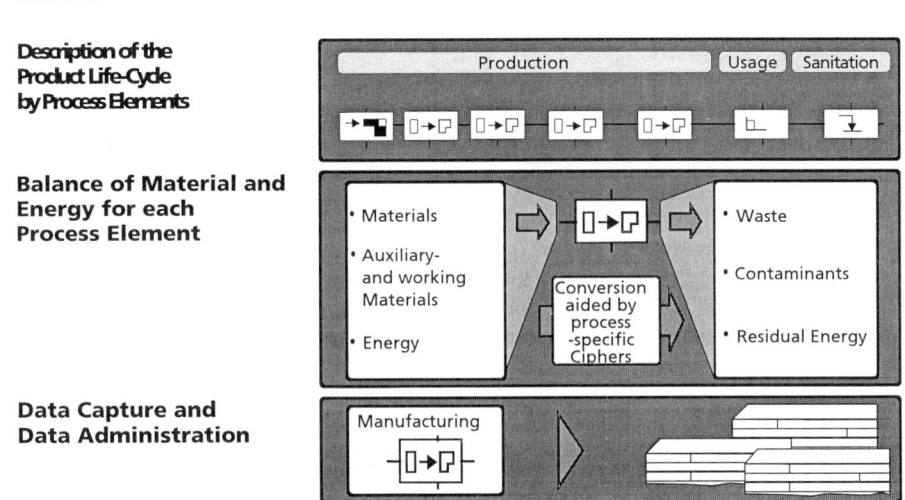

Figure 5 Data Capture by means of Employment of Process Elements.

The modules so far introduced do not include the piece number of the product in real terms though important for taking planning decisions. This renders an analysing system necessary standing behind the evaluation step but before the decision. This function is assumed by the fourth module "analysing" which, taking into account the other three modules, is responsible for finding out weak points and comparing alternative construction and manufacturing concepts. Based on the results of the application of this method it is possible to examine certain manufacturing concepts relating to the specific arising of demands of resources and to find out the weak points of these concepts. In addition to that it is possible within the scope of operative comparisons to prove alternative working process sequences in order to achieve the lowest demand of resources. Within the scope of strategical comparisons the user is able to study alternative production concepts from an ecological point of view and to choose the optimal concept (Figure 6).

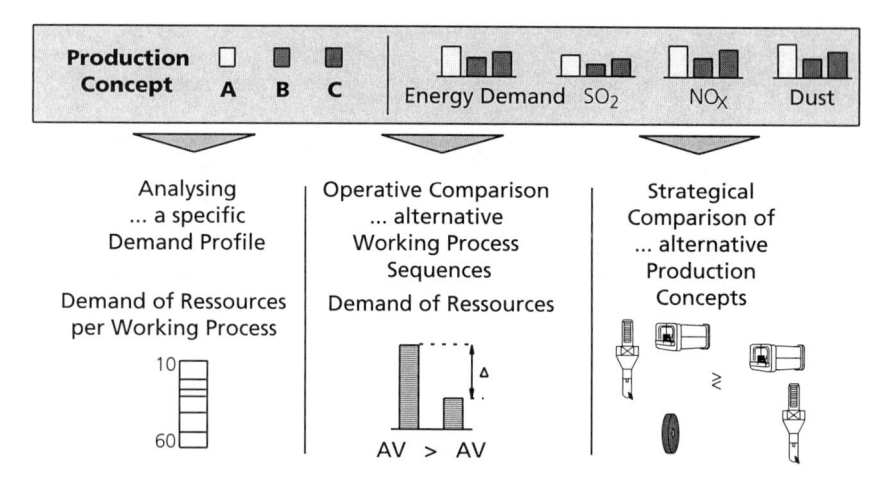

Figure 6 Actual Application Facilities for the Evaluation Method.

The application of the evaluation method performed in several projects confirmed its structure. Nevertheless, it proved to be quite large-scaled. In order to reduce the expenditure of the application the evaluation method was transformed into an „Electronic Data Processing system" (EDP, Figure 7). This transformation into the EDP-prototype CALA ensures a general application of the evaluation method and offers a structured decision tool to the user. With the evaluation system it is possible to analyse the demands of resources short-termed and to examine them within the scope of sensitivity analysis. As a result it is possible to assess the ecological effects of product realization in an early state of product planning.

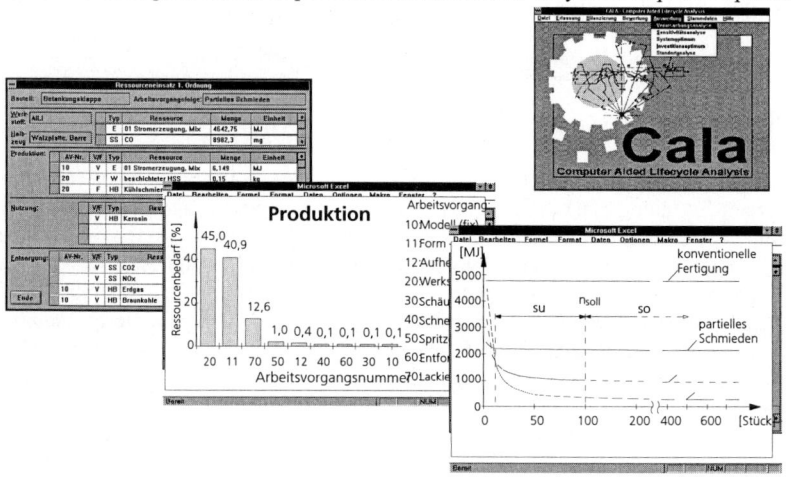

Figure 7 Evaluation Method for Performing Product Line Analysis.

Example

The following example is meant to illustrate the applicability of the evaluation method in general and of the developed prototype CALA in particular.

The evaluation task was to examine a designing concept for a gear shaft of an autombile motor with regard to ecological tolerance. With the conventional working process sequence it was planned to manufacture the component by applying different cutting procedures. With the development of an innovative alternative production concept employing massive reshaping the opportunity was given to fabricate a hollow gear shaft (Figure 8). The decision about the final production concept were given by the comparison of the energetic and ecological demand of recources.

Demand of Energy	Full Gear Shaft		Hollow Gear Shaft	
	46,12 MJ	Half Stuff	Half Stuff	26,70 MJ
	1,08 MJ	Sawing	Sawing	2,40 MJ
	2,30 MJ	Glowing		
	7,20 MJ	Reshaping	Reshaping	8,92 MJ
	0,41 MJ	Glowing	Friction Welding	0,72 MJ
	0,01 MJ	Centring		
	1,70 MJ	Rotate		
	1,38 MJ	Toothing		
	18,39 MJ	Hardening	Hardening	14,22 MJ
	0,76 MJ	Grinding of Rolling bearing seats	Hard-Rotating of Rolling Bearing Seats	28 MJ
	1,25 MJ	Grinding of Tooth Profile	Grinding of Tooth Profile	1,25 MJ
	80,60 MJ			54,49 MJ

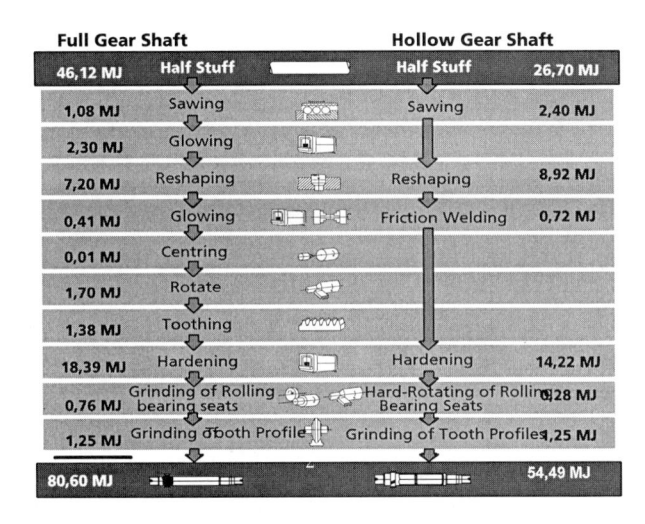

Figure 8 Comparing Productin Phase of alternative Production Concepts.

The application of the evaluation method lead to the conclusion that savings of energy demands for the production phase to the amount of almost a third of the original need could be achieved in connection with the new designing concept. Even the SO_2- and Co_x-emissions were cut by a similar amount.

On top of that the cost calculation performed in accordance with "Design for X" confirmed the improvement with regard to cost aspects. As a result to the application of the evaluation method it was possible to illustrate the optimization in money values. In this practical case it meant a production cost reduction from DM 12.05 to DM 8.48 before the production started (Figure 9). A further detailed analysis also confirmed that the hollow gear shaft caused a reduced demand of resources during usage and sanitation phase.

Piece Cost

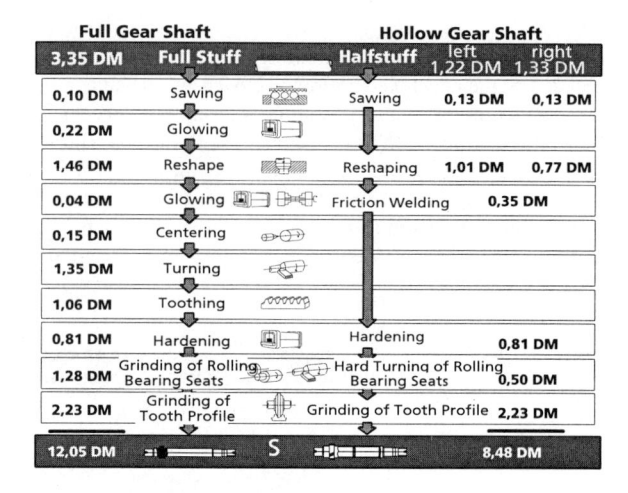

	Full Gear Shaft			Hollow Gear Shaft		
3,35 DM	Full Stuff			Halfstuff	left 1,22 DM	right 1,33 DM
0,10 DM	Sawing			Sawing	0,13 DM	0,13 DM
0,22 DM	Glowing					
1,46 DM	Reshape			Reshaping	1,01 DM	0,77 DM
0,04 DM	Glowing			Friction Welding	0,35 DM	
0,15 DM	Centering					
1,35 DM	Turning					
1,06 DM	Toothing					
0,81 DM	Hardening			Hardening	0,81 DM	
1,28 DM	Grinding of Rolling Bearing Seats			Hard Turning of Rolling Bearing Seats	0,50 DM	
2,23 DM	Grinding of Tooth Profile			Grinding of Tooth Profile	2,23 DM	
12,05 DM			S			8,48 DM

Figure 9 Comparing Balance Results for the Production Phase.

Conclusion

By isochronous application of the EDP-system CALA and conventional calculation devices it is possible to find out multidimensional demands of resources already in product design. By this proceeding a first step is done to realize the concept "Design for X", i.e. to consider isochronously all restrictions concerning the product design (Figure 10).

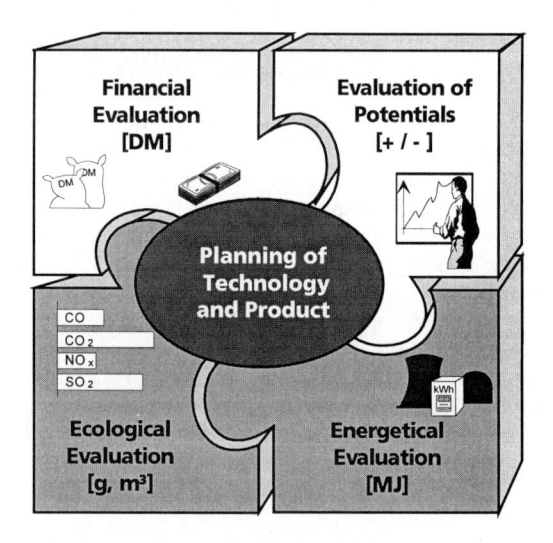

Figure 10 Future - oriented Planning of Technology and Product.

In order to ensure the competitiveness of enterprises at long sight it will be inevitable to consider isochronously financial, strategical, energetic and ecological aspects in product design.

3 REFERENCES

Alting, L. and Legarth, J. B. (1995) Life cycle engeneering and design, Keynote paper, Annals of the CIRP Vol.44/ 2/ 1995

Böhlke, U. (1994) Rechnerunterstützte Analyse von Produktlebenszyklen - Entwicklung einer Planungsmethodik für das umweltökonomische Technologiemanagement, Dissertation, Shaker Verlag Aachen

Boothroyd, G. and Alting L. (1992) Design for assembly and disassembly, Keynote paper, Annals of the CIRP Vol. 41/ 2/ 1992

Boothroyd, G. (1994) Product design for manufacture and assembly, Computer-Aided Design Vol. 26/ 7/ 1994

Ebach, H., Scherer, C. and Streibelt, H. (1992) Einfluß der Lifecycle-Betrachtung auf die recyclinggerechte Produktgestaltung, VDI Berichte 999, VDI-Verlag Düsseldorf

Eversheim, W., Böhlke, U., Hartmann M. and Katzy, B. (1993) Evaluation Strategies for Product Development, Production and Recycling Keynote paper "International Forum on Product design for manufacture and assembly" Newport, 14./15.06.93

Lehmann, S. (1990) Ökobilanz und Öko-Controlling als Instrumente einer präventiven Umweltpolitik, in: Unternehmen in: Vermeiden und Verwerten von Abfällen

VDI 2221 (1993), Methodik zum Entwickeln und Konstruieren technischer Systeme und Produkte, Beuth Verlag Berlin

VDI 2243 (1993), Konstruieren recyclinggerechter technischer Produkte - Grundlagen und Gestaltungsregeln, Beuth Verlag Berlin

Warnecke, G. and Sigl, M. (1994) Recycling ist Produktion, VDI-Z 136 Nr. 1/ 2

4 BIOGRAPHY

Prof. Dr.-Ing. Dr. h.c. Dipl.-Wirt.Ing. Walter Eversheim, born 1937, studies of mechanical engineering and afterwards additional studies of economics. Since 1973 holder of the Professorial chair of Production Systematics, RWTH Aachen, and member of the Directory of the Laboratory for Machine Tools and Production Engeneering (WZL). Since 1980 head of the Department „Planning and Organization" of the Fraunhofer-Institute for Production Technology (IPT), Aachen.

Dr.-Ing. Dipl.-Wirt.Ing. Uwe H. Böhlke, born 1964, studies of mechanical engineering and afterwards additional studies of economics. Since 1990 scientific employee, since 1994 chief engineer at the IPT.

Dipl.-Ing. Wilfried Kölscheid, born 1967, studies of mechanical engineering. Since 1995 scientific employee at the WZL.

7

Product Life Cycle Modelling for Inverse Manufacturing

Fumihiko Kimura and Hiromasa Suzuki
Department of Precision Machinery Engineering, The University of Tokyo
Hongo 7-3-1, Bunkyo-ku, Tokyo 113, Japan
Tel. +81-3-3812-2111 ex.6455, Fax +81-3-3812-8849
E-mail kimura@cim.pe.u-tokyo.ac.jp

Abstract

A new approach to product life cycle design, called as Inverse Manufacturing, is proposed, where products are designed for their entire life cycle from product conception to final reuse/recycling or disposal. A Right Quality Product (RQP) is considered where upgrading or degrading of product functionality can be achieved according to the users' requirements by applying flexible strategy of product maintenance and dynamic improvement with modular component replacement technology. Basic product life cycle modelling technology is discussed for supporting such activities.

Keywords

Product life cycle, Product design, Product modelling, Maintenance, Deterioration, Quality control

1 INTRODUCTION

Trends of recent manufacturing industry can be characterized by such factors as rapid changes of technology and market, severe competition due to the globalization of manufacturing activities, increasing social concern about clean manufacturing, and so on. In advanced industrial countries, major concern is no more directed to high productivity mass production, but rather to value-additive resource-saving flexible production adapted to customers' requirements. Various visions for the next generation manufacturing systems are being discussed. One of the most demanding factors for such consideration is environmental issues. From this standpoint, the efficiency of natural resource utilization might be the most important criteria for manufacturing activity evaluation.   In this paper, a new product life cycle concept is proposed for establishing a rational basis for precise evaluation of environmental burden or resource consumption of manufacturing activities and associated product behaviour throughout the total product life cycle.

Global environmental issues are very much complicated ones, even viewed only from the standpoint of manufacturing.   Due to the increasing activities related with production of industrial artifacts, especially in developing countries, many of the origins of environmental hazards or pollution, or waste of resources are attributed to manufacturing practices.   It is, therefore, the responsibility of manufacturing engineers to cope with these issues, and to find out the new industrial infrastructure which enables us to maintain today's level of industrial productivity and at the same time to keep our global environment in sustainable status.

As the first step, it is essential to clarify the influence of manufacturing processes and product usages on environment, and to quantitatively evaluate it.   There are a lot of works relating with this objective, such as LCA (Alting 1993).   But, actually it is very difficult to get scientifically satisfactory evaluation.   One of the reasons for this difficulty is that products and their production processes are well controlled within their manufacturing environment, that is, in their factory, whereas product life cycles are becoming very much vague or unrecognized after the shipment of products from the factory to market. According to the scope of interest for evaluation, quite different evaluation results can be obtained.

To cope with this difficulty, it is effective not only to consider the product design and manufacturing preparation concurrently, but also to design the total product life cycle as a whole from product planning, through product design and manufacturing, to product usage, maintenance and reuse/recycling/disposal.   And comprehensive strategy for product maintenance and improvement during product usage should be established, and all the life cycle processes are to be well controlled.   By such approach, reuse/recycling activities are also rationalized.   And the whole product life cycle can be made visible and controllable.   We shall call such approach as Inverse Manufacturing by stressing the controllability of reuse /recycling processes, where closed cycles of manufacturing activity are preplanned and controlled.   Related discussion are made in (Inoue 1995) and (Tomiyama 1995).

In the next section, product life cycle design concept is discussed.   In section 3, based on this concept, a new type of products named as a Right Quality Product is introduced as a basis for optimal product life cycle design.   In section 4, total product life cycle modelling technology is discussed for realizing a design support system.

2 PRODUCT LIFE CYCLE DESIGN

A simplified image of a total product life cycle is shown in figure 1.   By manufacturing activity, raw materials and energy sources are taken into the "Artifacts World", and material /energy waste is finally put back to the "Nature".   Each associated process also consumes

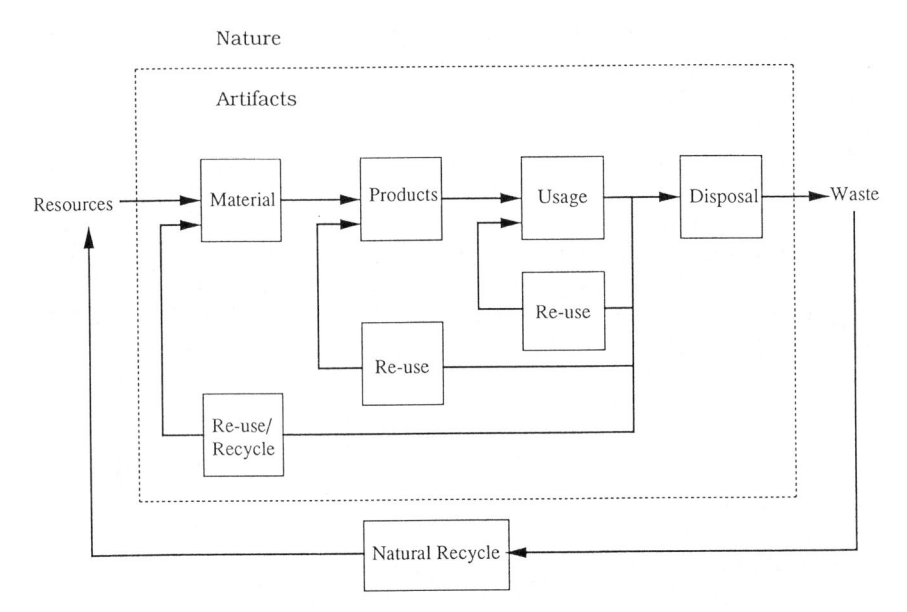

Figure 1 Total Product Life Cycle Modelling.

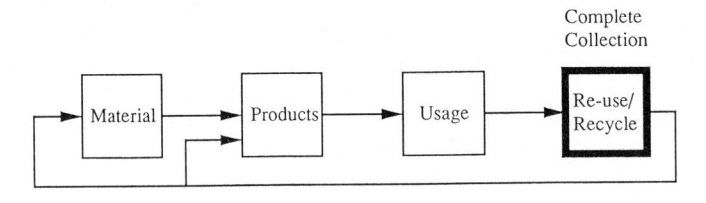

(a) High-Volume/Short-Life Products

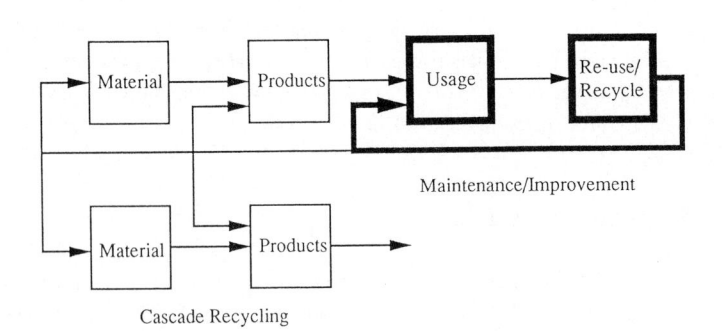

(b) Long-Life Products

Figure 2 Product Life Cycle Optimization.

resources and outputs waste. If the scale of the Artifacts World is small enough to be able to consider the Nature as unlimited in scale, we could think that the natural recycling capability is powerful enough to cancel all the resource and waste change, and neglect the effect of our industrial activity upon the Nature. But, in these days, this assumption does not seem to be true, and we need to know exactly the interaction between the Nature and the Artifacts World.

In the conventional industrial practice, we have spent much of our effort for good product design and efficient control of production processes. But, product usage /maintenance processes and recycling/disposal processes have not been well planned, and therefore those processes are rather invisible and not well controlled. Today we are confronted with a vast amount of materials/energy waste, and are forced to think about their elimination. However it is evident that efficient recycling/disposal processes do not necessarily optimize the total performance of the Artifacts World. Instead of making individual optimization of each process in figure 1, it is necessary to design the total product life cycle, as shown in figure 1, right first. It is still very difficult to make feasible engineering planning and evaluation of all the associated processes, but at least it is mandatory to explicitly describe those processes, and to have a planned control of those. It is the intent of the product life cycle design, and is a basis for Inverse Manufacturing.

In order to optimize a product life cycle with respect to certain environmental criteria, it is necessary to develop LCA processes with practical data. But, intuitively we could assume the following guideline for life cycle design. In figure 1, we can see several cycles of material flow. It seems that it is better to take inner cycles rather than outer cycles form the standpoint of resource consumption and environmental effects. Of course, this should be verified by LCA calculations. We could say that we should reuse products as much as possible at the level same as the original products, rather than destroying them to material recycling or disposal.

The previous discussion is purely based on engineering. In practice, we need to consider other factors as economical, social, and political factors. In industry, those factors are considered, and very much complicated product life cycles are actually taken. But, in many cases, those complicated life cycles are not designed, but just happened by local optimization effort.

The total product life cycle should be very much different, as shown in figure 2, according to the basic characteristics of products, such as product complexity, novelty, price, production volume, product life, etc. Simple products with volume production can take a simple product life cycle as shown in figure 2 (a). Paper and bottles are typical examples. More complicated single use-products, such as a single-use camera, are within this category. Durable products, such as buildings, aircraft/ships/trains, etc, take a very different product life cycle as shown in figure 2 (b). This cycle is characterized by the maintenance/improvement processes and associated cascade recycling processes. Probably aircraft is a good example, where the complete life cycle control is performed, and prolongation of product life cycle is economically justified by gradual improvement of performance by field replacement of modular functional components.

The issue for product life cycle design is to optimize product life cycle performance with respect to certain environmental criteria, and it requires further works, such as LCA. But, according to the intuitive guideline discussed above, the life cycle as shown in figure 2 (b) seems to be generally a good solution. The issue is then how to implement the same idea for commodity products where economic justification seems to be a major problem.

3 DESIGN OF RIGHT QUALITY PRODUCTS THROUGH DYNAMIC RENEWAL

In order to realize the product life cycle concept introduced in figure 2 (b), a concept of a Right Quality Product (RQP) is proposed. Total life cycle of RQP is shown in figure 3. A RQP is a product which has basically long-life, and can grow metamorphically by changing/adding/eliminating its functionality according to customers' requirements. Conventional products can be of long-life, but are not necessarily used for a long period due to the obsoleteness of functionality. If products can keep the Right Quality just fitted to users' needs without missing or excess functionality, they can stay competitive in the market.

For designing RQP, it is firstly important to design a total product life cycle as shown in figure 3. Especially it is essential to have good life cycle modelling capability dealing with product usage processes where product deterioration occur. Design of RQP includes not only precise evaluation of product quality/functionality/manufacturability, but also prediction of product quality degradation during product usage by sophisticated deterioration evaluation, and possible planning of proactive maintenance/improvement processes by modular renewal of functional components. During actual operations, disassembled components or parts from certain RQP are to be checked about their deterioration rate through model-based prediction and real inspections, and are reused for the same type machines or others.

By expanding the reuse of used components and parts, product recycling/disposal could be done in gradual and distributed manner, and can be made efficient by taking the advantage of various possibilities of product usage. Finally it may happen that clear distinction between completely new products and used products disappears because every product consists of appropriate new components and reused components. By such approach, new industrial society could be realized with much less resource consumption with the same level of industrial productivity as today.

Some of the important key characteristics of RQP are as follows:
- Modular configuration of product functionality and its associated product structure,
- Precise description and evaluation of product functionality/quality,
- Elimination of excess functionality/quality to be exactly fitted to demands,
- Prediction and control of product deterioration during operations,
- Modular improvement of product functionality under product deterioration status,
- Quality assurance and reuse of remanufactured components.

All those issues are extremely difficult in theoretical sense, and generic methods will not be available. Dedicated methods should be developed for classes of products respectively.

Product life cycle modelling for inverse manufacturing 85

We can see several industrial examples based on the above idea, although their scopes are limited and no theoretical justifications are given with respect to the total product life cycle efficiency. They include industrial manufacturing facilities, such as semiconductor manufacturing facilities, electronic devices, such as computers, and industrial equipments, such as air-conditioners. They are all industrially used products, and are under the good maintenance procedures. They are well modularized in their original design. It is a technical as well as economical challenge how to introduce this concept into commodity products with high volume production.

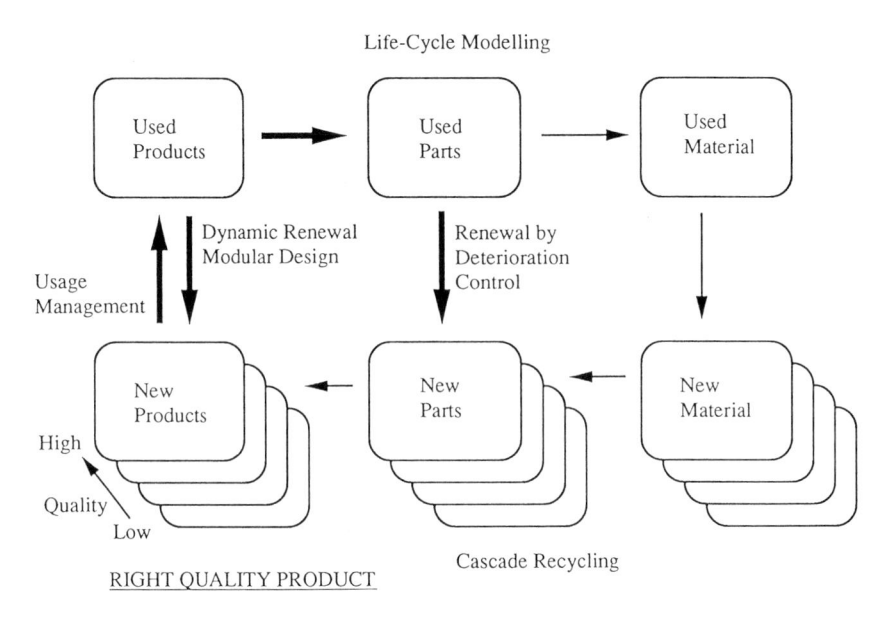

Figure 3 Total Life Cycle of Right Quality Products.

4 LIFE CYCLE DESIGN SUPPORT SYSTEM

Life cycle modelling is discussed for supporting the design and usage management of RQP. There has been a lot of work concerning with product modelling (Krause et al 1993a), but previous research primarily concerns with product design and manufacturing preparation, and very little attention has been paid for the product usage and maintenance stages. However, as shown in figure 4, conventional general purpose CAD/CAM system functionality with product modelling can be used as a platform for further development specific to RQP. General concept of virtual manufacturing (Kimura 1993a/1993b) is a basis for the total life cycle support system configuration. Modelling mechanisms, such as generic feature modelling

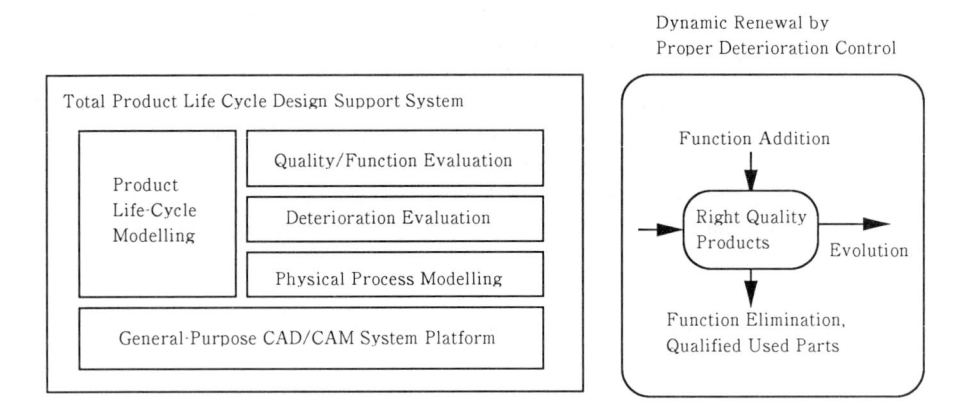

Figure 4 Life Cycle Design Support System.

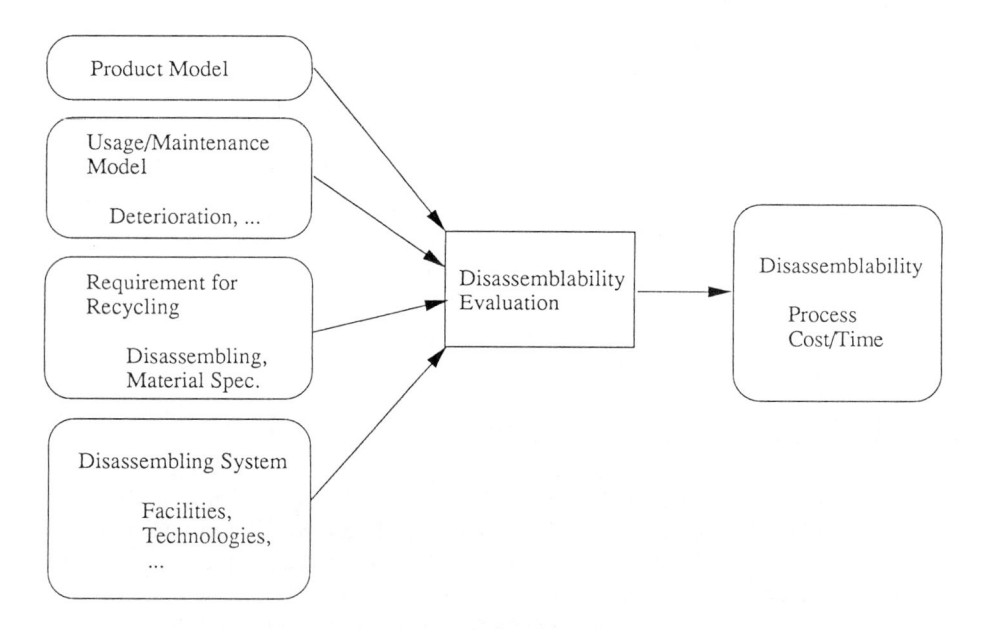

Figure 5 Disassemblability Evaluation Based on Life Cycle Modelling.

(Krause et al 1994), are applied to a life cycle modelling framework for RQP design.

Some of the key components for supporting life cycle modelling are as follows:
- Quality/Function Evaluation:

There have been many research works concerning with product quality/function description and evaluation. From the standpoint of RQP design, it is essentially important to be able to detect unexpected malfunctioning of products with certain deterioration effects. For this purpose, conventional methods, such as QFD, FMEA, robust design, etc., are not sufficient, but those methods should be augmented by comprehensive product modelling with deterioration and basic physical process evaluation, as described below. Some early examples are seen in (Ishii 1995, Krause et al 1993b, Takata 1995).
- Deterioration Evaluation:

Machine product deterioration will occur due to material degradation, wear, fatigue, fracture, shape deformation, dust, rust, etc., and it is practically impossible to precisely predict the status of deterioration quantitatively by theoretical calculation. Many of the data driven experience-based modelling of deterioration have been done in industry. Those results could be combined with general deterioration modelling based on the generic physical process modelling below.
- Generic Physical Process Modelling:

In relation with product modelling, modelling of basic physical processes has been pursued, such as shape deformation, dynamics, etc. One of the key issues is to evaluate the sensitivity of product physical behaviour with respect to physical property variations. Some of the examples are seen in (Inui 1994, Kimura 1992, Yoshizaki 1994). The extension of those works will be a basis for deterioration evaluation.

Importance of the total product life cycle modelling for RQP can be observed in the case of disassemblability evaluation as shown in figure 5. In RQP practice, disassembling is not simply a reverse operation of assembling. After several years of product usage, target objects for disassembling may have some defects or modifications, or suffer from rust. In the course of reuse strategy, requirements of disassembling will also vary. For instance, some parts can be destroyed because they are not reused, but some are to be carefully treated for right quality control for later reuse. Therefore comprehensive modelling of used products are mandatory for effective planning of disassembling.

5 CONCLUSION

Based on the concept of Inverse Manufacturing, a Right Quality Product is introduced, and some of the issues are discussed for technical implementation of the concept. For practical feasibility of the concept, social infrastructural support will be necessary to provide appropriate information during the product life cycle activities, as shown in figure 6. Any necessary information should be given for product maintenance and renewal as well as normal product design. The currently proposed information infrastructure, such as CALS, will be a vehicle for this purpose. It may not be practically feasible to keep all the information in the information database apart from products themselves. The concept of product ID/active memory will be necessary in conjunction with the life cycle database for effective implementation of RQP.

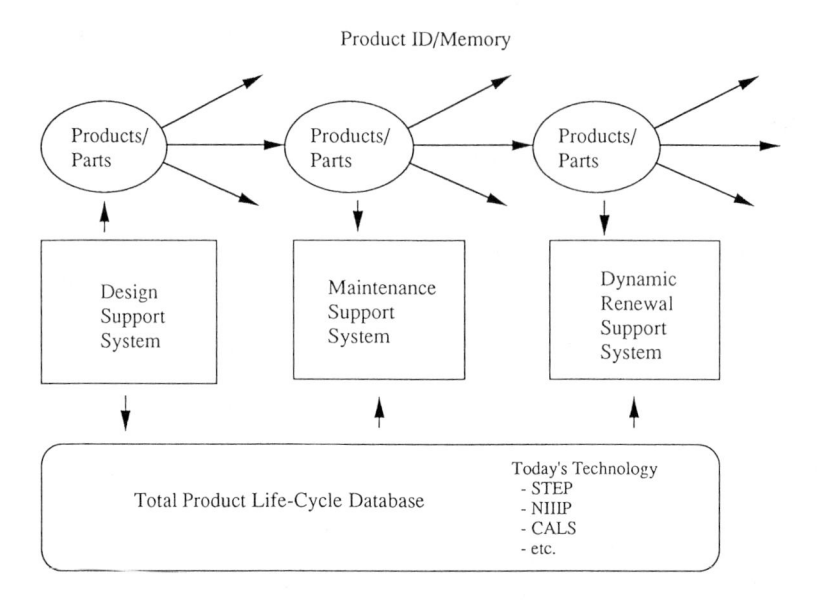

Figure 6 Information Infrastructure for Product Life Cycle Support.

6 REFERENCES

Alting, L. (1993) The Life Cycle Concept as a Basis for Sustainable Industrial Production, Annals of CIRP, 42, 163-167.
Inoue, H. (1995) Ecofactory: Ecologically Conscious Technology for the 21st Century, Int. J. Environmentally Conscious Design & Manufacturing, 4, 13-28.
Inui, M., Matsuki, N. and Kimura, F. (1994) Extended Formulation of Geometric Tolerances Based on Parametric Modifications of Surface Features, in *Feature Modeling and Recognition in Advanced CAD/CAM Systems* (IFIP WG5.3 Working Conference), 673-692.
Ishii, K. (1995) Life-Cycle Engineering Design, Trans. ASME, 117 (B), 42-47.
Kimura, F., Suzuki, H. and Takahashi, K. (1992) Product Design Evaluation Based on Effect of Shape Errors for Part Assembly, Annals of CIRP, 41, 193-196.
Kimura, F. (1993a) Product and Process Modelling as a Kernel for Virtual Manufacturing Environment, Annals of CIRP, 42, 147-150.

Kimura, F. (1993b) A Computer-Supported Framework for Concurrent Engineering Based on Virtual Manufacturing, in *Information Infrastructure Systems for Manufacturing* (IFIP WG5.3 Workshop), 345-359,

Krause, F.-L., Kimura. F., Kjellberg, T. and Lu, S.C.-Y. (1993a) Product Modeling, Annals of CIRP, 42, 695-706.

Krause, F.-L., Ulbrich, A. and Woll, R. (1993b) Methods for Quality-Driven Product Development, Annals of CIRP, 42, 151-154.

Krause, F.-L., Rieger, E. and Ulbrich, A. (1994) Feature Processing as Kernel for Integrated CAE Systems, in *Feature Modeling and Recognition in Advanced CAD/CAM Systems* (IFIP WG5.3 Working Conference), 693-716.

Takata, S., Hiraoka, H., Asama, H., Yamaoka, N. and Saito, D. (1995) Facility Model for Life-Cycle Maintenance System, Annals of CIRP, 44, 117-121.

Tomiyama, T., Sakao, T., Umeda, Y. and Baba, Y. (1995) The Post-Mass-Production Paradigm, Knowledge-Intensive Engineering, and Soft Machines, in *Proceedings of PROLAMAT '95*, Chapman & Hall.

Yoshizaki, K. (1994) Modelling of Machine Assembly Behaviour with Flexible Components, Ms. Thesis (Japanese), The University of Tokyo, 1994.

7 BIOGRAPHY

Fumihiko Kimura is a professor in the Department of Precision Machinery Engineering of the University of Tokyo. He was a research associate at the Electrotechnical Laboratory of the Ministry of International Trade and Industry from 1974 to 1979. He then moved to the University of Tokyo, and was an associate professor from 1979 to 1987. He has been active in the fields of solid modeling, free-form surface modelling and product modelling. His research interests now include the basic theory of CAD/CAM, concurrent engineering and virtual manufacturing. He is involved in the activities of ISO/TC184/SC4, and is a member of IFIP WG5.2 and 5.3, and a member of CIRP. He graduated from the Department Aeronautics, the University of Tokyo, in 1968, and received a Dr. Eng. Sci. degree in aeronautics from the University of Tokyo in 1974.

Hiromasa Suzuki is an associate professor of Department of Precision Machinery Engineering at The University of Tokyo. His research interests include geometric modeling and reasoning, and their applications to mechanical CAD/CAM systems. Suzuki received his doctor degree in precision machinery engineering from the University of Tokyo in 1986. He is a member of JSPE(Japan Society for Precision Engineers), JSME(Japan Society for Mechanical Engineers), IEEE, ACM etc.

8

Some Theoretical Issues and Methods of Life-Cycle Engineering

Valery Tarassov
Moscow State Bauman Technical University
2nd Baumanskaya St., 5 Moscow 107005, Russia
Tel.::+(095)263 69 95
E-mail: bulanov@mipk.bmstu.msk.ru

Abstract

At the beginning the filed of concurrent engineering could be seen as a relatively new customer-oriented design philosophy aimed at attaining better enterprise competitiveness that includes two main interrelated components - product competitiveness and production system competitiveness. It embraced a family of rather empirical methods and techniques of improving conventional design (such as the cross-disciplinary teaming, the design-for-manufacturing techniques) for raising product quality, reducing development costs, shortening lead times and enhancing overall productivity. Nowadays a wide implementation of emergent computer-based (and primarily intelligent) systems and technologies at enterprises requires the development of general enterprise integration theory based on the principles of life-cycle engineering [Douglas et al., 1993; Jo et al., 1991; Krause et al., 1994; Kusiak, 1993; Reddy et al., 1992; Tomiyama, 1991, etc.].

The main objective of this paper is to develop a systemic approach to enterprise modelling in the scope of life-cycle engineering. Such an approach has rich traditions in CAD theory. Here the works of German (Koller, Rodenacker, Roth, Spur) and Russian (Balashov, Matveyevsky, Polovinkin, Semenkov) authors are of special concern.

A basic idea of our approach to building the fundamentals of life-cycle engineering theory consists in specifying an exhaustive family of enterprise systemic attributes and in applying a reduced number of systemic principles to reveal different sides and to suggest non-conventional models of life-cycle engineering. For instance, on considering the enterprise as an open, dynamic, hierarchical, hererogeneous, distributed intelligent system and on taking into account the principles of systems hierarchy, goal analysis, functional decomposition, functions-structures unity, extensive-intensive structures unity and models concertation, we arrive to determine four crucial sides or problems of life-cycle engineering:

- enterprise data/knowledge engineering with paying special attention to building special knowledge accumulation, integration and reuse methods and tools, for instance, in the framework of life-cycle metaknowledge and enterprise metamodels;
- properly simultaneous engineering items involving the practice of transforming a sequential and linear nature of the product life-cycle into a highly parallelized, circular structure that means incorporating various product's life-cycle values at the most critical production time/space intervals, usually at early stages design;
- global life-cycle vagueness/uncertainty management with taking into various informational „No-Factors" proper to engineering activities and specifically to design process such as non-completeness, non-monotonicity, imprecision, fuzziness, ambiguity etc.;
- on-line coherence estimation/support and total quality contol.

Furthermore, we need a special methodology to face these problems in a modern computer-based environement. Ist implementation may include the following steps:
- non-linear life-cycle's representation, its detailed structuring and extraction of simultaneousity factors;
- development of mathematical methods and models to simulate and to control principal life-cycle parameters;
- development of appropriate functional and software architecture;
- validation and practical use of software tools.

A spiral life-cycle's representation suggested in Russia at the beginning of 80-s (see [Tarassov et al., 1994]) is discussed. Ist comparison with other close representations such as Jo's concurrent engineering wheel [Jo et al., 1993] and functionalities'cycle [Feru et al., 1994] based on Kimura's proposal is made. Following spiral model the product life-cycle is seen as a trajectory showing a time evolution of engineering information and product state from product specification to its dismantling (this final phasis of product dismantling is drawn at the center of the spiral). The spiral plane is divided into three sectors: design, manufacturing, exploitation (use and maintenance).

Such a representation enables us both to show in a clear and exhaustive manner the engineering data and knowledge circulation during the whole product life-cycle and to analyze basic interrelationships between different life-cycle stages and values for acquiring and manipulating life-cycle engineering knowledge. Specifically, here the number of spires may be attributed to a simultaneousity degree in lefi-cycle engineering: the less it is, the more simultaneous are the life-cycle values. This representation permits to analyse two main different types of simultaneousity encountered in enterprise activities.

The second part of the paper is devoted to the development of formal methods of life-cycle modelling. Specifically three groupes of models are considered:
- AI hybrid logics to model a simultaneous manipulation in multiple worlds and retractability features;
- Multiattribute reasoning/argumentation tools able to model various ways of practitioners'reasoning and argumentation, including common-sense reasoning;
- Neuro-fuzzy control models.

REFERENCES

[Douglas et al., 1993] Douglas R.E., Brown D.C., Zenger D.C.
„A concurrent engineering demonstration and training system for engineers and managers",
Revue internationale de CFAO et d'infographie, vol.8, no.3, 1993, p.263-301.

[Feru et al., 1994]   Feru F., Vat C., Cocquebert E., Deneux D., Rouchon C.
„Computer-aided functional design for manufacturing",
Revue internationale de CFAO et d'infographie, vol.9, no.5, 1994, p.665-683.

[Jo et al., 1991]   Jo H.H., Hamid R.P., Wong J.P.
„Concurrent engineering: the manufacturing philosophy for the 90's",
Computers in Industrial Engineering, vol.21, 1991, p. 35-39.

[Krause et al., 1994] Krause F.-L., Rieger E., Ulbrich A.
„Feature processing as kernel for integrated CAE systems"
Proc. IFIP Int. Conference on Feature Modeling and Recognition in Advanced CAD/CAM Systems, vol.2, Valenciennes, France, 1994, p.693-716.

[Kusiak, 1993]   Kusiak A.
„Concurrent engineering: automation, tools and techniques", J. Wiley and Sons, N.Y., 1993.

[Reddy et al., 1992] Reddy Y.V., Wood R.T., Cleetus Y.J.
„The DARPA initiative in concurrent engineering", Concurrent Engineering Research in Review, vol.1, 1991-1992, p.2-10.

[Tarassov et al.,]   Tarassov V.B., Kashuba L.A., Cherepanov N.V.
„Concurrent engineering and AI methodologies: opening new frontieres",
Proc. IFIP Int. Conference on Feature Modeling and Recognition in Advanced CAD/CAM Systems, vol.2, Valenciennes, France, 1994, p.867-888.

[Tomiyama, 1991]   Tomiyama T.
„Intelligent CAD systems", Advances in Computer Graphics, vol.1. Images: Synthesis, Analysis and Interaction (G. Garcia, T. Herman, ed.), Springer, Berlin, 1991, p.343-388.

9
Global Engineering Network - Applications for Green Design

H. Schott, H. Birkhofer
Departement of Machine Elements and Engineering Design
Technical University of Darmstadt
D 64289 Darmstadt
Magdalenenstr. 4
Telefon (06151) 16 6614
Telefax (06151) 16 3355
E-MAIL: schott@muk.maschinenbau.th-darmstadt.de

Abstract

Medium- and long-range environmental protection measures have to be preventative in nature. They should avoid potential environmental damage by stating suitable regulations at the product development stage, and implicitly minimise the potential damage over all phases of product life. However, the requirements posed by such an environment management system are to excessive to be met by small and medium-sized companies. The great variety of process analyses necessary to estimate the potential environmental effects throughout the overall product life and the environmental restrictions resulting from these effects cannot be managed by those responsible for environmental protection or the design or standards departments. Information systems that collect design relevant environmental information from ubiquitous sources and provide this information in a suitable form to the user groups are at least a partial solution to this problem.

Keywords

Design for Environment, Global engineering network, information conversion, Eco Design, Meta-information-services, hypermedia guidelines

1 Design for Environment - State of the Art and Need for Support

Design for Environment (DFE) is an optimisation process with the goal of minimising the detrimental impact of the product on the environment throughout its life cycle. The absolutely „green product" (i.e. a product that has no negative effect on the environment) does not exist.

We can only try to reduce the detrimental effects on the environment. The following constraints have to be taken into account while designing green products:

1.1 Constraints for Design for Environment

Constraints caused by the own company. The engineering design department is not only responsible for ecological criteria, but also for the technical function, economic feasibility and quality of their designs.

Constraints due to the decrease of net added value by the company. The increasing amount of outsourcing leads to a decrease of available options for a company to reduce negative environmental effects.

Constraints due to normative conditions. The multitude of legal and technical standards and specifications complicate the creation of useful design guidelines for the designer.

Constraints due to the extent of data to be collected for Life-Cycle-Assessment. To be able to judge the impact of a technical product on the environment over the whole Product-Life-Cycle, a vast quantity of data has to be collected.

1.2 Approaches for Design for Environment

In the last years numerous approaches to support Design for Environment have been created. Because the constraints due to the recycling-process can be easily formulated as design objectives, approaches for ease of recycling have found the largest dissemination. But the impact on the environment cannot be measured in terms of the recyclability of a product and/or its materials. All the approaches shown in fig. 1 have either the problem that the impact of a design solution on the environment can not be evaluated accurately or that the designer has to collect a vast quantity of data. To evaluate the environmental

Fig. 1: Approaches and objectives for design to environment

impact caused by the processes involved in the overall product life-cycle it is necessary to record the emissions, the raw material and energy consumption of all these processes. At the moment these data are not available.

1.3 Engineering Design as Information Conversion

Engineering Design is a process of information conversion [4]. The information needed in design for environment constitutes only a small part of the entire information that is received, processed and transmitted during the design process. The design process can be divided into the processes of creation and evaluation.

1.3.1 Information used to Search Environmentally Friendly Solutions (Creation)

The search for solutions can be based on intuitive methods, methods for searching existing and proven solutions and systematic solution development methods. It can be supported by various working aids like manuals and tools like CA-Tools or knowledge-based systems.

Methods for searching for solutions	Working Aids (and Sources)	Tools (and Sources)
Methods with an intuitive bias (Brainstorming, Delphi, stimulation: discussions, conversations, etc.)	*Guidelines* to awake the consciousness of designers (VDI-Guidelines for technology assessment) *Communication with organisations for environmental protection* to encourage intuition and to open new paths (Greenpeace, WWF, BUND)	none World-Wide Web (http://www.greenpeace.org)
Using solution catalogues (manuals, databases) (own products, patent files, market, design catalogues)	*Catalogues of evaluated (life-cycle-assessed) products* with referring processes. (meaningful and valid data is available only on the packaging of beverage (UBA)) *Design catalogues (manuals)* with information about efficiency and required auxiliary material and energy make it possible to evaluate the environmental impact. (such catalogues do not exist yet)	no tools and sources available yet (database-systems, case based reasoning systems, hypermedia-information-systems would be useful)
Systematic approach (function, physical effect, working principle, solution principle)	Working aids for the systematic approach are mainly used for checking and evaluating solutions. They help designers reducing the solution field.	no tools and sources available yet

Methods to design for ease of recycling and for waste processing	Guidelines and rules for Design for Environment (Working aids for design for recycling and design for disassembly: VDI-Guideline 2243, disassembly graphs, graphs of preferred disassembly order, lists of disassembly times, methods for evaluating disassembly operations Working aids referring to materials: tables of compatibility of materials, marking of materials)	knowledge-based systems (for recycling and disassembly: ReStar, DFA/DFD-software, Databases, LASeR, REKON, RECY-CLEAN) for materials: databases of material properties)

Tab. 1: Examples for information for the searching for solutions

Tab. 1 shows that the potential of scientific design methodology is not exhausted for Design for Environment.

1.3.2 Information used to Evaluate the Impact of Products on the Environment

Solutions are evaluated to verify them with respect to the requirements and constraints from the overall life-cycle.

Methods for selecting and evaluating solutions	Working Aids (and Sources)	Tools (and Sources)
Intuitive methods and methods without accounting impacts on environment	*Checklists* (Ecological quality of products, GH Kassel) Security data sheet (DIN) *Lists of forbidden and preferred materials* (depends on the company and country)	Spreadsheets, Databases, Word processors (Manuals, software)
Analytical Methods and methods accounting impacts *Ecobalances, life-cycle-assessment* (goal definition, inventory, environmental effects, evaluation) *Working aids and tools* (GaBi, Sima Pro, GEMIS, IDEA, Lifeway,...) For evaluation a number of different methods and criteria using different information and data exist: *judgement of experts* verbal, argumentative judgement from an expert *critical volumes* emissions and critical value of hazardous emissions (Process data, MIK-, MAK-Values (VDI, DFG)) *sustainability* emissions, raw material and energy consumption, deposit area, consumption of area,... (Process data, data about the „condition of the earth") *ecological shortage* mass flow, actual load, critical load (data about the mass flow, critical value of hazardous emissions, data about the load in the balanced area) *MIPS (material intensity per service)* Material intensity (total primary material consumption for processes), service unit referred (mass flow data, MI-tables, data about the usage of products) *toxicological evaluation* toxicological data and data about emissions (emissions, data about toxic effects on the environment and on man)		

Tab. 2: Examples for information for the evaluation of the ecological properties of solutions

Tab. 2 shows the tremendous amount of data, which has to be recorded for the quantitative evaluation of the environmental impact through products and involved processes. Only a small part of the necessary information can be recorded in the own company, and even this may exceed the possibilities of small or middle-sized companies.

1.4 General View on Sources used for Design for Environment

The information sources described in the section 2.3.1 and 2.3.2 can be divided into different levels:

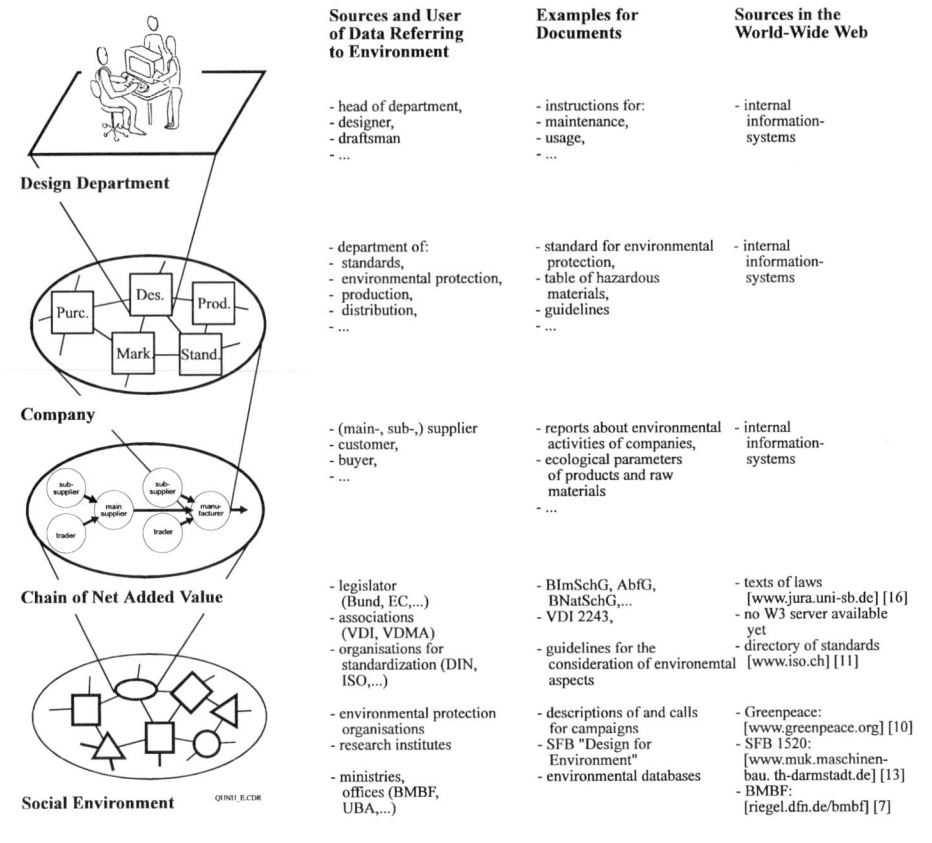

Level	Sources and User of Data Referring to Environment	Examples for Documents	Sources in the World-Wide Web
Design Department	- head of department, - designer, - draftsman - ...	- instructions for: - maintenance, - usage, - ...	- internal information-systems
Company	- department of: - standards, - environmental protection, - production, - distribution, - ...	- standard for environmental protection, - table of hazardous materials, - guidelines - ...	- internal information-systems
Chain of Net Added Value	- (main-, sub-,) supplier - customer, - buyer, - ...	- reports about environmental activities of companies, - ecological parameters of products and raw materials - ...	- internal information-systems
Social Environment	- legislator (Bund, EC,...) - associations (VDI, VDMA) - organisations for standardization (DIN, ISO,...) - environmental protection organisations - research institutes - ministries, offices (BMBF, UBA,...)	- BImSchG, AbfG, BNatSchG,... - VDI 2243, - guidelines for the consideration of environemtal aspects - descriptions of and calls for campaigns - SFB "Design for Environment" - environmental databases	- texts of laws [www.jura.uni-sb.de] [16] - no W3 server available yet - directory of standards [www.iso.ch] [11] - Greenpeace: [www.greenpeace.org] [10] - SFB 1520: [www.muk.maschinenbau.th-darmstadt.de] [13] - BMBF: [riegel.dfn.de/bmbf] [7]

Fig. 2: Examples for user and sources used for Design for Environment

Most of the sources named above, particularly on the level of the net added value chain and social environment are not suitable for developers and designers and difficult to access.

2 How can Information be Brought to the User?

The multitude of sources required for Design for Environment and the dynamic change of the contents of the corresponding documents require new approaches for the handling of the processes of information logistics. For the main tasks of information logistics (i.e. the structuring of the storage and flow of information) two trends can be observed:
- In information storage an increasing standardisation of product data and the provision of hypermedia product data can be observed.
- In the field of information flow the extensive dissemination of the Internet-service World-Wide Web (W3) has lead to a kind of standard for the exchange and the access of hypermedia documents.

At present there are about 30 Million people using the World-Wide Web (WWW or W3). The amount of people is increasing for about 10% each month. By means of the HTTP (hypertext-transfer-protocol) of the W3, documents can be accessed from all over the world in seconds. Although currently the amount of electronically accessible documents is small compared to the amount of printed documents, the number is increasing rapidly. The W3-Server of the Environmental Protection Agency for example is providing approx. 250 Gbyte of environmental information daily [9].

Hypermedia Documents on the World-Wide Web (W3)

Hypermedia documents have several attractive features:
- Several types of information (pictures, text, sounds, video), can be integrated and presented by one uniform user-interface.
- Connections between documents can be displayed and followed by links.
- Documents can be detailed and explained by linked documents.
- Documents can be directly accessed by methods of information retrieval (matching), or by following the links between the documents (browsing).

The advantages of hypermedia documents combined with the accessibility of information from all over the world is vital for the support of Design for Environment. One major task is the development of tools to help the user not to get lost in „hyperspace". The key to guided access is meta-information (information about information). By choosing the right information-system to transfer documents (e.g. the W3), and by developing adequate meta-information-systems it is possible to get the right information without delay. The storage of information and data is increasingly shifting from centralised to distributed information systems. The consequence of these points for engineering design is, that the sources of relevant environmental information on the levels described in fig. 2 can be easily accessed and information generated by designers can be provided and accessed by others.

To use and to extend the possibilities mentioned above, several research institutes and companies are working on a Global Engineering Network (GEN) (fig. 3). Their goal is to provide uniform access to distributed information independent of the physical location and to develop tools performing intelligent search and navigation. Due to the open architecture of the basic technologies and services like Internet and World-Wide Web it is possible to integrate several applications to the GEN. The GEN-Project co-ordinates these applications and presents the information under a uniform user-interface.

Global Engineering Network

Meta-Information-Services:

Yellow pages, catalogues of standard parts,
informations about product catalogues,
aids for selecting and evaluating semi-standard solution
elements and services, electronic tendering

Interactive Consulting and Brokering:

e.g. assistance and support in trouble-shooting, teleexpertise on design and manufacturing issues

Information-services:

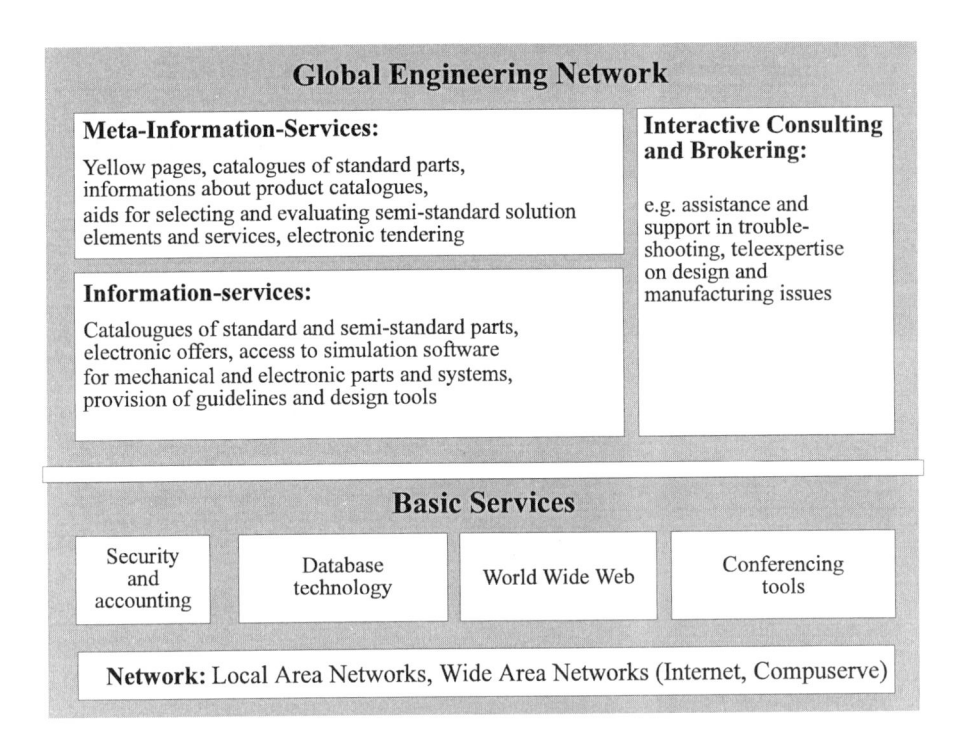

Catalougues of standard and semi-standard parts,
electronic offers, access to simulation software
for mechanical and electronic parts and systems,
provision of guidelines and design tools

Basic Services

| Security and accounting | Database technology | World Wide Web | Conferencing tools |

Network: Local Area Networks, Wide Area Networks (Internet, Compuserve)

Fig. 3: GEN Reference architechture

3 EDN - EcoDesign-Net

The broadness and variety of information needed for Design for Environment and the potential of the W3 for realising suitable concepts for information logistics are the motivation to develop an EcoDesign-Net (EDN). The EDN is meant to be a central information-tool for Design for Environment. The EDN collects information about topics that are relevant for Design for Environment. The information is adapted to the needs of designers and provided on the W3. Referring to fig. 2 EDN is an information-system on the level of the social environment of a company, accessible from individuals on other levels like the chain of net added value, companies or departments. We collect, adapt and provide the information that is at disposal from all the sources on the W3. Organisationally
EDN is part of the „Sonderforschungsbereich 1520 - Entwicklung umweltgerechter Produkte - Methoden, Arbeitsmittel und Instrumente (Design for Environment - Methods, Working Aids and Tools)", a research project at the Technical University of Darmstadt. The EDN is used for spreading the results of this research project. It is divided into *meta-information-services* and *information-services*. Especially the *information-services* are devised to provide the results of this research project.

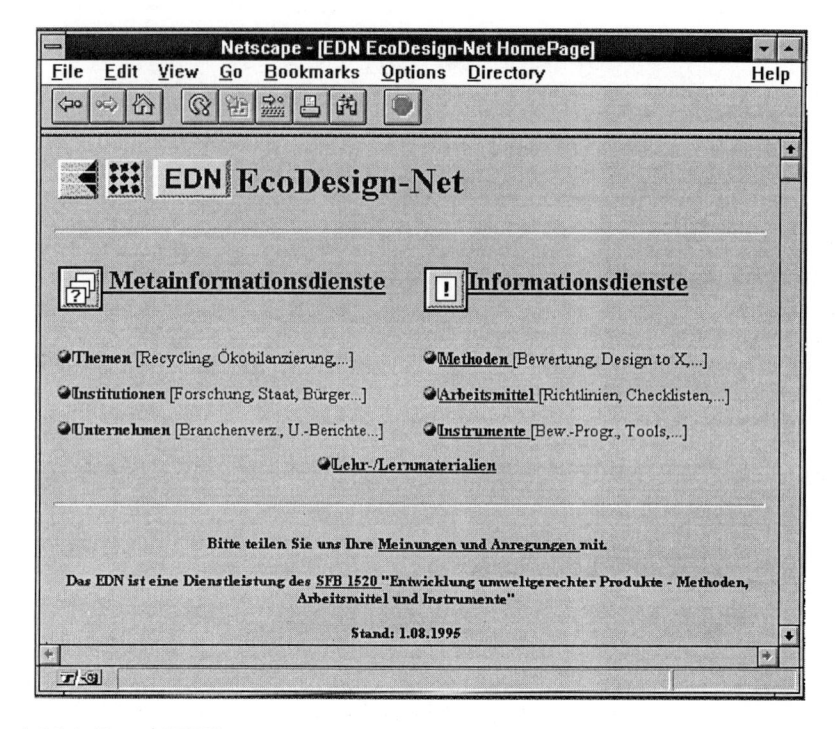

Fig. 4: Main Page of EDN

3.1 Meta-Information-Services

Using the meta-information-services it is possible to access information sources and relevant documents systematically. The users are able to search for Design for Environment related topics like recycling, disassembly, waste disposal or life-cycle-assessment. They are able to retrieve the relevant information on the level of the „social environment" from the W3. The most important ubiquitous sources and documents are selected, commented and provided by the EDN. The EDN enables the user to determine if a document might be valuable without having to retrieve it. A database, managing the meta-information belonging to the topics mentioned above, has been built up for this purpose. By using search engines (Web robots, wanderers, spiders) the database is continually updated. The EDN is using these systems with specific retrieval strategies adapted to the topics of design to environment.
Systematic search for topics and institutions. For the systematic search for documents (see fig. 5) specific topics can be selected. The query can be reduced to specific institutions like research institutions, companies, organisations etc. The user is able to complement the query by her/his own search keys. The search for the commented data records in the EDN database can be handed over to search engines like Lycos [12] or OpenText [15].

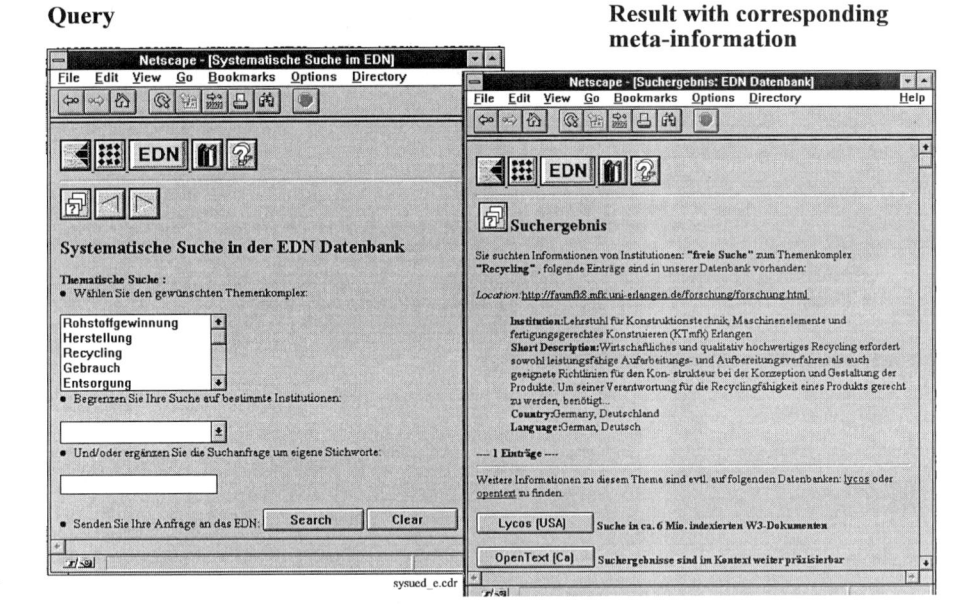

Fig. 5: Example of a query and corresponding result in the meta-information-service of the EDN

Access to information from companies. Reports about the activities of companies referring to environmental protection measures can be an aid for designers when selecting suppliers. At the moment there are no reports available, but for the future we expect a rising amount of reports being published electronically. To select suppliers by environmental criteria it is necessary to built up a standard enabling to compare the impacts of supplied products from different companies. In the middle of 1996 the ISO standard 14000ff for environmental management systems is expected [6]. This standard will describe criteria for life-cycle assessment. Up to now a first step towards a suitable selection of suppliers can be a file of companies being certified according to the ECO-Audit specification [1, 3] or the german „Umweltengel" [2].

3.2 Information-Services

To increase the contribution of research to the minimisation of detrimental impact of technical products on the environment, the EDN will provide the results of the research project „Design for Environment - methods, working aids and instruments" as soon as possible to the public. This way the results can be discussed very early with representatives of companies and the scientific community.

3.3 Using the Information Service „Guideline" to Set Up Integrated Hypermedia Guidelines for Green Design

The following example describes how the information- and meta-information-services can be used in a company. The goal of the integrated hypermedia guidelines is to reduce the effort necessary to set up guidelines in companies and to increase the efficiency by adding values. In our context a guideline is understood to be a working aid supporting designers while searching and evaluating solutions. The evaluation is based on qualitative aids like checklists and lists of hazardous materials.

Internal guidelines of companies: In comparison with general guidelines like the ones from the VDI, internal guidelines offer a company specific, summarised and adapted view of the topic. A large part of the guidelines is derived from external sources (see fig. 2).

Content Sources	goals	state of the art	references to methods and aids	design rules	checklists, lists of hazardous materials	further standards and laws
design department	-	-	standardisation department	engineering design department	-	standardisation department
company	management	department of research and development (R&D)	department of R&D, controlling	department of standards	production, person responsible for environ. prot.	person responsible for environ. prot.
chain of net added value	-	department of R&D of the supplier	-	associations	instructions of supplier and customers	instructions of supplier and customers
social environment	politics, public opinion	research institutions	research institutions	research institutions	research institutions	DIN, ISO, legislation

Tab 3: Examples of sources for environmental information relevant for guidelines

Integrated hypermedia guidelines: Due to the fact, that a major part of the company specific environmental guidelines are derived from external sources, the companies have only relatively small influence on them. Therefore it is useful to provide general guidelines on ubiquitous information systems for a broader circle of companies. These can in turn be integrated into specific guidelines of companies. The EDN can provide templates for company specific adaptations of the guidelines. However the integrity of the guideline has to be ensured. Following the (hierarchical) structure of print-documents, the integration of specific parts of guidelines can take place on the level of chapters, or paragraphs. The example below shows the adaptation of a specific guideline on the level of chapters. The chapters „goals" and „documented system procedure" are replaced by adapted chapters referring to the requirements of the company (see fig. 6).

Global engineering network 103

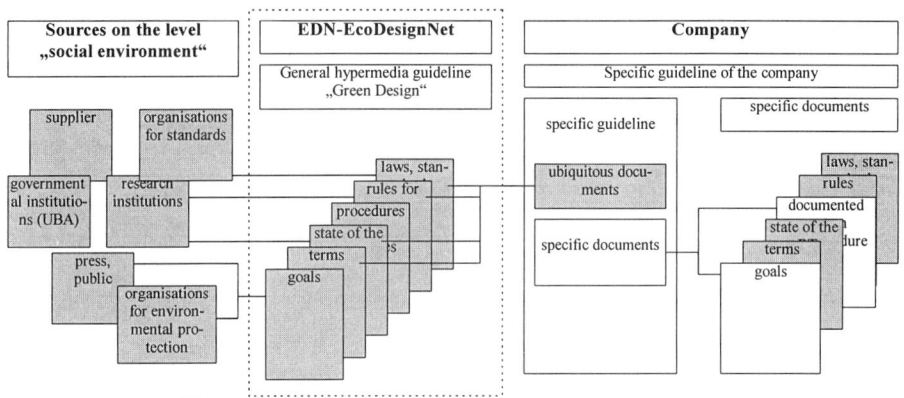

Fig. 6: Structure of integrated guidelines

Using the guidelines in the EDN it is possible for companies to access further sources concerning topics mentioned in the guidelines. Fig. 7 shows an example of an integrated hypermedia guideline and the possibilities of navigation.

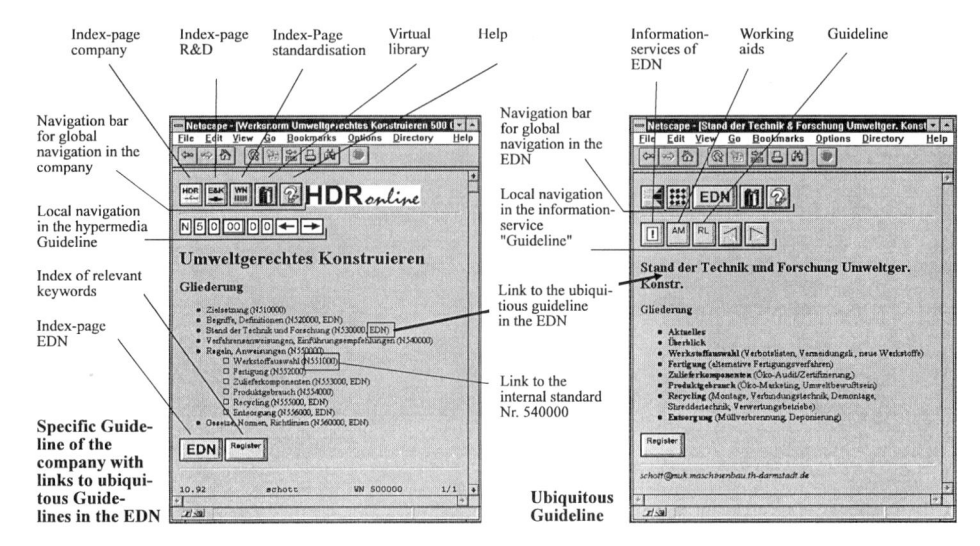

Fig. 7: Integrated hypermedia guideline for Design for Environment

Navigation in hypermedia guidelines: The available options for navigating hypermedia documents determine the possibilities of orientation and access to contents. We use two iconbars, one for local and one for global navigation. Global navigation is used to browse documents referring to the company (see fig. 6 and 7). The icon bar for local navigation is used to access the hypermedia guideline. Activating the numbers shown in fig. 8 the user is able to activate the document with the table of contents referring to the corresponding level (book of standards, directory of main topics, etc.). From every document there is a link to the index of the corresponding level. From this index every document is accessible through specific key words. The structure and way of navigating in the general guideline on the EDN follow the same scheme.

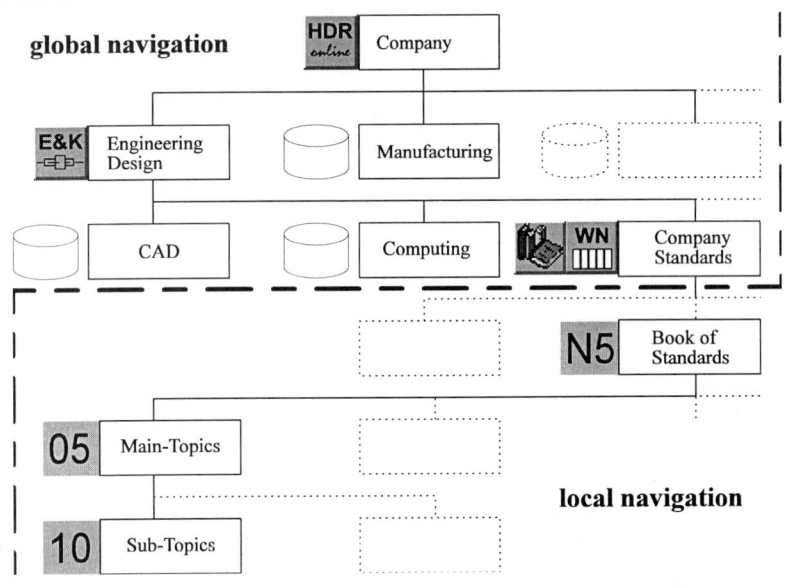

Fig. 8: Global and local navigation in hypermedia Guidelines

4 Conclusion and Future Work

Using the information- and meta-information-services of the EDN we have presented an approach to improve the provision of information for Design for Environment. The increasing commercialisation and dissemination of the World-Wide Web is promoting the broader use of this service in companies. The EDN can only be used successfully, when it is not the only (W3-) service used by the designer. Only when the W3 becomes an integral element of the tools used to design, the EDN can be used successfully. The GEN-Initiative is working on a broader dissemination of network-based information-services for engineering design. For the provision of product information and the interaction on so called virtual market places, the department of Machine Elements and Engineering Design at the Technical University of Darmstadt has developed and set up a system called CompoNet. The system is now marketed by a publishing company. To succeed in providing the information needed for Design for Environment, the EDN is depending on the close co-operation of companies, public authorities and research institutions. The described approach is feasible and we hope that many institutions will support the project to advance preventative environmental protection.

5 References

[1] **Alijah, R.**: Betriebliches Umweltmanagement: systematische Umsetzung der EG-Öko-Audit-Verordnung. WEKA, Augsburg 1995
[2] **Deutsches Institut für Gütesicherung und Kennzeichnung e.V. RAL (Hrsg.)**: Umweltzeichen - Produktanforderungen - Zeichenanwender und Produkte. RAL, Bonn 1992
[3] **N.N.**: VERORDNUNG (EWG) Nr. 1836/93 des RATES vom 29. Juni 1993 über die freiwillige Beteiligung gewerblicher Unternehmen an einem Gemeinschaftssystem für das Umweltmanagement und die Umweltbetriebsprüfung. Amtsblatt der Europäischen Gemeinschaften, Nr. L 168/1.
[4] **Pahl, G; Beitz, W.**: Konstruktionslehre - Methoden und Anwendung. 3.Aufl. Springer, Berlin 1993
[5] **VDI 2243**: VDI-Richtlinie 2243: Konstruieren recyclinggerechter Produkte. Beuth, Berlin 1993
[6] **Weber, J.**: Neue Öko-Norm: Wichtig fürs Exportgeschäft. VDI-N, Düsseldorf 32 (1995) S. 2

List of W3-Sources with their URL (Unified Ressource Locator)

[7] **bmbf:** Bundesministerium für Bildung, Wissenschaft, Forschung und Technologie. Bonn 1995. URL: http://riegel.dfn.de/bmbf
[8] **cern:** European Laboratory for Particle Physics, Genf, Schweiz 1995. URL: http://www.cern.ch
[9] **epa:** Environmental Protection Agency. USA 1995. URL: http://www.epa.gov
[10] **greenpeace:** Greenpeace, Amsterdam, Holland 1995. URL: http://www.greenpeace.org
[11] **iso:** International Organisation for Standardization, Schweiz 1995. URL: http://www.iso.ch
[12] **lycos:** Lycos-Projekt, Carnegie Mellon University, Center for Machine Translation, Pittsburg PA, USA 1995. URL: http://lycos.cs.cmu.edu
[13] **muk:** Fachgebiet Maschinenelemente und Konstruktionslehre. TH Darmstadt, Deutschland 1995. URL: http://www.muk.maschinenbau.th-darmstadt.de
[14] **ncsa:** National Supercomputing Agency, University of Illinois at Urbana-Champaign, USA 1995. URL: http://www.ncsa.uiuc.edu National Supercomputing agency
[15] **opentext:** Open Text Corporation, Waterloo, Canada 1995. URL: http://www.opentext.com
[16] **uni-sb:** Universität Saarbrücken Fachbereich Jura, Saarbrücken 1995. URL: http://www.jura.uni-sb.de

10

The Green Browser: A Proposal of Green Information Sharing and a Life Cycle Design Tool

Yasushi Umeda and Tetsuo Tomiyama
Department of Precision Machinery Engineering,
The Graduate School of Engineering, The University of Tokyo
Hongo 7-3-1, Bunkyo-ku, Tokyo 113, Japan.
Telephone: +81-3-3812-2111 (ext. 6481). Fax: +81-3-3815-7838
e-mail: {tomiyama, umeda}@zzz.pe.u-tokyo.ac.jp

Takashi Kiriyama and Yasunori Baba
Research into Artifacts, Center for Engineering(RACE),
The University of Tokyo
Komaba 4-6-1, Meguro-ku, Tokyo 153, Japan.
Telephone: +81-3-5453-5891. Fax: +81-3-3467-0648
e-mail: {kiriyama, baba}@race.u-tokyo.ac.jp

Abstract

The environmental issue is, without doubt, one of the most important and critical issues to solve urgently. In this paper, we propose the *Green Browser* which enables stakeholders (e.g., employees, shareholders, consumers, regulators, NGO, etc.) to have access to environmental information. Through information-sharing and the promotion of public discussion, raising green literacy and ensued market selection are expected to produce effective pressure for corporate and public decision bearing positively on green production and environmental protection. First, we propose a representational scheme called *green life cycle model* which organizes corporate information for the Green Browser. For the purpose of supporting design for life cycle of green products (*green life cycle design*), the scheme is built to illustrate a product's potential impacts from the raw material stage through use and eventual disposal or recycling. Firms are encouraged to process their firm-specific information based on the scheme. Second, we discuss how the Green Browser can support information sharing to enable stakeholders to obtain the detailed picture of products. We propose the coupling of the Green Browser with Internet for the sharing of green life cycle models and relevant data resources.

Keywords
Green Life Cycle Design, Green Information Sharing, Green Browser, Green Life Cycle Model

1 INTRODUCTION

The environmental issue is, without doubt, one of the most important and critical issues to solve urgently. However, the following characteristics of the environmental issue make it quite difficult to deal with:

- It is impossible to solve the issue only by using technologies. Rather, collaboration among technology, policy, and economy is essential. In this sense, a multidisciplinery research project is important to tackle the issue.
- The environmental issue includes a wide variety of problems ranging from global problems such as the green house effect to local problems such as the disposal of hazardous materials. Moreover, the issue is so complicated that relationships and trade-offs among these problems cannot be defined universally. As a result, we do not have an universal estimation method. For example, even if one can optimize the recycling process of a car from the viewpoint of material consumption, this process might be worse in terms of the green house effect.

How can AI contribute to this issue? Can it help introducing environmental concerns into corporate, policy, and decision-making process?

One option is to develop a tool for environmental decision-making. A tool that clarifies trade-offs among strategic targets such as growth, industrial competitiveness, and environmental impact is thought to enhance the ability of reasoning of decision-makers, especially in the conventional centralized (command-and-control) system. However, once the tool explores the complex problem of gaining global sustainability, negative outcomes may result when the tool carries an incomplete or wrong model of cause and effect.

The one alternative is to develop the *Green Browser* enabling stakeholders (e.g., employees, shareholders, consumers, regulators, NGO, etc.) to have access to environmental information. Through information-sharing and the promotion of public discussion, raising green literacy and ensued market selection are expected to produce effective pressure for corporate and public decision bearing positively on green production and environmental protection.

There has been already some networks about environmental information such as EcoNet[*], and EnviroWeb[†]. While we are planning to join these networks, our research focuses mainly on the manufacturing industry and aims at developing a methodology for supporting designers to design green products by using these networks.

This research proposes a representational scheme called *green life cycle model* which organizes corporate information for the Green Browser. For the purpose of supporting design for life cycle of green products (*green life cycle design*), the scheme is built to illustrate a product's potential impacts from the raw material stage through use and eventual dis-

---

[*]http://www.econet.apc.org/lcv/score100/econet_info.html
[†]http://www.gnn.com/gnn/wic/env.13.html

Process	Requirements	
	Design Related	Others
Manufacturing	no hazardous waste, least amount of material	low energy consumption
Operation	maintenability, long life, no hazardous wastes, low energy consumption	
Reuse and Recycling	reusable components, easy disassembling, recyclable material	no hazardous waste, low energy consumption
Disposal	no hazardous waste, least waste	

Table 1 Examples of Requirements in Each Life Cycle Process

posal or recycling. Firms are encouraged to process their firm-specific information based on the scheme. Then, stakeholders can obtain the detailed picture of products by browsing information with the Green Browser in the internet space. We call it *green information sharing*.

In sections 2 and 3, we discuss requirements and our approach for the life cycle design support and the green information sharing, respectively. We sketch the ongoing implementation of the Green Browser in Section 4. Section 5 concludes this paper.

2 GREEN LIFE CYCLE DESIGN

Designers should design a product and its life cycle system so as to meet the requirements of environmental friendness over the life cycle of the product. Many researchers so far have pointed out the importance of design for reusing and recycling products effectively (e.g., (Boothroyd and Alting, 1992, Ertel, 1994, Zust *et al.*, 1992)). Namely, reuse and recycling of products are ineffective and expensive unless they are purposely designed. For instance, design for disassembly and appropriate modular design are indispensable for economical disassembling and efficient reuse, respectively. Namely, in order to develop a green product, environmental requirements throughout life cycle of the product should be examined at the design stage. Table 1 shows some examples of requirements related to "greenness" in each life cycle process. As shown in this table, many requirements in each process are related to design. For example, in order to avoid producing hazardous waste in the manufacturing process, designers and manufacturing engineers should collaborate for designing the product so as to be manufactured in such a manner. Therefore, not only the product but also its appropriate life cycle system (e.g., manufacturing, operation support, maintenance, recycling, and disposal) should be designed at the design stage. We call this *green life cycle design*. The concurrent engineering (Sohlenius, 1992) is a hopeful approach to support the green life cycle design.

However, as we pointed out in Section 1, since the environmental issue is very complex, vague, and hardly well-evaluated, one of the most important needs for aiding the life cycle design is to support the designer to define the problem structure of the product life cycle at an early stage of design. This process includes picking out environmental and non-environmental requirements which the product should satisfy, clarifying the relations among these requirements, especially trade-offs, and selecting evaluation methods for the requirements.

In order to support this process, the Green Browser is designed to help a team of designers to form a consensus about "greenness" of the product. The Green Browser supports them to put together environmental requirements for the product into a model and thus to visualize trade-offs among them. The Green Browser is designed based on the following concepts:

1. Information generation
   While the green information sharing is important as described in Section 1, issues for it include by whom and how such green information is created. We believe that the created information about the life cycle of a product should be shared during and after the design. Therefore, the result of design should represent relations between the product and the environment over the life cycle clearly and explicitly.
   Here, two types of collaboration should be considered for supporting the information generation. One is collaboration in a process such as collaborative design work and the other is collaboration among different processes. For example, in the latter case, for executing design for disassembly, designers should collaborate with recycling engineers and be supplied basic data from the engineers.
2. Process modeling
   Environmental factors must be examined in each of the life stages such as raw materials, manufacturing the parts, assembly, shipping, duty time, reuse and recycling, and disposal. This means that the Green Browser should present the product information of different life stages. Furthermore, the planning of the production and the recycling process will be an important technical approach towards the environmental issue.
3. Qualitative representation
   Although impacts on the environment are important to be presented, it is not always possible to quantitatively evaluate them, as we pointed out in Section 1. For modeling such impacts, we use qualitative representations (e.g., (Weld and de Kleer, 1989)).

In order to support the life cycle design with these concepts, we here propose a representational scheme called *green life cycle model*. Figures 1 and 2 show the representational scheme of the green life cycle model and an example, respectively. The model of a product consists of three sub-models which are linked each other; namely, a strategy model, a process model, and object models.

Strategy Model
The strategy model represents how requirements for the product affect the achievement of overall goals of the development. The designers describe the following information in the strategy model.

- Requirements and goals

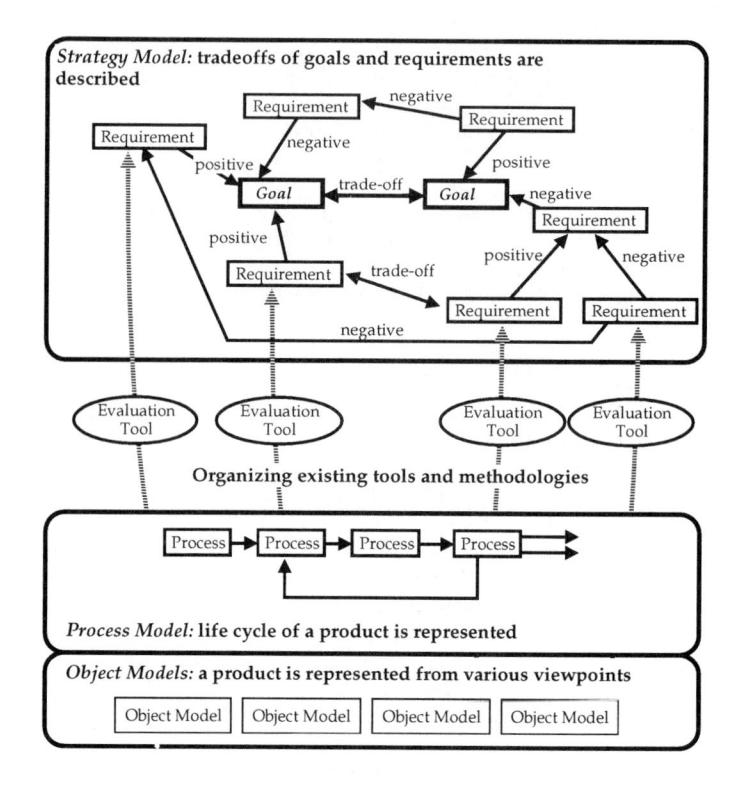

Figure 1 Scheme of the Green Life Cycle Model

Requirements includes environmental requirements such as "recyclability" and "no emission of hazardous materials" as well as general requirements such as "inexpensiveness" and "high speed." We call the most abstract requirements such as "competitiveness" and "greenness" *goals*.

- Relations among requirements

  The designer will find out relations among the defined requirements. For instance, a goal of greenness is positively affected by least materials, which is again achieved by selecting a recyclable material. For the same product, improving functionality will positively affect competitiveness. These two goals of greenness and competitiveness, however, may contradict in respect to materials because the selection of a recyclable material may reduce the functionality. Such positive, negative, and trade-off relations among requirements are represented with nodes and links as shown in Figure 1. Moreover, the designer attaches the reason to each relation why the relation is positive (negative). Among these relations, it is the most important for designing green products to specify the trade-off relations explicitly.

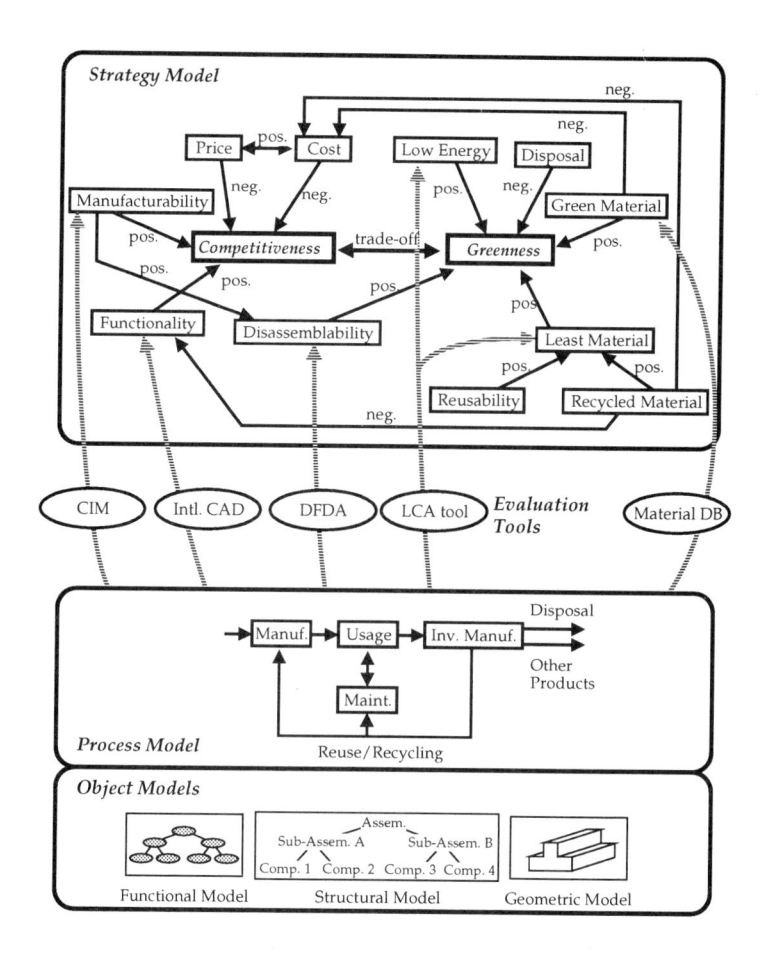

The green browser: a proposal of green information sharing

Figure 2 Example of the Green Life Cycle Model

- Weighing and evaluation criteria
  We assume that the designer can describe importance of each requirement as weight and evaluation criteria for concrete requirements. These kinds of information are put in each requirement node and will be used for evaluation of the product.
- Pointers to the process and the object models
  Each requirement should be related to some portion of the life cycle process and/or object models of the product. By using these pointers, the designer can organize existing tools and methodologies. This feature enables the designers to create green products and evaluate their whole life cycle.

Process Model
The process model represents the life cycle of the product. It depends on the life stage how a product requirements impacts on the environment. For instance, design for disassembly reduces impact in the recycling process, but may increase complexity in the manufacturing process. Links from the strategy model show on which stage of the life cycle the impacts of the requirements are considered.

Object Model
The object models represent the product from various viewpoints. Examples of the object models include functional model, geometric model, and structural model. These modelers of various viewpoints are being integrated into a framework called *Knowledge Intensive Engineering Framework* (Tomiyama *et al.*, 1994), which is currently developed by the authors. One of the features of the Knowledge Intensive Engineering Framework is to integrate various kinds of modelers including traditional modelers such as FEM modelers by providing a common knowledge base of ontology and relating concepts manipulated in each modeler to this ontology base.

3 GREEN INFORMATION SHARING

We believe that the basic information about "greenness" of an product must be included in the green life cycle model and shared among stakeholders using the Green Browser (see Figure 3). We designed the green information sharing facility of the Green Browser based on the following concepts:

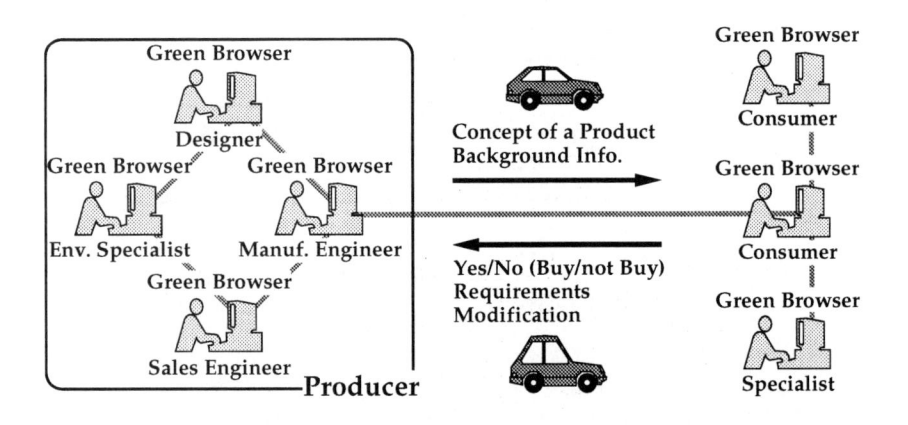

Figure 3 Green Information Sharing

1. Two layered structure
   As shown in Figure 3, the network of the Green Browser consists of two layers. One layer is inner-company network that supports a team of designers and other specialists to

clarify the concept of a product and, as a result, to construct the green life cycle model of the product. Another layer is outer-company network that bridges designers and customers. Through this network, the user collects information about greenness of the product and the company collects opinions of the customers. The information delivered into the second layer is made by aggregating the green life cycle model constructed in the first layer.

2. Linkage to external data sources

Recent advances of the computer network gives support to the effort of tackling the environmental issue. For example, as EcoNet and EnviroWeb provide, on the WWW one may find data of interest including life-cycle assessments, surveys, research papers, and reports. Such data resources may provide information relevant to the product. The Green Browser is designed so that the designer can obtain relevant data using it. Links to external data sources are associated with requirements in the strategy model. By selecting a requirement, the user can learn background information or relevant products associated with the data. For a team of designers, the browser allows to retrieve relevant data that have been linked up by other members.

3. Model sharing

To obtain a consensus about the concept of the product, it is important to learn the views of other designers. The browser is planned to be a common workspace for making a consensus in which the designers collaborate to construct a common green life cycle model of the product. The model represents explicitly different viewpoints of the designers in the strategy model; namely, while the strategy model representing difference is shared among the designers, the process and object models related to the strategy model might differ according to the difference of viewpoints.

It is supposed to happen that the same requirement is found in the strategy models of different products. For instance, for many products the requirement of least material is considered as a possible way to improve the greenness. So the requirements considered for one product might be used again for another product. For this reason we plan to collect strategy models of products, to extract requirements from them, and to put out a list of requirements on the network. Each requirement in the list has links back to the strategy model it was extracted from. This will allow the user to retrieve relevant strategy models as references.

4 IMPLEMENTATION

Currently, we are developing the Green Browser by mainly focusing on the strategy model and green information sharing between designers and customers. The system is being implemented in X-windows environment. Links to external data sources and tools are written in URLs (Universal Resource Locators) as practiced in the WWW. The Green Browser runs as a client of WWW, so that data requested by the user are obtained from remote servers in HTML. The Green Browser will be published with successful examples of life cycle design. The case of automobile is planned to be the example.

5 CONCLUSION

In this paper, we have proposed the *Green Browser* which supports green life cycle design and green information sharing. For supporting the green life cycle design, we have proposed the *green life cycle model* which consists of the strategy model, the process model, and the object models. In order to design green products, it is essential to support the designers to define the problem structure from the viewpoint of greenness. The green information sharing of the system encourages to share transparently the information about greenness of a product among designers, customers, regulators, NGO, and other stakeholders. We believe that, through information-sharing, green literacy is raised. It is expected that the green life cycle design support enables the designers to find out an answer for the green literacy of engineering.

Future work includes;

- continuing to develop of the Green Browser by integrating various modelers, evaluation tools, knowledge and data bases through the network,
- applying the system to many kinds of products in order to collect basic data, and
- providing the system as a public software for facilitating the green information sharing.

REFERENCES

G. Boothroyd and L. Alting. (1992). Design for assembly and disassembly. In *Annals of the CIRP'92*, volume 41/2, pp. 1–12, Berne, Stuttgart, 1992. CIRP, Technische Rundschau.

J. Ertel. (1994). Option of reusing electronic equipment. In K. Feldmann, editor, *RECY'94 (Second International Seminar on Life Cycle Engineering)*, pp. 234–237, Nuremberg, Germany, 1994. FAPS (Institute for Manufacturing Automation and Production Systems), University of Erlangen, Germany, and CIRP, University of Erlangen.

G Sohlenius. (1992). Concurrent engineering. In *Annals of the CIRP '92*, volume 41/2, pp. 645–655, 1992.

T. Tomiyama, T. Kiriyama, and Y. Umeda. (1994). Toward knowledge intensive engineering. In K. Fuchi and T. Yokoi, editors, *Knowledge Building and Knowledge Sharing*, pp. 308–316. Ohmusha and IOS Press, Tokyo, Osaka, and Kyoto, Amsterdam, Oxford, and Washington, 1994.

D. Weld and J. de Kleer, editors. (1989). *Readings in Qualitative Reasoning about Physical Systems.* Morgan-Kaufmann, San Mateo, CA.

R. Zust, R. Wagner, and B. Schumacher. (1992). Approach to the identification and quantification of environmental effects during product life. In *Annals of the CIRP'92*, volume 41/1, pp. 473–476, Berne, Stuttgart, 1992. CIRP, Technische Rundschau.

BIOGRAPHY

Dr. Yasushi Umeda has been Lecturer at the Inverse Manufacturing Laboratory, Faculty of Engineering, the University of Tokyo since 1995. He received his doctor's degree in precision machinery engineering from the Graduate School of the University of Tokyo in 1992. His research interests include soft machines (e.g., self-maintenance machines and cellular machines), green life cycle design, intelligent CAD for mechanical design, and functional reasoning.

Dr. Tetsuo Tomiyama has been Associate Professor at the Department of Precision Machinery Engineering, the University of Tokyo, since 1987. From 1985 to 1987, he worked at the Centre for Mathematics and Computer Science in Amsterdam. He received his doctor's degree in precision machinery engineering from the Graduate School of the University of Tokyo in 1985. His research interest includes design theory and methodology, knowledge intensive engineering, applications of qualitative physics, large scale engineering knowledge bases, and soft machines (self-maintenance machines and cellular machines). He is a member of the IFIP Working Group 5.2.

Dr. Takashi Kiriyama graduated from Department of Precision Machinery Engineering, Faculty of Engineering, the University of Tokyo in 1986. He obtained his Doctor's degree in Engineering from the Graduate School of the University of Tokyo in 1991. In the same year he started to work at the Department as Research Associate. In 1992 he became Lecturer at Research into Artifacts, Center for Engineering (RACE), the University of Tokyo, and has been Associate Professor since 1995. His research interest includes artifactural engineering, network-based collaborative design, qualitative physics, and design of micromechanisms.

Dr. Yasunori Baba has been Associate Professor of Manufacturing Science Division of Research into Artifacts, Center for Engineering (RACE) at the University of Tokyo since 1993. He received a BA in economics in 1977 from the Univ. of Tokyo and MPA in 1981 from International Christian University, Tokyo. After obtaining a Ph. D. from the Univ. of Sussex, U.K., he worked as a Research Fellow at the Science Policy Research Unit (SPRU) at the university from 1986 to 1988. He was a senior researcher at the National Institute of Science and Technology Policy (NISTEP) of Science and Technology Agency (STA) from 1989 to 1991 and had a joint appointment at the Saitama University at the Graduate School of Policy Science from 1991 to 1992. His research field is the economics of technical change and its application to science and technology policy. He is currently working on the information of "Artifactual Engineering" with a focus on global environmental issues.

11

Towards a new specification method for an Automated System

JACQUET L., SALLEZ Y.,SOENEN R.
Laboratoire d'Automatique et de Mécanique Industrielles et Humaines,
URA CNRS N°1775, Equipe Génie Industriel et Logiciel,
Université de Valenciennes et du Hainaut-Cambrésis,
Le Mont-Houy, BP 311, 59304 VALENCIENNES Cedex, France
Tel : (33) 27 14 13 45
Fax : (33) 27 14 12 88
E-Mail : lgil@univ-valenciennes.fr

Abstract
The design of an automated system is shared among several actors (mechanical engineer, automation engineer) and a design method must be used. Therefore, the objective of this paper is to propose a new specification method according to the concurrent engineering concept. After a presentation of the successive phases, an aircraft assembly system is studied.

Keywords
Concurrent engineering, Automation, Design, Method, Specifications

1 INTRODUCTION

Different design methods (Suh,1990) permit to conceive an automated system. One of them is the system approach. This top-down approach is structured and can be supported by the "V" life cycle. However disadvantages of the "V" life cycle are the sequential aspect and the lack of information about steps that constitute each phase.

In this context, the first part of this paper propose a new functional specification method that takes into account simultaneously the mechanical and automatical aspects. This iterative method takes progressively information from the needs of the client. Its first aim is to obtain a product in accord with the requirements, and the second goal is to reduce the design time under the concurrent engineering concept.

The second and third parts are respectively focused on the first and second phases of the specification method.

Then, the last part validates the previous method through the study of an aircraft assembly system.

2 PROPOSED DESIGN METHOD

This design procedure, presented on figure 1, is iterative and its aim is to obtain a product more conformable to the client needs. In the same way, its purpose is to reduce the delay of the design with simultaneous engineering concepts (Jagou,1993). Firstly, this design method, defines the functional aspect through three specification levels that model respectively:

- Service functions that explain the need according to a client vision (AFNOR,1989). For that, we define all material or no material elements that can interfere on the product. This phase permits us to verify that all the information given by the client is coherent and is enough for a correct definition of the client need. In this step, we answer the question: "What are the client needs?".
- Operational functions that must be defined for each service function. This phase consists in explaining the functional chain that satisfies the considered service function. The decomposition finishes when the elementary operational function appear. Then we select the possible solution principles for each elementary operational function. In this phase we answer the question: "How to satisfy each client need?".
- Technical functions constitute the lowest level decomposition in accordance with the studied solution principle.

Furthermore, the design procedure defines the technical aspect (preliminary design phase). In relation to the constraints previously explained in the first phase (performance, safety, existing material parts), we can define the support of technical functions. Previously a kinematic chain has been studied for each mechanical solution principle. The classification step of technical solutions is the end of the top-down phase and the beginning of the separate study of mechanical (process) and control/command part.

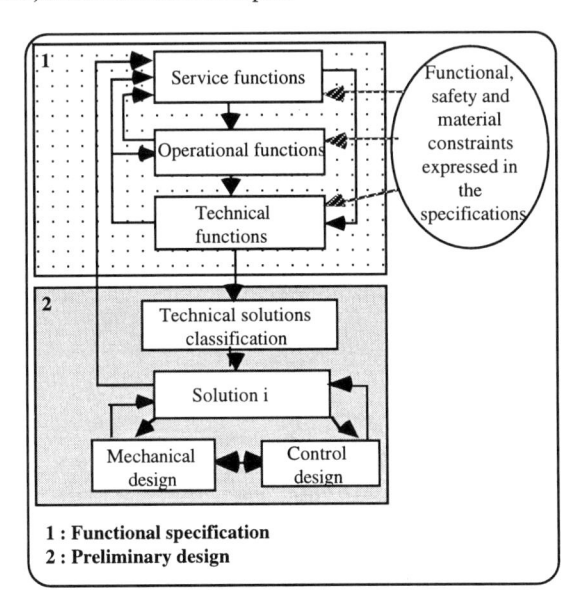

Figure 1 Design procedure

After this global presentation of the design procedure, we explain more precisely the first and the second procedures of the functional specification phase.

3 SERVICE FUNCTIONS

The definition of service functions (Jacquet,1995 a) permits to understand clearly the client needs. This phase is essential because it structures information given by the client and it's permits to verify their coherence. The service functions are defined through five steps presented on the figure 2. Each step will be explained in details in the following sections.

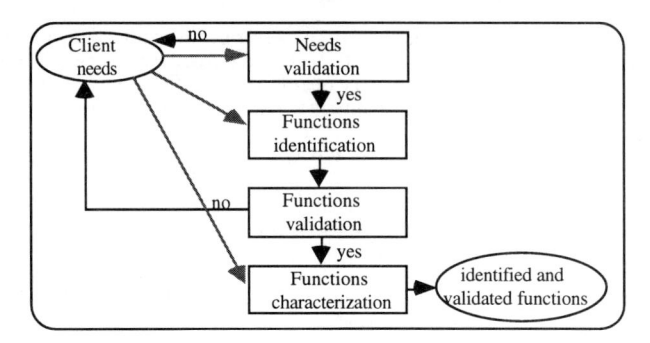

Figure 2 Definition procedure of service functions

3.1 Needs validation

This first step of the procedure certifies the validity of client needs. Indeed, there is no interest to define a system that satisfy just temporarily the client. So, the aim of this step is to understand clearly the present and future client needs. The method that permits to validate the client needs consists in answering the following questions :

- Why do the client needs exist ? (In what aim ?, For what reasons ?).
- Can it evolve ? (if yes, how ?).
- Can it disappear ? (if yes, how ?).

The answers to the previously questions permit in the first case to validate the client needs and permit to continue the service functions definition procedure. In the worst case, if we can not validate the client needs (inconstant needs) we define again the needs.

3.2 Functions identification

The product functions are defined by the services that it is supposed to do. Thereby, it is very important to forget the material solutions when we define this type of function. Even, we must try to explain them as clearly as possible. When it is possible we must use an infinitive verb with names following it. In the majority of cases, a product (or system) must realise several functions. These functions can be classified in two types :

- **Service functions** are functions for which the product is effectively built. They are "*principal*" when they traduce the needs for which the product is built. They are

"*complementary* " when they traduce the use we want to do with the product in its future environment. In our method, the two external elements that generate one service function can not be used to generate a constraint function. These external elements induces constraints on the service function but these constraints are explained in the function characterisation phase. Beside, no difference is effected between the principal and complementary service functions.
• *Constraint functions* are dictated by regimentation that the designer must respect to obtain a product in accordance with standards. In our method, these functions are global constraint functions that all the service functions must respected.

Used method
In this context, the more attractive functional analysis method (AFNOR,1986), (Lachnitt,1994) is the SAFE method. First of all, this method defines the external elements that will permit to exhibit the service and constraint functions that have to satisfy the system. This method analyses successive sequences of using system. In our method, the definition of the functions is executed after the selection of external elements that can interfere with the future system in each life phase (Installation, Using, Maintenance, End of activity). These elements can be classified in several types :
• Human : users, operators,...
• Material elements : adjust tools, inspect tools,...
• Environment : humidity, temperature,...
• Standards : Human safety, material safety,...

Representation model
The designer uses the matrix representation model to show the automated system functions. This model presents the identified external elements in its abscise and ordinate. Existing functions are materialised by one sign on the line column intersection. The constraint functions, presented by "C" sign, are positioned on the diagonal matrix. The others (service functions), presented by "F" sign, can be positioned in all of the matrix part. The advantage of this representation model is its legibility whatever the number of identified external elements.

3.3 Functions validation

The aim of this step is to validate each of the previously identified functions (previous step). The method which permits to support this step (Chabert,1992) consist in answering the following questions :
• Why does the function exist ?
• What is its aim ?
• What are its existence reasons ?
• What could make it disappear ?
• What could make it evolve ?
• What is the risk of its disappearance ?

3.4 Functions characterisation

This step consists in assigning and weighting, for each function, the functional performances required by the client in the specification. These characteristics are related to functions and not materials. They must only precise client's realisable exigencies (Lachnitt,1994).
There is a method that permits to characterise each service function (AFNOR,1986), (AFNOR, 1989). On the first time, this method defines **test criteria** which will permit to verify if the service functions are well satisfied and if the constraint functions are well respected. In the second time, we must define one **test level** for each test criterion. If the test level explains a quantitative measure we speak of performance level. In the last step, we must precise the **flexibility** of each test level. The value of this flexibility will be **"0"** if the value

of the test level can not be changed; **"1"** when the value can be modified (modification is cost-effective); **"2"** when a discussion between the designer and the client is advised.

The weight of each functional performance is obtained by using the crossed sort method.

The flexibility test level permits to know functions that designer will modify if there are not technical solutions able to answer problem (preliminary design phase).

Then, in accordance with simultaneous engineering concepts that try to reduce study time, we classify previously identified functions in several groups. This classification will permit to define concurrently operational functions (second phase of the procedure). Indeed, it is possible to constitute several engineerist teams. Thereby, each of them, can study one type of service function family and thus we can reduce study time of an automated system. The different function families are : using, safety, informational, degraded and constraint.

4 OPERATIONAL FUNCTIONS

After a correct definition of the problem, the goal of the functional specification procedure is to define **how** could the needs identified by the service functions (Jacquet,1995 b) be satisfied. This phase has two goals. The first is to define the operational chains. Each operational function is decomposed in operational sub-chains until obtain the elementary operational functions. The second aim consists in suppressing the solution principles inappropriate with the constraints associated to the studied elementary operational function. This phase is supported by six steps presented on the figure 2. Each step will be explained in details in the following sections.

4.1 Definition of operational chains

The goal of this step is to define the different operational chains that permit to satisfy each service function explained in the first phase of the procedure (Service function). It is a strictly creative step, where imagination is not constricted. However, this step's result depends on the specific knowledge of the designer and on his ability to integrate it.

The principal used creative methods (Petitdemange,1987), (Bellut,1990) are the following:
- **The brainstorming** is a collective method of ideas research. In this method, the partners must give without restriction all ideas to solve the problem. The problem is studied on different points of view and the solution ideas are generated spontaneously without control. Thereby, the solutions set is increased by an ideas association procedure.
- **The achronique board** consists in defining the initial (initial states) and the final (final states) characteristics of the product that will be transformed by the process in an aim of defining correctly the working type to realise. The working scheduling and the material process are not considered in this method.
- **The generic model** consists in choosing the corresponding model of the need in an existing list. The major disadvantage of this method is that we can apply it only in a specific field. Indeed, we must have defined before the library of all generic model that answer to the studied problem.

In our approach, we use the best part of these different creative methods and we define the operational chains through the following steps :
- Definition of stable intermediate state of the flow.
- Definition of the activity (operational functions) which permit to obtain each state.
- Definition of the possible operational chain.

The steps of "Definition of operational chain" and "Choice of one operational chain" are iterative. Each iteration consists in decomposing functions to subfunctions that precise the solution of the problem. The decomposition stop criterion is reached when we can associate an elementary material part to each operational function.

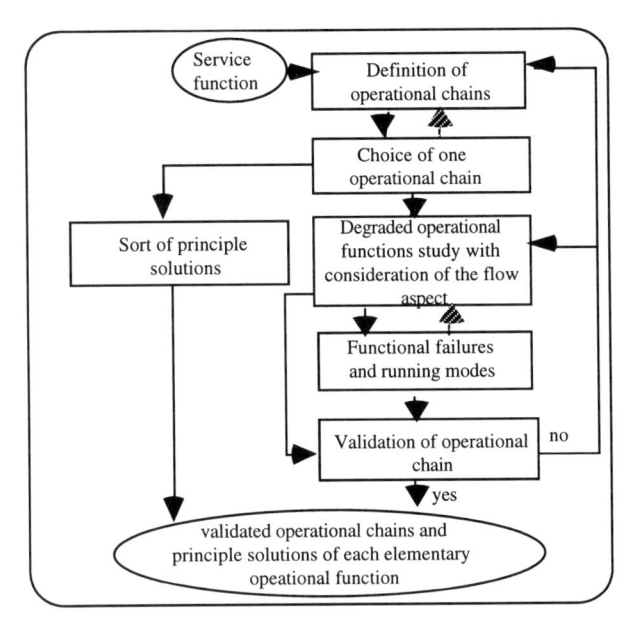

Towards a new specification method for an automated system 121

Figure 3 Definition procedure of operational functions

4.2 Choice of one operational chain

This step has the aim to avoid the systematic study of the different operational chains that have been defined in the previous step. In relation to the criteria of selection like informational safety, just one operational chain will be chosen. Indeed, it is better to have a **stable state** (flow state can not change) before realising another action.

By iteration with the previous step, each operational function is decomposed in operational sub-chains until obtain the elementary operational functions ("move", "maintain", "control", "change", ...).

4.3 Degraded operational functioning study with consideration of flow aspects

This step is dedicated to the material flow through the automated system. All cases of degradation flow that could appear in the control operational functions, will be studied and the operational solutions will be explained. The objective of these solutions is to avoid the generation of new operational functions. Thus, the designer just creates new links between the operational functions already defined.

4.4 Functional failures and running modes

Firstly, the goal of this step is to study the functional failures of each operational function and to define an efficient solution. This step is fundamental when we study problems with human flow or continuous process. The four possible functional failures (AFNOR,1986) are the following :
• **Function missing:** the function is not performed when it is required,

- **Function stop:** the function stops running,
- **Function degradation:** performance decreasing,
- **Unreasonable function starting:** the function is realised when it is not required.

However the solutions of the functional failures can become quickly complex according to the number and type of failing functions. Therefore, in the aim of time and motion study, we will consider the hypothesis that only one operational function can be failing at the same time. In the case of more than one effect can be attributed to the same default, we will take the most dangerous. If the human aspect must be studied (if it is different of the flow), it will be considered when we will explain the functional failure aspect.

The methods that permit to study the functional failures is the **FMECA** (RIOUT,1994). It is a method that at the first time consists in searching the potential defaults that can be generated by the process on the product. Then we define the effects and the causes of the defaults. After that we calculate their criticality in an aim of default hierarchy, that permits to know the necessity of finding one solution to the functional failures.

Therefore, from the study of the functional failures, **at first,** we will try to solve each failure effect with **operational way**. This fact consists in defining new links between operational functions. Those links are activated from one calculated delay whose aim is to unjam the system and inform it that one failure has occurred (degraded mode). However, the failure effects can not be solved in an operational way. Therefore, **in a second time**, we must calculate the criticality of each operational function to define the most dangerous. This is calculated from the following weighting :
- **"0"** if the failure has no effect on the system or on the flow,
- **"6"** if the failure disturbs the system or flow working,
- **"25"** if the failure stops the flow or the working system,
- **"100"** if the failure generates an opposed working to the initial working system or if it is dangerous for the flow (destruction).

For each function, from the calculation of the criticality and the maximum criticality level that has been specified by the client, we must define functional solutions. These solutions will permit to answer the most critical functions. In fact, the solutions are often functional redundancies. According to the redundance type, we can have two distinct running modes that are safe and degraded. The first corresponds to a passive redundance and the second to an active redundance. However, when the functional redundance does not permit to suppress the failure effect, the designer must define a new operational function to do it. Then, he must study the functional failure of this new function.

This step will be used to generate the structure of the running modes model (Parayre,1992).

4.5 Validation of operational chains

Operational chains have been explained progressively through the previous steps. The Petri nets are the used tools to validate them .The Petri nets permit to take into account the synchronisation between operational functions by modelling the flow with tokens. Those nets can present continuous and discrete aspect. This formal tool permits to verify net characteristics (safe, limited, deterministic, etc.).

4.6 Sort of principle solutions

The second goal of the operational function phase is performed by this step. From the elementary operational functions defined in the 4.1-4.2 iterative steps, we take into account all principle solutions able to perform each elementary function. For example, the elementary function "move" can be supported by the following principle solutions "electromagnetism", "with fluid", "mechanic by weight", "mechanic by human energy", "mechanic by other energies".

Then we suppress some of these principle solutions from solving constraints (geometric, kinematic, production, environment, ...). In the case of the "move" elementary function, the result of this step is one trajectory with its kinematic criteria in accordance with the selected principle solution.

5 VALIDATION ON AN INDUSTRIAL EXAMPLE

The goal of this part is to study an automated system that assemblies components in aircraft industry. A plane is build with several sub-structures. The assembly of each sub-structure is performed by riveting several components. These last are located on a support, hold in position, and riveted. Holes of Ø 4mm are realised before the riveting.

5.1 Functions identification

The functional matrix (See Figure 4) shows all service and constraint functions that will have to satisfy the aircraft assembly system.

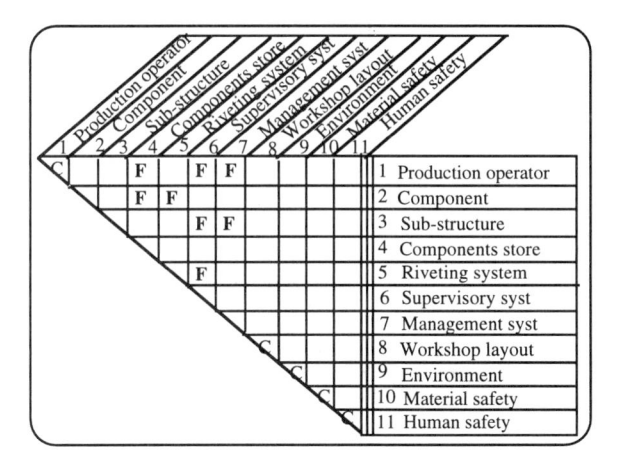

Figure 4 Functional matrix of the using phase

On this matrix, we have defined :
• Service functions as "Assembly the components with riveting system" or "Permit to the operator to inform the management system"....
• Constraint functions as "Respect human safety and material safety standards"...

5.2 Functions validation

After answering the question the array of Figure 5, it is possible to validate each function. In all other cases the function is suppressed.

The remedy answers can be binary or moderate. However, in an objective of simplicity we take the binary case.

Function :	Assembly the components with riveting system		
Why does the function exist ?	What is the cause of function existance ?	What is the thing that could do to disappear the cause ? (remedy)	Is the remedy stable ?
To obtain an assembled sub-structure with components	It's diffcult to assembly a plane section once only Some positions of components are constraining	Few components to assembly Lack of components constraining positions	Unprobable Impossible
Conclusion	stable		

Figure 5 Function validation

5.3 Functions characterisation

The characterising criteria of each function are in the specifications. This step permits to verify if the information given by the client are coherent and enough to characterise the need.

Function	Family	Characteristics	Value	Flexib.
Assembly components with system riveting	Using	Material : - Weight comp. i - Width comp. i - Riveting system - Width system Functionnal : - Time - Cost - Riveting inspection	 10 m2 max 10 mn 100 kF max manual	0 0 1 0 2 1 0

Figure 6 Functions characterisation

5.4 Definition of operational chain

The assembly order of the components is the following : (1-2)->3->4
The stable intermediate states are :
- Components 1-2-3 located   **(0.1)**
- Component 4 located   **(0.2)**
- Components 1-2-3 assembled   **(1)**
- Components 1-2-3-4 assembled   **(2)**

When the operational functions are grey (see Figure 7), this represents the fact that the stop criterion is not reached. Therefore, we must decompose operational function into subfunctions until we can associate an elementary operational function to each of them (stop criterion). When this decomposition level is reached, as we can see on the Figure 8, the operational function become white.

```
┌─────────────┐                          ┌─────────────┐
│To assembly the│                        │To assembly the│→ Components 1-2-3-4
│ components  │→ Components 1-2-3        │ components  │  assemblied  (2)
│ 1-2 and 3   │  assemblied (1)          │ 1-2-3 and 4 │
└─────────────┘                          └─────────────┘
     Components 1-2-3 located (0.1)          Component 4 located (0.2)
```

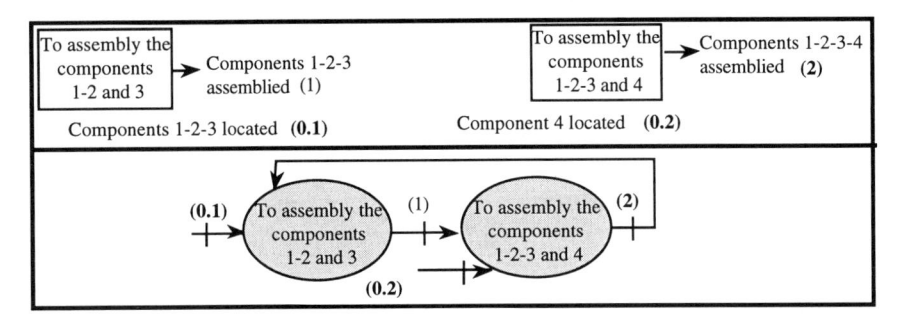

Figure 7 Operational chain (level 1)

Each assembly function must be decomposed into the following operational functions :
- "To assembly" that models a more elementary assembly level.
- **"To maintain or to release"** that models the maintain or the suppression in position of at least one component.
- **"To move"** that models the component moving function.
- **"To inspect"** that models the transformation inspection.
- "To transform (no elementary)" that models a set of elementary functions.
- **"To transform the shape"** that models the shape modification of one component.
- **"To transform the external structure"** that models the object structure modification performed by addition or suppression of mater between at least two components.
- **"To transform the internal structure"** that models the modification of the internal structure of the object.

The **bold** operational functions are **elementary**.

The decomposition of the operational function "To assembly the components 1-2 and 3" gives the following intermediate states and the functional chain of the Figure 8:
- Component 1 located   (**0.1.1**)
- Component 2 located   (**0.1.2**)
- Component 3 located   (**0.1.3**)
- Component 1 hold in position (3)
- Component 2 hold in position (4)
- Component 3 hold in position (5)
- Components 1-2-3 hold in position (6)
- Components 1-2-3 Ø 4mm   (7)
- Components 1-2-3 riveted   (8)
- Components 1-2-3 assembled   (**1**)

In accordance with the paragraph 4.2, the choice of a set of operational chains is effected rejecting the solution of the operational parallelism (sequential constraint).

The feedback "*10* " on the figure 8 between the inspection and the realisation of the Ø 4mm hole is relative to the degraded operational functioning study as explained in 4.4. In our aircraft assembly case, if a defective rivet is detected, we must make again the hole at this location and put an other rivet.

As explained in 4.6, the lowest level of decomposition is reached on the Figure 8 because we obtain only elementary operational functions.

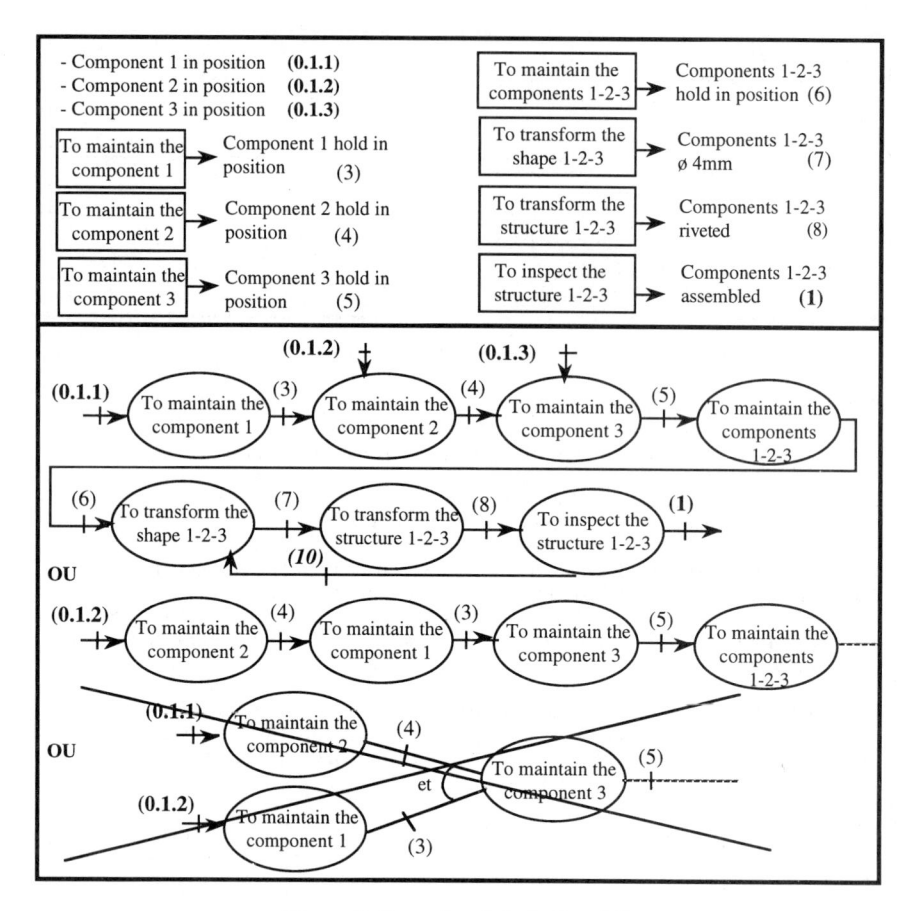

Figure 8 Operational chain (level 2)

6 CONCLUSIONS AND FUTURE WORKS

The design of an automated system is complex and a method must be used. We have proposed a new one which is based on concurrent engineering concepts. Then, we have focused on the first and second specification phases respectively dedicated to the definition of service and operational functions. We have explained the steps, which have permitted the definition of operational functions. This procedure decomposes operational chains in elementary operational functions, which constitute the lowest level.

Finally, we will have to define the technical functions that will support each elementary operational function. These functions will be defined for the principle solution that will be chosen among the set of possible principle solutions.

The method will be supported by the software development platform CAS.CADE (Nguyen, 1994).

7 REFERENCES

AFNOR X60_510 (1986) Fiabilité maintenabilité disponibilité, *Techniques d'analyse de la fiabilité des systèmes, procédure d'analyse des modes de défaillance et de leurs effets* (AMDE), édition AFNOR.

AFNOR X50_153 (1989) Gérer et assurer la qualité, *Analyse de la valeur, Recommandations pour sa mise en oeuvre*, édition AFNOR.

Bellut S. (1990) La compétitivité par la maîtrise des coûts, *collection Afnor gestion*, édition AFNOR

Chabert C. (1992) Cahier des charges fonctionnel d'un système informatique, technologie **50**,55-63

Jacquet L. (1995 a) An autamated system specification method according to concurrent engineering concepts, ASI'95, 25-28 June; Lisbon.

Jacquet L.(1995b) Toward a specification procedure of operational functions for an automated system In: Proceedings of the 1995 IEEE International Conference on system, Man and Cybernetic, Vancouver, Canada.

Jagou P. (1993) Concurrent Engineering, la maîtrise des coûts, des délais et de la qualité, collection Hermes, Paris.

Lachnitt J.(1994) *L'analyse de la valeur*, collection Que sais-je ?, Paris.

Nguyen T.(1994) CAS.CADE : une architecture avancée pour l'ingénierie simultanée et concourante, IFIP conference : feature modeling and recognition in advanced cad/cam systems,95-109, Valenciennes.

Parayre T.(1992) MESAP : vers une méthodologie d'exploitation des SAP, Thèse de doctorat en automatique, Université de valenciennes et du hainaut cambrésis.

Petitdemange C.(1987) *Créer et développer vos produits*, édition Afnor.

Riout J. (1994) *Le guide de l'AMDEC machine*, CETIM.

Suh Nam P.(1990) *The principles of design*, col: oxford series on advanced manufacturing, oxford.

PART FOUR
Decision Support

12

An approach to planning of textile manufacturing operations: a scheduling method

D. Mourtzis*, N. Papakostas*, G. Chryssolouris**
Laboratory for Manufacturing Systems
University of Patras
Greece
Tel.: ++30-61-997262, Fax: ++30-61-997744
E-Mail: xrisol@mech.upatras.gr
* Research Assistant
** Professor, Director of the Laboratory

Abstract
This paper describes a planning method and its application to Textile Industry. The field of application of this method covers the production planning and scheduling processes of a typical textile factory. The scheduling method performs production planning, scheduling and control, based on a multiple-criteria decision-making technique, suitable for any kind of manufacturing system. A number of combinations of criteria, provided by this scheduling method, are applied to the industrial test-case at hand, and the related results are discussed. In addition, other conventional scheduling policies are applied and their results evaluated and compared with the proposed method. The proposed work shows that the multiple criteria technique provides better results.

Keywords
Production planning, scheduling, manufacturing, textile industry

1 INTRODUCTION

This paper describes the application of a scheduling method to production planning and scheduling and analyses the planning problem of a Typical Textile Factory (TTF). The method is applied to a Greek industrial firm, in the field of textile products. One of its production lines is dedicated to carpet production and consists of three different sections namely dyeing, spinning and weaving. The scheduling method has been applied to the selected production line. The data and shop-floor configuration come from the above mentioned TTF.

2 THE TEXTILE INDUSTRY PLANNING PROBLEM

The work presented in this paper is part of a requirements analysis study from a Textile factory, on the automation of the overall industrial process, from product and process design to production planning and control, for one of the Textile factory plants, the Combed **Y**arn **C**arpet **P**roduction (CYCP) area. This paper is concerned with the CYCP area production scheduling at various levels of the factory hierarchy (Ghryssolouris et al. 1985). The overall objective is to make these production planning phases at the different levels, totally integrated and flexible enough, to allow for increased operational and decisional capability. The next sections describe the application of the scheduling method to the TTF test-case both at job-shop and work center levels.

2.1 Production system description

The under study Production System characterized as a TTF, consists of three separate sections or production areas, which are the Dyeing, the Spinning, and the Weaving sections. It is well known that a manufacturing system can be defined as a combination of humans, machinery, and equipment, which are bound by a common material and information flow. This definition will be used to describe the three above mentioned sections.

Dyeing Section: The first step to produce a carpet is to dye the required quantities of raw material/fibers in the appropriate colors. This operation is performed in the Dyeing section. The material input in this section is wool, acrylic or polypropylene fibers. The information input is the type, color and quantity of the fibers as well as the time the fibers have to be available in the next (spinning) section. The raw material inventory control information is also an information input for the Dyeing section. The main processes performed within this section, are the dyeing, drying and opening of fibers (Figure 1). The output of the Dyeing section, the dyed fibers, is an input for the spinning area.

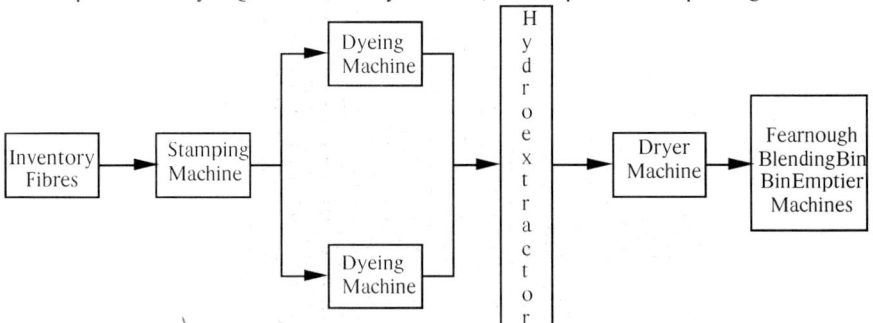

Figure 1 The manufacturing processes in the dyeing section

Spinning Section: The second step in producing a carpet is to spin the required quantities of yarns. In order to perform the operation of weaving, three systems of yarns are needed namely warp, weft and tuft. The warp and weft yarns are not of interest because they are purchased from external suppliers, while the tuft yarns are produced in the spinning section of this TTF. The information input is the type, color and quantity of yarns and the time these yarns have to be available in the next (weaving) section. The semifinal product inventory control information, is also an input for the Spinning section. In order to produce these yarns, the process sequence is as follows:
Sliver forming (Carding and Gilling), Roving forming, Yarn forming, Yarn stabilization, Yarn Cleaning and Package forming, Yarn Plying and Final Package forming (Figure 2).

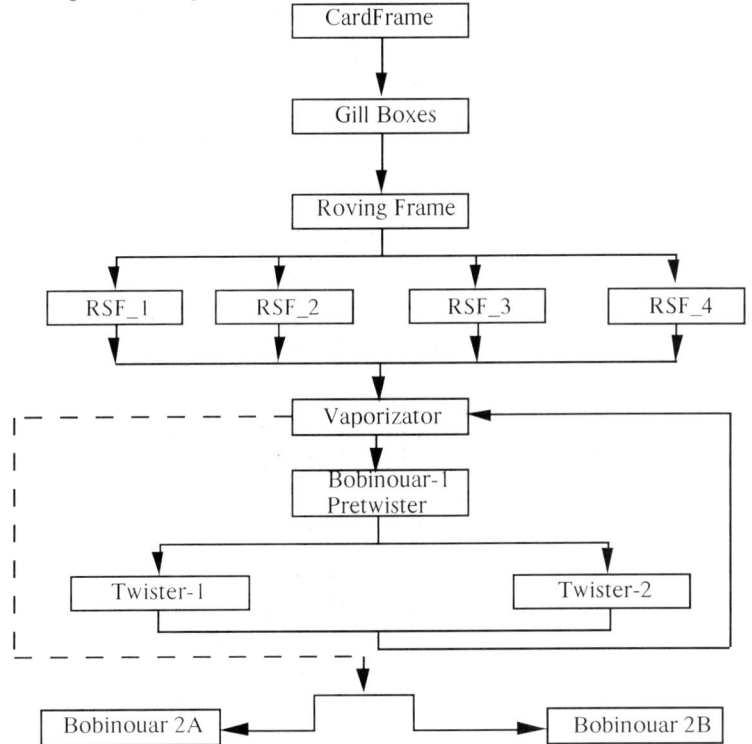

```
                    CardFrame
                        |
                        v
                    Gill Boxes
                        |
                        v
                   Roving Frame
          _____|_____
         |        |         |        |
         v        v         v        v
       RSF_1   RSF_2     RSF_3    RSF_4
         |        |         |        |
         |        |         v        |
         |        |     Vaporizator <-----+
         |        |         |             |
         |        |         v             |
         |        |  Bobinouar-1          |
         |        |   Pretwister          |
         |        |    |       |          |
         |        |    v       v          |
         |        | Twister-1  Twister-2  |
         |        |                       |
         |        |                       |
                  v
         Bobinouar 2A <---+   +---> Bobinouar 2B
```

Figure 2 The manufacturing processes in the spinning section

Weaving Section: The third step to produce a carpet is the weaving. In this section the carpet is produced by interlacing the three systems of yarns Warp, Weft and Tuft. Each carpet type contains five different colors, depending on the color set, meaning that five different quantities of dyed yarns must be available simultaneously, in order to begin the execution of an order in the weaving section. The information input is the type, design, color set and quantity of carpets and the time they have to be produced. The final product inventory control information is also an input for the Weaving section. In order to produce these carpets the single operation to be performed is Weaving (Figure 3).

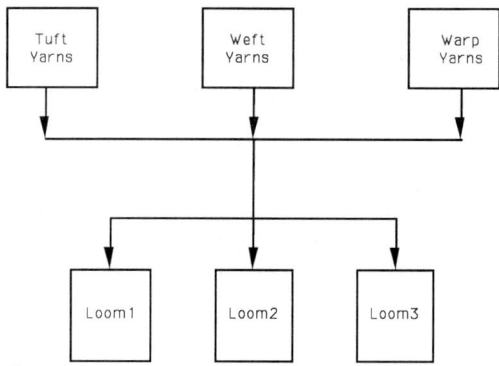

```
        ┌───────┐    ┌───────┐    ┌───────┐
        │ Tuft  │    │ Weft  │    │ Warp  │
        │ Yarns │    │ Yarns │    │ Yarns │
        └───┬───┘    └───┬───┘    └───┬───┘
            │            │            │
            ▼            ▼            ▼
        ┌───────┐    ┌───────┐    ┌───────┐
        │ Loom1 │    │ Loom2 │    │ Loom3 │
        └───────┘    └───────┘    └───────┘
```

Figure 3 The manufacturing processes in the weaving section

2.2 Planning requirements and problem description

The point of interest is what a plant manager must achieve, when planning a manufacturing system like that.
- Customer satisfaction in relation to product attributes.
- Accurate determination of due dates which must be followed.
- Small inventory at all stages of the production.
- Low production cost.
- Excellent product quality.
- Flexible production planning so that orders may change at an acceptable cost and quality especially without provoking manufacturing process corruption.

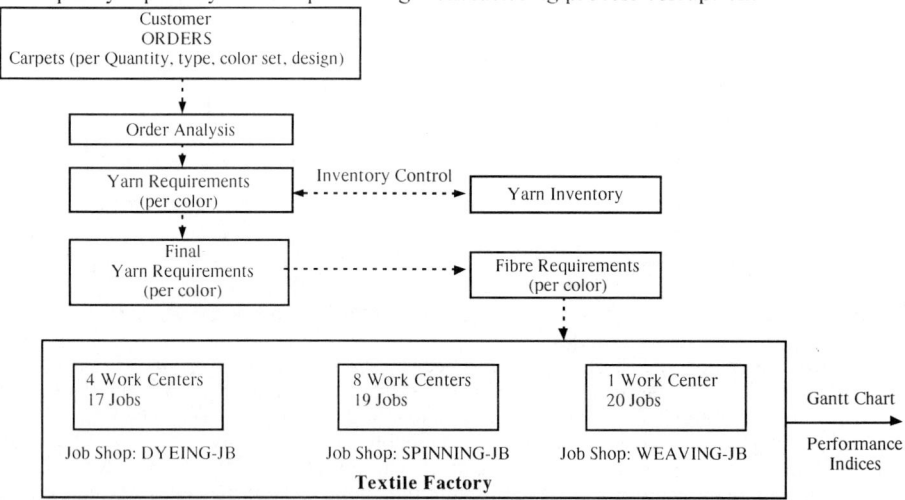

```
    ┌─────────────────────────────────────────────┐
    │              Customer                        │
    │               ORDERS                         │
    │ Carpets (per Quantity, type, color set, design)│
    └──────────────────┬──────────────────────────┘
                       ▼
           ┌───────────────────────┐
           │    Order Analysis     │
           └───────────┬───────────┘
                       ▼
           ┌───────────────────────┐   Inventory Control    ┌───────────────────┐
           │  Yarn Requirements    │◄─ ─ ─ ─ ─ ─ ─ ─ ─ ─ ─ ►│  Yarn Inventory   │
           │     (per color)       │                        └───────────────────┘
           └───────────┬───────────┘
                       ▼
           ┌───────────────────────┐                        ┌───────────────────┐
           │        Final          │                        │ Fibre Requirements│
           │  Yarn Requirements    │─ ─ ─ ─ ─ ─ ─ ─ ─ ─ ─ ─►│    (per color)    │
           │     (per color)       │                        └─────────┬─────────┘
           └───────────────────────┘                                  │
                                                                      ▼
    ┌─────────────────────────────────────────────────────────────────────┐
    │  ┌─────────────┐     ┌─────────────┐     ┌─────────────┐            │──► Gantt Chart
    │  │4 Work Centers│    │8 Work Centers│    │1 Work Center│            │
    │  │ 17 Jobs     │     │ 19 Jobs     │     │ 20 Jobs     │            │──► Performance
    │  └─────────────┘     └─────────────┘     └─────────────┘            │    Indices
    │                                                                      │
    │  Job Shop: DYEING-JB  Job Shop: SPINNING-JB  Job Shop: WEAVING-JB    │
    │                        Textile Factory                               │
    └─────────────────────────────────────────────────────────────────────┘
```

Figure 4 The production Planning Problem Description

In order to implement the above requirements, the question is how one should plan the entire manufacturing system (the three described sections), using the scheduling method described in section 3, making at the same, time the necessary inventory control (Figure 4). In this case the scheduling is both a resource allocation problem and a material requirements planning problem.

Problems with Current Planning Practice

Currently, production planning is done based on the experience of the plant manager and on some information about the state of the textile factory, with no computer assistance. The planning procedure is the following :
- The sales manager empirically decides, when, in what quantities and what types of carpets will be produced as inventory over a long period of time.
- This work load contains a set of orders and it is an information input for the manufacturing system at the weaving section.
- This work load needs to be scheduled to each section of the textile factory (Dyeing, Spinning and Weaving section), after performing a usually insufficient inventory control.
- The plant manager must also build the bill of materials (BOM). Using BOMs for all the work load planned, he is able to plan material requirements.
- After that the production engineer must implement the decided scheduling (allocation of Jobs/Tasks to available resources) with no possibility for performance estimation of this scheduling.
- Since the system is both inventory and make-to-order oriented, another stream of orders is possible to arrive at the textile factory at any time. These orders must be executed in the textile factory too, thus, being an additional input for the weaving section.
- After a new inventory control, the plan manager must try to reschedule the work load in all three sections, including the new orders. This task is very difficult, because there is no real time information about the shop floor status.
- The production engineer tries to execute the new production plant by scheduling the allocation of resources with no valid estimation of the consequences of his decision. Steps 6,7,8 are repeated any time a new order arrives.
- This planning method may have a decision horizon, ranging from a few hours to some days. The performance of the decided scheduling will be estimated only after the planning execution. The mean production cost, the mean tardiness, the mean flow time and the mean quality are used as performance indicators for estimating the applied scheduling

3. PROPOSED METHODOLOGY

The resource allocation problem, referring to the assignment of a set of resources to a set of tasks over time, is of utmost importance to many industrial activities, particularly to the planning and control of manufacturing systems. In actual facilities the number of resources and tasks is large enough to make the problem combinatorial explosive. In this work, the following hierarchical model will be used. A manufacturing system (factory) consists of a set of job shops, each producing a family of similar products or subassemblies. Each job shop is further partitioned into work centers, consisting of resources with similar manufacturing functions. A resource is defined as an individual production unit, which can represent either a single machine or a manufacturing cell (machines grouped together with auxiliary devices). The imposed scheduling method assigns the available resources to pending production tasks following, a number of steps that a human undertakes, when making a choice.

Determine a set of relevant decision making criteria: The choice of decision-making criteria at the work center level, must be guided by the need to reflect overall system objectives. Since work centers represent a relatively low level of control within a manufacturing system hierarchy, the connection between the criteria for decision making at the work center level, and the overall system objectives may be difficult to ascertain. For these cases, a suggestion is made for the construction of the hierarchy of objectives, which will help to establish the correspondence between these objectives and the low-level, decision-making criteria/attributes (Chryssolouris et al.1991). This approach is adapted to the manufacturing environment and followed throughout the work presented in this paper.

Determine a set of alternatives: An alternative is defined as a set of possible assignments of available resources to pending tasks. The times at which the decision-making activity occurs, are referred to here, as decision points, and the spacing between two subsequent decision points is referred to as decision interval. The decision-making activity is triggered by a change of the status of the system, namely either by the completion of a new task. The decision horizon is a time interval that begins at the decision point. There are a number of issuers that have to be taken into account to address the problem of determining alternatives (Chryssolouris et al. 1991).

Determine the consequences of the different alternatives with respect to the different criteria: The consequence of an alternative is defined as the values of the different criteria for this alternative. Since an alternative consists of an ordered set of resource/task pairs, the value of a criterion for an alternative, is the aggregation of the values of this criterion for each individual pair. These values are estimated with the help of data, such as the start/completion date of a task or parameters associated with resource performance. These data may be made available through adequate use of database concepts and data manipulation mechanisms (Chryssolouris et al. 1992).

Apply decision making rules in order to select the best alternative: Once the consequences of the different alternatives have been established, the problem of selecting the best alternative, is reduced to evaluating a decision matrix. In general, the evaluation of decision matrices involves the execution of a procedure that specifies how attribute information is to be processed to arrive at a choice. A utility function is applied, which is a linear combination of the consequence values with weights that represent the relative importance of the criteria. The alternative with the best utility is selected (Chryssolouris et al. 1992)

4 THE PRODUCTION SYSTEM MODEL

Based upon the hierarchical approach previously defined, one can view a manufacturing system as a number of job shops that consist of a number of work centers. The textile factory, being under study, consists of different production lines and as defined in section one this paper is concerned with the CYCP line scheduling. The scheduling tool data entry phase is divided into three categories:
- Facilities which include Factory model, Job-shop model, Work center definition and Resource definition.
- Work load which includes Job definition, Work load definition and Arrival profiles.
- Operating policy.

Operating Policy is the decision-making logic used to assign manufacturing system resources to various production tasks. The scheduling method allows the definition and use of a criteria combination or a conventional dispatch rule as the operating policy. The criteria defined in the scheduling method used are: Mean Flowtime, Mean Tardiness, Mean Cost and Mean Quality. The implemented dispatch rules are: first in first out (FIFO), last in first out (LIFO), shortest processing time (SPT).

4.1 Factory model

The CYCP line consists of three sections, the Dyeing Section the Ring Spinning Section, and the Weaving Section. Every section of this typical textile factory can be modeled as a job shop, which consists of a number of Work Centers (Table 1).

Table 1 The factory model

Job shop - ID	Process - ID	Work center number
Dyeing - JB	Dyeing fibers	4
Spinning - JB	Yarn forming	8
Weaving - JB	Carpet forming	1

4.2 Job shop model

In sequence, every section of this factory can be modeled with the help of a Work Center and Resource hierarchy.

Dyeing Section: As table 2, shows the dyeing section includes the following work centers, such as Stamping-WC, Dyeing-WC, Drying-WC and Blending-WC. Each of them includes only one resource, with the exception of the Dyeing-WC and Blending-WC. Dyeing-WC, including two resources, D1 & D2 (Dyeing machines), which have different capacities/productivities (D1 has the lowest one). Blending-WC includes five identical resources (Blending Bins).

Table 2 Job Shop model - Dying Section

Work Center-ID	Process-ID	Resource number	Resource-ID
Stamping-WC	Stamping	1	S1 (Stamping Machine)
Dyeing-WC	Dyeing	2	D1, D2 (Dyeing Machine 1,2)
Drying-WC	Hydroextraction Drying Opening	1	DR1 (Hydroextractor, Dryer, Fearnough)
Blending-WC	Storing-Feeding	5	BB1-BB5 (Blending Bin 1-5)

Spinning Section: As table 3 shows the spinning section includes the following work centers, CardGill-WC, Roving-WC, Spinning-WC, Vapor-WC, Bobin1-WC, PrTw-WC, Twist-WC, Bobin2-WC. Each of them includes only one resource with the exception of the Spinning-WC, Twist-WC and Bobin2-WC. Spin-WC including four identical resources (Ring-spinning-frames) and Twist-WC including two identical resources (Twisters). Bobin2-WC including two different resources, B2a and B2b (Bobinouars), which have different capacities / productivities (B2a has the lowest one).

Table 3 Job Shop model - Spinning Section

Work Center ID	Process-ID	Resource number	Resource-ID
CardGill-WC	Cleaning-Carding	1	CG1 (Cardframe, Gill Boxes)
Roving-WC	Roving Forming	1	RF1 (Roving Frame)
Spinning-WC	Yarn Forming	4	RSF1,2,3,4 (Ring Spinning Frames)
Vapor-WC	Yarn Stabilization	1	V1 (Vaporizator)
Bobin1-WC	Yarn Cleaning & Package Forming	1	B1 (Bobinouar 1)
PrTw-WC	Yarn Plying	1	PT1 (Pretwister)
Twist-WC	Yarn Twisting	2	T1 & T2 (Twister 1 & 2)
Bobin2-WC	Package Forming	2	B2a&B2b (Bobinouar 2a&2b)

Weaving Section: As table 4 shows, the weaving section includes only one work center, Carpet-WC. This work center includes three resources (Looms). Each type of carpet may be produced at any loom

Table 4 The job shop model - weaving section

Work Center-ID	Process-ID	Resource number	Resource-ID
Carpet-WC	Weaving	3	LOOM1 LOOM2 LOOM3

Some restrictions have been assumed: material handling systems have not been considered; buffers before and after each work center/resource have infinite capacity.

4.3 Work load model

The manufacturing system under study, produces carpets for inventory as well as executes orders for different customers. These orders may arrive at the factory at any time. The operating scenario assumed is as follows:
-Orders/jobs arrive at the weaving section of the textile factory. The main characteristics of a job are quantity, carpet type, color set, carpet design. In order to perform the operation of weaving, three systems of yarns, Warp, Weft and Tuft are needed. The availability of the tuft yarns is of special interest, because they are produced as semifinal products by the under study textile factory, while the warp and weft yarns are purchased by external suppliers and it is assumed that they are available when needed. Each carpet type contains five different colors, which means that five different quantities of dyed yarns must be available simultaneously, to begin the execution of an order in the weaving section.
- An inventory control must be performed in order to know if some of the required yarn quantities are available from the inventory.
- The rest of the yarn quantities must be produced in the Spinning Section, in predetermined colors and qualities.
- To produce these yarns one must have readily available the required quantities of dyed textile fibers. This job will be performed in the Dyeing Section.

Job definition
The nature of the work load (parts, jobs, Tasks), in every separate section of the under study Textile Factory, is different. Three classes of carpets are produced in this Production Line. Each job arrives at every section of the textile factory accompanied by a process plan (namely, a set of instructions that determine the sequence of the different tasks as well as their technological constrains).

Work load in Dyeing Section
In this section each job models a quantity of textile fibers, which must be dyed in a predetermined color, with the different operations to be performed for dyeing the textile fibers modeled as tasks.
JD-Ckk: Stamping, Dyeing, Drying, Feeding, where **kk** is the color code.

Work Load in Spinning Section.
In this section each job models a quantity of yarn which must be spun in a predetermined color, with the different operations to be performed for yarn spinning modeled as tasks.
JS-Ckk: Carding, Roving forming, Yarn forming, Yarn stabilization, Yarn Cleaning & Package Forming, Yarn Plying, Yarn Twisting, Package forming, where **kk** is the color code.

Work Load in Weaving Section.
In this section each job models a quantity of the final product (Carpet) which must be produced in a predetermined design, with the only one operation to be performed for carpet weaving modeled as a task.
JWx-yyyy-ORz: Carpet Weaving, where: **x** is carpet type, **-yyyy-** is Color set, **z** is order code.
Data on each process (processing times, set-ups, etc.) are given in (Working Paper Chryssolouris et al. 1995).

Work load definition.
Work load consists of four different customer orders, which contain carpets to be produced in different types, color sets, dimensions, designs and quantities (**suborders**). These suborders must be grouped to achieve a more efficient production planning and scheduling. The way these orders are grouped, depends on the color set. The result of this grouping, is 20 groups of carpets, 17 groups of fibers to be dyed and 19 groups of yarns to be spun. The precise definitions of the implemented work loads are given in (Working Paper Chryssolouris et al. 1995)

4.4   Problem approximation

In order to solve the above described production planning and scheduling problem, regarding the under study typical Textile Factory, one could approach the problem in the following way:
Planning for the entire manufacturing system, using as input in the weaving section the customer's orders (quantities of carpets). Furthermore, continuing production planning

by using the result of the weaving section scheduling, as input, to the spinning and dyeing sections. These are broken down into required quantities of yarns and fibers, making at the same time the necessary yarn and fiber inventory control. This way, the due dates for the tasks of a job shop are chosen in relation to the start times of the related jobs in the next job shop. In order to keep the problem close to reality, a number of assumptions are made:
- non zero yarn inventory, zero fiber inventory
- all tasks are released at predetermined times in the system
- infinite buffer capacity regarding the buffers between the three job shops (sections).

The scheduling of the under study manufacturing system has been implemented using a number of different policies. These different polices are combinations of four different conflicting criteria, namely mean tardiness, mean flowtime, mean cost and mean quality.

In addition, four conventional dispatching rules, shortest processing time (SPT), first in first out (FIFO) and last in last out (LIFO), have been implemented. Simulation results from all these cases have been obtained in Gantt chart and in alphanumeric form.

5. RESULTS AND DISCUSSION

At the end of a run or at any point, during a simulation, statistically calculated performance measures, can be reported about jobs, work flow, resource status, utilization, and so on. Performance measures form the basis for any conclusions drawn from the playing out of a scenario, based on the predetermined choice of facilities, work description and work center policies.

As far as job orders are concerned, detailed schedules for each job order are listed. Each schedule includes information about each task in the job order, such as the resource it was processed on, times (arrival date, start date, completion date, due date, etc.). Summary data on times for an entire job order, are also available. Some indicative results, concerning the Weaving section, are presented in Figure 5 and Table 5.

Results regarding the utilization and the mean job tardiness for the weaving work center, are also presented in Figures 6-9, in comparison to other conventional dispatch rules.

Table 5. Results for the Weaving Job Shop using the proposed scheduling method

Resource Name	Work Center	Name of Task	Start Time	End Time
LOOM1	CARPET-WC	W1-0018-OR0	0.000	9.897
		W1-0018-OR1	9.897	117.167
		W1-0015-OR2	117.167	205.749
		W1-0015-OR1	205.749	237.855
		W1-0018-OR2	237.855	348.706
		W1-0018-OR3	348.706	525.413
LOOM2	CARPET-WC	W2-0018-OR0	0.000	25.808
		W2-1557-OR1	25.808	83.692
		W2-2018-OR1	83.692	145.999
		W2-0018-OR1	145.999	183.299
		W2-OO18-OR2	183.299	348.475
		W2-0018-OR3	348.475	465.056
		W2-2018-OR2	465.056	644.803
LOOM3	CARPET-WC	W3-0001-OR0	0.000	13.932
		W3-0001-OR1	13.932	37.141
		W3-0501-OR1	37.141	115.984
		W3-0001-OR3	115.984	191.109
		W3-0501-OR2	191.109	341.792
		W3-0501-OR3	341.792	412.946
		W3-0001-OR2	412.946	557.592

Mean Job Tardiness (hours): 10.01
Mean Job Flowtime (hours): 244.94
Mean Job Quality: 1.00
Mean Job Cost: 161 080.80

6 CONCLUSIONS

This work demonstrates a new approach to production planning and scheduling of a typical textile plant. The results from simulations applied to the weaving work center are presented using the new method and some dispatch rules (FIFO, LIFO, SPT).

The new method suggests the combination of different criteria for the production planning in the textile industry. Every time a decision is required, a set of feasible alternatives is produced and the best alternative is selected.

The criteria applied to the work center can also be applied to the job shop level, where short and medium term production scheduling decisions should be taken.

The results regarding the utilization and the mean job tardiness for the weaving work center, seem to be very satisfying, when the proposed method is implemented in comparison to the other rules.

7 REFERENCES

Chryssolouris, G., K. Wright, and W. Cobb (November 12-15, 1985) *Decision Making Strategy for Manufacturing Systems.* Proceedings of the International Conference on Cybernetics and Society, IEEE, Tucson, Arizona, 64-72.

Chryssolouris, G. (Spring 1987) *MADEMA: An Approach to Intelligent Manufacturing Systems.* CIM Review, Vol. 3, No. 3, 11-17.

Chryssolouris, G., and I. Gruenig (December 1988) *On a Database Design for Intelligent Manufacturing Systems.* International Journal of Computer Integrated Manufacturing, Vol. 1, No. 3, 171-184.

Chryssolouris, G., S. Graves, and K. Ulrich (January 9-11 1991) *Decision Making in Manufacturing Systems: Product Design, Production Planning, and Process Control.* Proceedings, 17th NSF Design & Mfg. Systems Grantee's Conference, University of Texas, Austin, 693-701.

Chryssolouris, G., J. Pierce, and K. Dicke (1991) *An Approach for Allocating Manufacturing Resources to Production Tasks.* Journal of Manufacturing Systems, Vol. 10, No. 5, 368-382.

Chryssolouris, G., (1992) *Manufacturing Systems: Theory and Practice.* Springer-Verlag, New York.

Chryssolouris, G., J. Pierce, and K. Dicke (1992) *A Decision-Making Approach to the Operation of Flexible Manufacturing Systems.* International Journal of Flexible Manufacturing Systems, Vol. 4, No. 3/4, 309-330.

Chryssolouris, G. and M. Lee (1994) *An Approach to Real-Time Flexible Scheduling.* International Journal of Flexible Manufacturing Systems. Vol. 6, 235-253

Chryssolouris, G., and M. Lee (June 1992) *An Assessment of Flexibility in Manufacturing Systems.* Manufacturing Review, Vol. 5, No. 2, 105-116.

Chryssolouris, G. and M. Lee, (1994) *An Approach to Real-Time Flexible Scheduling.* International Journal of Flexible Manufacturing Systems, Vol. 6, 235-253.

Chryssolouris, G., D. Mourtzis, and N. Papakostas (1995) *Study, Design and implementation of a planning and scheduling system for a textile company,* Working Paper.

An approach to planning of textile manufacturing operations 143

Resource Names	Job ID	Duration	Scheduled Start	30/7	6/8	13/8	20/8	27/8	3/9	10/9	17/9	24/9	1/10	8/10	15/10	22/10	29/10	5/11
LOOM1,LOOM2,LOOM3	ORDER0 DueDate	0h	7/8/95 8:01pm		◆													
LOOM1,LOOM2,LOOM3	ORDER1 DueDate	0h	22/8/95 2:00pm				◆											
LOOM1,LOOM2,LOOM3	ORDER2 DueDate	0h	19/9/95 2:00pm								◆							
LOOM1,LOOM2,LOOM3	ORDER3 DueDate	0h	2/10/95 8:01pm										◆					
LOOM1	*****W1-0018-OR0*****	9.9h	4/8/95 6:00am		▬													
LOOM1	W1-0018-OR1	107.3h	4/8/95 3:55pm		■													
LOOM1	W1-0015-OR2	88.6h	15/8/95 11:12am			■												
LOOM1	W1-0015-OR1	32.1h	22/8/95 7:49pm				■											
LOOM1	W1-0018-OR2	110.9h	24/8/95 7:55pm				■											
LOOM1	W1-0018-OR3	176.7h	4/9/95 6:49pm						■									
LOOM2	*****W2-0018-OR0*****	25.8h	4/8/95 6:00am		▬													
LOOM2	W2-1557-OR1	57.9h	7/8/95 3:49pm		■													
LOOM2	W2-2018-OR1	62.3h	11/8/95 9:42am			■												
LOOM2	W2-0018-OR1	37.3h	17/8/95 8:00am			■												
LOOM2	W2-0018-OR2	165.2h	21/8/95 1:18pm				■											
LOOM2	W2-0018-OR3	116.6h	4/9/95 6:31pm						■									
LOOM2	W2-2018-OR2	179.7h	14/9/95 7:06am								■							
LOOM3	*****W3-0001-OR0*****	13.9h	4/8/95 6:00am		▬													
LOOM3	W3-0001-OR1	23.2h	4/8/95 7:55pm		■													
LOOM3	W3-0501-OR1	78.8h	8/8/95 11:06am			■												
LOOM3	W3-0001-OR3	75.1h	15/8/95 9:54am			■												
LOOM3	W3-0501-OR2	150.7h	21/8/95 9:01pm				■											
LOOM3	W3-0501-OR3	71.2h	4/9/95 11:42am						■									
LOOM3	W3-0001-OR2	144.6h	8/9/95 6:55pm							■								

Figure 5 Results in Gantt Chart form for the weaving job shop (2 shifts/day, 5 days/week).

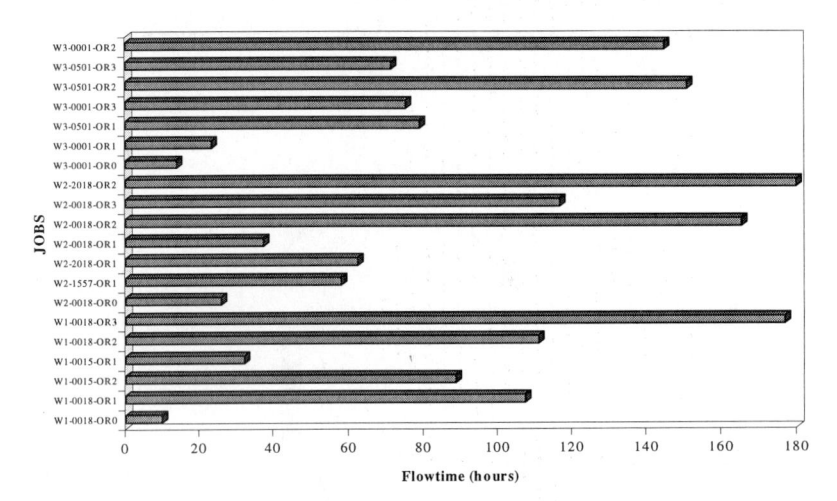

Figure 6 The flowtime per job in the weaving job shop using the proposed scheduling method.

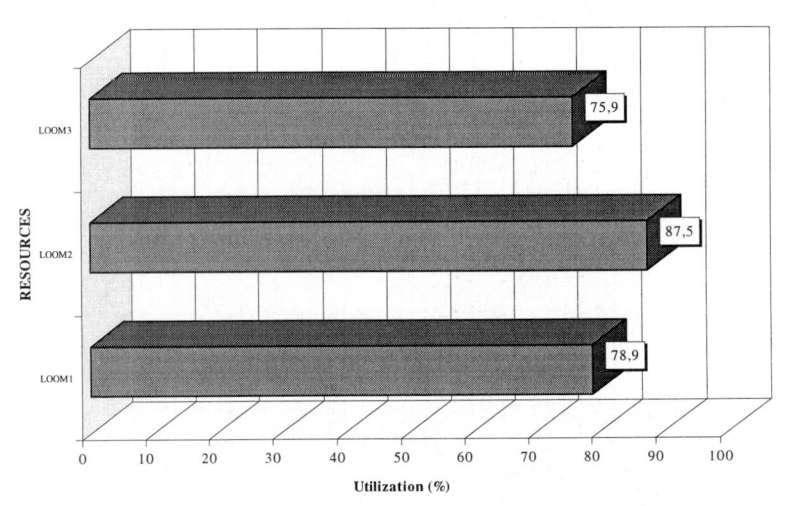

Figure 7 The machine utilization in the weaving job shop with the proposed scheduling method.

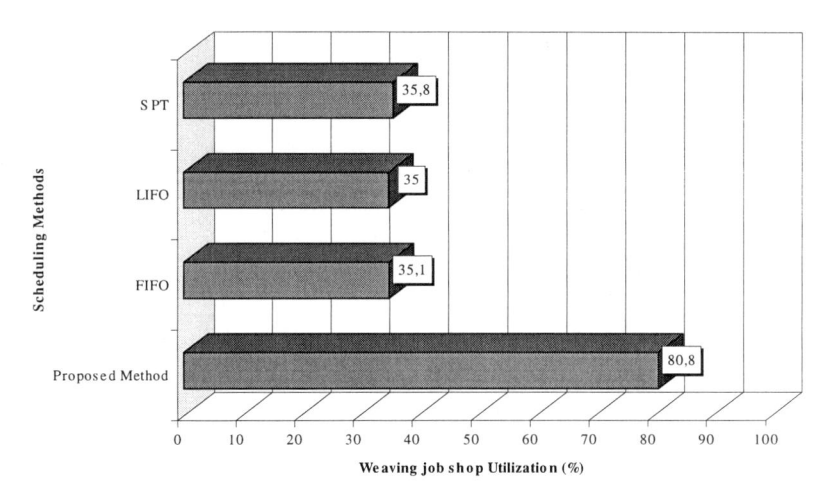

An approach to planning of textile manufacturing operations 145

Figure 8 The mean machine utilization in the weaving job shop using the proposed scheduling method vs conventional scheduling methods.

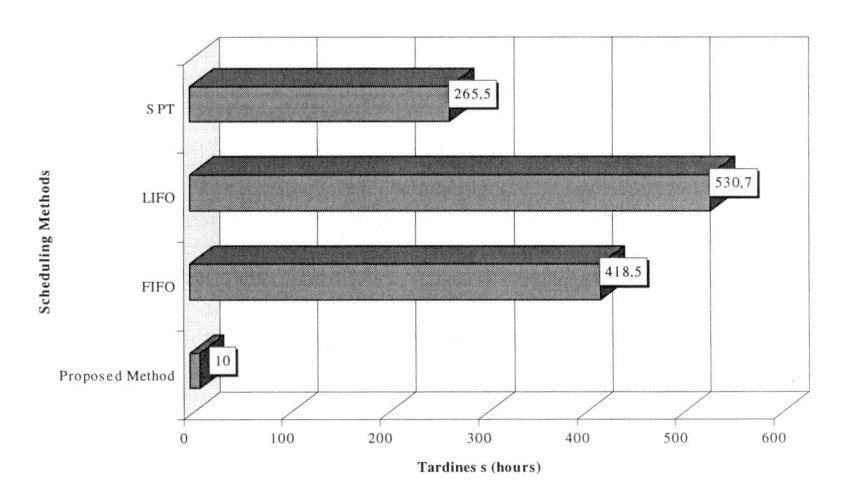

Figure 9 The mean job tardiness in the weaving job shop using the proposed scheduling method vs conventional scheduling methods.

13

A unified decision support tool for product management

H. E. Cook
C. J. Gauthier Professor
Department of Mechanical and Industrial Engineering
University of Illinois
Urbana, Illinois 61801

Phone 217-244-7992
Fax: 217-333-0721
Email: cook@ux1.cso.uiuc.edu

Abstract

An integrated structured methodology is described for aiding product management decisions across all elements of the enterprise. Decisions on new product alternatives are based upon demand and profit forecasts determined from forecasts of the value of the proposed new product to the customer, manufacturing costs, investment levels, and forecasted actions of competitors. The model used to support the structured methodology is designed to be as simple as possible to treat the problem but having sufficient rigor to support trustworthy decision making. The major issues related to marketing, finance, design, engineering, manufacturing, service, environment, and quality management are encompassed by the methodology.

Keywords

New products, customers, profits, strategic goals, product realization, structured methodologies, Total Quality Management, Quality Function Deployment, Taguchi Methods, decision support tool, value benchmarking, strategic quality deployment.

1 INTRODUCTION

Solving Unstructured Problems
Product decisions are complex because the basic problem is unstructured due to the fact that there are many different ways to design, manufacture and market most products. Moreover, outcomes in the marketplace are not just a function of what you do to your product but also highly dependent on what your competitors do to theirs. Because of the complexity and uncertainty, tools are needed to bring order to the product realization process and to manage its innate uncertainty so that good and timely decisions are the norm and not the exception.

Such tools are known as *structured methodologies*, one of the best known being Total Quality Management (TQM) (Gitlow, Oppenheim, and Oppenheim, 1995). Other well-known examples are Quality Function Deployment (QFD) (Akao, 1990; American Supplier Institute, 1993), Manufacturing Resource Planning (MRP) (Chase and Aquilano, 1995), Design of Experiments (DoE) (DeVor, Chang, and Sutherland, 1992), and just-in-time (JIT) (Schonberger, 1987) which deal with subsets of the general problem of product management.

Need for Continuous Improvement
Products must be continuously improved to survive in highly competitive markets. This in turn requires that continuous advances be made in the methods used by the enterprise in support of the product realization process. Although the techniques in use today are much improved over those in place in the early 1980's, significant shortcomings remain. TQM, for example, now strongly supports such sound practices as statistical process control and design of experiments but many elements of TQM's unifying structure today are platitudes. The QFD discipline has caused enterprises to get closer to the customer and thereby gain a better understanding of customer needs before embarking on a new product. However, once the customer needs have been identified, the QFD deployment process rests on subjective judgments and fails to utilize more rigorous concepts from economic theory and statistics. Although Taguchi Methods (Taguchi and Wu, 1980) have been shown to be very effective at the component level of design, a system level support tool akin to Taguchi Methods has not received wide usage. Finally, none of these tools are well integrated into an overall guiding structure which all of the stakeholders within the enterprise can embrace.

Purpose
The purpose of this paper is to first outline the key elements that a structured methodology for product management should have and then to review a structured methodology which contains those elements based upon a simple model of the marketplace collectively referred to as the S-Model (Cook and DeVor, 1991; Cook, 1992; Cook and Kolli, 1994; Kolli and Cook, 1994).

2 TIME-INVARIANT CUSTOMER NEEDS LOOP

Structured methodologies for product management need to be developed around those elements of the product realization process that are deemed as fundamental. The loop shown in Figure 1 which begins with customer needs is taken as being fundamental because it is time-invariant. Customers will always purchase products to fill their needs. Although a multitude of attributes are specified by the designer and manufactured into the product, only a select few -- the system level attributes -- will be of direct importance to the customer. For example, a potential customer for a new vehicle does not care about the myriad details of the parts that form the braking subsystem, only that the vehicle stops reliably, smoothly, quietly, and in a short distance.

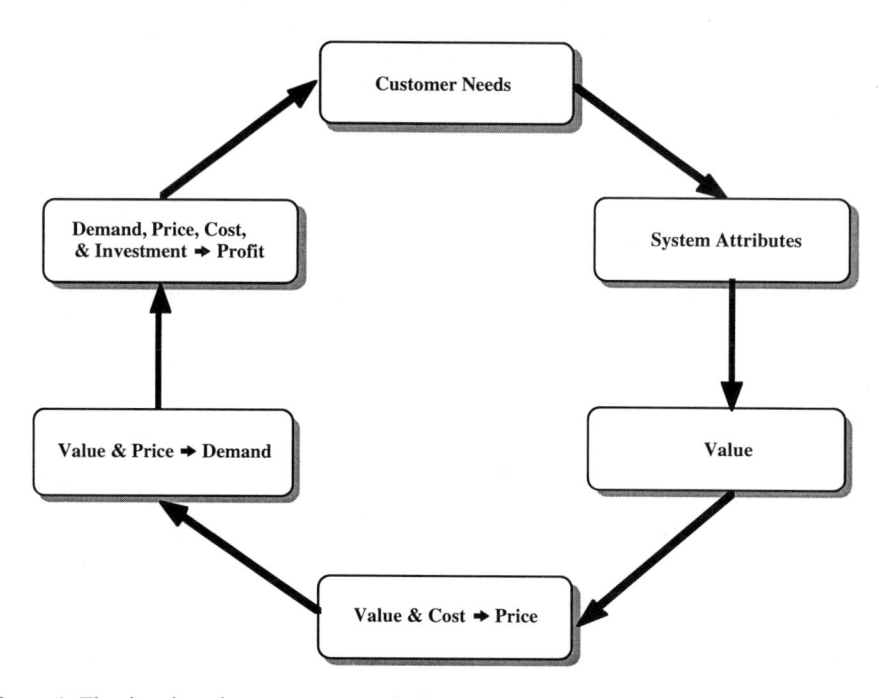

Figure 1 The time-invariant customer needs loop.

If a buyer found that an unexpected squeal was always emitted during braking, this would result in dissatisfaction and a loss of value. If the brake squeal intensity at every application was at 120 dB, the threshold of pain, the loss of vehicle value might be 100% in that the buyer would not drive the vehicle until it was fixed. The problem might have resulted from the brake pad chemistry being out of specification, a component level attribute, but what affected the customer was the system level attribute, noise. Thus as we move around the loop in Figure 1, we see that customer needs are translated into product value through the system level attributes.

Product value and cost place upper and lower bounds on price. If price is higher than value no one will purchase the product and, if price is lower than cost, money will be lost on every sale. Value and price drive demand. Profit, which is determined by demand, price, cost, and investment, creates the working capital needed by the enterprise to develop new products to meet the growing expectations of its customers.

A major challenge in managing products through the customer needs loop is to reliably connect system attributes to product value and to quantify how the demand of a product is affected by changes in its price and value and in the price and value of the products competing against it. Another challenge in developing a structured methodology is to find the right blend between simplicity and rigor. The approach must be straightforward and transparent if practitioners are to use it but sufficient rigor should exist so that users can depend on the results.

3 DEMAND MODEL

The S-Model attempts to strike the balance between simplicity and rigor by using a demand expression (Cook and Kolli, 1994; Donndelinger and Cook, 1995):

$$n_i = K\left\{V_i - P_i - \frac{1}{N}\sum_{j \neq i}\left[V_j - P_j\right]\right\}, \tag{1}$$

obtained from a Taylor's expansion about a so-called "cartel point" where the price and value of all the competing products are the same. The parameter N is the number of competitors and the summation is over the $N-1$ products competing against product i. The term K is the negative slope of demand as a function of price. The linear expression should be valid in the vicinity of the cartel point, a reference state given by the average price and value of the competing products. The restriction of linearity limits the application of Eq. 1 to products within the same market segment.

4 VALUE BENCHMARKING

We see from Eq. 1 that a demand forecast requires an estimate of the values and prices of all of the products that will be competing with each other. Value benchmarking is the first step in estimating value as shown schematically in Figure 2 for the S-Model (Cook and Kolli, 1994; Donndelinger and Cook, 1995). The process involves three steps. The first step is to determine the values and value differences of the products currently competing in the market place. The second step is to consider the differences in the attributes of the products and see if the value differences found can be explained reasonably in terms of the attribute differences. If this is successful, then a forecast of the change in value of a product resulting from a proposed design change can be made by translating the attribute changes into value changes. The process of selecting the best combination of attributes for a future product using the S-Model has been termed Strategic Quality Deployment (Kolli and Cook, 1994), the box labeled SQD in Figure 2.

DP Analysis

The values of the products currently competing in the marketplace, (noted as DP Analysis in Figure 2) are determined by solving the linear system of equations given by Eq. 1 in terms of value for known demands and prices (Donndelinger and Cook, 1995):

$$V_i = \frac{N[n_i + n_T]}{K[N+1]} + P_i, \qquad (2)$$

where n_T is the total demand for the N products. The difference between the values of two competing products is given by:

$$V_i - V_k = \frac{N[n_i - n_k]}{K[N+1]} + P_i - P_k. \qquad (3)$$

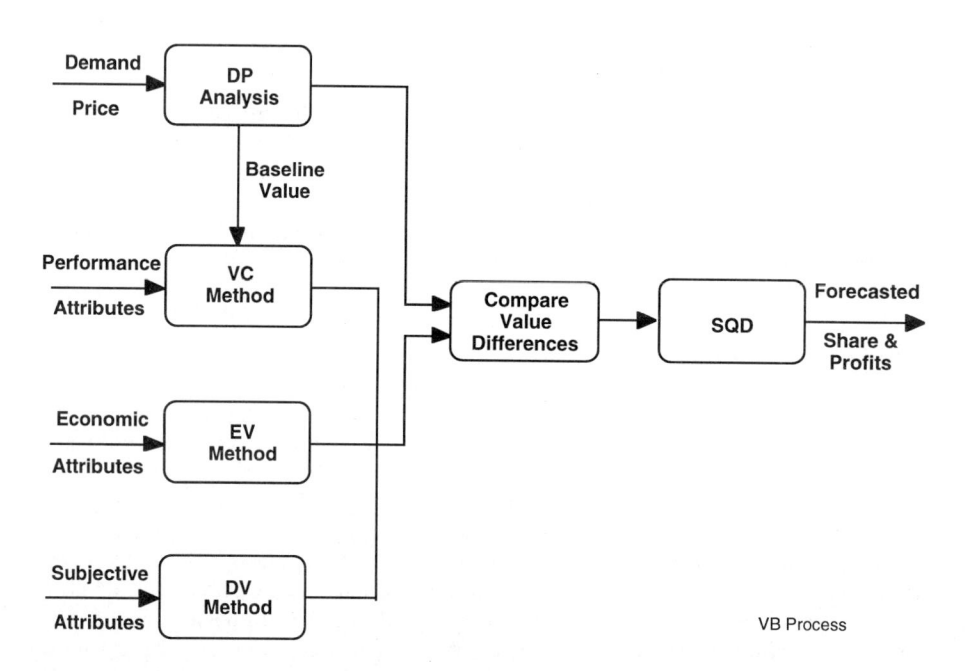

Figure 2 The S-Model value benchmarking process (Donndelinger and Cook, 1995).

Attribute Analysis

Three different procedures are used to estimate how the attribute differences generate value differences between the products. The VC method uses value curves, Figure 3. A rapid method for arriving at an estimated value curve for an attribute is to use an exponentially weighted parabolic function:

$$v(g_j) = \left\{ \frac{[g_{jC} - g_{jl}]^2 - [g_j - g_{jl}]^2}{[g_{jC} - g_{jl}]^2 - [g_{j0} - g_{jl}]^2} \right\}^{\gamma_j}, \tag{4}$$

for value divided by the baseline value, V_0, as a function of the attribute g_j. The parameters in Eq. 3 are defined in Table 1. The value of attributes such as the fuel economy of automobiles or the operating costs of construction equipment can be estimated by straightforward economic calculations, noted as the EV method in Figure 2. Certain attributes of a product such as the style of a vehicle or the value of the name of the manufacturer of the product can be highly subjective in regard to their value. For these attributes, surveys are used in which respondents are asked their willingness to pay for a change in the attribute versus a baseline level. This is noted as the DV method in Figure 2.

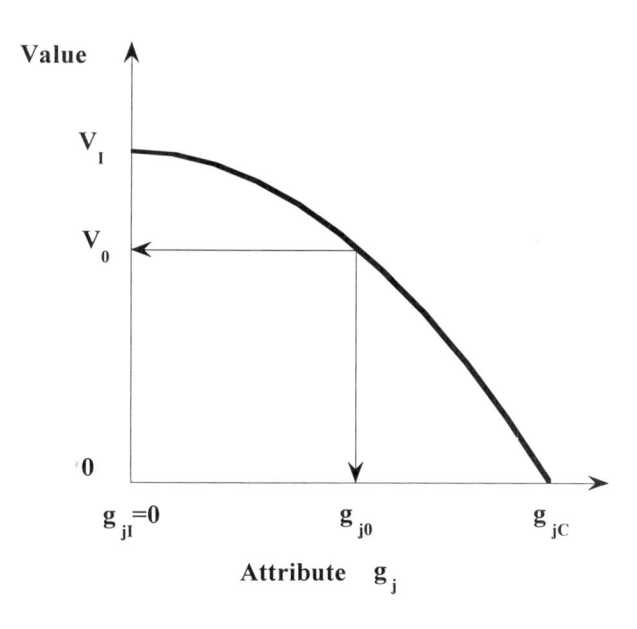

Figure 3 A schematic value curve.

The techniques can also be combined to arrive at an estimate of the value of an attribute. McConville and Cook (1995) combined the VC and DV methods to arrive at an estimate of the value of 0 to 60 mph acceleration times for a family sedan. Donndelinger and Cook (1995) averaged the results from DV and EV methods to arrive at an estimate of the value of fuel economy. Simek and Cook (1995) used the DV method in conjunction with the VC method to arrive at the value of leg room and head room in an automobile.

Automotive Applications
The value benchmarking process has been applied to mid-sized (Donndelinger and Cook, 1995) and compact automobiles (Schildt, 1995) in the U.S. market. The comparison between the value differences determined from the DP and attribute analyses found for mid-sized automobiles is shown in Figure 4. The vehicle brands examined have been disguised using the letters A, B, C, D, and E. The expression used for estimating the value of multiple attributes was the heuristic form:

$$V_i(\mathbf{g}) = V_0 v(g_1) v(g_2) v(g_3) ... v(g_j) + \Delta V_{Opt} , \qquad (5)$$

which was chosen because it has the property that the value of the product goes to zero (except for salvageable options, the last term in Eq. 6) when $g_j = g_{jC}$, where $v(g_{jC}) = 0$.

Attributes found to have the largest impact on the value of mid-sized vehicles were the lack of an airbag for vehicle D (the other vehicles had a drivers air bag as standard equipment) and the value of the name of the manufacturer which correlated with the reliability of the vehicles reported by *Consumer Reports*. Schildt (1995) found for compact cars that the high customer satisfaction with the purchase and service experience at the dealership was worth $1000 to $1500 in value.

Table 1 Points on value curve

g_j	Attribute for factor j.
g_{j0}	Baseline level for attribute j.
g_{jI}	Ideal level for attribute j.
g_{jC}	Critical level for attribute j.
γ_j	Weighting Factor for attribute j.

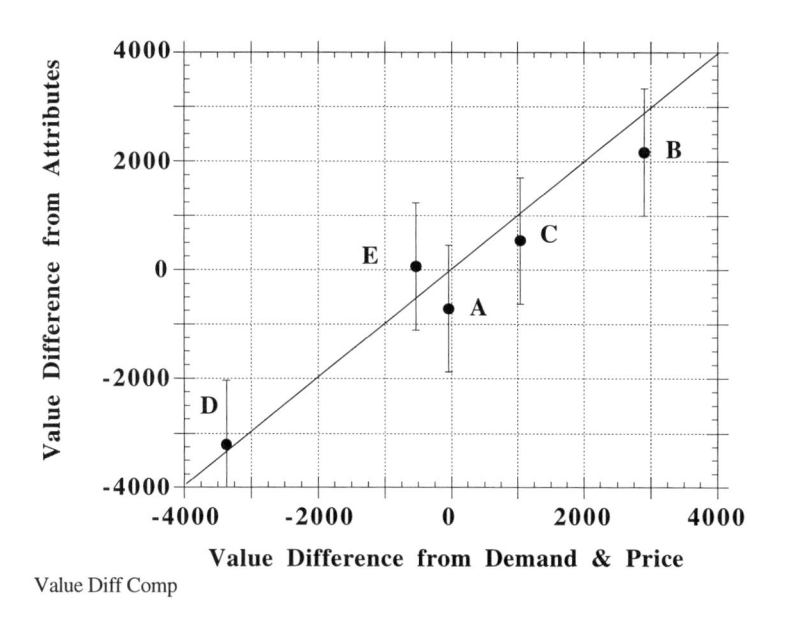

Figure 4 Comparison of value differences for five mid-sized vehicles calculated from their attribute differences and from demand and price analysis (Donndelinger and Cook, 1995).

SQD Waterfall of Experiments

The SQD process can be represented by a waterfall of experiments from the system to subsystem (SS) to component (C), Figure 5 (Kolli and Cook, 1994). At the system level different subsystems are evaluated versus the baseline product. The example illustrated in Figure 5 evaluates alternates for three subsystems using four experimental trials at each level. A one in a column for an alternative signifies that it is being evaluated for that trial. Trial 1 is an evaluation of the baseline product. Trial 2 evaluates a prototype which has the alternatives for SS2 and SS3 replacing their baseline counterparts, etc.

For each trial, the multiple system level attributes are evaluated and converted into value. The system level costs and investment for each prototype are also determined and the profitability (net cash flow) is forecast using the expression:

$$A_i(g) = n_i(g)[P_i(g) - C_i(g)] - F_i(g) - M_i(g), \tag{6}$$

where g is a vector of product attributes, $n_i(g)$ is annual demand, $P_i(g)$ is the price, $C_i(g)$ is the variable cost, $F_i(g)$ is the fixed cost, and $M_i(g)$ is the investment. The price used for each trial (q) is computed according to pricing scenario of the form:

$$P(q) = P_0 + \alpha_V \delta V(q) + \alpha_C \delta C(q), \tag{7}$$

where P_0 is the price of the baseline product and the coefficients α_V and α_C are both approximately equal to 1/2 (Monroe, Silver, and Cook, 1995). The two coefficients can be changed by the system task to reflect the strategic pricing plan for the product.

System Level

Trial	Base	SS1	SS2	SS3	Attributes	V	C	A
1	1	0	0	0	$g_1, g_2, g_3, \ldots$	V(1)	C(1)	A(1)
2	1	0	1	1	$g_1, g_2, g_3, \ldots$	V(2)	C(2)	A(2)
3	1	1	1	0	$g_1, g_2, g_3, \ldots$	V(3)	C(3)	A(3)
4	1	1	0	1	$g_1, g_2, g_3, \ldots$	V(4)	C(4)	A(4)

Subsystem SS2 Level

Trial	Base	C1	C2	C3	Attributes	V	C	A
1	1	0	0	0	$g_1, g_2, g_3, \ldots$	V(1)	C(1)	A(1)
2	1	0	1	1	$g_1, g_2, g_3, \ldots$	V(2)	C(2)	A(2)
3	1	1	1	0	$g_1, g_2, g_3, \ldots$	V(3)	C(3)	A(3)
4	1	1	0	1	$g_1, g_2, g_3, \ldots$	V(4)	C(4)	A(4)

Component C2 Level

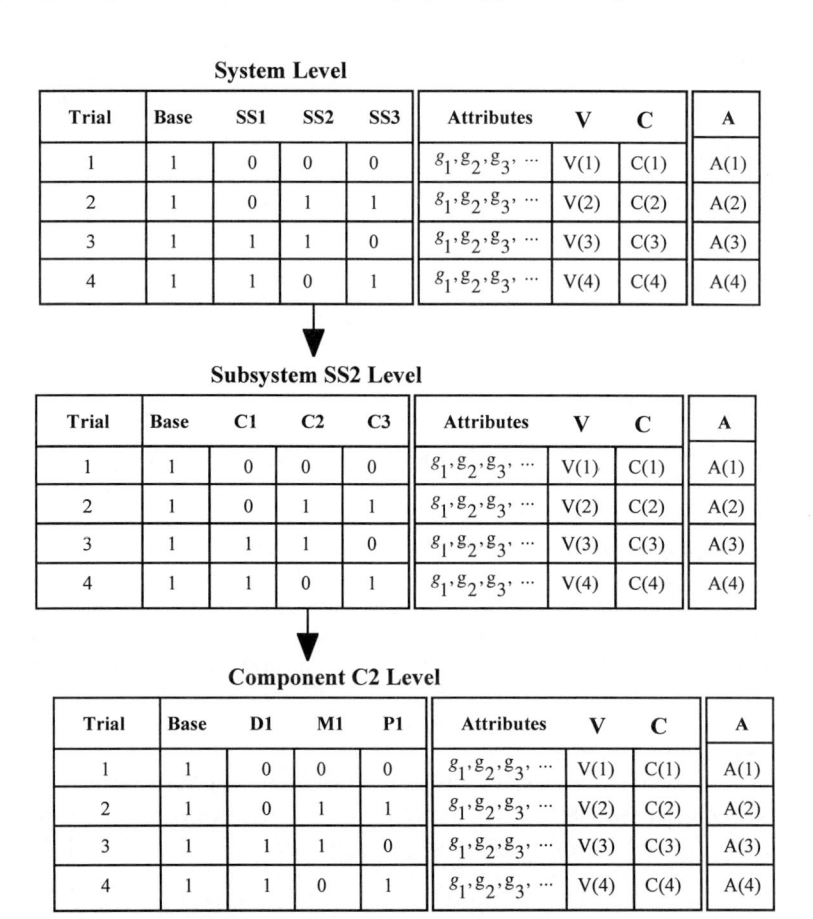

Trial	Base	D1	M1	P1	Attributes	V	C	A
1	1	0	0	0	$g_1, g_2, g_3, \ldots$	V(1)	C(1)	A(1)
2	1	0	1	1	$g_1, g_2, g_3, \ldots$	V(2)	C(2)	A(2)
3	1	1	1	0	$g_1, g_2, g_3, \ldots$	V(3)	C(3)	A(3)
4	1	1	0	1	$g_1, g_2, g_3, \ldots$	V(4)	C(4)	A(4)

Figure 5 Waterfall of experiments.

If, for example, system level analysis confirms that the proposed alternative for SS2 is to be used for production, then those responsible for SS2 will evaluate the possible component alternatives for this subsystem in order to meet or exceed the key requirements specified by the system level task (the internal customer of the subsystem units). A key set of component requirements is then sent to those responsible for designing and manufacturing the components. Persons within the component tasks then determine the best dimensioning, materials, and manufacturing process for meeting their internal customer's requirements at the lowest possible cost. The objective function for each trial at every level is the same, forecasted profitability. This results in the same system level attributes being evaluated for each trial at every level for conversion into value.

5 DISCUSSION AND SUMMARY

The waterfall of experiments shown in Figure 5 examine the impact of the alternatives considered against the multiple system level product attributes of importance to the customer. At the component level, the experiments are related to Taguchi's parameter design experiments, the difference being that a loss function is used by Taguchi Methods practitioners; whereas, forecasted profitability is used in SQD. Because Taguchi's loss function can be derived from the S-Model theory of quality, the two approaches are in harmony (Cook and DeVor, 1991). The S-Model was developed to treat the full range of design issues from the system level to the component level which requires consideration of multiple attributes. Taguchi's formulation focuses on the parameter and tolerance design single attribute problem.

Product experiments are traditionally made using prototype hardware but computer simulation is a promising and growing alternative which may soon replace much of the labor and cost of prototype evaluations. It is important in the development of computer simulations of product behavior that the engineers take the simulation "all the way to the customer" instead of ending the simulation at the point where a traditional engineering result is obtained. For example, a new suspension design for an automobile may be under consideration. The traditional engineering calculation might end at the computation of the critical wheel hop frequency and maximum possible lateral acceleration. Taking the simulation to the customer requires that the impact on the system level product attributes be computed and those translated into value and costs and ultimately into profit.

Although the S-Model is relatively new, it is already being used as a decision support tool on an experimental basis in developing new automotive and construction equipment products. As the usage increases, we expect, of course, to find areas where refinements are needed. Opportunities for improving the SQD methodology lie in developing more accurate models for 1) translating customer needs into value and 2) for converting value and cost into forecasted demand.

6 ACKNOWLEDGMENTS

The author is deeply grateful to Caterpillar and Ford Motor Company for support of this research through unrestricted grants.

7 REFERENCES

Akao, Y. (Editor) (1990) *Quality Function Deployment QFD*. Productivity Press, Cambridge, Massachusetts

Chase, R.B. and Aquilano, N.J. (1995) *Production Operations Management Manufacturing and Services*. 7th edition, Irwin, Chicago, Illinois, 589-633.

Cook, H.E. and DeVor, R.E. (1991) On Competitive Manufacturing Enterprises I: The S-Model and the Theory of Quality. *Manufacturing Review*, **4**(2), 96-105.

Cook, H.E. (1992) New Avenues to Total Quality Management. *Manufacturing Review*, **5**(4), 284-292.

Cook, H.E. and Kolli, R.P. (1994) Using Value Benchmarking to Plan and Price New Products and Processes. *Manufacturing Review*, **7**(2), 134-147.

DeVor, R.E., Chang, T., and Sutherland, J.W. (1992) *Statistical Quality Design and Control*. Macmillan, New York.

Donndelinger, J. and Cook, H.E. (1995) Benchmarking Product Value: Mid-sized Automobiles. *Report No. UILU-ENG-95-4008*.

Gitlow, H., Oppenheim, A., and Oppenheim, R. (1995) *Quality Management*. 2nd ed., Irwin, Burr Ridge, Illinois

Kolli, R.P. and Cook, H.E. (1994) Strategic Quality Deployment. *Manufacturing Review*, **7**(2), 148-163.

McConville, G. and Cook, H.E. (1995) Estimating the Value Trade-off Between Automobile Acceleration Performance and Fuel Economy. *Tech. Report No. UILU-ENG-95-4010*.

Monroe, E.M., Silver, R.L., and Cook, H.E. (1995) Segmentation of the Market for Family Automobiles by Price and Value. *Tech. Report No. UILU-ENG-95-4031*.

Schildt, T. (1995) *Value Benchmarking In The Compact Car Market*. Diplomarbeit, RWTH, Aachen.

Schonberger, R.J. (1987) *World Class Manufacturing Casebook, Implementing JIT and TQC*. The Free Press, New York.

Simek, M.E. and Cook, H.E. (1995) A Methodology for Estimating the Value of Interior Room in Automobiles. *Tech. Report No. UILU-ENG-95-4009*.

Taguchi, G. and Wu, Y. (1980) *Introduction to Off-line Quality Control*. Central Japan Quality Association, Nagoya, Japan.

Transactions from the Fifth Symposium on Quality Function Deployment (1993) American Supplier Institute, Dearborn, Michigan.

8 BIOGRAPHY

Dr. Cook's current research and academic interests are in the broad aspects of product realization and the management of technology focusing on manufacturing competitiveness, advanced quality systems, and product management. Before joining the University, Dr. Cook was Director of Automotive Research and Technical Systems and later General Manager of Scientific Affairs at Chrysler. He helped spearhead the development of the first cooperative research partnership within the U.S. based automotive industry -- the formation of a pre-competitive research joint venture between Ford, GM and Chrysler for structural polymer based composites. He is a member of the National Academy of Engineering and a Fellow of both the American Society for Metals and the Society of Automotive Engineers. He is also a member of the National Materials Advisory Board.

14

Scenario-Management during the early stages of product development

Prof. Dr.-Ing. J. Gausemeier
Dipl.-Wirt.-Ing. A. Fink
Dipl.-Wirt.-Ing. O. Schlake
Heinz Nixdorf Institute, University of Paderborn, 33102 Paderborn, Germany, Tel. +49 5251 606267 Fax: +49 5251 606268.
E-mail: gausemeier@hni.uni-paderborn.de.

Abstract
During the early stages of product development - product planning and drafting - the fundamental points for the success of a new industrial product are fixed. With the increase of competition, enterprises primarily tend to concentrate on turning out a profit at the end of the financial year. For the short term an increase of efficiency and cost-reduction are the decisive factors of their thinking and acting. How intense the awareness in costs and efficiency in product development might ever be: this will not be enough to guarantee the enterprise's future. The decision-makers have to create ideas of a possible future and resulting chances and risks for new products /1, 2/. The scenario-planning helps to systematically create these ideas of a possible future based on the principles of "network thinking" and a "multiple future". Today many enterprises take advantage of the scenario-planning but without using its full potential. Therefore the Heinz Nixdorf Institute has advanced the scenario-planning to the Scenario-Management.
The article describes how scenarios are created and how they are used during the early stages of product development. First it describes the idea of Scenario-Management. The article concentrates on the scenario-based development of successful product strategies, the choice of basic solutions and the wording of future visions for products and technologies.
The final target of Scenario-Management is to reach future-robust decisions in the early stages of product development, that means to perfectly match the decisions on possible future situations.

Keywords
Scenario-Management, product development, scenario-planning, multiple-future, network-thinking, future-robust visions

1. SCENARIO-MANAGEMENT

When strategy entered economics in the 1960s and 1970s it was judged as an general remedy for upcoming uncertainties on markets, in technique and social environments. In product planning strategic thinking lead to more importance. Today, thirty years later, this euphoria has gone. Many decision-makers don't rely on strategic management anymore, because "strategic plans proofed wrong in retrospective too often" or because they think that "strategic management can not be controlled in everydays business". The renunciation of strategic planning is always a loss of future opportunities. The visionary view, the constant search for new success-potentials and

Scenario-management during the early stages of product development 159

market conform transformation are key factors for success in competitive and turbulent markets. Today, the enterprises have to detect and examine the influence factors on business, their possible future developments and their systematic interaction more than ever before. They have to create visions, strategic targets and flexible and robust strategies (figure 1). For this Scenario-Management is an ideal tool /1/.

We understand a scenario as a description of a future situation, which occurrence can not be predicted with certainty and the description of the developments which led from the present to this situation /1, 3/.

1.1 Multiple futures and network thinking

The American futurologist Herman Kahn has first introduced scenarios into economics and sociology in the 1950s at Rand Corporation. The new idea about Kahn's scenarios was, that they describe what could happen instead of what will happen. With the rise of strategic planning the scenario-planning became of interest for several enterprises. Today many enterprises take advantage of the scenario-planning but without using its full potential. /4/. The Heinz Nixdorf Institute advanced the scenario-planning to Scenario-Management. Scenario-planning is based on multiple futures and network thinking. This eliminates two main deficiencies of strategic planning:

Multiple futures

Future becomes more and more unpredictable. But as a first important deficiency of common planning most enterprises rely on just one single, general forecast instead of taking several projections into account. With scenarios it is possible to eliminate this problem, because scenarios take several future projections into account, which are all possible but not sure. We call this "multiple future" (see figure 2). While a single-forecast-based planning inevitably will go wrong, a scenario-based-planning tries to make decisions relying on different future possibilities. This is why Scenario-Management leads to future-robust decisions.

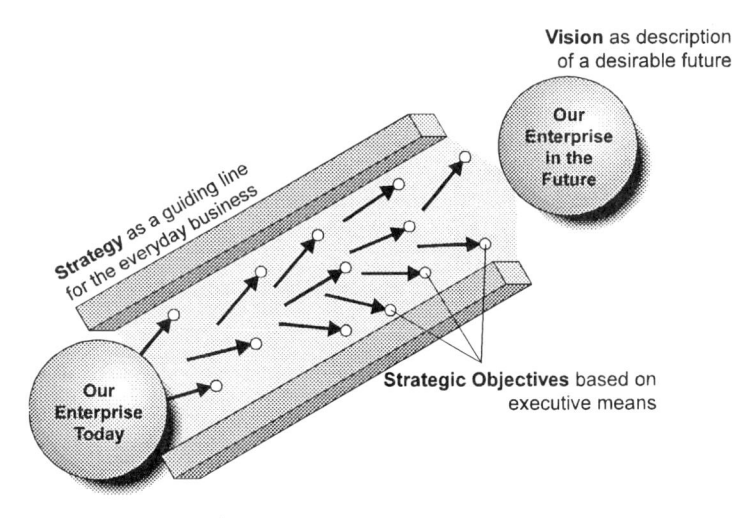

Figure 1 Visions and strategies

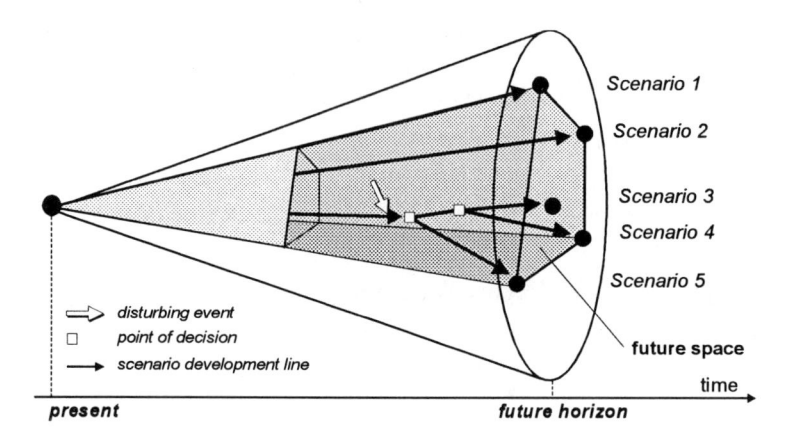

Figure 2 Scenario-funnel

Network thinking

Another deficiency of common strategic planning is dealing with complexity. Dörner impressivly describes in his book "Die Logik des Misslingens", that humans are limited in grasping complex coherences /5/. With the increase of complexity traditional management theories must fail as well. They rely on disconnected environments and influence spheres e.g. markets, production and sales. But today the influences between former "disconnected" environments gain of importance /6, 7/. The Scenario-Management counts on the detection and analysis of relevant influences between enterprises and their environment.

1.2 Phases of Scenario Management

To support entrepreneurial decisions by a scenario-project it is first necessary to define the so-called "examination field". This is the target of the project, e.g. an enterprise, a product or a technology. For example a manufacturer of ATM's (automatic telling machines) could support his decision for a new product line with the help of a scenario-project. In this case the examination field is the product "ATM" (see figure 3). A scenario-project has to provide the means to support these decisions. These alternative scenarios describe the future development of the non-influencible environment of the examination field. Environments of the "ATM" are e.g. market evolution (computer banking), technique evolution (semiconductor development) and global trends (multimedia). By separating these environments from the examination field it is possible to derive instructions for the examination field from non-influencible environments that strongly determine the future developments of this field.

Scenario-Management consists of five phases (figure 4) /1/. The core-phases are phases two to four, called "scenario creation" which are followed by the scenario transfers dealing with the further application of the scenarios.

Scenario-Preparation

During the first phase the creation and application of the scenarios is planned. For this purpose, an examination field (e.g. a product, a technology) and a scenario-field will be defined, and the organisational form of the scenario-project will be defined. In a second step, the examination field in its actual situation will be described: this is called "examination field analysis". Therefore the examination field is split up into its single components (see figure 3). Means of description are: strength-weakness profiles, portfolios and other management instruments. The result of the scenario-preparation is called "scenario-base".

Scenario Field Analysis

The scenario field analysis starts with the creation of scenarios (see figure 5). The scenario field is split up into appropriate spheres of influence: markets, politics, etc. Spheres of influence of an ATM are, e.g. banking, trade and industry, society and information technology. Each one of these environments is described by single influence factors. In most cases too many influence factors will be defined, so that relevant key factors must be selected with the help of influence analysis. In this step the direct and indirect influences between all of the influence factors will be detected and analysed. The software "SCENARIO-MANAGER", developed at the Heinz Nixdorf Institute is used to identify the key factors. These key factors are the input for the prognosis process, the scenarios prognostics (phase 3).

Scenario Prognostics

This is the heart of the Scenario-Management. Here the "look in the future" is carried out. Several projections will be developed and outlined for each key factor. These projections can be extreme images of the future (extreme
projection) or probable images (trend-projection). Possible extreme projections for the key factor "speed of innovation" are e.g. "boost of innovations by information-technology" and "slow down by technological hostility". This also shows an important advantage of scenarios: future forecasts don't have to be quantified but can be described as qualified developments.

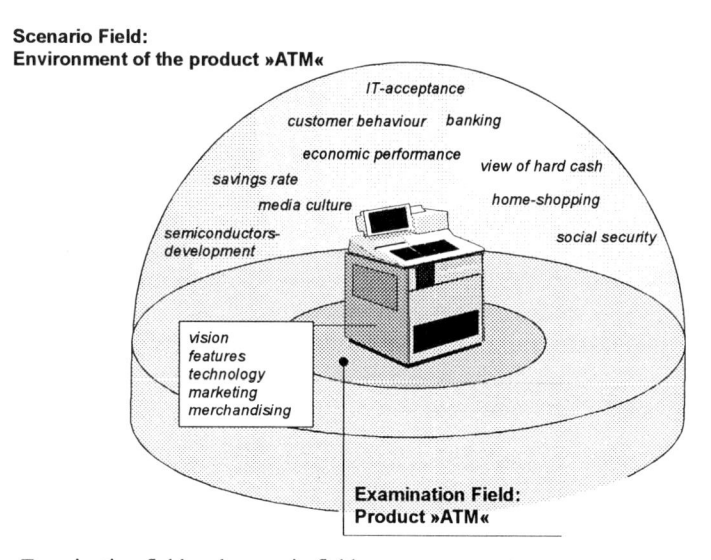

Figure 3   Examination field and scenario field

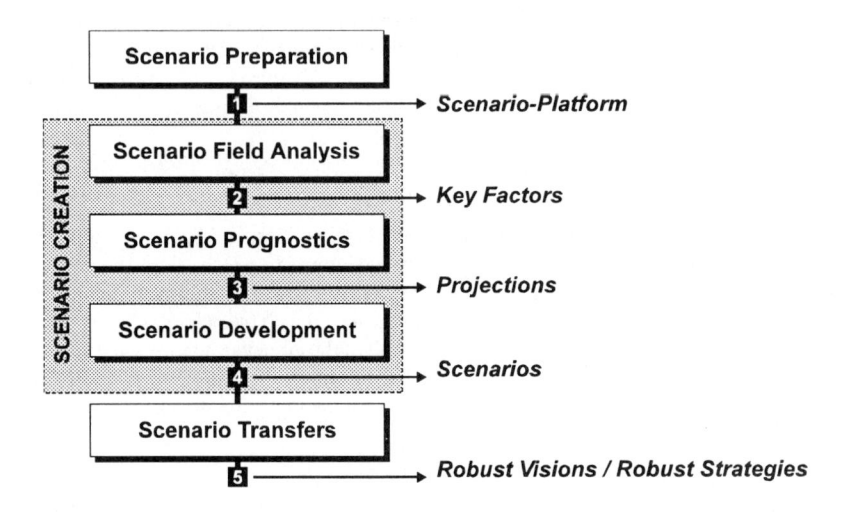

Figure 4 Phases of scenario-management

Scenario Development
In phase four all possible combinations of projections are assessed with regard to contradictions and plausibility. Plausibility here means "is it possible that these two projections of two different key factors appear in a common scenario". The valuation of consistency usually reduces the possible combinations of projections from some millions to less than fifty thousands. These consistent projection-bundles, which resemble each other are than tied together by the aid of cluster analysis for two to four crude-scenarios. A terse and detailed description turns these alternative crude-scenarios into easily understandable scenarios. The person who creates the scenario will be able to image the future space by using statistic-aided graphic programs. She/he will use the "two dimensional maps of the future" in order to interpret the crude-scenarios in an creative manner, and afterwards describe the scenarios in prose.

Scenario Transfers
The scenario-transfers start with an analysis of the effects. Chances and risks can be found out by a matrix of effects - each scenario and each component of the examination field can be treated separately. Using the results, different measures to seize the chances and minimize the risks can be developed. These contingency schemes are alternative ways of action and postly contradictors. Within the scope of a robust planning, the contingency schemes can be combined so that a future-robust strategy can be developed. The strategy is named "robust" because it can be successfully employed with different scenarios.

2 SCENARIOS DURING THE EARLY STAGES OF PRODUCT DEVELOPMENT

The success of an enterprise is decisively influenced by the quality and efficiency of its product development process. The early stages of product development - product planning, product mar-

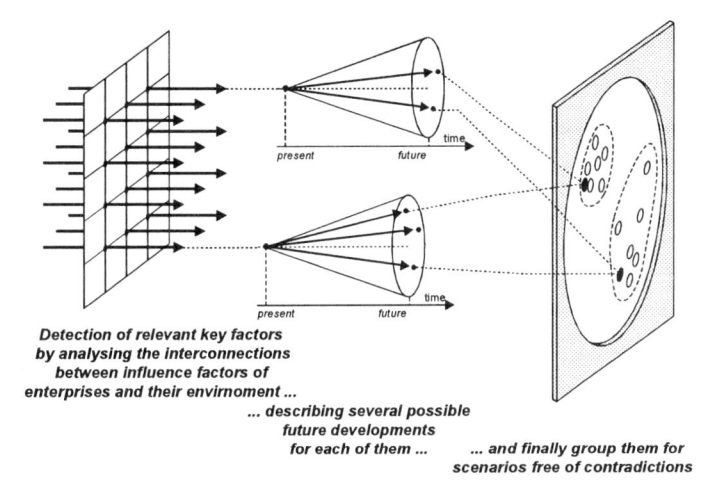

Figure 5 Principle of scenario creation

keting and product conceiving are of essential importance. Here momentous decisions are made. Although the early stages of product development are very important there are still deficits in a future-oriented procedure.

With scenarios brought into action the early stages could be tied much closer to the development of markets and technologies /8/. Three possible uses for scenarios in product development are:

1. With scenarios it is possible to design product strategies in a way that they are resistent to several possible developments in markets and other surroundings.
2. Technologies embedded in new products could be selected and designed to catch up with different developments. We call this future-robust solutions.
3. Product and technique visions are qualified to estimate and consider the demands of social relevant groups (unions, environmental groups etc.) on new products and technologies. These visions of a "desirable future" become more important. Therefore we present a method to develop future-robust visions.

2.1 Development of future-robust product strategies

Most of the major enterprises face the increasing complexity and variety of their interests with a specific structurisation of their business e.g. in strategic business areas (SBA). This is done by a matrix (see figure 6) built by components of market performance (products, product groups, service packages) and the relevant market segments (homogeneous customer groups). A business area is represented by an occupied space on this matrix. Means of a business area are turnover, profit and the strength of the business potential. Strategic business areas imply a possible sustained success for the enterprise. To develop a product strategy the market performance of this SBA will be defined as examination field (see phase 1).

The development of future-robust product strategy is modelled very close to the five phases of the Scenario-Management (see figures 4, 7).

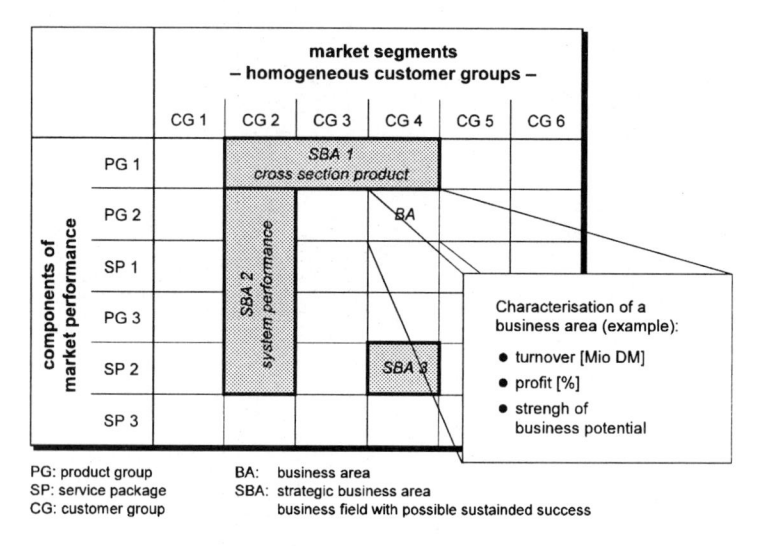

Figure 6 Visualizing of strategic business areas

Product analysis (phase 1 in Scenario-Management)
The ATM will be defined as examination field and will be described in its actual situation. To do so it is subdivided in examination field components (EFC) e.g. functionality, used technologies, energy consumption etc. To determine the strength-weakness-profiles the integrated technology-market portfolio by McKinsey will be used /1, 9, 10/.

Development of product scenarios (phases 2 to 4)
The influence spheres of ATM e.g. banking and trade (closer spheres) and economy, society or information technology (global spheres) will be defined as scenario field. Extreme scenarios will be set up for this scenario field and understandable images of a possible future situation will be described /1/.

Effects Analysis (phase 5)
To begin with the consequences of single scenarios for single EFC will be determined and collected in an effect matrix. A scenario that describes e.g. "Tele-banking" as a penetrating technique that will have considerable effects on several components of the ATM. The use of effects matrix ensures that no fundamental effects will be left out because decision-makers concentrate only on certain scenarios or certain examination field components.
After that scenario specific chances and risks for the EFC's can easily be read from the matrix. For the above scenario and the EFC "features" it could be a risk, that the ATM does not need additional features because most money transactions are done much more efficiently via "Tele-banking" at home.

Eventuality Planning (still phase 5)
At this stage of scenario transfers, scenario specific measures will be derived from the chances and risks. E.g. the ATM should be exclusivly designed to fulfil the needs of a "Tele-banking"

Scenario-management during the early stages of product development

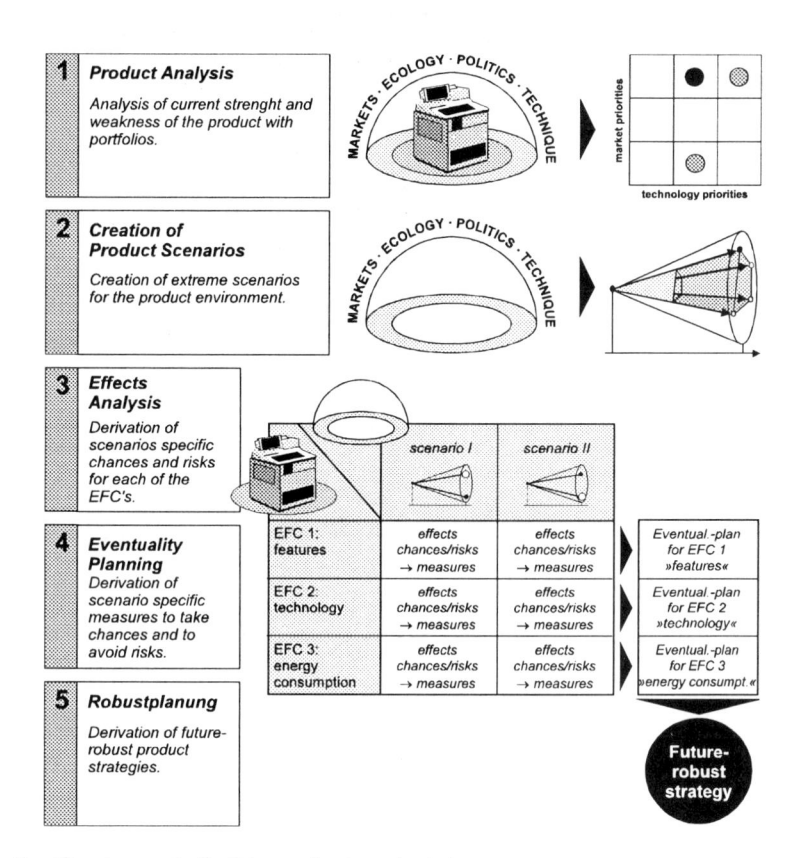

Figure 7 Development of a future-robust product strategy

customer, concentrating on hard cash in- and output. All measures that concentrate on a certain scenario and on one single EFC, are called eventuality plan.

Robust Planning

During robust planning eventuality plans are tied together for a robust plan. Robust plans are combinations of measures that meet the demands of several scenarios. E.g. the features of ATM's could be extended and at the same time adjusted to the needs for "Tele-banking" customers. Another example of robust planning is the power producing industry that invests in nuclear power and at the same time does research in the area of solar and wind power.

2.2 Development of a future-robust solution

A lot of problems concerning the conceptual design and planning of products come from the fact that decision-makers when "looking at the future" only regard projections and forecasts which are universally valid. They should rather take several possibilities of development into account and make their strategies "future-robust" instead.

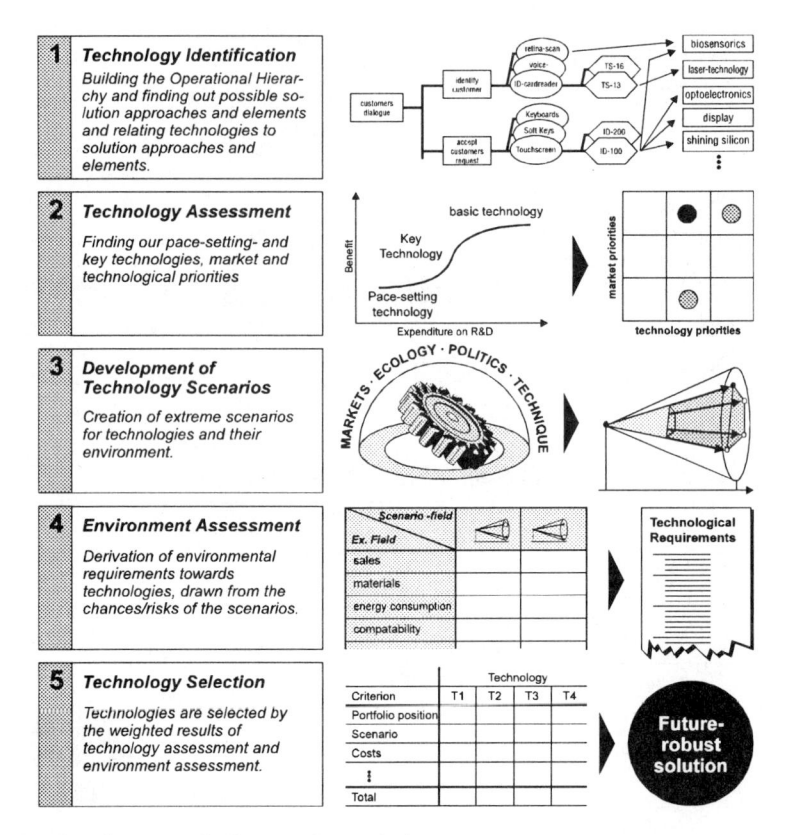

Figure 8 Development of a future-robust solution

For a sustained securing of market chances the basic technologies of a product have to be replaced in time by key and pace setting technologies. These processes of substitution are neglected in practice very often. Therefore we present a method to develop a future-robust solution. This method is based on the five phases of the Scenario-Management (see figures 4, 8).

When developing the operational structure, the design engineer can realize a function using solutions approaches or solution elements (modules). A function called "identify client" e.g. could be realized by the solution approaches "IO card", "voice recognition" or "retina scanning". The solution approach "IO card" contains e.g. the solution elements "EG card" and a choice of credit cards. The final solution should be fixed with respect to the development potentials and future prospects of the chosen technologies.

After setting up the operational hierarchy, the technologies used for solution approaches will be identified. Every solution approach can be related to at least one technology.

In phase two, these technologies will be analysed and assessed. The aim is to find out key technologies and pace-setting technologies and to define the technological and market priorities. The result of technology assessment is a ranking of technologies and, thereby, of solution approaches.

Scenario-management during the early stages of product development 167

In phase three the scenario-planning is applicated, which makes it possible to consider possible developments of technologies and their environments. Usually no more than three scenarios free of contradictions will be set up. They should be as different as possible. For these scenarios, chances and risks for the examined technologies are found out and will serve to derive environmental requirements towards these technologies. E.g. the energy demand of a technology can be both chance and risk. If a scenario forecasts the tendency to a rigorous increase of energy costs, the dominating requirement will be the one of having a low-energy technology.

In phase four each of the technologies identified and analysed during the second phase will now be assessed with regard to the requirements.

In the fifth phase - technology selection the solution approach or element to be realized is defined. The results of the third and fourth step are weighted and combined. Then technology combinations for possible solution approaches are found out. Thus, the design engineer has got a basis for selecting the solution approach. Not only the present state but also their potentials and future prospects are shown.

2.3 Development of future-robust product visions and technology visions

Visions are of greater importance for product development. They could be compared with the north star: It is not the destination of the voyage, but it leads your way in the long term /11/.

We understand product visions as desirable future market performances. Technology visions are of importance for the success of an enterprise as well. New technical solutions can only be developed if, at first images of the possible futre are created in the minds of engineers, managers, employees and customers. Technology visions are essential first to support the search for new technological knowledge and technological solutions and second to call for innovative entrepreneurs /12/. There are two ways to use technology visions in enterprises:

Many products and technologies commonly known today where first thought of as visions. Later their "creators" further developed them to competitive articles. A excellent example is the "vision of electronic data-processing" by Herman Hollerith. The young mining-engineer created this technology vision during the long-winded manual evaluation of the US-American census in 1880, which took more than seven and a half years. He constructed a machine, which punched holes into cardboard according to incoming data. With sensors these punch cards are then sorted and easily counted by electronic equipment. 1896 he founded the TABULATING MACHINE COMPANY, the forerunner of IBM, the most powerful computer enterprise in the world today.

The development of future-robust visions can be combined with the method of stakeholder-scanning /1, 13, 14, 15/. We present a procedure of six steps closely modelled to the five phases of Scenario-Management (see figure 4). This method helps to develop visions for enterprises as well as product and technology visions (figure 9).

Stakeholder Scanning (phase 1 Scenario-Management)

The first step is to identify groups or associations which may influence a certain enterprise, product or technology or its environment. These groups are called stakeholders. Besides shareholders, customers, employees it is important to take groups into account that appear to have no influence at first e.g. environmental groups, consumer associations.

Stakeholder Assessment (still phase 1)

All identified stakeholders will assessed by means of their short and long time targets and of their influential power (e.g. number of members, political lobbies, financial power).

1 Stakeholder Scanning

Identification of influential groups with a possible impact on the future development of a product or technology.

2 Stakeholder Assessment

Assessment of identified stakeholder by means of goals and power.

3 Creation of Stakeholder Scenarios

Set up of the Stakeholder Field and definition of its environment as Scenario Field. Development of extreme

4 Effects Analysis

Diversion of system caused, scenario specific demands on product and technology visions

5 Set up of scenario specific visions

Diversion of specific product and technology visions from specific demands of single scenarios.

6 Set up of future robust visions

Integration of specific visions within future-robust product and technology visions

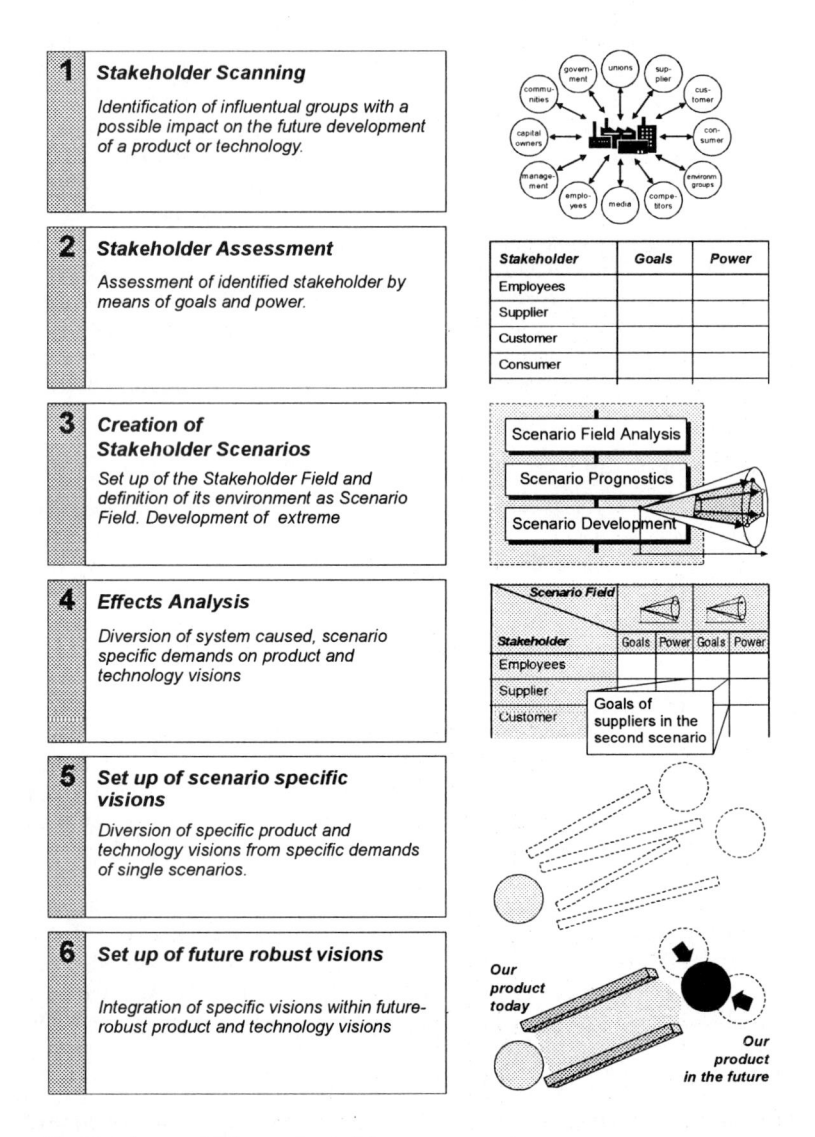

Figure 9 Development future-robust visions

Creation of stakeholder scenarios (phases 2 to 4)
The environment of the enterprise, product or technology will be defined as scenario-field. It includes all stakeholders which can't be influenced by the decisions-makers. According to phases 2 to 4 of the Scenario-Management, possible future developments of the scenario-field will be described as extreme scenarios.

Effects Analysis (phase 5)

The scenario-transfers start with the evaluation of consequences for goals and power of the individual stakeholder derived from the scenarios. The results are called "scenario specific demands" on the visions, e.g. the vision of the product "automobile" changes caused by the growing influence and power of environmental groups for an ecological orientated vision of individual local traffic.

Creation of scenario specific visions

In this step all scenario specific demands of one single scenario will be grouped for a scenario specific product or technology vision. If the decision-makers rely on theses visions they imply and accept a certain scenario as a future fact. Even though scenario specific visons are not an integral part of the strategy of an enterprise they are useful for a much better understanding of future opportunities and for a quicker response to future changes.

Creation of future-robust visions

In this last step the scenario specific visions will be integrated for one single future-robust product or technology vision. With this vision the enterprise tries to deal with as many as possible developments within its environment. The advantage of such vision is that it does not have to be altered as many times as a traditional vision has to.

3 REFERENCES

/1/ **Gausemeier**, J.; **Fink**, A.; **Schlake**, O. (1995) Szenario-Management – Planen und Führen mit Szenarien; München, Wien : Carl Hanser Verlag.
/2/ **Hamel**, G.; **Prahalad**, C.K. (1995) Wettlauf um die Zukunft – Wie sie mit bahnbrechenden Strategien die Kontrolle über ihre Branche gewinnen und die Märkte von morgen schaffen : Wien, Ueberreuther.
/3/ **Reibnitz**, U. von (1991) Szenario-Technik, Wiesbaden : Gabler.
/4/ **Meyer-Schönherr**, M. (1992) Szenario-Technik als Instrument der strategischen Planung; Ludwigsburg, Berlin : Verlag Wissenschaft & Praxis.
/5/ **Dörner**, D.(1992) Die Logik des Mißlingens – Strategisches Denken in komplexen Situationen; Reinbek bei Hamburg; Rowohlt.
/6/ **Ulrich**, H.; **Probst**, G.J.B. (1991) Anleitung zum ganzheitlichen Denken und Handeln : Ein Brevier für Führungskräfte – 3. erw. Aufl. – Bern; Stuttgart: Haupt.
/7/ **Senge**, P. (1990) The Fifth Discipline – The Art and Practice of the Learning Organization; New York; Doubleday/Currency.
/8/ **Gausemeier**, J.; **Paul**, M. (1994) Szenariobasiertes Konzipieren – Der Blick in die Zukunft der Technologie beim Konzipieren ; International Conference on Computer Integrated Manufacturing, Zakopane 10.-13.5.94, Mechanika Z. 118, Gleiwitz.
/9/ **Krubasik**, E.G. (19982) Strategische Waffe : Wirtschaftswoche, 36. Jg., No. 25, 28-33.
/10/ **Servatius**, H.-G. (1985) Methodik des strategischen Technologie-Managements – Grundlagen für erfolgreiche Innovationen – 2., unveränd. Aufl. – Berlin; Erich Schmidt Verlag.
/11/ **Hinterhuber**, H.A. (1990) Wettbewerbsstrategie : 2., völlig neu bearb. Aufl. – Berlin, New York; de Gruyter.
/12/ **Koolmann**, S. (1992) Leitbilder der Technikentwicklung : Das Beispiel des Automobilbaus – Frankfurt/Main; New York : Campus Verlag.
/13/ **Freeman**, R.E. (1994) Strategic Management : A Stakeholder Approach – Marshfield; London; Melbourne; Wellington: Pitman.
/14/ **Scholz**, C. (1987) Strategisches Management; Berlin, New York; de Gruyter.
/15/ **Gausemeier**, J.; **Fink**, A.; **Schlake**, O. (1995) Entwicklung zukunftsrobuster Leitbilder durch Stakeholder-Szenarien; in: io management, 10/95

15

The FABERCOAT Decision Support System: a design tool for plasma sprayed coatings

M. Foy, M. Marchese and G. Jacucci
Università di Trento, Dipartimento di Informatica e Studi Aziendali
Laboratorio di Ingegneria Informatica, Via Zeni 8, Rovereto (TN), Italy
Ph: +39 - 464 - 443132, Fax: +39 - 464 - 443 141
e-mail: marchese@lii.unitn.it

Abstract

The FABERCOAT Decision Support System (DSS) is a design tool for the production of plasma-sprayed coatings developed in a CEC project among European universities, laboratories and manufacturing industries of plasma-sprayed coatings. Specifically the FABERCOAT System utilizes a genetic algorithm to find a set of spray settings which can be used to achieve one particular set of desired final (and measurable) properties of the coating. At present, these desired outcomes are based on specific industrial requirements based on total porosity demands and most frequent porosity aspect ratio. The FABERCOAT System includes a model for the plasma spray manufacturing process developed and validated within the CEC project. The model has been used to "harvest" the information gained in the extensive production and characterization programme of the CEC project and has provided an innovative way of feeding back this information to the end-users (mainly technicians at the production laboratories).

Keywords

Decision support systems, expert systems, genetic algorithms, coating manufacturing

1. INTRODUCTION

In the following the concepts and the implementation of a prototype Decision Support System (DSS) for the manufacturing of plasma-sprayed coatings named FABERCOAT will be presented and discussed. The name is taken directly from the name of the European research project (Fabercoat, 1995) under the BRITE/EURAM programme on the characterisation and modelling of the manufacturing process of plasma-sprayed coatings.

Briefly, the FABERCOAT System can be used to find a set of spray settings in order to achieve one particular set of desired final (and measurable) properties of a coating. The system is based on a model for the deposition process that provides microstructural information (such as total porosity and/or porosity aspect ratio distributions) that are needed for the evaluation of a coating's effective properties.

The aims of the FABERCOAT System are:
- to simplify the use of deposition models in order to simulate the spraying process in a fast and effective way;
- to use the information of theoretical and experimental research in order to provide computerised decision support to people who perform plasma spraying manufacturing.

The final aim is to allow operators to specify what coating characteristics are desirable (porosity/temperature profile/residual stress profile), and ask the DSS to give suggestions on how to configure the spraying operation.

Section 2 focuses on the definition of the concepts and the user requirements for the FABERCOAT System as they emerged in the course of the project. Section 3 is centred on the description of the implemented prototype and on it's application in specific cases.

2. FABERCOAT SYSTEM CONCEPTS

The FABERCOAT System is centred around a model for the plasma spray manufacturing process developed and validated within the FABERCOAT project (Fabercoat, 1995). A deposition model has been developed and used to "harvest" the information gained in the extensive production and characterisation programme of the project and has provided an innovative way of feeding back this information to the end-users (engineers and technicians at the production laboratories).

Specifically the FABERCOAT System consists of a deposition model for the plasma spray manufacturing process, together with two components: (1) a graphic user interface (GUI) to interface/communicate between operators and the model; (2) an expert system component that co-ordinates the execution of the model.

The expert system component is a so called a "front-end system" (O'Keefe, 1986). In fact, in our system, this component acts as a "front-end" for defining a numerical simulation which is subsequently run on its own. The solution space of a great number of numerical simulations are automatically analysed and searched for the desired properties. The expert system component is used:
- to combine project information and knowledge into one tool;
- to allow a user to perform intelligent and automatic searches of the spray parameter space;
- to assist final users in performing reliable correlations between initial spray parameters and final coating properties.

The overall scheme of the FABERCOAT System is summarised in Figure 1.

2.1 Basis for the Development of the FABERCOAT System

Before describing the FABERCOAT System functionality, we recall here the main results of the investigation that have been relevant in the definition of the FABERCOAT System.

In regard to the experimental investigations, the FABERCOAT System relied on the following results (Fabercoat, 1995):
i) collection of experimental data on the plasma spray manufacturing processes at the three different length scales of the process: a) data on the single droplet of ceramic particle (in flight and after impact on the substrate), b) data on the mechanism for the assembly of droplets, c) information on the macroscopic properties of the final coatings;
ii) development and use of new techniques to measure plasma-particle interactions to provide the necessary data to the model of the deposition process;

FABERCOAT SYSTEM CONCEPT

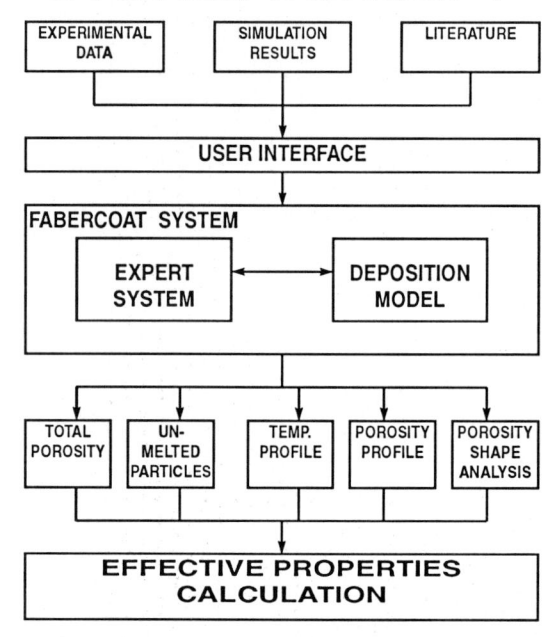

```
  ┌──────────────┐  ┌──────────────┐  ┌──────────────┐
  │ EXPERIMENTAL │  │  SIMULATION  │  │  LITERATURE  │
  │     DATA     │  │   RESULTS    │  │              │
  └──────────────┘  └──────────────┘  └──────────────┘
                           │
  ┌─────────────────────────────────────────────────┐
  │                 USER INTERFACE                  │
  └─────────────────────────────────────────────────┘
  ┌─────────────────────────────────────────────────┐
  │ FABERCOAT SYSTEM                                │
  │   ┌──────────────┐         ┌──────────────┐    │
  │   │    EXPERT    │◄───────►│  DEPOSITION  │    │
  │   │    SYSTEM    │         │    MODEL     │    │
  │   └──────────────┘         └──────────────┘    │
  └─────────────────────────────────────────────────┘
       │         │         │         │         │
  ┌────────┐┌────────┐┌────────┐┌────────┐┌────────┐
  │ TOTAL  ││  UN-   ││ TEMP.  ││POROSITY││POROSITY│
  │POROSITY││ MELTED ││PROFILE ││PROFILE ││ SHAPE  │
  │        ││PARTICLES││        ││        ││ANALYSIS│
  └────────┘└────────┘└────────┘└────────┘└────────┘
                           │
  ┌─────────────────────────────────────────────────┐
  │         EFFECTIVE PROPERTIES                    │
  │             CALCULATION                         │
  └─────────────────────────────────────────────────┘
```

Figure 1: FABERCOAT System Concept

iii) development of image analysis tools to characterise coating microstructures (porosity, crack network fractal dimension);
iv) X-ray analysis to measure phase composition and residual stresses.

In regard to the modelling and simulation investigations, the following results have been used in the Decision Support System (DSS):
1) a deposition model (2-d) for the plasma spray manufacture of ceramic coatings;
2) theoretical and numerical models for the description of the mechanical and the thermal material behaviour of ceramic layers;
3) a residual stress model to estimate strain and stresses developed in the coating during fabrication.

The deposition model has been identified as the central component in order to fulfil the DSS goals, namely to try to correlate process parameters and final coating characteristics. The main capabilities of the deposition model (Cirolini et al, 1991 and 1995) consist of predicting a number of final coating microstructural characteristics (such as total porosity, porosity profile and shape distributions, and temperature distributions) that can be used to determine, in a second stage, final coating mechanical and thermal properties (among others, elasticity modulus and thermal conductivity). In the following we review briefly the input data needed for the deposition model; in turn they represent the necessary input data for the FABERCOAT System:

(1) correlation between torch settings and plasma and particle properties, i.e.:
- data on plasma gas (mainly temperature) as a function of torch settings and gun distance;
- data on in-flight particle characteristics such as particle radius, velocity and temperature distributions as a function of torch design, settings and distance.

Such detailed data are complex to measure, but have been obtained in a complete form for one specific torch in the course of the Fabercoat project (Fabercoat, 1995) as well as in other research work (Fauchais, 1989). It is useful to underline that significant effort has been applied in the last 10 years in the plasma-spray community to model the plasma gas and the transport of particles in it (Vardelle, 1986). By now several models have been developed and validated for some plasma torches, and are commercially available. They could be easily integrated in the FABERCOAT System to extend the needed correlation between torch settings and plasma and particle properties for other apparatus. For this prototype version of the deposition model we have limited our simulation to consider only the acquired experimental data for one torch.

(2) Correlation between in-flight and after-impact particle properties, i.e.:
- data on the behaviour of particles after impact as a function of initial conditions.

Theoretical and numerical models of the impact process have been developed in the past 5 years in order to obtain information on the particle flattening degree as a function of a particle's initial conditions and material properties (Trapaga et al, 1991 and 1992, Fukai, et al, 1993, Bertagnolli et al, 1995). The available information has been integrated in the present version of the FABERCOAT System.

2.2 FABERCOAT System functionality

The FABERCOAT System provides multiple levels of functionality so different types of deposition model simulations can be easily carried out, and results can be examined as to their quality.

Multiple simulation executions

The first necessary functionality allows the user to execute multiple simulations automatically. Users often would like to run multiple simulations (possibly varying one spray parameter to see how that effects one or more of the overall characteristics of the final coating), but this task can often be laborious, as the user has to continuously monitor the results, change input files, and then reiterate simulations. The FABERCOAT System alleviates this difficulty by allowing the user to quickly specify a multiple simulation cycle, and have this cycle automatically performed without user intervention.

Expert system component for intelligent search

Furthermore, beyond this simple iterative simulation functionality, this system integrates an expert system component, which intelligently performs more complicated simulation cycles. This expert system component intelligently directs multiple simulation cycles, specifically, it performs intelligent search. Explicitly, the component executes a search within the space of spray parameters, with the goal to locate a particular set of coating characteristics. For example, a user is able to specify a certain set of desired outputs (such as a desired total porosity of 10%), and then direct the system to find a set of spray parameters (within certain ranges) that most closely matches the set of desired outputs. The system then responds as an expert would, by giving advice about how to set the spray conditions (gun settings, substrate conditions, etc.) in order to achieve the desired properties in the final coating.

Simulation history archiving

Within these iterative simulation cycles (either directly or through search), the system is also be able to archive all results from all simulations executed. For example, when performing a search for a set of spray parameters, the system maintains a "history file" containing a list of all the simulations which have been performed. This is useful for at least three reasons. First, this permits the user to go back and browse the simulation history, so that trends and correlations between model inputs and model outputs (i.e., between spray parameters and final coating characteristics) can be discovered. Second, by maintaining a simulation history, if the search were to terminate before finishing, all results would not be lost, and it may be possible to start the search off from where it left off. This is an important issue to some users because the deposition model can take considerable computation time, and therefore, every simulation is valued and they may want to look at the results from all simulations that have been performed. Lastly, a large set of simulations results can be useful to model developers because it can help them detect bugs in the simulation model computer code, so that these bugs can be fixed.

N-best members produced during search

In addition to archiving a complete history, when the system is performing a search, a set of n-best solutions is maintained, and available for review at the end of the search. That is, the n-best set of spray parameters which give results most closely matching those specified as most desirable by the user are recorded. Therefore, at the end of a search, not only is the best set of spray parameters presented, but the n-best (e.g., 20-best) that have been encountered during the entire search. This is useful so the user can see a number of spray settings that will match her needs, rather than getting only one "answer", and also allow her to see how much "spread" (i.e., separation or distance) there is between good sets of spray settings.

Graphical user interface

All of these functionalities are contained within a graphical user interface (GUI) that easily allows the user to execute these tasks. In particular, the user is able to specify parameters and preferences within the GUI (using type-in boxes, etc.), and then carry out actions with the use of buttons.

Overall, these functionalities should provide users useful information and decision support by permitting them to structure simulations and searches to assist in the understanding of the deposition model. Therefore users should gain a better understanding of the mechanics of the deposition model, and plasma spray processes, because trends in the model, and correlations between inputs and outputs, will be much easier to observe.

2.3 The deposition model parameters

The deposition program has a long list of input parameters that can be divided in three main groups: "operator" - "physical" - "computational".

The "operator" group includes, for example, the gun distance during the spray, the ceramic powder granulometry, and the final thickness of the coating. These parameters are supposed to be directly established by users and users should specify them in standard "engineering" units.

The "physical" group includes, for example, the particle's velocity and temperature at splat time. These values describe the current state of the physical process that is being simulated; they refer to quantities not directly established in real life but that depend on those in the first category by known relations, either analytical, experimental or produced by other simulations.

The "computational" group includes parameters related to the inner workings of the deposition program like grid sizes, random generator seeds and so on. These parameters affect the "precision" and the run time of the program and may depend on those in the "operator" and "physical" classes. For example, the grid size depends on the size of the particles and on the final thickness of the coating (for more information see Cirolini et al, 1991 and 1995).

Originally the second class (i.e., the "physical" group) was the most numerous one because the deposition program was correlating basic physical quantities. The availability of experimental measurements (in-flight velocity and temperature as a function of torch settings) and the results of other simulations (particle splashing and curling for different temperature and velocity combinations) allows for filling the gap between real life parameters and physical parameters thus increasing the size of the "operator" class and reducing the "physical" one.

2.4 Selection of the searchable spray parameters

A subset of the above deposition model input parameters were chosen for inclusion in the set of spray parameters which may be "searched over" by the expert system component. This selection focused on the most important operator-controlled spray parameters which have the most important effect on the final coating characteristics.

Functionality has been included which allows the user to choose which of these parameters should be search over (including what range they should be searched) and which should not be searched, but left at particular constant values. This allows the user the flexibility to decide what parameters are available to be changed in the spraying process, and which ones should remain fixed due to some limitation of the spraying process. This is an important point because the final goal is to make the program usable by operators, not only by physicists or programmers.

The searchable spray parameters, with their corresponding typical values or ranges, are:
GUNDISTA	0.075 - 0.15 m	distance between gun and substrate
GUNVELOC	0.02 - 0.25 m/sec	velocity of the gun
PARTDIAM	20 - 45 microns	diameter of particles
PARTDIAM_DEV	1-5 microns	standard deviation of diameter of particles
SPRAYRAT	0.05 - 0.4 g/sec	powder feed rate
SUBSROUG	0 - 50 microns	substrate roughness
SUBSTEMP	300 - 800 °K	substrate temperature

2.5 Selection of the objective function

The objective function is the physical property of the coating the operator wants to control, i.e. the quantity in the output of the deposition program for which the operator wants to choose a desired value and have the Expert System component search for the unknown parameters.

The outputs of the deposition program at present include: a simulated micrograph of a section of the coating, the intermediate and final temperature profile (through the coating thickness), the intermediate and final porosity profile, the final pore aspect ratio distribution and the final unmelted particle percentage.

The simulated section of the coating can be compared to a microscope image of a section of the coating, and allows one to see pores and individual splats. It gives an overall view of the coating, but its information is hard to translate into a quantitative one, and it is better used as a visual check to indicate if something wrong happened in the simulation.

The through-thickness temperature profile is available at different times during the deposition, the last one being especially important for the successive development of residual stresses during cooling. Temperature is usually fairly constant in the substrate, with the exception of a sharp change at interface and quite a big gradient in the coating. The information contained in the temperature profile could be reasonably summarised in one number by taking the gradient in the coating, or measuring the temperature step at the interface.

The porosity profile is probably the most interesting result. Porosity influences the mechanical properties of the coating and can be correlated with measurements done on real coatings. As with the temperature profile there is the problem of compressing its information to just one number. A reasonable first approximation is to compute from the porosity profile the total mean porosity of the coating section.

Another output parameter is the distribution of the aspect ratio of the pores: its importance relies on its strong influence on the mechanical and thermal effective properties of the coating (Rickerby et al, 1987 and McPherson, 1989). The typical feature of these types of distributions are the abundance of thin, elongated pores (like cracks) with an aspect ratio less than 0.3 in the simulated coating. These cracks are found to have the greatest effect on the mechanical and thermal effective properties of the ceramic.

The last output parameter is the unmelted particle percentage. Apart from correlating strongly with the final mean porosity present in the coating, the amount of unmelted material needs to be controlled during fabrication to assure the overall quality of the coating.

In the first development phase of the FABERCOAT System, it was decided to use the total mean porosity as the objective function for its smooth behaviour. The other outputs have been provided in this phase as a check on the correctness of the results from the prototype system.

3. FABERCOAT SYSTEM PROTOTYPE

A prototype FABERCOAT System that utilises the concepts presented in the last sections has been developed and is under use in the plasma-spray laboratory of project's partners.

3.1 Search method

As described in a previous section presenting the FABERCOAT System functionalities, the system has the capability to perform intelligent searches of the spray parameter space (i.e., a search through all possible sets of spray parameters). This is directed by an expert system component. Due to the requirements of this component, a specialised search method was selected to play a part in the expert system component.

In this particular domain of ceramic thermal coating processes, where knowledge of the domain is hard to codify (i.e., 'rules of thumb' are vague and difficult to construct), the selection of a search method for the expert system component is a good choice. This is due to the fact that, in general, search methods do not rely on 'rules of thumb', rather, rules are not required and an intelligent search algorithm can actually facilitate the user in identifying 'rules of thumb'.

The selection of the actual search method was made among the following possible methods: hill-climbing, simulated annealing, and genetic algorithms. In the end, genetic algorithms (GAs) were selected as the most desirable method because:
(a) they can perform unbiased search,
(b) they make no assumptions about the search space (i.e., the search space does not have to be smooth or regular),
(c) they carry out a more effective search of an irregular, multi-dimensional space because they search from a population of points rather than a single point,
(d) their search is not random, but intelligent (they utilise operators which are patterned after natural genetics), and
(e) they have been shown effective at finding optimal or near-optimal solutions to dynamic real-world problems (Holland, 1975 and Goldberg, 1989).

For a complete description of GAs, how they function, etc., refer to Holland (1975) and Goldberg (1989).

3.2 Linking the deposition model and the genetic algorithm (GA)

To allow the GA to search the space of the deposition model spray parameters, the deposition model is linked to the GA, and the GA uses the deposition model as the evaluation function. Therefore, whenever the GA wants to evaluate the "worth/fitness" of a set of spray parameters, the deposition model is called, and the final outcome is returned to the GA so that a fitness can be computed. This new hybridised-system component is called the "deposition model-GA" or the DEP-GA. Figure 2 illustrates how the deposition model (DEP model) and GA are linked to form the DEP-GA component.

For the DEP-GA to find a near-optimal spray parameter set for given desired final coating characteristics, it goes through three primary steps.

First, the DEP model and the GA are initialised. The initialisation of the GA involves establishing an initial-random population of spray parameter sets which includes only the spray parameters that are being searched (these are called population members, and are represented within the GA as bit strings).

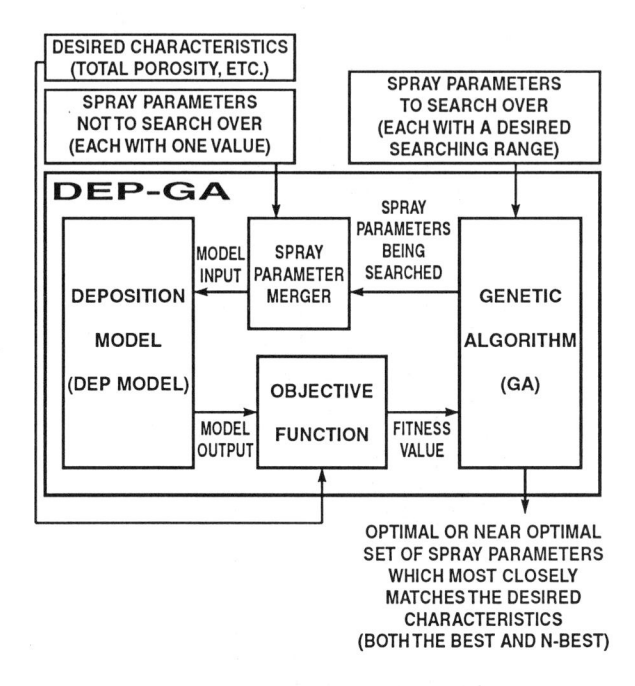

Figure 2: Structure of the DEP-GA

The second main step in the DEP-GA is the fitness computation. This involves taking each GA population member (one set of searchable spray parameters) filling in the spray parameters which are not being searched with base values, and executing a DEP model simulation with this full set of spray parameters. The outputs that come from this execution are then passed through the objective function (which can be structured by the user - see Section 2.5), matched against the desired characteristics, and finally a fitness value is produced which is then returned to the GA. This fitness evaluation step is executed many times because new population members are generated by the GA at the end of each generation cycle.

The last main step is the evolution of the GA population. This involves manipulations on the bit strings (i.e., operations on the population members). The three manipulations, or operators, used in the DEP-GA are reproduction, crossover, and mutation (descriptions of these operators can be found in Goldberg, 1989). GA evolution is usually continued until the GA has converged on an optimal or near-optimal set of spray parameters which best matches the desired characteristics (total porosity, etc.) specified by the user. Lastly, the best and n-best set of spray parameters are produced for inspection by the user.

3.3 DEP-GA performance

Multiple sets of DEP-GA test runs have been done within the FABERCOAT System. As a starting point for evaluating the effectiveness of the DEP-GA, we chose an objective function which only included the total porosity (even though we could have also used the final coating temperature or the unmelted particle percentage as well - see Section 2.5 for more information). Additionally, we limited the spray parameters to search over to three: GUNDISTA, PARTDIAM, and SPRAYRAT (as described in Section 2.4). The below sub-sections describe the results obtained within this scenario.

GA convergence on an optimal set of spray parameters
The DEP-GA has performed similarly to most GAs, in that during a run, the average fitness of the populations, over time, has increase. That is, the members in the later populations have converged on maximum members in the space (i.e., on optimal sets of spray parameters). Figure 3 shows that the DEP-GA does in fact produce this typical performance. Specifically, this figure illustrates the evolution of fitness values coming from a DEP-GA run where the user has specified to find a set of spray parameters which will give 10% overall porosity. Note that as generations proceed, the GA weeds out members which do not produce the desired results (i.e., members that have low fitness) and focuses on members that produce coatings near the desired 10% porosity (i.e., members that have high fitness). Also note that by the fifth generation, the GA has probably already converged on the optimal spray parameters because the maximum fitness does not increase after the fifth generation, and because the average fitness does not make any significant upward trends after the tenth generation (in fact our full space investigations have shown that the maximum member in the fifth generation was the best member in the entire space). This behaviour supports the hypothesis that the GA run need not be run for 60 generations (as shown in Figure 3), but could be terminated between the fifth and tenth generation and still obtain very good results. Preliminary statistical runs have confirmed this hypothesis.

Locating n-best sets of spray parameters
The above discussed performance is a positive indictor of how well the GA has been able to find a good set of spray parameters that match the user's specified needs (and how well we expect it to work in the future under different objective functions and search parameter scenarios).

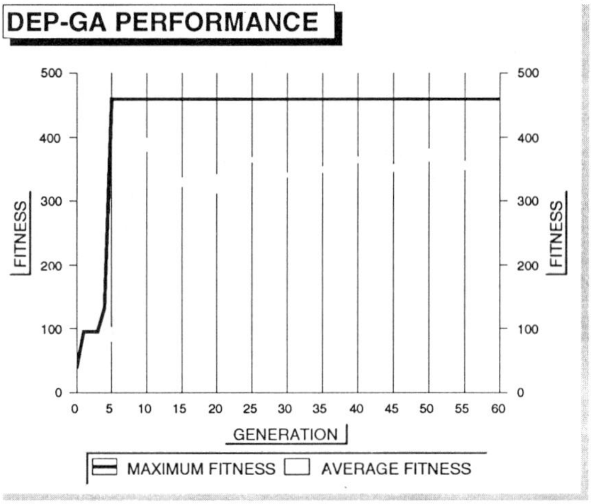
DEP-GA PERFORMANCE
FITNESS
500
400
300
200
100
0
0 5 10 15 20 25 30 35 40 45 50 55 60
GENERATION
MAXIMUM FITNESS
AVERAGE FITNESS

Figure 3: DEP-GA Performance

For more specific results a number of runs have been performed on the chosen subset of the parameter's space, namely the subset referring to three spray parameters: GUNDISTA, PARTDIAM and SPRAYRAT. For this case, a complete run of the deposition model has been made in order to visualise the whole space (i.e., all possible combinations of these three spray settings within reasonable ranges have been run, and the results have been combined). Figure 4 displays the total porosity results, as produced by the deposition model, as a function of SPRAYRAT (different sections, range 0.04 - 0.4 g/sec), GUNDISTA (x-axis, range 0.075 - 0.15 m) and PARTDIAM (y-axis, range 20 - 45 microns). Contour lines refers to different porosity levels ranging from 5 to 19 %; each shade indicates a 2% interval of porosity. Some important features of this space can be underlined: (a) porosity tends to increase with increasing gun distance, and tends to decrease with decreasing spray rate, (b) porosity distribution is quite similar for most spray rate values, and (c) there exists different disconnected regions where total porosity values are equal (i.e., the model predicts different disjoint sets of parameters that result in the same value of final total porosity) (note for example the tile in the middle-right of Figure 4, there is a 9-11% porosity region in the upper-right of this tile, and also in the lower part of this tile, and these regions are separated by a region with 11-13% porosity).

The black dots in Figure 4 indicate the n-best members (in this case, 20-best members, i.e., the best, 2nd best, 3rd best, etc.) found by the DEP-GA run described above (i.e., targeting 10% total porosity, and searching over only 3 spray parameters). As can be seen, because a diverse set of n-best members has been produced, the GA has not searched just one part of the space, but has been able to search different parts of the space. Additionally, because this n-best tracking has been included, the GA can indicate more than just one set of parameters that will give the desired porosity, but a number of sets of parameters (which may be spread out over the whole space) that will give results very near to the desired porosity.

This allows the expert system component to provide a widely distributed set of near-optimal spray parameter sets, thereby giving the user the opportunity to examine many different solutions, and focus on the one that most closely matches her needs (e.g., the user may like to avoid solutions which give particular types of porosity shapes, or a particular percentage of unmelted particles).

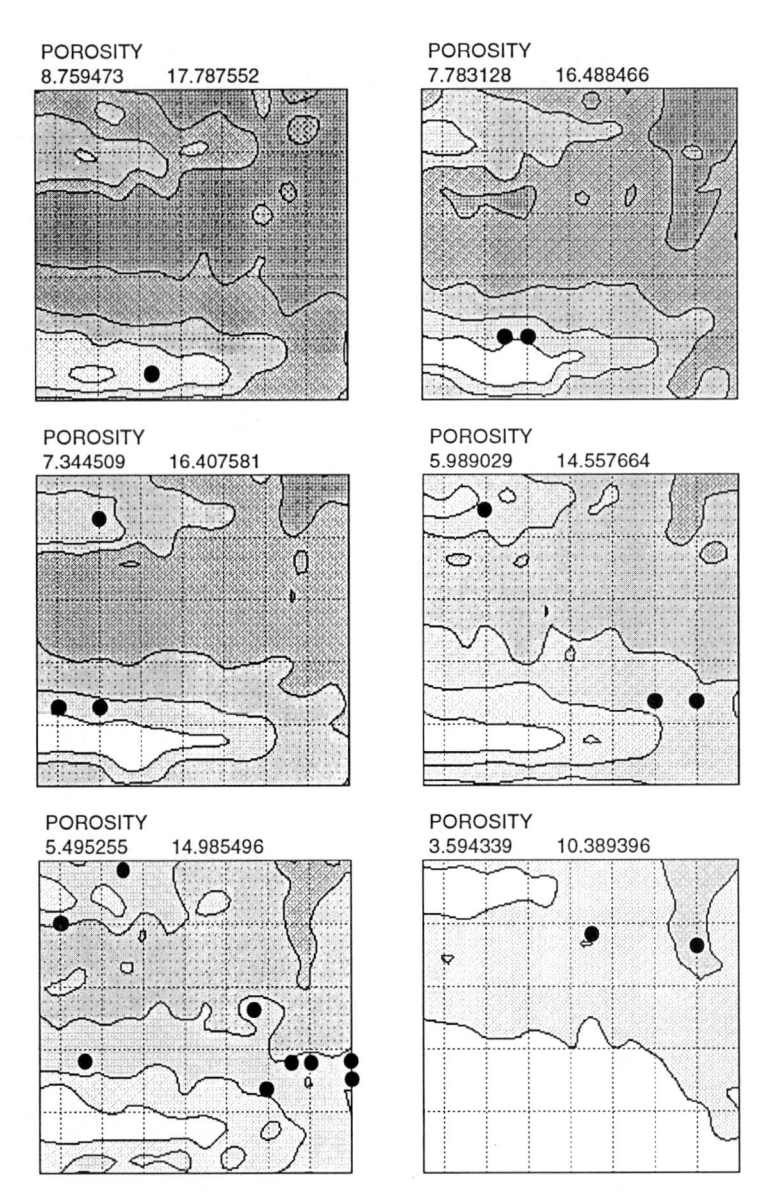

Figure 4: Sections of the three spray parameter's space indicating total porosity levels - GUNDISTA (x-axis), PARTDIAM (Y-axis), SPRAYRAT (different sections). Contour lines refer to porosity between 5 and 19 %, each shade indicating a 2% interval of porosity. Black dots refer to the 20-best members found by the DEP-GA when searching for 10% porosity.

Overall, taking into consideration the results from the above described scenario, and additionally other test runs, we have found the FABERCOAT System with the DEP-GA to perform very well. Furthermore, we expect that when a real-user utilises the DEP-GA, it will produce good sets of possible spray parameters which the operator can then use to determine the actual spray parameters to use in a real spraying process. This is provided by the searching capabilities of the GA.

3.4 Computer aspects

Graphical user interface
The FABERCOAT System integrates a graphical user interface (GUI) around the deposition model and genetic algorithm, thereby making the system easier to use. This GUI is implemented using a public domain tool kit called SUIT. One of the largest advantages of SUIT is that it makes the FABERCOAT System portable, that is, it will run on 4 different computer platforms: UNIX X-Window machines, IBM-PC DOS machines, IBM-PC MS-Windows machines, and Macintosh machines (as long as there is a deposition model compiled on the platform with the FABERCOAT System).

Run time
At present, it is most desirable to run the FABERCOAT System on a fast UNIX X-Window workstation because these machines have the greatest ability to run the deposition model in a reasonable amount of time (1 to 2 minutes for one simulation; about 10 hours for a full DEP-GA run), but in the future, when all machines become faster, it should even be reasonable to run the DEP-GA component on an PC type machine.

4. CONCLUSIONS

The prototype FABERCOAT System has proved that it has the capability to provide decision support in the area of the manufacturing of plasma sprayed coating. At the present time the prototype DSS is under use in the plasma-spray laboratories of the partners involved in the project.

The FABERCOAT System utilises a genetic algorithm (GA) to find a set of spray settings which can be used to achieve one particular set of desired final (and measurable) properties of a coating. These desired outcomes can be based on total porosity and/or final temperature of the top of the coating and/or final unmelted particle percentage requirements. The FABERCOAT System is centred around a model for the plasma spray manufacturing process developed and validated under the CEC project. The model has been used to "harvest" the information gained in the extensive production and characterisation programme of the project and has provided an innovative way of feeding back this information to the end-users (mainly technicians at the production laboratories).

5. REFERENCES

Bertagnolli M., Marchese M. and Jacucci G., (1995) Thermomechanical simulation of the splashing of ceramic droplets on a rigid substrate. *Journal of Plasma Spray*, **4(1)**, 1--9.

Cirolini, S., Harding, J.H., and Jacucci, G. (1991) Computer simulation of plasma--sprayed coatings: I. Coating deposition model. *Surface and Coating Technology*, **48**, 137--145.

Cirolini S., Marchese M., Jacucci G. and Harding J. (1995) Modelling the deposition process of thermal barrier coatings. *Journal of Plasma Spray*, **4(1)**, 10--18.

Fabercoat Brite/Euram Project No.BREU-0418 (1995) *Fabercoat: Modeling and characterization of the manufacturing process of ceramic thermal barrier coatings*, Final Technical Report, European Union, Bruxelles.

Fauchais, P., Grimaud, A., Vardelle, A., and Vardelle, M. (1989) La projection par plasma: une revue. *Annales de Physique*, **14**, pp 261--310.

Fukai, J, Zhao Z., Poulikakos D., Megaridis M. and Miyatake O. (1993) Modeling of the deformation of a liquid droplet impinging upon a flat surface. *Phys. Fluids A*, **5(11)**, 2588-2599.

Goldberg, D.E. (1989) *Genetic Algorithms in Search, Optimization, and Machine Learning*. Addison-Wesley, Reading, MA, USA.

Holland, J.H. (1975) *Adaptation in Natural and Artificial Systems.* University of Michigan Press, Ann Arbor, MI,USA.

McPherson, R. (1989) A review of microstructure and properties of plasma sprayed ceramic coatings. *Surface and Coating Technology*, **39-40**, 173--181.

O'Keefe, R. (1986) Simulation and Expert Systems - A Taxonomy and Some Examples, *Simulation*, **46(1)**, 10-16.

Rickerby, D.S., Eckold, G., Scott, K.T., and Buckley-Golder, I.M., (1987) The interrelationship between internal stress, processing parameters and microstructure of physically vapor deposited and thermally sprayed coatings. *Thin Solid Films*, **154**, 125 --141.

Trapaga, G., and Szekely, J. (1991) Mathematical modeling of the isothermal impingement of liquid droplets in spraying processes. *Metallurgical Transactions B*, **22B**, 901--914.

Trapaga, G., Matthys, E.F.,Valencia, J.J., and Szekely, J.(1992) Fluid flow, heat transfer, and solidification of molten metal droplets impinging on substrates: comparison of numerical and experimental results. *Metallurgical Transactions B*, **23B**, 701--718.

Vardelle, A., Vardelle, M., and Fauchais, P. (1986) Les transfers de quantit\'e de mouvement et de chaleur plasma particules solides dans un plasma d' arc en estinction. *Revue Internationelle des Hautes Temperatures et Refractaires,* **23**, pp 69--85.

16

Modeling high precision assembly processes using discrete event simulation - an alternative solution for decision making in manufacturing

Il. Astinov
P. Hadjijski
Laboratory 'Simulation Modeling in Industry' (SMI)
MTF, TMMM, Technical University - Sofia (VMEI), 1756 Sofia, BULGARIA, Tel: (+359 2) 636 3784, FAX: (+359 2) 683 478,
E-mail: ila@bgtus4.vmei.acad.bg

Abstract
The paper presents a non-traditional application of discrete event simulation as a decision aiding tool for determining the most suitable manufacturing method to be applied in high precision assembly processes. The production of bearing components and their assembly is being modeled using discrete event simulation techniques. Three manufacturing approaches are being simulated - random part selection assembly, selective assembly and compensator feedback manufacturing. Statistics are collected and compared on the number of bearing components and failed bearings expected to be produced in order to manufacture the required quantity of bearings meeting the quality criteria. The major conclusion that can be reported is that discrete event simulation provides the manufacturing decision makers with more realistic and precise data aiding the process of meeting quality criteria in high precision assembly operations.

Keywords
Assembly, quality, modeling, simulation, decision making

1. INTRODUCTION

The trend towards increasing the quality of products is common to manufacturing. Nowadays quality demands increase much faster due to technology development, market competition, human and ecological factors. Figure 1 illustrates a viewpoint of a production process dividing the latter in two stages - manufacturing of components and assembly of the final product. Each stage has specific characteristics and contributes to the quality of the end product depending on the process plan.

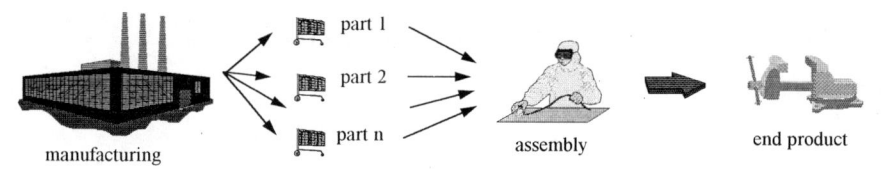

Figure 1 Stages in a production process.

In mechanical engineering dimensions accuracy is a major quality criteria, specially dimensions of parts or clearances which determine the functionality of the product. Provided the manufacturing stage is in a position to produce the parts of the product within the predefined tolerances, then the major contribution towards the quality of the dimensions comes from there. The assembly stage in this case is organized on the **random assembly method** where all parts included in the product are picked up from the batches at a random and assembled in the predefined in the process plan sequence.

If dimensions accuracy can not be met with the existing manufacturing infrastructure, there are two general solutions to the problem - the innovation of the manufacturing infrastructure with new, high-precision hardware or utilization of production methodologies which can guarantee the requested quality. The second approach is most suitable specially for cases, where clearance tolerance values are crucial. A typical example for products with increased quality requirements are ball bearings. They will be used further in this paper as a representative for products, requiring high precision manufacturing and assembly operations. One of the major quality criteria is the clearance between the balls (spheres) and the outer ring (Tc on Figure 2).

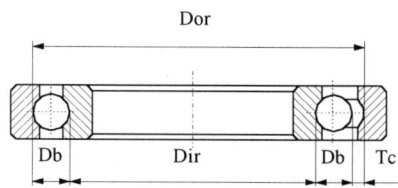

Figure 2 Ball bearing major dimensions and clearance.

In the treated case, the major contribution toward meeting the dimension quality comes from the assembly stage. Two general methods are currently utilized:

- **Selective assembly method**: parts are manufactured with large tolerance values. Subsequently the tolerance value of each part is divided in subintervals and parts are sorted

within these subintervals in groups according to their actual dimension. The effect is narrowing the tolerance of parts in each group. Upon assembly the final product is composed by parts originating from equivalent groups.
- **Compensator feedback manufacturing**: parts are picked up at a random from batches except one, called the compensator. In the case of bearing manufacturing the compensator is the set of balls. The product is assembled without the compensator. After measuring the subassembly, the required dimension and tolerance of the compensator is estimated. A feedback request is made for a compensator part having the estimated dimensions. The latter is picked-up from the appropriate group of premanufactured batches.

The general advantage of both methods is that they allow the production of parts with extended tolerances which decreases the accuracy and quality requirements of the manufacturing stage. However the obvious major disadvantages are:

- Additional operations in grouping within the selective assembly method for all parts and similar activity in the compensator feedback concerning the compensator part.
- The compensator feedback method requires permanent control on the amounts of compensators in all groups and as soon as these amounts pass beyond a predefined threshold, compensators with such dimensions should be manufactured. This means that the manufacturing facility must be able to produce in a reduced time compensators with very high requirements on dimension accuracy.
- Both methods in many cases need specialized manufacturing hardware typical for the mass production and because of that they are difficult to be utilized in small and medium sized enterprises (SME).

Decisions on which of these methods to utilize for achieving a high quality demand are difficult to make. The general constraint in any case is the accuracy that can be realistically achieved in the manufacturing stage.

2. PROBLEM DEFINITION

The problem treated in this paper, based on the presumptions stated in the Introduction is illustrated on Figure 3.

- given the accuracy that can be realistically achieved in the manufacturing stage of a specific high-precision product ;
- given the quantity of the latter that is to be produced.

⇒ which of the three assembly methods will be more suitable to implement?

Figure 3 Problem definition.

Existing solutions to this problem are generally based on analyzing the dimensional built-up of the product. For the case of ball bearings, similar to the one, given in Figure 2 the graphical outline of the dimensional built-up is shown on Figure 4.

(a) Bearing drawing (b) Dimensional built-up

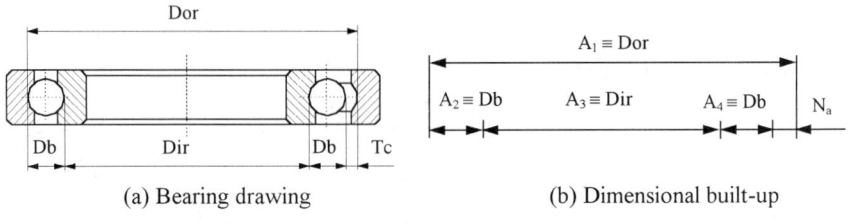

Figure 4 Dimensional built-up of ball bearing.

The formal expressions for calculating the clearance dimensions and tolerance value (Tc in Figure 4) have the following general form :

$$\left. \begin{array}{l} T_c = \sum_{i=1}^{n} T_{Ai} \\ es_{Na} = \sum_{i=1}^{k} esA_{sk} - \sum_{i=1}^{l} eiA_{tl} \\ ei_{Na} = \sum_{i=1}^{k} eiA_{sk} - \sum_{i=1}^{l} esA_{tl} \end{array} \right\}, \text{ where} \qquad (1)$$

- Tc   -   clearance tolerance ;
- T_{Ai} -   tolerances of the dimensions of part i ;
- Na   -   clearance dimension ;
- es_{Na} -   maximum value of the Na dimension ;
- ei_{Na} -   minimum value of the Na dimension ;
- A_{sk} -   dimension of part k leading to the increase of Na in the dimensional built-up ;
- A_{tl} -   dimension of part l leading to a decrease of Na in the dimensional built-up ;
- es   -   designates maximum values ;
- ei   -   designates minimum dimensions ;
- n   -   number of parts in the assembly ;
- k   -   number of parts increasing Na ;
- l   -   number of parts decreasing Na.

The number of groups in the selective assembly method and compensator feedback is estimated by expressions, having the following general form:

$$q = \frac{\sum_{i=1}^{n-1} TA_{ji}}{T_c}., \text{ where} \qquad (2)$$

- TA_{ji} -   tolerances of part i which can be realistically achieved ;
- q   -   number of groups.

Both expressions (1) and (2) are deterministic models for estimating the clearance value and tolerance of the assembled product. In real life dimensions of parts and clearance values vary from part to part and product to product, which determines them as having stochastic properties. Thus expressions (1) and (2) do not take into account such stochastic aspects. So in addition to the problem definition, an alternative solution is described which **considers the stochastic characteristics** in manufacturing high-precision products. A powerful methodology in modeling stochastic systems and applicable in this case is discrete event simulation - Astinov (1990), Pritsker (1986).

3. THE MODELING APPROACH

3.1 Modeling principles

Manufacturing stage

Dimension accuracy is the main point of interest in the current development. The process of achieving actual dimensions of parts in the manufacturing stage is determined by the values of the dimension and tolerances preassigned in the part design. An example of a preassigned dimension and tolerance field for the outer ring of a ball bearing is given on Figure 5.

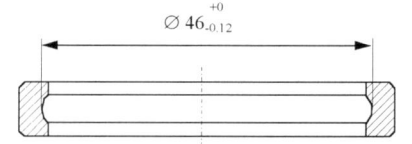

$\varnothing\ 46^{+0}_{-0.12}$

Figure 5 Typical preassigned dimension and tolerance for the outer ring of a ball bearing.

The actual dimension of each part is distributed within the preassigned tolerance field. This is illustrated in Figure 6, where the tolerance field and a typical distribution of actual dimensions are shown (Figure 6 (b)).

(a) Preassigned dimension and tolerance

(b) Typical dimension distribution

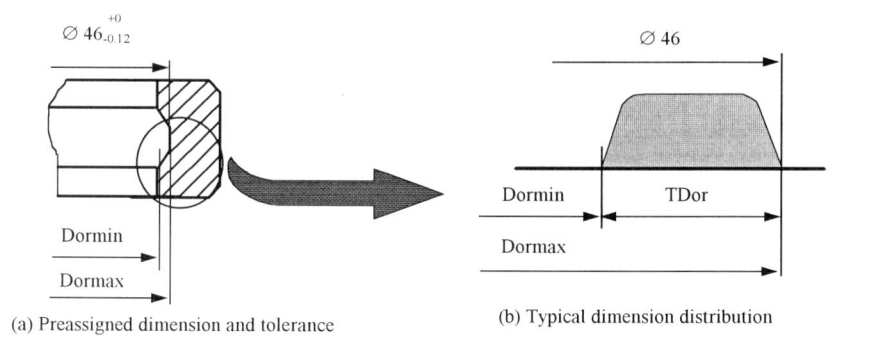

Figure 6 Tolerance field and typical dimension distribution.

The term 'distribution' in this case indicates the probability density function (PDF) of the dimension. The PDF could be a standard continuous distribution, such as Gauss, uniform or in very limited cases the triangular distribution, as well as a composite distribution including all three mentioned. The latter can be represented by an empirical PDF. In this case, the dimension of any part j of type i can be expressed with the following formal notation:

$Aij = ?[Di, \mathbf{pi}].$, where (3)

- Aij - the actual dimension of part j of the type i e.g. the actual dimension of the jth outer ring ;
- ?[...] - formal notation indicating that Aij has stochastic properties ;
- Di - the PDF type of the Aij. This can be a Gauss, uniform, triangular, empirical or other distribution ;
- **pi** - an array, containing the parameters of the PDF. In the case of a PDF equivalent to the Gauss distribution, $\mathbf{p}=[A_1 \equiv \mu, A_2 \equiv \sigma]$.

Actual values for Aij in expression (3) can be generated with random variate generators. Procedures for generating random variates distributed within a large variety of PDF-s are incorporated in every discrete-event simulation software (Astinov, 1990). The distribution type and parameters (i.e. Di and **pi**), most suitable as a model for the manufacturing process of each part can be obtained in one of the following ways:

- curve fitting techniques based on observation data such as χ^2, Kolmogorov-Smirnov or other goodness-of-fit tests (Bratley, Fox, Schrage, 1983);
- assume Gauss distribution with 6σ equal to the tolerance of the dimension or uniform distribution with upper and lower limits equal to the minimum and maximum value of the dimension, where such distributions have proven to be typical and suitable for the utilized manufacturing method ;
- assume triangular distribution based on expert opinion (Law, Kelton, 1991).

Thus consecutive generation of values for the dimensions of each part, based on expression (3) is actually a model, representing a random selection of the parts from the batches as they are produced in the manufacturing stage. Moreover, each set of parts will have random values for their dimensions which are within the tolerance limits, just as the case is in real life.

Assembly stage

Once actual values of dimensions for all parts are modeled and available, the value of the clearance which will appear after the assembly can be calculated with the following expression:

$$N_a = \sum_{i=1}^{n} A_i * k_i$$, where (4)

- Na - clearance dimension ;
- A_i - the dimension of the ith part ;
- k_i - coefficient, having the values of 1 or -1 and indicating whether A_i contributes to increasing or decreasing of the value of Na.

Having the defined mechanism for assigning part dimensions and calculating the clearance value, the effect of using each of the three assembly methods for the current set of parts can easily be studied (Astinov, Hadjiski, Pashov, Stoichkov, Milanov, 1995). Figure 7 shows the principle algorithm of the assembly stage, simulated by the model developed and implemented.

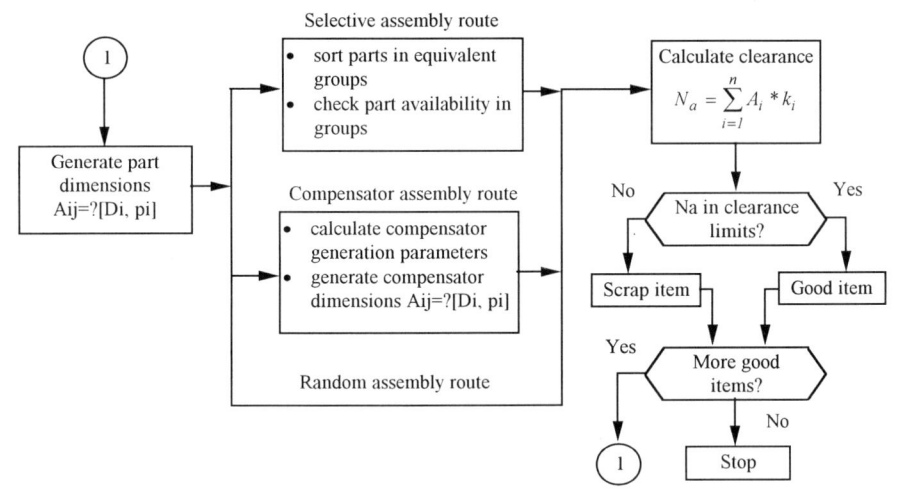

Figure 7 Principle algorithm of the assembly stage.

Events analysis
The overall process of manufacturing and assembly was analyzed from a discrete event point of view, regarding the dynamic characteristics of the latter. The chart on Figure 8 shows the events identified, their sequence and modeled duration.

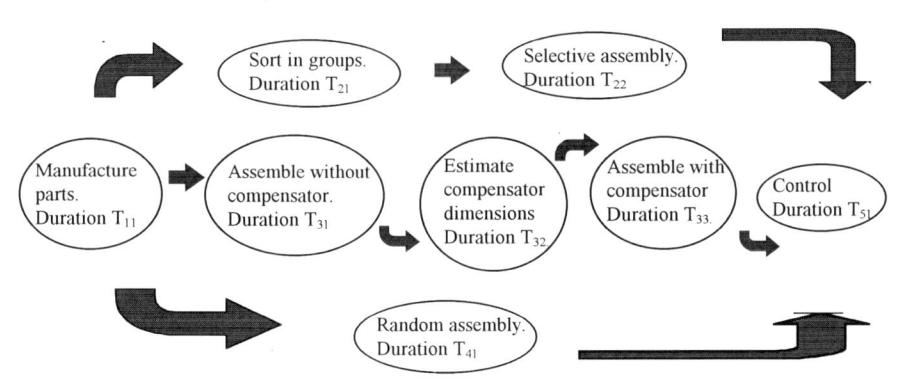

Figure 8 Events, sequence and durations in assembly stage.

It should be noted, that all durations given in Figure 8 in the general case have also stochastic properties. This means, that actual values for every duration can be modeled using following expression:

$$T_{ij} = ?[D_i, \mathbf{p}_i].,\ where \qquad\qquad (5)$$

- T_{ij} - the actual interval duration of the occurrence of each event ;
- ?[...] - formal notation indicating that T_{ij} has stochastic properties ;
- D_i - the PDF type of the T_{ij} ;
- $\mathbf{p}_i$ - an array, containing the parameters of the PDF.

3.2 Input data and performance measures

Input data
The following input data is considered important and required to run the model according to the problem definition (see Figure 3) :

- PDF distribution type and parameters for each part reflecting the accuracy that can be realistically achieved by the existing manufacturing stage ;
- PDF distribution and type for durations of the events (see Figure 8) ;
- maximum and minimum values of the dimensions for each part. These parameters represent the quality control implemented in the manufacturing stage ;
- quantity of items to be produced ;
- number of groups for selective assembly ;
- number of groups for the compensator part.

Performance measures estimated
As shown in Figure 7, the model developed would initiate a loop and would keep performing the calculations until the required quantity of items is produced by each of the assembly method. Statistics are collected on the following variables being the performance measures of the system:

- total parts of each type including scrap and good parts for each of the assembly methods ;
- total scrap items produced for each of the assembly methods ;
- total parts unused and remaining in all groups for the selective assembly method ;
- total compensators in each group necessary to produce the required amount of items ;
- total durations for each event.

The statistics include the mean, standard deviation, coefficient of variation, minimum and maximum values for each of the performance measures given above.

The model was developed using SLAMSYSTEM general purpose simulation package (Pritsker, 1986, *SLAMSYSTEM User's Guide,* 1993) linked with specially designed additional procedures written on FORTRAN 77 programming language. The scenario concept and automatic statistic collection in SLAMSYSTEM are extensively used. Computer animation of the production process is also generated.

4. CASE STUDY - THE PRODUCTION OF A BALL BEARING

The following example will be used to illustrate the practical aspects of the present development. Consider that the management of an enterprise has to take a decision whether or not to accept a contract for manufacturing a quantity of 1000 non-standard ball bearing with some basic dimensions as given in Figure 9.

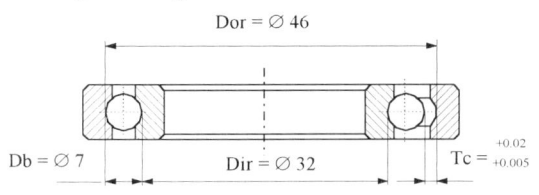

Figure 9 Dimensions and requirements of a non-standard ball bearing.

The manufacturing facility of the enterprise is in a position to achieve the dimensions of each part as shown in Table 1. Additionally the parts will pass a control operation before continuing to the assembly stage which will narrow the final tolerance field.

Table 1 Dimensions realistically achievable in the manufacturing stage

Part dimension	μ (mm)	σ (mm)	Minimum (mm)	Maximum (mm)
Dor	45.94	0.02	45.88	46
Db	31.958	0.01	31.928	31.988
Dir	6.985	0.005	6.97	7

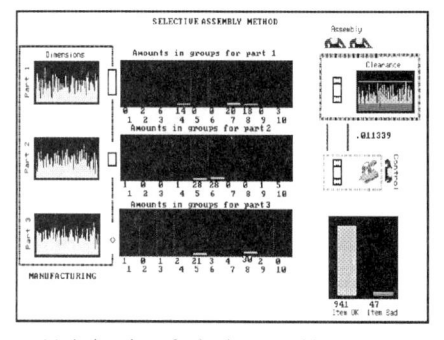

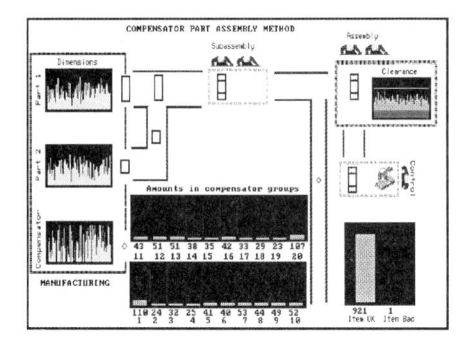

(a) Animation of selective assembly

(b) Animation of compensator assembly

Figure 10 Animation output of the model for two of the assembly methods.

Figure 10 illustrates the animation generated by the model for two of the assembly methods - selective and compensator feedback. The animations shows the dynamic behavior of the production process in the sense of actual part dimensions, clearance values achieved and the accumulation of good and scrap bearing. The manufacturing process for the bearing was simulated in five independent runs each terminating when 1000 bearings, meeting the clearance restrictions were produced by all three methods of assembly. As it is obvious from Table 1, the random assembly method is not applicable in this particular case since the manufacturing stage can produce parts with larger tolerance values. Therefore selective and compensator feedback methods are to be applied. The results, computed by the model for these to methods are as follows. Figure 11 is the generated bar graph, showing the mean value of the produced scrap bearings in the case of selective assembly (a) and compensator feedback assembly (b) as a result of one simulation run.

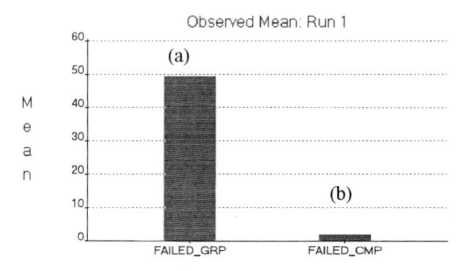

Figure 11 Mean values of produced scrap bearings.

Figure 12 is a printout of the statistics collected and available after the total simulation runs.

```
                S L A M   I I   S U M M A R Y   R E P O R T
SIMULATION PROJECT ASSEMBLY                 BY IL. ASTINOV
DATE   8/15/1995                            RUN NUMBER    5 OF    5

            **STATISTICS FOR VARIABLES BASED ON OBSERVATION**

                MEAN      STANDARD   COEFF. OF  MINIMUM   MAXIMUM   NO.OF
                VALUE     DEVIATION  VARIATION  VALUE     VALUE     OBS
FAILED_GRP      .728E+02  .141E+02   .193E+00   .490E+02  .860E+02  5
S_OUTRINGS      .115E+04  .195E+02   .169E-01   .112E+04  .117E+04  5
S_INRINGS       .115E+04  .180E+02   .156E-01   .113E+04  .117E+04  5
S_SPHERES       .115E+04  .179E+02   .155E-01   .113E+04  .117E+04  5
FAILED_CMP      .280E+01  .164E+01   .587E+00   .100E+01  .500E+01  5
C_OUTRINGS      .101E+04  .182E+01   .181E-02   .100E+04  .101E+04  5
C_INRINGS       .101E+04  .235E+01   .233E-02   .100E+04  .101E+04  5
C_SPHERES       .101E+04  .235E+01   .233E-02   .100E+04  .101E+04  5
CMP_1           .129E+03  .140E+02   .109E+00   .111E+03  .148E+03  5
CMP_2           .270E+02  .548E+01   .203E+00   .190E+02  .330E+02  5
CMP_3           .316E+02  .477E+01   .151E+00   .250E+02  .370E+02  5
CMP_4           .350E+02  .510E+01   .146E+00   .290E+02  .400E+02  5
CMP_5           .446E+02  .518E+01   .116E+00   .360E+02  .490E+02  5
```

Figure 12 Printout of the statistics collected during simulation.

High precision assembly processes using discrete event simulation 193

In this specific case, in order to produce the required amount of 1000 bearings having a clearance within the limits, indicated on Figure 9 the following major facts can be drawn:

Case - applying selective assembly

If selective assembly is applied, the mean value of the amount of outer rings that is to be produced is 1150 with a standard deviation of 19.5 parts going to a maximum of 1170 (see the row named S_OUTRINGS on Figure 12). From these parts 1000 will be assembled in the non-scrap bearings, a mean of 72.8 with a standard deviation of 14.1 will go in scrap bearings (see the row named FAILED_GRP on Figure 12) and the rest will either be unused and remain the groups or will not pass the dimension quality control in the manufacturing stage. Or in other words it is expected that by applying the selective assembly method some additional 120-170 outer rings are to be manufactured and unused in order to produce 1000 bearings. The same conclusions can be drawn for the other parts - the inner rings (row S_INRINGS on Figure 12) and sets of balls (row S_SPHERES on Figure 12).

Case - applying compensator feedback

In the same manner, a mean value of 1010 with a standard deviation of 1.82 outer rings are to be produced when applying the compensator feedback method (see row C_OUTRINGS on Figure 12). It is expected that a mean of 2.8 with a standard deviation of 1.64 bearings will fail the clearance control (see row FAILED_CMP on Figure 12). Similar conclusions can be drawn for the other parts. As the compensator dimensions are curtail for applying this method, statistics are collected for the amounts of compensator parts falling in each group. In Figure 12 a printout is given only for the first five groups, from a total of twenty. For example, a mean of 129 with a standard deviation of 14 sets of balls, with dimensions falling into group one are to be produced in order to fulfill the requirements of the assembly method (see row CMP_1 on Figure 12). Figure 13 shows the bar graph generated by the model and indicating the distribution of mean amounts of compensator parts amongst groups.

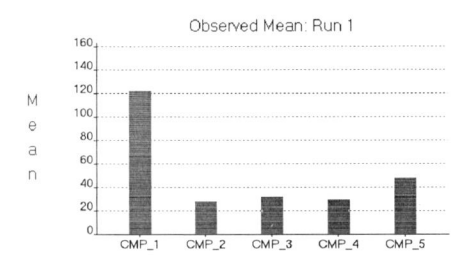

Figure 13 Distribution of the mean amounts of compensators in groups.

The estimates computed through the simulation runs provide quantitative results of the application of each assembly method taking into consideration the stochastic properties of the manufacturing and assembly stages. They will support the final decision which is to be made by the manufacturing decision maker towards rational trade-offs between contract deadlines, costs and other major specific factors.

5. CONCLUSIONS

The following major conclusions can be drawn based on the results achieved so far:

- Applying discrete event simulation in analyzing high precision assembly does no require a complex formalization. The formalization of the problem represents more realistically the real-life phenomena of the treated problem.
- The influence of the stochastic factors is taken into account very precisely. The deterministic models of the type shown in expressions (1) and (2) are not in a position to do so. This capability of the present model leads to a more precise estimation of the major performance measures and will help decision makers in taking more accurate decisions.
- Despite the fact, that bearing manufacturing is the base of the current development, the present model can be utilized in any manufacturing case, specially where stochastic factors play a vital role.
- The non-standard application of discrete event simulation presented facilitates a justified implementation of assembly methods in enterprises which do not have suitable specialized manufacturing infrastructure. This is especially valid for the case of the compensator feedback method, where the amounts of compensators in each group can be initially estimated and competitors manufactured without having the traditional feedback to the manufacturing facility.
- Care should be taken on the choice of distributions describing the stochastic factors as a wrong estimation or approximation may lead to generating output data which will not represent the real situation thus leading to wrong decision making.

The major conclusion that can be reported is that discrete event simulation as a methodology in this case provides the manufacturing decision makers with more realistic and precise data aiding the process of meeting quality criteria in high precision assembly operations.

6. REFERENCES

Astinov, Il. (1990) Simulation modeling of machine tools and systems. *PhD thesis*. Technical University Sofia, Sofia.

Astinov, Il. Hadjiski, P. Pashov, Tc. Stoichkov, B. Milanov, M. (1995) Simulation modeling for providing accuracy for precise products by the method of group self-replacement. *Advanced Manufacturing Technology Conference Proceedings (AMTECH'95)*, **Volume 1**, 147-55.

Bratley, P. Fox, B. Schrage L. (1983) Rational choice of input distributions. *A guide to simulation*. Springer-Verlag, New York, Berlin, Heidelberg, Tokyo

Law, A.M. Kelton D.W. (1991) Selecting a distribution in the absence of data. *Simulation modeling & analysis*. McGraw-Hill, Inc, New York

Pritsker, A.A.B. (1986) Applications of simulation. *Introduction to simulation and SLAM II*. Halsted Press, John Wiley & Sons, New York, Chichester, Brisbane.

SLAMSYSTEM User's Guide (1993) Pritsker Corporation

17

A market model of manufacturing control*

T. Kis, A. Márkus and J. Váncza
Computer and Automation Research Institute
Hungarian Academy of Sciences
H-1518 Budapest, POB 63, Hungary
Phone: 36-1-1810-194
Fax: 36-1-1667-503
E-mail: {tamas.kis;markus;vancza}@sztaki.hu

Abstract
In the paper there will be suggested a novel approach to cope with disturbances and variability inherent in present day manufacturing systems: production control based on a market model. In a setting of distributed scheduling, Machine and Management agents pursue their own interest by operating under bounded rationality in a changing, hardly predictable environment. Local decisions of agents are coordinated via negotiation and bargaining. The interplay of agents can compensate for their narrow perspective and this is what ensures the viability of the overall system.

Keywords
cooperative manufacturing, production control, market model, multi-agent systems, negotiation, distributed scheduling

1 INTRODUCTION

There is not much doubt about that today's market conditions force the evolution of manufacturing control systems toward cooperative manufacturing. This new trend complements the efforts made for the realization of (computer) integrated manufacturing.

Cooperative manufacturing is based on the recognition of ignorance; on accepting that in an inherently unstable and unpredictable environment it is difficult, or even impossible, to be prepared with pre-programmed responses to any abrupt production disturbances (Hatvany 1985, Buzacott 1995). In order to be able to absorb sudden changes both in the internal and external conditions, the systems must be more closely linked to the environment, and have an organizational structure with overlapping and redundant functions. Both conditions require the allocation of decision authority to those agents in the manufacturing game who access the pertinent information and have also the specific knowledge to process it. Co-locating decision rights and distributed information leads to a flatter management hierarchy and to a more decentralized control.

---

* This work has been supported by the grant T-7341 of the National Research Foundation of Hungary (OTKA).

Though, in a distributed multi-agent setting cooperation and collaborative work cannot easily be guaranteed. As for cooperation, the question emerges: how to meet system-wide criteria? For instance, how can the system comply with due dates? Or, how can the global goal of profit maximization be asserted? In general, how can both individual and global goals be reconciled and pursued as dual goals?

In a distributed framework, a central driving factor for cooperation is uncertainty: agents do need to share some information just because it is of incomplete, incorrect and inconsistent nature. Uncertainty can be resolved only in the course of a cooperative dialogue, i.e. via negotiation. For coordinating the decisions and actions of agents some rules are needed whose observance is mandatory. This is actually what the agents give in return of the decision rights they receive. An important set of rules concern information exchange and provide a protocol for negotiation, while other rules couple and aggregate the agents' behavior. However, how can we ensure an acceptable overall behavior of the system when its autonomous components are supposed to be neither altruistic nor benevolent? And how to guarantee the overall performance in situations which we are admittedly unable to anticipate?

In this paper we try to give answers to the above questions within the framework of a particular, though by now more and more typical, manufacturing control problem. We address the problem of distributed dynamic scheduling (Sycara et al. 1991, Burke and Prosser 1994, Duffie and Prabhu 1994). Following a statement of the scheduling problem in a dynamic environment, agents who take part in the evaluation of orders and generation of schedules are defined. Then we define a mechanism that governs the interaction of agents on market-based principles.

The motivations of introducing the market model into manufacturing control are as follows: markets provide a well-proven mechanism for controlling agents that pursue their own interest by operating under bounded rationality (i.e., with limited information and fallible reasoning capacity), in a dynamically changing, hardly predictable environment. Emerging conflicts can be resolved by negotiation and bargaining on the very simple common terms of services (goods) and prices. Hence, heterogeneity of agents causes no problem here and the presence of competitive agents can even be an advantage. Conflicts, which inevitably crop up in scheduling, will in fact drive the system toward generating efficient patterns of adaptive behavior - just as in the case of real market economy (see Horváth et al. 1993).

2 SCHEDULING WITH LIMITED CAPABILITY AGENTS

In the control of a manufacturing system one has to perform two basic kinds of activities: those belonging to management and to production. In our model, while the management-related activities are performed by a single agent called *Management*, production is performed by several, so-called *Machine* agents. In a nutshell, the Management accepts or rejects incoming orders that arrive from the outside world and ensures that the accepted orders (called jobs) reach their required finished state. Machines perform the manufacturing operations on the jobs; the selection from among the technologically feasible machines results from a bidding process. Each Machine maintains its own opportunistic schedule. The outlines of the system are shown in Fig. 1.

The overall objective of the manufacturing system is to earn, in a long range time scale, as much profit as possible. We consider the whole manufacturing system as a single piece of property (i.e., no subcontracting is being considered now), so profits earned by management and machines are summed up when evaluating the performance of the system. However, there are further aspects of the distribution of the profit: On the one hand, since we do not feed back the profit earned into the system (no investments are made, no agent may go bankrupt), it is of minor importance whether the profit is shared *fairly* or not. On the other hand, maximization of its own profit is the only tractable objective function of each agent: good coordination of these conflicting objectives is the major concern of our approach.

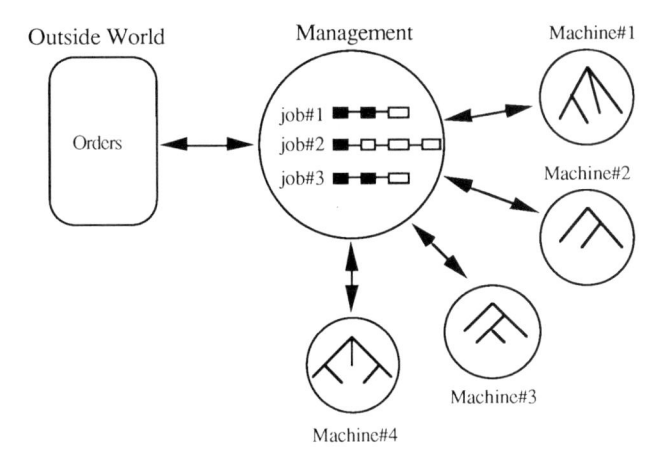

Figure 1 The basic setting of Management and Machines.

In case when each agent has unlimited computing power and the cooperation among them is ideal, the manufacturing system finds a steady state, such that
- the Management accepts a mix of orders that are appropriate with respect to the technological capabilities and the present workload of the machines,
- Machines work in a schedule that ensures that the total profit of the system be as high as possible.

Our aim is to control the a manufacturing system under more realistic assumptions: here the agents have limited computing power and their cooperation is far from the ideal. These assumptions are rooted in the manufacturing practice:
- The agents have to make decisions in limited time (e.g., an order has to be accepted/rejected as soon as possible, else another manufacturer will take it, or some other orders will be left unnoticed).
- The agents have limited knowledge on the state of the overall system: Management knows the technological and cost parameters of the Machines but it does not know how they make their detailed schedule. Similarly, each Machine knows only about those tasks that are waiting in its own input buffer.
- In real life, manufacturing agents are self-interested as far as the rules of the factory allow. For instance, in order to achieve higher profit or build up better working conditions for themselves, they may deliberately bias the estimates on their future workload.
- The scheme for information exchange does not allow Machines to form coalitions.
- Under realistic working conditions the outside world is changing, too: the manufacturing system, in the first line the Management, has to adapt itself to new situations (e.g., a new technological bottleneck may develop when well-paying orders show an increased need of a particular machining resource).

In order to keep the agents' activities within limits of common-sense manufacturing rationality even under these weak assumptions, we will introduce a bidding protocol and a set of rules that the agents must obey, both in their bidding and manufacturing activities. Supposing, as we have told, that our agents all belong to the same company, such rules will be indeed kept by all the agents.

3   THE OBJECTS

3.1   Orders, jobs and tasks

Orders arrive from the outside world. While the future stream of incoming orders is unknown, we suppose that the major parameters are steady or do not change quickly. (E.g., 5% of the orders are well-paying rush orders, 10% are very small sized ones, 75% and 25% of all machining requests are milling and drilling etc.) Actually, one of our main concerns is that the system has to show a robust behavior against moderate though sudden changes of the order stream parameters.

An *order* consists of a definite sequence of tasks. An order, once accepted by the Management, may leave the manufacturing system only when all of its tasks have been executed on the machines.

To each order an *ArrivalTime*, a *DueTime*, a *ContractPrice* and a *TardinessPenalty* function are assigned. Price is a fixed number of monetary units, the tardiness penalty is a monotonously increasing function of time. If the order is finished in due time then the outside world pays its contract price to Management; later the price is decreased by the tardiness penalty. In the minimal model, finishing earlier than *DueTime* has no consequence to the income of Management.

As soon as the Management accepts an order, a *job* is created. Rejected orders disappear from the horizon. Actually, while jobs exist solely for the Management, the tasks do so for the Machine agents.

A job is described with its task sequence, arrival time and due time; these parameters of the job are the same as those of the order. However, to each task the Management computes an estimated manufacturing price and an estimated time frame within which the task should be made. The calculation of these estimates is, as a matter of fact, the main task of Management: both the acceptance of orders and the whole bidding process are built on these data (details will be discussed later).

Usually, the sum of the estimated manufacturing prices is less than the contract price, and the difference is the planned profit at the Management. While the profit of the management is defined only on the level of jobs, the Machine agents deal with tasks only.

Whenever a job is getting ready after its due time, the outside world decreases the Management's income by the tardiness penalty of the job. The lateness of a job may be caused either by wrong decisions of Management (a relatively cheap or a too urgent order accepted in times when workload is high) or by the selfishness of one or more Machines (e.g., the most suitable machine waits idle for more profitable tasks and the management could not help but give long tasks to slower machines).

While in the manufacturing system, the jobs have no own interests or decisions; they simply have to obey the rules.

Each *task* is given as a 4-tuple: (*Service Volume ArrivalTime DueTime*). A service is a particular kind of machining operation. E.g., (drilling 20 100 150) means that 20 units of drilling operations are to be made in the time frame of 100 to 150 minutes.

In the minimal model, no tasks may be shared between machines, delayed or abandoned after having started on a particular machine.

3.2 Machines

The manufacturing system operates with a number of different machines that have partially overlapping technological capabilities: some of the machines may be more suitable for specific kinds of services, some may be generally better than other machines.

Each machine is given as a set of its *resources* that are (*Service Speed MinuteCost*) triplets: e.g., the task of drilling 20 units, if made on a machine with a resource (drilling 5 8) takes 4 minutes and its technological cost is 32. Costs of maintenance, tooling etc. are all considered as parts of this technological cost.

The same service (on the same machine) may be accomplished with different speeds and costs. If a machine works on low speed, its minute cost is supposed to be smaller. Although in this way the machine's profit is likely to be smaller, too, the other choice may be to stay idle and earn nothing; accordingly, working slow may be a reasonable choice. On the other hand, from the Management's point of view, whenever a job is ahead of its schedule, so that it would finish too early, slow and relatively cheap use of a machine may be an advantage.

In the minimal model, machines are supposed not to break down while making a task and need no maintenance. In addition, the technological parameters of the machines do not change, new machines do not start to work. These assumptions are rooted not as much in the scheduling method used here, but come from our inability to attach monetary equivalents to these factors in a sensible way. As a matter of fact, if we do not want to deal with the financial consequences of a breakdown event or of a maintenance period, the scheduling method can easily be modified with inserting idle periods.

With their given technological capabilities, the machines are free to use whatever price calculation scheme for accomplishing a task. Each machine has its own account. Its sole income is the payment from the tasks made there, and its profit is the income minus the technological cost.

4 THE PROCESSING OF ORDERS

4.1 Outlines of the market mechanism

The Management takes care of each job in the system: as soon as a task becomes the first one on the job's undone task sequence, the Management makes an *announcement* containing all the relevant data of the task. As it will be shown later, it may happen that the same task should be announced again.

Now it is the Machines' turn to prepare *bids* to the announcement. Each bid contains a *StartTime*, an *EndTime* and a *Price* of executing that task. Each machine may send more than one bid for a task, or may leave the announcement unanswered. Obviously, the bids must not hurt the cost and time constraints of the announcement.

The Management, having received the bids, selects a subset of them, and notifies the Machines which of their bids have been accepted and become so-called *assignments*. As an extreme, the Management may choose to create for each announcement only one assignment, however, our intention is to exploit the flexibility that can be achieved by alternative assignments. If no assignment is made, the Management has to make a new announcement of the same task (probably, now a higher price will be offered for the accomplishment of the task).

For each task and machine, the obligation is only on the *set* of the alternative assignments, however the machine is free to choose any member of its assignment set when it starts to work on a task. Obviously, when a machine starts to work on an assignment, all the alternative assignments for the same task cease to be real ones, so a message is sent to other Machines, and each updates its own schedule tree (details will be discussed later).

When starting to work on an assignment, the machine has to withdraw all assignments that overlap with the work in progress. This may lead to a situation when no assignment remains for some task: that is why the Management has to be informed of the withdrawn assignments. In a case when the last alternative of some task has been discarded, the Management makes a new announcement of the same task.

In addition, the Management is informed that the task will be ready at *EndTime* of that assignment; from now on, the next task of the job may be announced with that *ArrivalTime*. When a Machine finishes to work on an assignment, no administration is needed.

Timing and some other technical issues of the above bidding protocol will be outlined later. In the following points it will be discussed how some major decisions of the bidding protocol are being made in our experiments. However, most of the decision strategies, bidding and manufacturing rules may be replaced with other ones, whenever it can be shown that the basic assumptions are not hurt.

4.2 How does Management accept or reject orders?

As soon as a new order arrives, Management decides whether to accept (and make an appropriate job) or to reject it. The absolute requirement of acceptance is that the manufacturing system be able to provide all the required services. Then Management has to judge whether the new order could comply with the constraints that were generated by the orders already accepted, and it has also to calculate the expected cost of executing the order.

Management estimates the machining time and cost of each task in the order by considering both the specification of the tasks and the actual workload of the system. With a margin of safety, it can decide then whether there is any chance of finishing the order for the given due time with an acceptable profit, still under the given *ContractPrice*.

However, the total machining cost estimate may have also extra components which depend on the characteristics of the order. For instance, such a cost factor expresses the risk of accepting an order that contains tasks calling for unusually long machining services, or an order that consists of many tasks.

4.3 When and how does Management make task announcements?

As soon as a new job is created or a Machine begins to work on a particular task, Management announces the first task from the sequence of the job's undone tasks. The announcement contains *ArrivalTime*, *DueTime* and *PriceLimit* beyond the task description.

ArrivalTime of the task is set to *ArrivalTime* of the job if a new job is at hand, and to the *DueTime* of the previous task otherwise. This can be done because a rule guarantees that a task - when already under machining - will be completed without any disturbances and delay.

Due time and price limit for the task of a job are suggested on the basis of Management's evaluation of the job's overall progress. Once an order has been accepted, Management has an estimation of the flow of the events that relate to the job. Accordingly, at each announcement Management knows whether the current task of the job is ahead or behind its estimated schedule, and whether, up to the given point, it has spent more or less money than planned. It can guess whether the job as a whole is *early* or *tardy*, *under-* or *overspent*.

Management has to give reasonable intermediate time and price constraints for a task that is only one item in a job's task sequence. Hence, it takes the estimated processing time and cost of the task and modifies them according to the job's temporal and budget status. For instance, if the job is early, then the task may be delayed (see Fig. 2). The rate of modification depends on the measure of the earliness (lateness) in a progressive way. The same holds for the budget data. When a task of a tardy and overspent job is to be announced, Management should decide whether it is better to loss time or money.

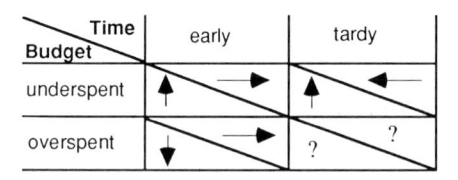

Time Budget	early	tardy
underspent	↑ →	↑ ←
overspent	↓ →	? ?

Figure 2 Directions of modifying due time and price limit in a task announcement according to the job's estimated temporal and budget status.

4.4 How does a Machine make its bid?

A basic assumption of our approach is that Management will accept more than one bids for (some of the) announcements. In this case, when there are two or more assignments for a single task announcement, machines in concern may not be sure what their actual schedule will be. The schedule alternatives of each machine are handled in form of *schedule trees*; each tree is known and updated exclusively by its owner machine.

Nodes of the schedule tree are assignments (accepted bids) of the tasks. Brother nodes are alternatives of the assignments, and a branch of a tree corresponds to a schedule. Each branch has to be executable. Some branches may be shorter than others, i.e., contain no assignment for an announcement.

Whenever a new announcement arrives, the machine starts to insert new (hypothetical) assignments into its schedule tree: first it considers its resource alternatives, then, for each alternative, it inserts the task's execution into several places: (1) starting after the arrival time of the task, (2) immediately preceding and following any assignments of the tree, and (3) ending just at the task's due time. Obviously, all these alternatives must form collision free schedules.

It is the machine's choice whether it indeed sends all of these bids to the Management. The only exception is the following bidding rule: if a machine has a long enough time frame for executing the task, then it is obliged to make a bid that ends just at the given due time of the task. This is called a *last resort bid*. This specific bidding technique ensures that all jobs indeed leave the system and the Management's eventual losses are distributed.

The price tags of the assignments are determined in the following way: first, the machine's usual profit is added to the technological cost, then other factors are considered. With some of these factors the Machine agent wants to guess the Management's price tolerance (e.g., if the assignment falls within an early segment of the announced *ArrivalTime - DueTime* interval, then price is further increased by a small factor). Some other factors might compensate the machine's implicit losses when this assignment becomes reality: if the assignment leaves free a small, useless period of time at its start or end point, then price is increased.

In the broader perspective of a learning scheduler, the machine's usual profit rate needs to be updated, too: it may happen that, with respect to the stream of new orders, the machine has to devaluate its importance (unless the machine wants to stay idle for most of the time).

4.5 How does Management decide which bids to accept?

First of all, for some pairs of the bids one of them is definitely better than the other: e.g., with the same start-end time frame offered, if one is cheaper than the other, no rationally acting Management will accept the more expensive, so-called dominated one. (As a matter of fact, this is true only when there is no hidden agreement between management and the expensive but favored machine to compensate this decision at some later time.) In order to prevent the agents' cheating and outguessing of each other, one of our bidding rules tells that, from among comparable bids, the dominated ones have to be discarded.

A less trivial problem is the selection from among non-comparable bids: when the one is cheaper, the other becomes ready earlier (and leaves more freedom for the scheduling of the other tasks). As for tardy jobs, the offered *EndTime* will be the distinctive factor. The same mechanism is used when the announcement of a task is repeated: tasks of tardy jobs incur higher price limits.

In addition, Management may take into consideration a large variety of further aspects; here we give some illustrative examples only:
- investigate how threatening is the tardiness penalty function of the job,
- disperse the accepted bids among machines,
- give different treatment to one of the first tasks of a job than to one of the last ones.

A special bidding rule says that Management is obliged to accept all of the last resort bids.

Anyway, in one extreme strategy, Management may accept all non-dominated bids, in the other, only a single bid from among all non-dominated ones. The first extreme leaves far more decision power at the machines; we have followed this path in our experiments.

4.6 How does a machine decide which assignment to work on?

At any time, each machine knows its schedule tree containing all the assignments corresponding to its accepted and not invalidated bids. (Invalidated bids are those that were on pruned branches of the tree.)

When finishing with an assignment, there are two cases: (1) if the Machine has no assignment starting at that point then it can do nothing else than go idle until the next possible start time. (2) If there is at least one assignment that may be started, then, according to a manufacturing control rule, the machine is obliged to start to work on one of them (i.e., it must not stay idle). In addition, if one of the assignments is of a last resort bid, then the machine is obliged to select that one.

When deciding what to do, there are two conflicting objectives: (1) earn as much as possible *now*, or (2) consider the consequences of a decision, and take into account the future earnings as well. Due to the following reasons, the second objective is rather difficult to attain:
- a machine may deem a branch of its schedule tree very promising, but later it may be unable to work along that branch since other machines will have selected and done some of those tasks in the meantime;
- one should attach a value to the free time segments of the schedule branches, too; this estimation may change in the future (e.g., if orders are getting better, free time is getting more valuable).

Up to now, we have not found any discounting or weighting scheme for the future earnings, so we adopted a simple greedy selection that considers only the immediate earning. However, there are alternative objectives even in this limited setting: the best assignment might be defined either as one with the largest profit, or as maximizing profit per time.

4.7 How does Management make repeated task announcements?

Management must announce a task again whenever it receives no bid from the Machines to an announcement. Due to the last resort bidding rule, such a situation happens only when under the current workload no machine is technically able to fit the task - as announced by Management - into its schedule. Another case for making a new announcement is when all assignments of a task have been withdrawn.

As for the second case, an announcement is sent immediately after the withdrawal of the last task-related assignment. When negotiation fails, the task will be announced again only after an appropriate change in the technical conditions: when there is a new time-slice at least on one of the machines that could provide the required service. This may happen either because a machine completes a task, or because one of its assignments has been withdrawn in the meantime. In such a moment, Management announces the task again by using its usual mechanism. Since time is passing, the new announcement will most probably differ from the previous one (e.g., a job thought to be early becomes tardy in the meantime).

5 A SIMULATION ENVIRONMENT OF THE MARKET SYSTEM

We have implemented a simulation of the above market system in a simple, instantaneous message-passing software architecture. In order to model the limited computing power of the agents, each agent records its summed up computing time. Our message passing system uses time-stamps, and an aging mechanism: old, unprocessed messages are deleted from the mail-boxes of the agents (e.g., an announcement not considered since the machine was busy with processing its bids to an earlier announcement). In addition, a priority mechanism of the message types ensures the proper handling of the content of the mail-boxes (e.g., a message telling that a task execution has been started on a machine deserves higher priority than new announcements in the mail-box).

On a more technical level, the program works with a monitor that gives time slots to the agents in turn. While those programs of the agents that make some simple administrative work are considered as consuming no computing power (e.g., message handling), the core of the scheduling work is made by so-called anytime algorithms which are able to give some result even if abandoned at any time.

The message aging mechanism is implemented with the time slots of the monitor, too: e.g., all the agents know that the evaluation of bids to an assignment will be started in the k-th round after the announcement.

In the experimental implementation as a simplifying assumption we have stated that machining and computing power can not be converted into each other.

Since ties frequently occur (e.g., two machines could start to work on a task at the same time), a random unbiased selection is severely used. As usual in such cases, randomization adds variance to the experiments, so special care is needed to check what results are indeed significant.

6 RELATED WORK

The idea of negotiated factory scheduling emerged long before this work. The early attempts concentrate on the negotiation mechanism and handle the bargaining problem apart from the actual scheduling problem (Maley 1988, Parunak 1988). The main concepts and protocol of negotiation have been usually borrowed from of the so-called Contract Net Protocol (Smith 1980).

Investigations of computational ecosystems have shown that in a multi-agent setting, where agents are designed to achieve a single common goal, the rational behavior of individuals may force the overall system toward suboptimal overall behavior (Hogg 1995). For instance, the shared use of public resources can easily cause so-called social dilemma when both the performance of individual agents and of the global system are far away from the possible optima. The overuse of relatively cheap common resources is such a typical situation known in the field of sociology as the "tragedy of commons" (Hardin 1968).

Human societies suggest several ways for resolving social dilemmas: the most important mechanisms are (1) taxation schemes (2) markets, and (3) small-sized communities with long-term goals.

In this work we made an attempt to define a distributed scheduling framework where individual and common interests are reconciled via a market mechanism. Here we have to emphasize that our market model was tailored to the specifics of a dynamic job-shop scheduling problem. Other domain-dependent market-based control schemes are known in distributed processor (workstation) scheduling (Stankovic et al. 1985, Waldsprunger et al. 1992).

Notwithstanding, one may already have at hand general-purpose market-based computational tools which have a solid underlying theoretical basis provided by general equilibrium theory (Wellman 1993, 1995). So-called market-oriented programming has successfully been applied to several resource allocation problems which had no temporal aspect. Though, in the context of our particular dynamic scheduling problem we had two severe arguments against an equilibrium-based approach: (1) it gives little insight into the temporal (e.g., transient) behavior of the system, and (2) we cannot suppose that a reduced consumption for one machining service does not cause a reduced consumption of others as well: this property makes the bargaining process unstable.

As noted above, there is a method for resolving social dilemmas which is free of any direct economic considerations: this is the formation of relatively small communities by individuals aimed at a common long-term goal (Glance and Huberman 1994). Cooperation within such groups is self-initiated and it dominates, at least in the long run, myopic and selfish defection. In our opinion, research of holonic manufacturing (Valckenaers et al. 1994) is heading in this direction. Though, at the moment it is not clear what general system properties and what domain- and task-dependent conditions allow together the operation of task-oriented cooperative agents (holons) in manufacturing environments.

A pragmatic mix of ideas from taxation schemes, markets and coalition formation appears in the framework of random manufacturing designed for dynamic distributed scheduling (Iwata et al. 1994). Here incoming orders are announced with their technological constraints (tasks, due dates) and economical promises: fixed rewards for performance and penalties for non-performance. After forming coalitions, making a common schedule and agreeing on the distribution of profits, machine groups send a tender for getting the order. Bargaining around prices is, actually, one-sided: machines can express their disinterest in the running prices only by sending no tender for an order. Though commitments are made quite early (whole orders are assigned to machine groups), there is no emphasis on retracting unfavorable decisions and stepping back from hopeless situations.

The opportunistic self-scheduling of orders (or parts) onto machines appeared in several past works. In (Upton et al. 1991), queuing theory is applied to action-based dispatch rules: bids are offered on the basis of actual buffer load and expected sojourn time, and then jobs select machines which offer the lowest expected processing time.

7 CONCLUSION

In this paper we have presented a distributed framework for solving dynamic scheduling problems. The decisions of Management and Machines, who take part in the scheduling process as autonomous agents, are coordinated by a market mechanism.

The proposed control model has to be corroborated by extensive experimentation. We make experiments with the market-based control mechanism (1) by studying how the system behaves under stationary working conditions, (2) by investigating the reactions of the almost saturated system to changes of the incoming order stream or processing capabilities of machines, and (3) by confronting the system with paradoxical situations that are typical in distributed processing: when the addition of a new resource or the removal of a temporal constraint degrades the global performance (see Stankovic et al. 1995).

Based on studies of computational ecosystems it has been pointed out that flexible manufacturing control schemes could give rise to unpredictable or chaotic behavior on the shop floor (Parunak 1991). Our negotiation protocol and incentive system are designed so that the overall system be able to generate feasible schedules. Further on, we expect that the system can absorb changes both in its external and internal conditions by restructuring the pattern of accepted orders and the workload pattern of the individual machines.

8 REFERENCES

Burke, P. and Prosser, P. (1994) The distributed asynchronous scheduler. in *Intelligent Scheduling* (eds. Zweben, M. and Fox, M.S.), Morgan Kaufmann, 309-339.

Buzacott, J. (1995) A perspective on new paradigms in manufacturing. *Journal of Manufacturing Systems*, **14**(2), 118-125.

Duffie, N.A and Prabhu, V.V. (1994) Real-time distributed scheduling of heterarchical manufacturing systems. *Journal of Manufacturing Systems*, **13**(2), 94-107.

Glance, N.S. and Huberman, B.A. (1994) Dynamics of social dilemmas. *Scientific American*, **270**(3), 76-81.

Hardin, G. (1968) The tragedy of the commons. *Science*, **162**, 1243-1248.

Hatvany, J. (1985) Intelligence and cooperation in heterarchic manufacturing systems. *Robotics and Computer-Integrated Manufacturing*, **2**(2), 101-104.

Hogg T. (1995) Social dilemmas in computational ecosystems. in *Proc. of the IJCAI-95 Conference*, 711-716.

Horváth, M., Márkus, A. and Váncza, J. (1993) Conflicts in manufacturing systems - A problem setting. In Yoshikawa, H. and Goossenaerts, J. (eds.), *Information Infrastructure Systems for Manufacturing*, North-Holland, 265-279.

Iwata, K., Onosato, M. and Koike, M. (1994) Random manufacturing systems: A new concept of manufacturing systems for production to order. *Annals of the CIRP*, **43**(1), 379-383.

Maley, J.G. (1988) Managing the flow of intelligent parts. *Robotics and Computer-Integrated Manufacturing*, **4**(3-4), 525-530.

Parunak, H.V.D. (1988) Distributed artificial intelligence systems. in *Artificial Intelligence - Implications for CIM* (ed. Kusiak, A.), IFS Ltd./Springer Verlag, 225-251.

Parunak, H.V.D. (1991) Characterizing the manufacturing scheduling problem. *Journal of Manufacturing Systems*, **10**(3), 241-259.

Smith, R.G (1980) The contract net protocol: High-level communication and control in distributed problem solving. *IEEE Trans. on Computers*, **C-29**(12), 1104-1113.

Stankovic, J.A., Ramamritham, K. and Cheng, S. (1985) Evaluation of a flexible task scheduling algorithm for distributed hard real-time systems. *IEEE Trans. on Computers*, **C-34**(12), 1130-1143.

Stankovic, J.A., Spuri, M., Natale, M.D. and Buttazzo, G.C. (1995) Implications of classical scheduling results for real-time scheduling. *Computer*, June 1995, 16-25.

Sycara, K. Roth, S., Sadeh, N. and Fox, M.S. (1991) Distributed constrained heuristic search. *IEEE Trans. on Systems, Man, and Cybernetics*, **21**(6), 1446-1461.

Upton, D.M., Barash, M.M. and Matheson, A.M. (1991) Architectures and auctions in manufacturing. *Int. Journal of Computer Integrated Manufacturing*, **4**(1), 23-33.

Valckenaers, P., Van Brussel, H., Bonneville, F., Bongaerts, L. and Wyns, J. (1994) IMS test case 5: Holonic manufacturing systems. in *Preprints of the IMS'94 Conference*, Vienna, Austria, 19-24.

Waldsprunger, C.A., Hogg, T., Huberman, A., Kephart, J.O. and Stornetta, W. (1992) Spawn: A distributed computational economy. *IEEE Trans. on Software Engineering*, **18**(2), 103-117.

Wellman, M.P. (1993) A market oriented programming environment and its application to distributed multicommodity flow problems. *Journal of Artificial Intelligence Research*, **1**, 1-23.

Wellman, M.P. (1995) Market-oriented programming: Some early lessons. in Clearwater, S. (ed.), *Market-Based Control: A Paradigm of Distributed Resource Allocation*, World Scientific.

9 BIOGRAPHIES

Tamás Kis is a PhD student at the Eötvös University Budapest where he received his MS degree in computer science in 1994. At the same time, he is a research assistant at the Computer and Automation Research Institute. His interest includes distributed artificial intelligence, classical and applied planning and evolutionary game theory.

András Márkus is a research scientist at the Computer and Automation Research Institute of the Hungarian Academy of Sciences. His research interests include artificial intelligence and its engineering applications. He received his MS degree in mathematics from the Eötvös University Budapest in 1975, and the PhD from the Budapest Technical University in 1986.

József Váncza is a research scientist at the Computer and Automation Research Institute of the Hungarian Academy of Sciences, where his research interests include artificial intelligence and its applications to planning and control problems of manufacturing. He received his MS degree in electrical engineering from the Budapest Technical University in 1984, and the PhD from the Hungarian Academy of Sciences in 1994.

PART FIVE

Assembly and Disassembly

18

Integration of assembly considerations in product design

Michiel A. Willemse, Ton Storm
Laboratory for Flexible Production Automation
Delft University of Technology
Landbergstraat 3
NL 2628 CE Delft
tel +31-15-2786502
fax +31-15-2783910
m.a.willemse@wbmt.tudelft.nl
t.storm@wbmt.tudelft.nl

Abstract

The paper reports on a four-year research project carried out at the Laboratory of Flexible Production Automation of Delft University of Technology. The project goal is to create economically feasible and technically reliable possibilities for flexible automated assembly for small series of mechanical products. Based on experiences with flexible assembly automation, it was concluded that the main bottleneck is the existence of many "Not-Designed-for-Assembly" situations in products. Use of various DFA-methods doesn't guarantee a proper product design. The feedback cycle of the assembly considerations is too long to make the necessary adjustments in the design without affecting the majority of the completed steps in the design.

In order to shorten the feedback cycle, the design activities have to be approached systematically. A framework for this view on product design is provided in the Design Coordination Framework. An important step is the separation of the design process in three types of actions: design coordination, design activities and design results. The paper describes the three parts of product design for these issues: the design of part relations, design for gripping, accessibility analysis and stability analysis.

Keywords
Assembly, design coordination, design support, design for assembly

1 INTRODUCTION

Background

The Laboratory for Flexible Production Automation has been studying techniques for flexible automated assembly since 1986. One of the major results is the Delft Intelligent Assembly Cell (DIAC), a 75 manyear research project. DIAC was focused on development of techniques for flexible automation of assembly. Attention was given to assembly planning, logistics, sensory control, and several other aspects of assembly in 15 PhD theses. A testbed for the implementation of the ideas developed in the theses was provided in an experimental assembly cell. Figure 1 shows a picture of the cell as it is realized by 1995.

Figure 1 Picture of Delft Intelligent Assembly Cell

The DIAC project was formally completed in 1993. Continuing research in the Laboratory is based on the conclusions and results of the DIAC project. One of the conclusions was that application of flexible assembly requires not only a flexible assembly system, but also a product that is suited for assembly by this system. This means that the product design must be adapted to the possibilities and limitations of the assembly system.

Research goal

The research goal was formulated as the creation of techniques to utilize flexible automated assembly for small to medium product volumes in a economically justifiable and technically reliable way.

Research areas

Two areas of research can be discerned: design methodology and assembly processes. With respect to the design process it is important to define the product range. The products considered in this research must fit within certain ranges that makes them suitable for flexible automated assembly. The annual volume must approximately be between 1.000 and 100.000. The dimensions and weight of the parts and the completed product can be limited arbitrarily to a 50 cm cube and 10 kg.

With respect to the assembly process the abstraction level has to be defined. Heemskerk (1990) has developed a hierarchic decomposition of assembly as a reference model for planning purposes. The model consists of four levels:

- the batch level, which deals with strategies and optimization for a series of products;
- the product level, which deals with sequence planning for a product;
- the part level, determining resources, strategies, sensory and control activities for single parts;
- the primitive level, describing the single steps needed for assembly of one part.

This research deals with tasks related to the assembly of individual parts, which corresponds with the part and primitive level. A separate research project at the Laboratory is directed towards the higher abstraction level of product structuring (Tichem, 1995/2).

Two remarks on the limitations of this research can be made:
The main scope is flexible automated assembly, however, most of the proposed methods will also benefit manual assembly.
This research shows design optimization with respect to assembly only. This might conflict with other life cycle aspects. Techniques that help to balance between these different aspects are not considered in this research, but are necessary as well.

The paper is structured in four main parts. The first part analyses and defines the assembly process. The next part develops a view on product design process, and consequences for design support. Then a study of part relations and a description of support in different areas is provided. The paper ends with conclusions and future work.

2 ANALYSIS OF THE ASSEMBLY PROCESS

For the assembly of parts two basic models have been developed at the Laboratory. The first model is described in Baartman (1995), and is shown in figure 2.

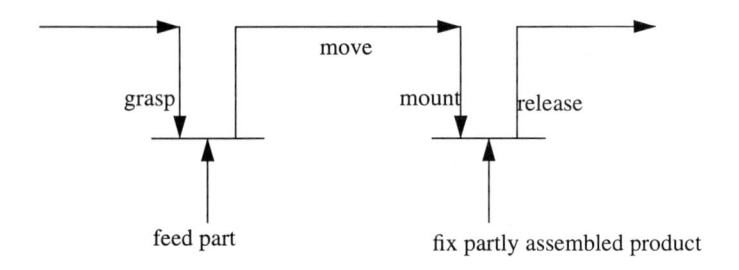

Figure 2 Model of the assembly process

In this model five phases are distinguished in the assembly of a part:
feed - grasp - move - mount - check.
These are the fundamental steps for the assembly of one part in a product. In case of automated assembly, the horizontal lines represent the coarse motion of the robot, where no interaction between the robot and other objects is expected, and the vertical lines represent the fine motion, where the motion can be influenced by contact between the part and the product. Especially the mount phase is difficult to control, since the interaction between moving parts in contact is not understood. However, there are solutions that help to make the process robust, that work without a thorough knowledge of this interaction. An example is the DFA rule to add chamfers to pegs and holes in order to facilitate the first stage of insertion.

The second model is described in Martens (1992), and is a refinement of the mount phase. Figure 3 shows the definition of this phase in the connection model. The model represents a connection in which only one degree of freedom is left without a geometrical form closure after the mount operation. A close study of existing products shows that this model is applicable for a considerable number of the part relations.

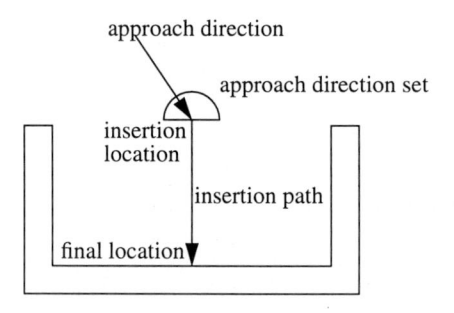

Figure 3 Connection model

In this paper assembly is defined as the realisation of a geometrical relation between a part and a partly assembled product. From now on, the term 'partly assembled product' will be abbreviated by the word 'product'.

Important planning aspects of assembly are determination of the assembly sequence, the fixturing of the product and the selection of the base part, that is the first part to be assembled that makes contact with the fixture. Where necessary, assumptions will be made about these planning tasks. The research goal can now be rephrased as definition of the features and properties of a part and a product that create the conditions for a reliable execution of assembly tasks.

3 VIEW ON PRODUCT DESIGN

Design can be regarded as a structured decision making process, aimed at a goal described in the specifications. The structure of the decision sequence is dependent on the intermediate results and the unsolved aspects that get the attention of the designer.

In the design of a product the major part of the life cycle properties of the product are defined. Understanding of the importance of this fact has led to various design review methods such as Design for Manufacturing, Assembly, Logistics, or Ecology, Quality Function Deployment, Failure Mode and Effect Analysis. Using these methods, the design is analysed at certain phases in the progress. Characteristic for this kind of methods is the long feedback cycle. Negative consequences of design decisions emerge in the evaluation, and not at the moment the decision is taken. Improving that decision will in most cases also affect other completed steps, since every decision is determined by the state of the design as it is at that moment. This view on the activities in product design is illustrated in figure 4. Note that the trigger 'gives rise to' in this figure indicates that the sequence of design activities is based on the previous design results. Design can not be described as a fixed sequence of activities.

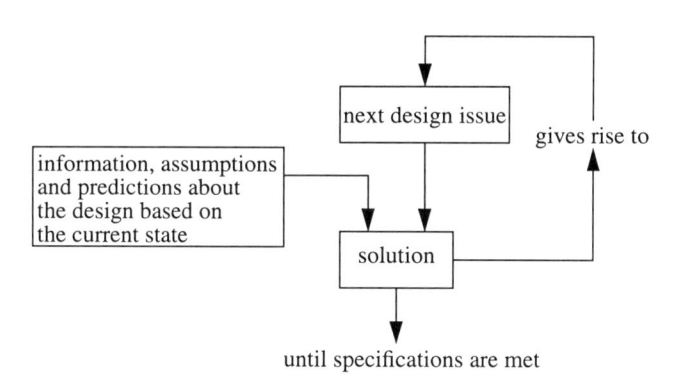

Figure 4 Design as a decision making process

Some of the situations in which the design is based on wrong decisions can be avoided by integration of several disciplines into a design team. Of course this approach is applicable only in major design tasks. Furthermore, putting several experts together doesn't guarantee a balanced and structured participation of all the disciplines, nor does it help in solving the tasks in a more structured order.

An attempt to overcome these problems is described in a framework for design coordination (Andreasen ea, 1994). This framework consists of an extensive set of models of the activities and results in the design process. Important is the fact that the coordination of the activities in design is separated from the execution and the results of the design. A simplified version is proposed in (Tichem, 1995/1) and will serve as reference model.

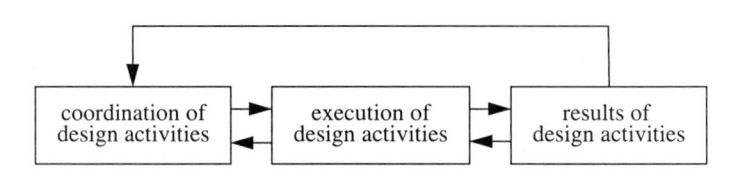

Figure 5 Separation of design tasks; a reference model

Based on this model, design support can also be separated in support in coordination, support in execution and support in the description of the results. The contents of the results must be described, the execution must explain how to apply the method, and the coordination must describe in what situations the method should be applied.

4 THE DESIGN OF PART RELATIONS

Introduction
The part relation is defined by the faces of the part and the product that can be in contact during or after the assembly. The mount operation is mainly determined by the geometry of the part relation. The functional reasons that resulted in the definition of the relation can be treated as constraints on the relation, and given these constraints the relation has to be designed in such a way that the assembly of the relation can be executed reliably.

Reasons for part relations
The individual geometrical relations between parts are designed with a certain intent, based on one or more of the following principles: functional reasons are the relative motion possibilities and different material properties, other reasons are the limitations of manufacturing or (dis-)assembly techniques and the use of standard parts and modules.

Demands on part relations
The relation has to be realized by means of an assembly operation. This operation is mainly determined by the geometry of the relation. The result of the assembly operation has to be a situation that can be expected to be stable under the subsequent assembly operations. This requires that the forces that act on the part and the product will not cause a relative motion of the part or the product. Therefore, the result of an assembly operation must be a secure obstruction of all the degrees of freedom (DOF) of the part relative to the product. The obstruction is determined by the geometrical relation

between the part and the product, and can be based on form closure or force (gravity, friction) closure.

Classification of part relations

Therefore a classification of part relations must be developed, including the functional DOF as well as the assembly operations necessary to establish the relation. A classification of the functional DOF can be developed by systematically combining DOF, as shown in figure 6. The figure combines translations and rotations in three principal directions.

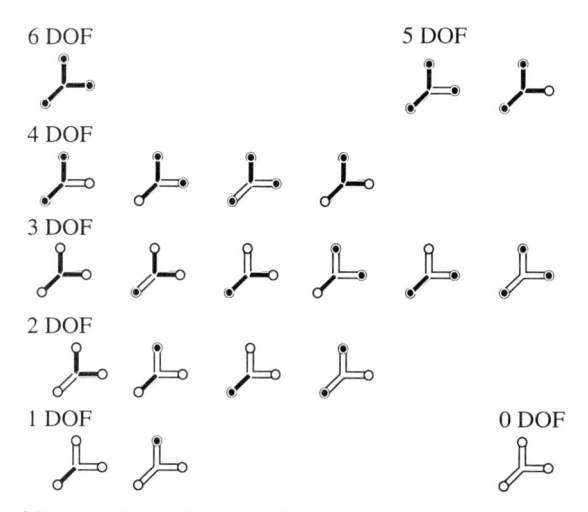

Figure 6 Systematic combination of DOF

Next, constructional solutions must be generated that satisfy the resulting motion possibilities. Since the objective is to relate the functional DOF to the assembly motions, only solutions consisting of two parts are allowed. In this paper only a few solutions will be described. Examples of some relations are provided in figures 7, 8 and 9.

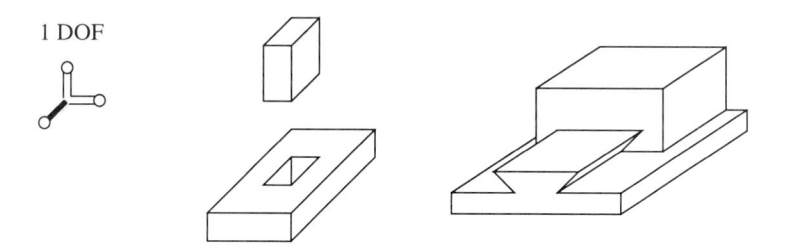

Figure 7 Relations with 1 translational DOF

The constructions of figure 7 both have 1 translational DOF. The assembly of the dovetail is a notorious problem. A constructional solution that is applied in several machine tools is the separation of the dovetail in two tapered parts. The difficult assembly is now replaced by an additional manufacturing operation and an assembly process consisting of an easy mount operation and an adjustment operation.

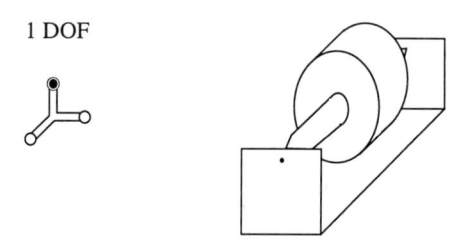

1 DOF

Figure 8 Relation with 1 rotational DOF

Figure 8 shows an example of 1 rotational DOF. Assembly of this construction requires either elastic deformation or separation of one of the parts.

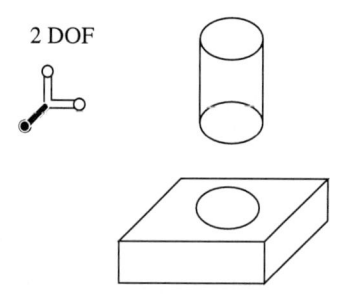

2 DOF

Figure 9 Relation with 2 DOF

Figure 9 shows the simple peg-in-hole assembly. This assembly operation has been studied in various research laboratories. Compliance is the ability of an assembly system to adapt the motion to the forces that act on the system in such a way that the assembly will be successfull. Mechanisms as well as sensor-based control systems have been developed for this purpose. Another achievement in this field is the wedging and jamming diagram of Nevins and Whitney (1989). This diagram relates the occurrence of jamming or wedging during assembly to geometrical properties of the parts. Use of this diagram during design can guide the designer to define a relation in which wedging and jamming will be avoided.

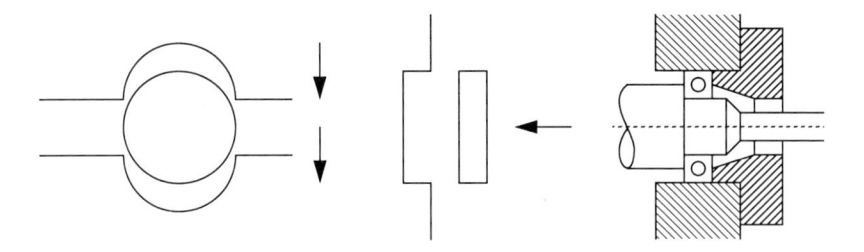

Figure 10 Roller bearing in housing

A more complicated example is shown schematically in figure 10. This figure shows two fundamental solutions for the construction of a rolling bearing in the housing of for example a gear box. The left part shows a construction where the housing is separated into two parts. The two right sketches show a solution where the housing is not separated. Assembly of the axle will go through the hole of the bearing, which is a constraint for the diameter of all the elements on the axle.

5 DESIGN FOR ASSEMBLY SUPPORT

The previous paragraph described the geometrical relations between parts and the link to the mount operation. This paragraph describes some other aspects that are necessary for a successfull and reliable assembly process. A more extensive description can be found in Willemse (1995).

5.1 Design for gripping

Introduction
Gripping is defined as the establishment of a temporary relation between a part to be assembled and an assembly effector. Gripping is a necessary action for every part. The problems in gripping can be divided into three areas. A relation between the gripper and the part must be possible. The conditions of the assembly motions that act on the part-gripper relation must be considered. Finally, the motion of the gripper and the part towards the final assembly position must be possible. Support in this area should be aimed at selection of the gripping principle, selection of the gripper and including the gripping properties into the part.
Part of the knowledge necessary for design for gripping is specific for the available assembly system, especially the grippers.

Selection of gripping principles
Gripping principles can be divided into three categories:
- mechanical
- magnetic
- vacuum

The category of mechanical grippers is extensive. A basic distinction can be made between form closure and force closure gripping. Generally speaking, form closure can ensure a higher gripping accuracy but lacks flexibility.

The selection of a gripping principle depends on gripper availability and part characteristics such as material, shape, elasticity, weight.

When a gripping principle is chosen, the relation between gripper and part has to be detailed. The faces of the part that have contact with the product or with the feeding mechanism are not available for gripping. Baartman (1995) used these constraints to determine the faces of a product that are suitable for gripping (see figure 11).

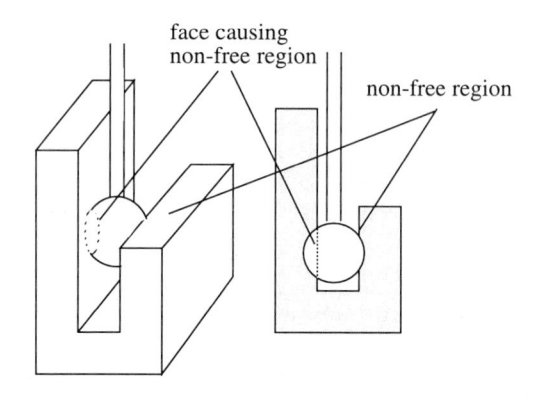

Figure 11 Non-free region caused by a face of a part

When the free faces are determined, features can be integrated in the product that facilitate gripping with the selected principle. Features can be a specific configuration of faces where the fingers of the gripper should grip, or specific protrusions or holes that form an interface between the part and the gripper.

Another important aspect is stability. The grip has to be stable under normal assembly conditions. Two types can be distinguished, as illustrated in figure 12. Stability of the contact points, the left part of the picture, is dependent on the shape and the configuration of the contact faces. The position of the centre of gravity relative to the grip position determines the gravitational stability.

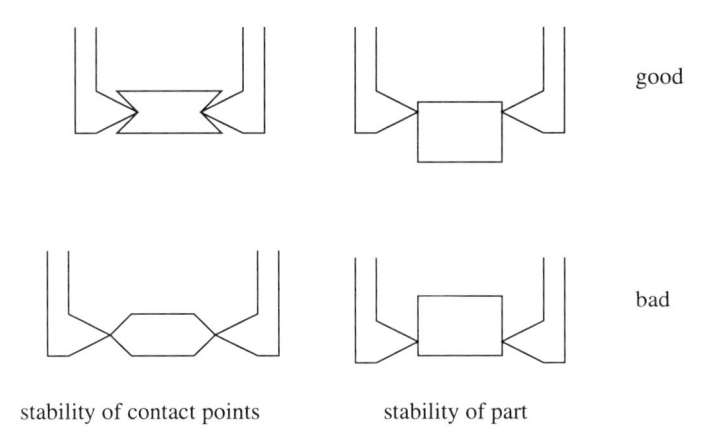

Integration of assembly considerations in product design 219

stability of contact points   stability of part

Figure 12 Grip stability

This section only described the gripping of a single part. Much optimization is possible when a sequence of parts is considered. This optimization can be directed towards minimization of the number of grippers necessary for assembly of one product, and minimization of the number of gripper changes in order to reduce the assembly time. These aspects are not elaborated in this paper.

5.2 Accessibility analysis

A necessary condition for the assembly of a part is the existence of a collision free assembly path. Collision can occur between the part and the product, but also between the gripper or the robot and the product. It is obvious that the assembly sequence highly influences the available collision free space. Techniques for collision detection have already been studied in the field of robot motion planning (Latombe, 1991). The challenge is to find methods that determine which type of solution is advantageous in case of collision problems. Basically two types of action can be distinguished: selection of an alternative assembly sequence or adaptation of the geometry of the product.

5.3 Stability analysis

Stability of the product is a necessary condition in case of automated assembly, but it will benefit manual assembly tasks as well. For stability analysis two types of information have to be analysed. The forces expected to act on the product or part being considered have to be summarized, and the DOF of the part or product have to be analysed with respect to form or force closure. By comparing the forces and DOF with corresponding orientations, stability can be predicted.

Stability has to be ensured in several stages of the assembly process. The product has to be stable after completion of the assembly process, the part assembled in the product must have a stable relation with the product and the product must resist the forces acting on the product during the assembly process.

6 CONCLUSIONS AND FUTURE WORK

The paper presents an analysis of the assembly properties of part relations, and a framework for design coordination. This knowledge is intended to support the design process in order to integrate the assembly properties into the product design. Future work will be focused on elaboration of the proposed techniques and development of designer support based on the framework. Implementation in a CAD environment will serve as a validation of the methods.

7 REFERENCES

Andreasen MM, Bowen J ea (1994) Design coordination framework, paper for CIMMOD/CIMDEV Workshop, Torino, September 22-23
Baartman JP (1995) Automation of assembly operations on parts, Delft University of Technology, ISBN 90-370-0119-X
Heemskerk CJM (1990) A concept for computer aided process planning for flexible assembly, Delft University of Technology, ISBN 90-370-0041X
Latombe J-C (1991) Robot motion planning, Stanford University, ISBN 0-7923-9129-2
Martens P (1991) CAD/CAM for assembly planning, Delft University of Technology, ISBN 90-370-0057-6
Nevins JL, Whitney DE (1989) Concurrent design of products and processes, McGraw-Hill Publishing Company
Tichem M, Storm T (1995/1) Design support for product structuring, to be published in Proceedings of Cim at Work, Joint final conference of Esprit working groups CIMMOD/CIMDEV, Eindhoven, August 27-30
Tichem M, Willemse MA, Storm T (1995/2) Product structuring and design coordination, paper for WDK Workshop on Product Structuring, Delft University of Technology, June 22-23
Willemse MA, Storm T (1995) Design support for assembly decisions, to be published in Proceedings of Cim at Work, Joint final conference of Esprit working groups CIMMOD/CIMDEV, Eindhoven, August 27-30

8 BIOGRAPHY

Michiel Willemse is a PhD candidate at the Laboratory for Flexible Production Automation, Delft University of Technology. He obtained his MSc degree in mechanical engineering in 1992, when he graduated on a concept for a learning CAPP system for parts manufacturing. His current research is in the field of assembly processes and design for assembly.

Ton Storm is associate professor at the Laboratory for Flexible Production Automation, Delft University of Technology. He obtained his doctors degree cum laude on the thesis "Design of a high speed cylindrical grinding machine". He was manager of the department "development of numerically controlled turning machines" at Hembrug in The Netherlands, then worked as a consultant in the area of production and production automation, and after that became manager of Development and Production at Gefra B.V.. His current specialty is flexible assembly systems.

19

Modeling and Planning of Disassembly Processes

E. Zussman[*], B. Scholz-Reiter[**], H. Scharke[**]
[*] Mechanical Engineering Department, Technion - Israel Institute of Technology, 32000 Haifa, Israel, eyal@HiTech.technion.ac.il
[**] Faculty of Mechanical, Electrical, and Industrial Engineering, Brandenburg Technical University, POBox 101344, D-03013 Cottbus, Germany, bsr@itt.tu-cottbus.de

Abstract
A certified technology for recovery of obsolete products is disassembly. Disassembly can lead to cost minimization, hazardous materials isolation, and opportunities to re-use or re-utilize materials and components. This paper presents an approach for modeling disassembly processes with Disassembly-Petri-Net (DPN) and probabilistic data with Bayesian Networks, which quantify beliefs about object states and disassembly operations. The combination of these models gives a planner the ability to react efficiently to different product states and unexpected results of disassembly operations.

Keywords
Petri Net, Bayesian Network, Disassembly for Recovery

1 INTRODUCTION

Due to new environmental legislation and continually increasing landfill costs, manufacturers are forced to consider new technologies to cope with obsolete products. Currently, the most promising solution seems to be automated product disassembly which can lead to cost minimization, hazardous materials isolation, and opportunities to re-use or re-utilize materials and components. Unlike existing manufacturing and assembly processes, disassembly processes are characterized by a high variety of products and manufacturers, uncertain product condition after usage, and a not rigidly defined process goal. Therefore, a new methodology is required for advanced modeling and planning of the disassembly process.

The present work aims at expanding well-known methods from the manufacturing area, and artificial intelligence, by offering a set of tools for integrating aspects of uncertainty

into the product's disassembly process. These tools include the following:
- A new Petri Net, Disassembly-Petri-Net (DPN), which contains all feasible disassembly sequences and relevant information.
- A Causal-Network, Bayesian-Network, which quantifies beliefs about the object states and disassembly operations. Capturing new data from the product's conditions, system beliefs are updated by means of backward and forward probability propagation.
- A linked mechanism which combines the DPN with probabilistic values from the Bayesian Network.

Our new model enables planning and reacting to different product states caused by different product conditions or different results of disassembly operations. It is the first model which includes all feasible disassembly sequences and the relevant uncertainty data. It is easily integrated into the large task of designing a product for disassembly (predictive planning) or to the actual disassembly planning process (reactive planning).

The paper is organized as follows: Section 2 introduces a Disassembly Petri Net, derived from an ordinary Petri Net, to model disassembly sequences and the system resources. Section 3 presents the Bayesian Network for modeling the probabilities along the disassembly sequences. Section 4 presents the integration of the DPN and the Bayesian Network to achieve reactive and predictive planning environment.

2   MODELING DISASSEMBLY PROCESSES

Modeling the disassembly sequences is an important task in creating the reactive disassembly planner and an intelligent decision support system for the disassembly process. To cope with this problem, several methodologies for representing product conditions and constraints as well as assembly sequences have been utilized and can also be used for representing disassembly sequences. In principle one can distinguish between methods describing the product conditions and precedence relations, connections and contacts, and methods mapping assembly tasks in the elements of representation. The former include connectivity graphs [Homem de Mello and Sanderson,1991] [Huang, 1993], precedence graphs or block graphs [Pagello and Qian, 1994], and the latter are based on And/Or graphs [Homem de Mello and Sanderson,1991], [Zussman et al., 1994], or Petri-Nets [Dicesare et al., 1993], [Baumgarten, 1990].

Using Petri nets to model the disassembly system has many advantages. One is able to describe the product, the disassembly process and the disassembly system with a single representation. The graphical representation is easy to understand and is used as a standard in industry. In addition, the Petri nets provide an abstraction in different levels of disassembly planning.

To construct a Petri net for disassembly we can choose different ways (by "Petri net for disassembly" we mean a Petri net for the planning level). In the literature the decomposition method is often used to gather assembly sequences [Kanehara et al., 1993]. If there already exists an Assembly Petri Net (APN), we can follow the same method and represent the disassembly process by reversing this APN. Generally, an APN is defined as a 4-tuple:

$$APN = \langle P, T, Pre, Post \rangle \qquad (1)$$

where:
P - places,
T - transitions,
Pre - pre arcs,
Post - post arcs.

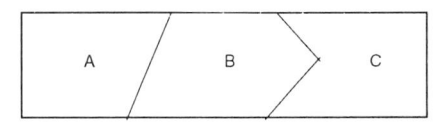

Figure 1 A three-part assembly

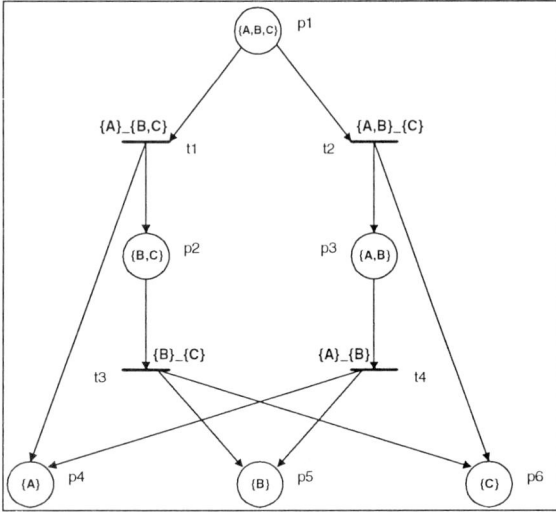

Figure 2 An inverted assembly Petri net for the {A,B,C} assembly

For example the reversed Assembly Petri Net of the assembly {A,B,C} (Figure 1) is shown in Figure 2. If the information on the APN does not exist then it can be extracted from the product structure.

Nevertheless, such a representation reflects a static process, i.e., a process with a fixed order of operations. This means we assume that every product has the same properties or conditions and that the disassembly operation has the same results for certain every time. Since the product properties and the results of the disassembly process are mostly unknown, we introduce in this research work a new type of Petri Net which we termed as Disassembly Petri Net (DPN).

Similar to the APN, the DPN consists of places, transitions, pre arcs and post arcs. To explain the differences, imagine, the dispatcher ordered a machine to disassemble part {A} from {A,B,C}. A normal Petri Net would contain a place p_1, the operation t_1 and the following state p_2 and p_3.

In the DPN a token passes from place p_1 through the transition t_1 to place p_{1a}, were place p_{1a} marks an unknown state (Figure 3). This means that we are not sure whether the machine will have, or has had, success with this operation or not. At this stage we still do not have enough information to make a certain statement. With the existing knowledge and the known product conditions we are only able to formulate beliefs in success or failure. We believe that the operation was or will be successful but do not know the exact events. Additionally, after the operation is done we know nothing about the actual product state until we sense the product state and properties (transition t_{1a}) to define the conditions for certain.

As Figure 3 shows, we have in the example two possible states after the disassembly. Either we were successful, in which case the result is part {A} and a subassembly {B,C}, or the disassembly operation was a failure and we are still left with the original assembly {A,B,C}. For more complex products it is conceivable to have more than two different options. Also due to the product form and connection points it is possible to have different disassembly opportunities, for example should we at first loosen Part {A} from {B,C} or should we start with loosening Part {C} from {A,B}? One can also use different fixtures to hold the product and different tools or machines for one and the same disassembly operation. To choose the different ways to disassemble the product and to distinguish the results from the sensing operation we attached values to the relevant arcs. The arc values after a decision place give us information about the optimal disassembly path.

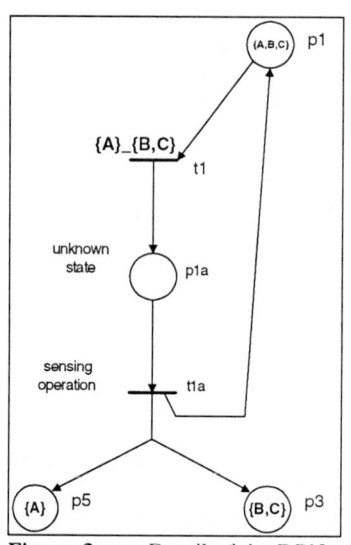

Figure 3 Detail of the DPN of the ABC assembly

Figure 4 describes a fully specified disassembly Petri net for the assembly {A,B,C} from the viewpoint of geometrical object constraints. The DPN introduced presents only an abstract level and sub-Petri-nets are used to model the machines, robots, conveyers, tools or fixtures in the disassembly system.

A disassembly Petri net (DPN) is defined as a 8- tuple:

$$DPN = \langle P, T, \Pr e, Post, \mu, \sigma, U_p, U_T \rangle . \qquad (2)$$

where:
P - places,
T - transitions,
Pre - pre arcs, where each pre arc $\Pr e(p_i, t_j)$ is associated with a decision value $0 \leq \mu(\Pr e(p_i, t_j)) \leq 1$ calculated from an internal algorithm,
Post - post arcs, where each post arc $Post(t_j, p_k)$ is associated with a probability value $0 \leq \sigma(Post(t_j, p_k)) \leq 1$ determined by external sources,
U_p and U_T - utility functions assigned to each place and transition.

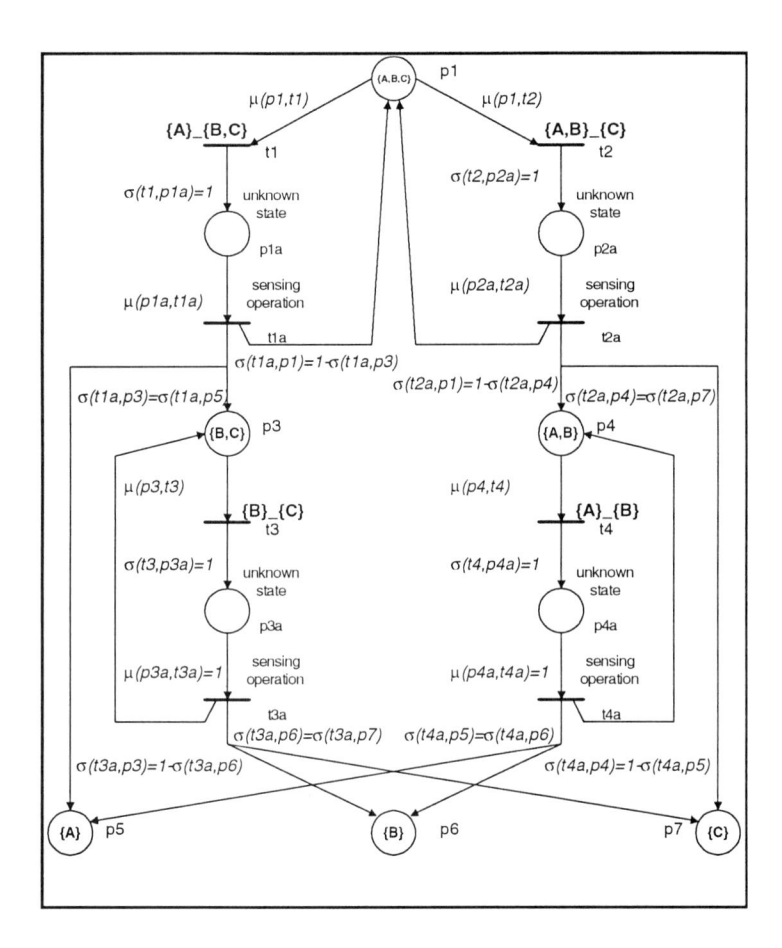

Figure 4 A DPN of the ABC assembly

To summarize the explanations, a DPN is a special case of ordinary APN's. We have to add different features to make them suitable for the demands of reactive disassembly processes:
- Due to different initial product conditions the same disassembly operation on different products of the same product type can result in different states. So, in the worst case, one always has the opportunity to receive the input state after a disassembly operation.
- After each disassembly operation we added a place for an unknown state and a sensing operation to define the results and the actual product state. Therefore, a place or a transition in the DPN can correspond to two different types. A place can be a feasible subassembly state of the product, or a unknown subassembly state. A transition can be

a feasible disassembly operation or an sensing operation.
- To express beliefs about the resulting product state we needed to add probabilities for each post arc coming from a sensing operation.
- To make a decision, we added to the pre arcs after each place arc values corresponding to the preferred disassembly path of the product.

3 UNCERTAINTY REPRESENTATION

This section is dedicated to Bayesian Networks [Perl, 1988], [Neapolitan, 1991] embodying a probabilistic expert system to describe the uncertainties in the actual product conditions and in the disassembly operations. These uncertainties are quantified and further attached to the Disassembly Petri Net.

Generally it is difficult for humans to structure and represent knowledge about a domain. Therefore a special representation model, called Semantic Networks, is often used to get a fast overview of the space and to demonstrate the causalities. In such networks, nodes represent propositional variables and arcs local dependencies. In Artificial Intelligence special applications to Semantic Networks are Neural Networks for learning mechanisms [Nadi et al., 1991], Influence Diagrams for decision making under uncertainty [Levitt et al., 1990], or Bayesian Networks for probability modeling [Heckermann, 1995].
In recent years many scientists have taken an interest in problems connected with uncertainty handling in Expert Systems. The Bayesian theory and the graphical modeling language, the Bayesian Networks, is in this area an often used and robust method of handling uncertainties using the probability theory. Many real-world applications using Bayes axioms have been developed [Heckermann and Wellman, 1995], such as in diagnosis, forecasting, automated vision, planning, sensor fusion or approaches dedicated to manufacturing control.
Since expert knowledge, sensor data, complete or incomplete statistical data, and uncertain product conditions have to be considered we have selected the probability approach. Instead of classifying the product or initial product conditions, like Fuzzy Logic, we quantify observed product properties and propagate events in the future [Drakopoulos, 1994]. This results in the opportunity to express beliefs about the product conditions and success or failure of operations. Bayesian Networks are a tool to represent uncertain knowledge in an intuitive and expressive way. The network is a directed or multi-directed graph where nodes represent variables $X = \{x_1,...,x_n\}$ about states in the world and these nodes are linked with arrows when they are dependent. The direction of the link represents local causal relations. Hence, a Bayesian Network represents a global joint probability distribution for a domain of variables.

The most important theorem and the heart of *Bayes theory* is illustrated in the following formula:

$$P(A_k|B_j) = \frac{P(A_k \cap B_j)}{\sum_i P(A_i \cap B_j)} = \frac{P(A_k) \cdot P(B_j|A_k)}{\sum_i P(A_i) \cdot P(B_j|A_i)}.\qquad(3)$$

This theorem represents a special form of *conditional probability*:

$$P(A_k|B_j) = \frac{P(A_k \cap B_j)}{P(B_j)} = \frac{P(A_k) \cdot P(B_j|A_k)}{P(B_j)}\qquad(4)$$

and has been developed under the exploitation of the product rule (*multiplication theorem* of joint probabilities):

$$P(A_i \cap B_j) = P(A_i) \cdot P(B_j|A_i)\qquad(5)$$

and sum rule (*marginal probability*):

$$P(B_j) = \sum_i P(A_i \cap B_j).\qquad(6)$$

If we take a closer look at Bayes theorem, we can see something like an inversion, meaning the revision of the *posterior probability* $P(A_i|B_j)$ after obtaining additional information about the *prior probability* $P(B_j)$. The Bayesian idea allows us to calculate the probability that there was an event A_i given that we observed the event B_j. This equations now provides the basis for propagating probabilities or beliefs in cause-effect models. Eq. (6) described the prediction (forward propagation) and Eq. (3) allows us to make a diagnosis (backward propagation) concerning an event.

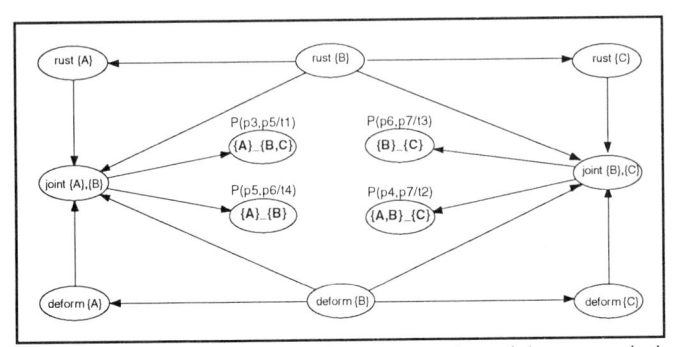

Figure 5 Bayesian Network representation of the uncertainties in the condition of the {A,B,C} assembly

To illustrate the above explanations, let us construct a Bayesian network for the assembly *{A,B,C}* according to the following procedure:
- First, find out what are the system variables and states.
- Second, the expert or the group of experts have to encode their knowledge about causes and effects.

- With the results of the previous steps the experts can construct a directed or multi-directed acyclic graph.
- In the final step they have to assign the initial probabilities for the root nodes and the local conditional probabilities distributions for the other nodes.

Our variables in this domain can be rust or deformation on part *{A}*, *{B}*, or *{C}* and the states of the joints between *{A}* and *{B}*, and between *{B}* and *{C}* (Figure 5). To simplify we consider only rust in part *{B}*, the two joints, and only discrete binary variables. Consequently we define three variables and their states:

- part *{B}* is rusted or not R_B $\{r_{B1} = yes, r_{B2} = no\}$,
- joint *{A},{B}* is good or bad J_{AB} $\{j_{AB1} = good, j_{AB2} = bad\}$,
- joint *{B},{C}* is good or bad J_{BC} $\{j_{BC1} = good, j_{BC2} = bad\}$.

Furthermore one has to think about the causalities. Considering the product geometry, it is obvious that the rust in part *{B}* has a big influence on both joints *{A},{B}* and *{B},{C}*. The representation of this information is illustrated in the causal network in Figure 6.

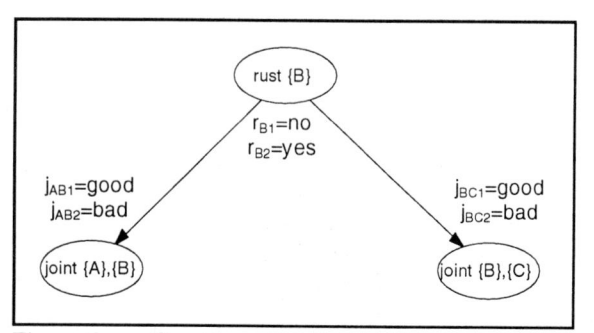

Figure 6   Graphical presentation of the causal relations for the assembly {A,B,C}

In addition it is necessary to specify the probability data for the nodes. Assume the probability of getting a product to be disassembled with a rusted part *{B}* is the same as that of getting a product where part *{B}* is not rusted, i.e. we set both to 0,5. Assume further that from a statistical database or from an expert with experience we got the following conditional probability values. The probability that joint *{A},{B}* is in good condition given the information that part *{B}* is not rusted is:

$P(J_{AB} = j_{AB1} | R_B = r_{B1}) = 0,9$

and the probability that joint *{B},{C}* is in a good state presupposed part *{B}* is not rusted is:

$P(J_{BC} = j_{BC1} | R_B = r_{B1}) = 0,8$.

Simultaneously we can calculate the values:

$P(J_{AB} = j_{AB1} | R_B = r_{B2}) = 0,1$ and $P(J_{BC} = j_{BC1} | R_B = r_{B2}) = 0,2$.

Given the initial probabilities, the network in Figure 7 allows us to make some probability computations. In general, computing interesting probabilities in causal networks is called *probabilistic inference* and is based on the theorems in Eq. (3) to (6).

The probability that part *{B}* has no rust, joint *{A},{B}* is in a good state, and joint *{B},{C}* is in a good state is the *joint probability*:

$P\left(R_B = r_{B1} \bigcap J_{AB} = j_{AB1} \bigcap J_{BC} = j_{BC1}\right)$

$= P(R_B = r_{B1}) \cdot P(J_{AB} = j_{AB1} | R_B = r_{B1}) \cdot P(J_{BC} = j_{BC1} | R_B = r_{B1}) = 0.5 \cdot 0.9 \cdot 0.8 = 0.36$

Or the probability that joint {A},{B} is in a good state is the sum over all outcomes. This means good conditions for joint {A},{B} if there is rust on part {B}, or good conditions for joint {A},{B} if we observe no rust on {B}.

$$P(J_{AB} = j_{AB1}) = P(J_{AB} = j_{AB1} \cap R_B = r_{B1}) + P(J_{AB} = j_{AB1} \cap R_B = r_{B2})$$
$$= 0.9 \cdot 0.5 + 0.1 \cdot 0.5 = 0.5$$

As mentioned, a property of Bayesian networks is the opportunity to make predictive and diagnostic statements along the network structure.

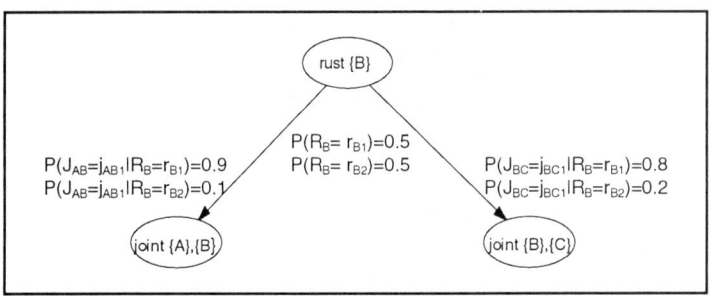

Figure 7 Detailed Bayesian network of the {A,B,C} assembly

Assuming that we measure the rust on part {B} with a sensor and we observe no rust, we can update the probabilities according to our measurement to $P(R_B=r_{B1}) = 1$ and $P(R_B=r_{B2}) = 0$. Now we can make a prediction and calculate the new probability for the joint {A},{B}, which results in:

$$P(J_{AB} = j_{AB1}) = P(J_{AB} = j_{AB1}|R_B = r_{B1}) \cdot P(R_B = r_{B1}) + P(J_{AB} = j_{AB1}|R_B = r_{B2}) \cdot P(R_B = r_{B2})$$
$$= 0.9 \cdot 1.0 + 0.1 \cdot 0.0 = 0.9$$

Another way is the diagnostic estimation. Imagine we have no sensor and try to unscrew joint {A},{B}. If this operation was successful, which means joint {A},{B} is in a good state, we can say something about the rust in {B}. Our interest is now the conditional probability $P(R_B=r_{b1}|J_{AB}=j_{AB1})$ of how our beliefs on rust in part {B} will change knowing that joint {A},{B} was in a good state. With Bayes theory (Eq. (3)) this is:

$$P(R_B = r_{B1}|J_{AB} = j_{AB1}) = \frac{P(R_B = r_{B1} \cap J_{AB} = j_{AB1})}{\sum_i P(R_B = r_{Bi} \cap J_{AB} = j_{AB1})} = \frac{0.5 \cdot 0.9}{0.5} = 0.9$$

After we have obtained a new value for the belief that part {B} is not rusted, we also want to know the new probability that joint {B},{C} is in a good state. Taking the new values, we have the probability that joint {B},{C} is in a good state:

$$P(J_{BC} = j_{BC1}|J_{AB} = j_{AB1}) = P(J_{BC} = j_{BC1})$$
$$= P(J_{BC} = j_{BC1}|R_B = r_{B1}) \cdot P(R_B = r_{B1}) + P(J_{BC} = j_{BC1}|R_B = r_{B2}) \cdot P(R_B = r_{B2})$$
$$= 0.8 \cdot 0.9 + 0.2 \cdot 0.1 = 0.74$$

4 TOWARDS A PREDICTIVE / REACTIVE DISASSEMBLY PLANNING SYSTEM

The goal of this approach is to model systems for predictive/reactive disassembly processes and to provide a decision support of selected methods and tools for such processes. The work here is limited to predictive/reactive planning systems, without scheduling methods, and depicts just the control of operation sequences in a single process.

In the previous sections we introduced tools for modeling uncertainties and process plans. At this section we explain the special case of a DPN linked to a Bayesian network to provide predictive/reactive decision support. This combination provides many advantages, so we are able to simulate different product and process conditions and states.

As previously mentioned we use probability values to represent our uncertainties about the product conditions. The probabilities are not of a static nature. Each product has different initial properties and conditions and, in addition, at the beginning of the disassembly process we may not have all the necessary information to describe the product conditions completely. This means that during the disassembly we continually have to try to improve our knowledge about the product to support an optimal decision of the disassembly methods and tools. Therefore the probabilities are updated, as do our beliefs, with every new piece of information. As usual in manufacturing new information referring to the product state, properties and condition is generated by sensors after a sensing operation. To be always up to date about the product one has to take into account the new information after each disassembly step. It is now clear that such a data exchange can only result if one links the structure of the DPN with the structure of the Bayesian network. A link means that relevant parts of the structure of the DPN communicate with parts of the Bayesian network. In the DPN we connected each probability value in the post arcs after a sensing operation with the corresponding probability value in the Bayesian network for the success or failure of the operation (Figure 8) as following:

for(j:=1; number(t_j); j++)
      if (t_j == sensing operation)
           link $\sigma\bigl(Post(t_j,p_k)\bigr)$ with the corresponding probability value
           in Bayesian Network;
      end;
end;

Thus each individual probability influences the other probabilities in the DPN. For example, a failure to disassemble part *{A}* from *{A,B,C}* decreases the probability of later disassembling part *{A}* from *{A,B}* for certain, and we can also suppose that the whole product is in bad condition.

If the sensor detects rust or deformation on part *{A}* this will change the probability distribution in the Bayesian network. Through the link, the updated probabilities for the disassembly operations appear immediately in the DPN and give us information about the product conditions and resulting "success rates" for different states. Assuming we decided to try a disassembly operation and sensed the actual state, the probabilities for the states after this operation will change either to one for the obtained state or to zero for all others, since an arc with zero decision values means no relation between the place concerned and the transition. We also have a feedback to the Bayesian network through the link. The new event, success or failure of the disassembly operation, changes our belief in the whole Bayesian network. The inference machine in the Bayesian network recalculates now by means of forward and backward propagation a new probability distribution. To close the loop, the new probability distribution in the Bayesian network again affects the DPN and a review of the optimal disassembly path in the DPN is necessary.

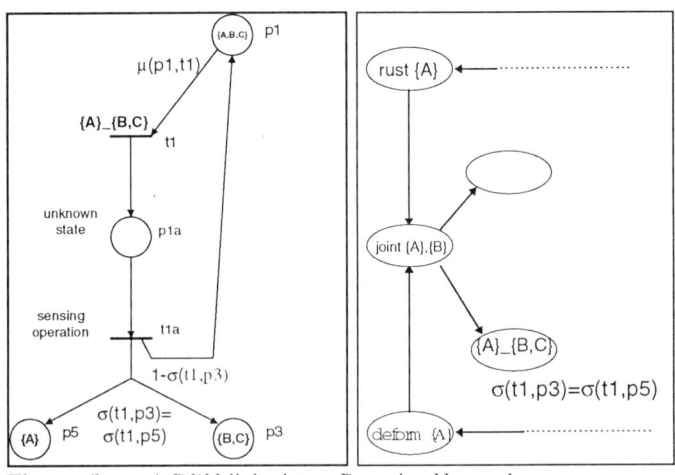

Figure 8   A DPN linked to a Bayesian Network

5 CONCLUSIONS

The presented approach introduces a new concept for modeling disassembly systems as well as uncertainties in product conditions and disassembly operations. Modeling is based on a Disassembly Petri Net (DPN) which holds information regarding disassembly environment, and the possible and feasible disassembly paths. Various uncertainties resulting from different product types, usage influences and the alterations of the original product's conditions are represented by means of a Bayesian Network.

It has been shown that the new method of linking the DPN and the Bayesian Network creates an appropriate environment for predictive and reactive planning. This can be used either at the Design for Environment (DFE) stage of the product, or directly at the recovery process planning stage.

6 ACKNOWLEDGMENT

This research was supported by the fund for the promotion of research at the Technion. Mr. H. Scharke was supported by the "Moshe Greidinger Scholarship Fund", administered by the Rotary International, District 2490, Israel. Finally thanks to Mr. Sven Vestergaard for supporting us with the Bayesian Network simulation software HUGIN (copyright Hugin Expert A/S, Niels Jernes Vej 10, Denmark).

7 REFERENCES

Baumgarten, Bernd (1990) *Petri-Netze Grundlagen und Anwendungen*. Wissenschaftsverlag, Mannheim.

Dicesare, F., Harhalakis, G., Proth, J. M., Silva, M., Vernadat, F. B. (1993) *Practice of Petri Nets in Manufacturing*, Chapman & Hall, London.

Drakopoulos, J. (1994) Probabilities, Possibilities, and Fuzzy Sets. *Stanford University, Department of Computer Science, Knowledge Systems Laboratory*, 701 Welch Road, Palo Alto, CA 94304-0106.

Heckerman, D. (1995) A Tutorial on Learning Bayesian Networks. *Technical Report MSR-TR-95-06*, Microsoft Research Advanced Technology Division, Microsoft Cooperation.

Heckermann, D. and Wellman, M. P. (1995) Real-world applications of Bayesian Networks, *Communications of the ACM*, Vol. 38, No.: 3, 24-57.

Homem de Mello, L. S., Sanderson, A.C. (1991) A Correct and Complete Algorithm for the Generation of Mechanical Assembly Sequences. *IEEE Transactions on Robotics and Automation*, **7**, **2**, 228-240.

Huang, K. (1993) Development of an Assembly Planner Using Decomposition Approach. *Proc. of the IEEE in Robotics and Automation*, 63-68.

Kanehara, T., Suzuki, T., Okuma, S. (1993) On Algebraic and graph structural Properties of Assembly Petri Net. *Proceedings of the 1993 IEEE/RSJ International Conference on Intelligent Robots and Systems*, 2286-2293.

Levitt, T., Agosta, J. and Binford, T. (1990) Model-based influence diagrams for machine vision. *Uncertainty in Artificial Intelligence 5*, L. N. Kanal and J. F. Lemmer, Eds. North-Holland, N. Y. 371-388.

Nadi, F., Agogino, A. and Hodges, D. (1991) Use of influence Diagrams and neuronal networks in modeling semiconductor manufacturing process. *IEEE Transactions on Semiconductor Manufacturing*, **4**, 52-58.

Neapolitan, R. E. (1991) *Probabilistic Reasoning in Expert Systems: Theory and Algorithm*. Northeastern Illinois University, Chicago, Illinois, John Wiley & Sons, Inc. Pagello, E., Qian, W.-H. (1994) On the Scenario and Heuristics of Disassemblies. *Proc. of the IEEE in Robotics and Automation*, 264-271.

Perl, J. (1988) *Probabilistic Reasoning In Intelligent Systems: Network of Plausible Inference*, Morgan Kaufmann.

Zussmann, E., Kriwet, A., Seliger, G. (1994) Disassembly Oriented Assessment Methodology to Support Design for Recycling. *Annals of the CIRP* **43**,1.

20
Recycling and Disassembly of Electronic Devices

K. Feldmann, O. Meedt
Institute of Manufacturing Automation and Production Systems
Egerlandstr. 7-9, Germany, D-91058 Erlangen
Tel.: ++49 (0) 9131 / 857710, Fax.: ++49 (0) 9131 / 302528
Internet: meedt@faps.uni-erlangen.de

Abstract

In this contribution possibilities for the improvement of the disassembly of electronic devices are presented and discussed. In the first part a software tool for fast planing and simulating disassembly in view of different frame conditions is shown.

On the other hand new partial destructive tools for the improvement of actual and future disassembly processes, which have been developed at FAPS, will be presented. This tools which are specially designed for disassembly are highly flexible and can be a major improvement for manual disassembly as well as a good starting point for further research on automated solutions.

Keywords
Disassembly, Disassembly Tools, Disassembly Planning, Recycling

1 INTRODUCTION

The environmentally friendly treatment of products at their lifes end becomes, not only because of environmental but also for economical reasons, more and more important.

Recycling concepts very often require disassembly processes to assure i.e. an efficient separation of hazardous materials or the accumulation of worthy ingredients for further recovery. Because of this fact and because of the low automation rate disassembly processes are a main cost factor in the process of recycling. Therefore tools and methods have to be found to improve the disassembly of electronic devices in order to make them more cost–efficient.

2 DISASSEMBLY –AN IMPORTANT STEP FOR THE RECYCLING OF ELECTRONIC DEVICES

Regarding disassembly of products, we have to distinguish between the treatment of conventional products which are currently returned, and products which will be designed for recycling and disassembly, returning in future times (Figure 1).

The vast majority of products returned now was designed years ago. Thus, for reducing the waste problem in the near future, first of all we have to develop technologies and techniques for recycling and disassembly in respect to old products.

Simultaneously design guidelines which support the development of environmental friendlier products have to be developed and improved. In addition for these new products technologies and techniques for the different recycling and disassembly strategies have to be found.

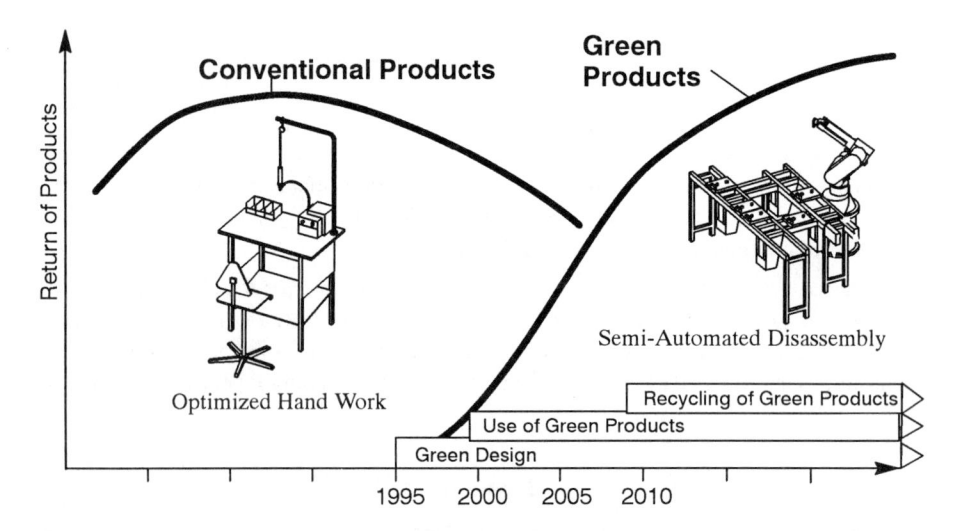

Figure 1 Alternativ strategies for the disassembly of actual and future products /1/

The major possibilities to improve disassembly processes are shown in Figure 2 . Only a combined interaction between Design for Disassembly, stored product data and flexible disassembly–tools (hardware and software) can assure efficient disassembly processes. Regarding the disassembly of currently returned products which are not designed for recycling, besides little support of product data, only the disassembly–tools can be improved in order to achieved efficient disassembly processes. For future products this knowledge should be used and included into the design process.

In the following, flexible tools and methods for the improvement of the disassembly of electronic devices (Figure 2 right) will be presented.

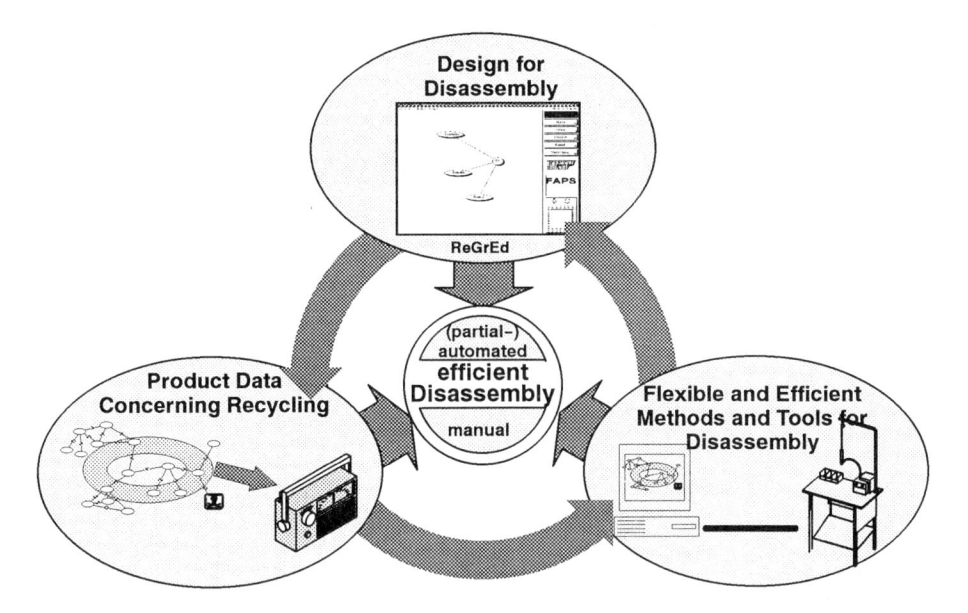

Figure 2 Preconditions for efficient disassembly processes

3 COMPUTER-AIDED DISASSEMBLY-PLANING

Regarding the disassembly of electronic devices in the industry shows, that there is no systematic planing of the disassembly processes. Up to now in almost all enterprises disassembly is only made for the separation of hazardous materials /5/. The aim of the system presented in this chapter is to support the disassembly simulation and planing in order to show the economically and ecologically optimum. The results can be used for disassembly enterprises as well as for the design of disassembly and recycling friendly products.

A rough model of a disassembly enterprise can be described as follows: There is an input, usually different product types and an output which is in general a spectrum of different fractions of materials and components for further recovery. The main function of a disassembly enterprise is the economically and ecologically optimal dismantling of the incoming products so that the components can be related to the output fractions. This can be reached considering the optimal disassembly depth for the different products, where the resulting costs for the gained fractions and the costs for needed disassembly processes are minimal /2, 8/.

These considerations are the starting point for the development of the software-tool for disassembly planning (DisPlay) which is currently being developed at FAPS.

The software tool is based on the Recyclinggraph Editor (ReGrEd) /2, 3, 4/, which is a software module for modeling products with information regarding disassembly and recycling (e.g. structure of the product, components, connections) about a certain product

(Figure 3). This data can be used for improving the design of a product as well as for disassembly planning.

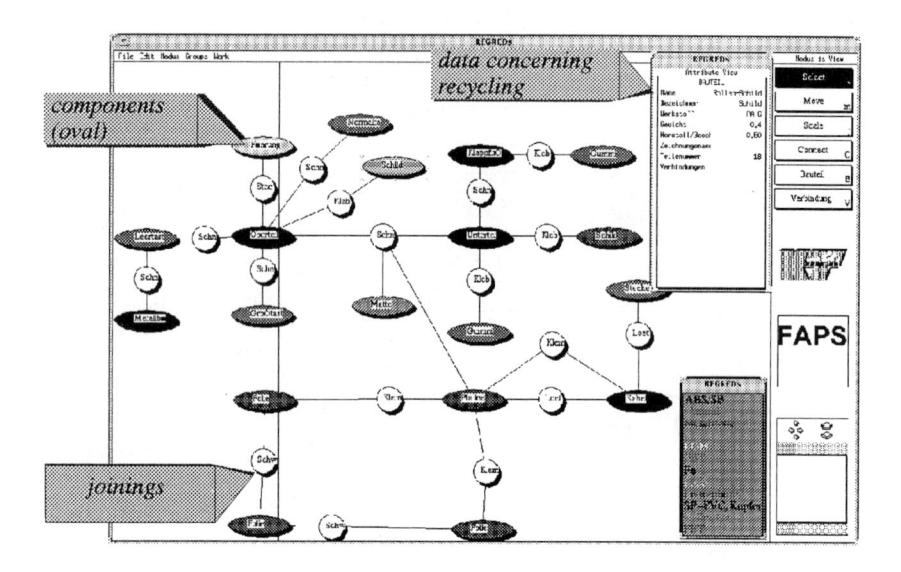

Figure 3 The Recyclinggrapheditor (ReGrEd) exemplary for a keyboard

The concept for the disassembly planing module DisPlay is shown in Figure 4. There are two groups of data which are needed for the simulation and calculation of disassembly processes.

- **Product data** can be gained from ReGrEd (Figure 3). The source of the data can be product marking in combination with a database (at the beginning of the product–lifecycle) or disassembly samples for the generation of the necessary product data.
- **Data concerning the frame conditions** are gained through modeling of a real or virtual disassembly enterprise. This database contains the actual or virtual frame conditions for disassembly. The output fractions for example are described by a special system and are related to the costs or gains. Disassembly tools, too, and the costs for different disassembly processes as well as labour costs are stored in the database.

The main function of DisPlay is the calculation of the optimal disassembly depth under consideration and combination of the data mentioned above. Therefore in a first step all technical possible disassembly steps are calculated. This algorithm is based on the product data stored in ReGrEd using the information about the structure of the product. The detailed description of the algorithm is very complex /6/ and not subject of this paper.

In a second step the clumps of all resulting disassembly possibilities have to be related to the modelled output fractions. Because the modelled output fractions are oriented on the real or virtual market they are not completely compatible to the material combinations in the resulting clumps. Therefore it is possible that material combinations fit into several output–fractions (i. e. to a fraction for reuse, to fraction containing mixed plastics, or a fraction with hazardous materials). The problem of relating the clumps to the right output fraction will be solved by a material interface in combination with a rule- interpreter. In addition to the right output-fraction the costs for the disassembly processes too, have to be related to all disassembly combinations.

In the next step the disassembly combination with the lowest costs is calculated and shown as the optimal disassembly depth to the user. In addition to that, the resulting disassembly sequence is calculated.

The results of this calculations can be used for the following actions:

- Determination of the optimal disassembly processes and sequences for the manual or automated disassembly
- Determination of a complete disassembly documentation containing a recovery plan, the costs and the amounts of the different output fractions.
- Proposals for building recovery groups concerning different disassembly processes.

As mentioned above the calculation is based on two types of data. For the simulation of the disassembly both categories of data can be changed. So it is possible to change the frame conditions for disassembly for getting different scenarios. On the other hand also the product can be changed with regard to recycling and disassembly. This is an important information for product designers. Therefore the results can be used for efficient disassembly processes for building up optimal output fractions as well as for improving the design of the products. A more detailed description of the software tool for disassembly planing and for design for recycling is given in /2/.

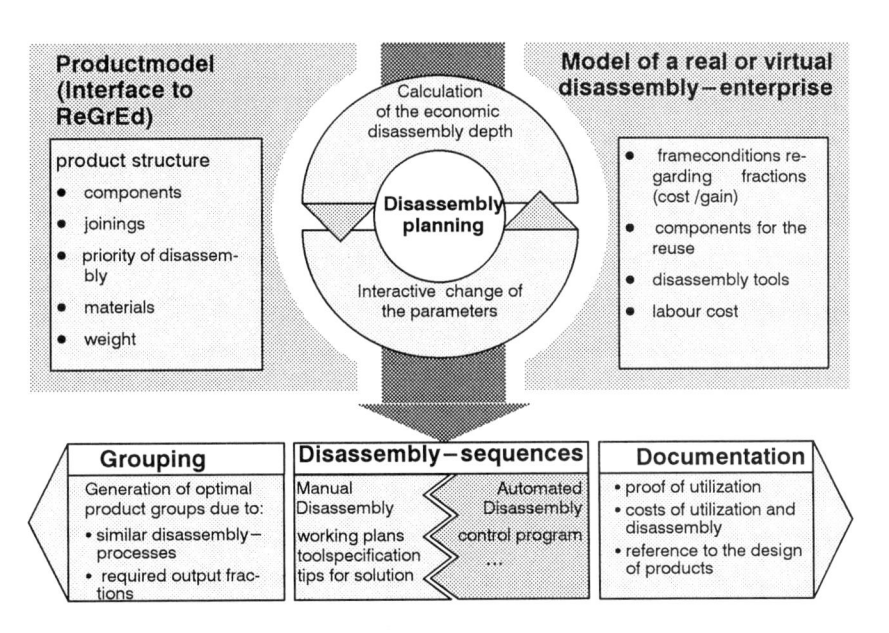

Figure 4 Concept of the system for disassembly planning DisPlay

4 TOOLS AND CONCEPTS FOR EFFICIENT AND FLEXIBLE DISASSEMBLY OF ELECTRONIC DEVICES

In addition to the planing of optimal disassembly and the calculation of the most economical disassembly depth the major point of improvement concerning the disassembly of existing devices, is increasing the efficiency of the disassembly processes. Up to now most of the tools used in disassembly enterprises are designed for assembly or maintenance of products /7/. Because the frame conditions for disassembly are different, these tools don't fulfil the requirements. Highly flexible tools for example are needed. On the other hand a partial or complete destruction of components does not matter in order to improve the efficiency.

Disassembly processes can be described in an abstract form as shown in Figure 5. The function of disassembly tools is based on a chain of elementary processes as for example forces, torques, temperature, which are transmitted to the product. This can lead to different types of disassembly:

- destructive processes in order to reach the destruction of components or joinings
- destructive processes in order to generate working points for further destructive or non destructive processes
- non destructive processes to inverse the assembly processes

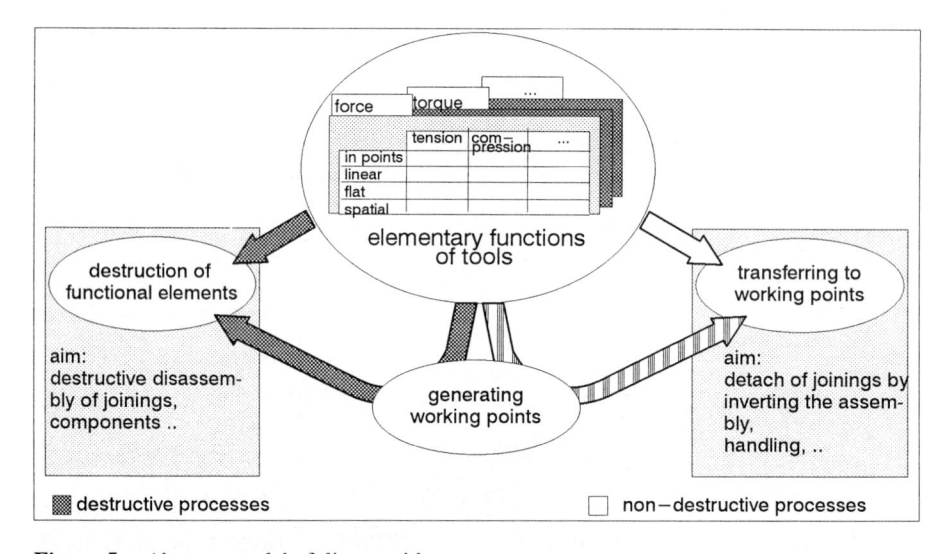

Figure 5 Abstract model of disassembly processes

Taking these the frame conditions of the disassembly of electronic scrap into account, there are the following requirements main on disassembly tools:

- Flexibility concerning products and variants
- Flexibility concerning different processes
- Failuretolerant systems

Recycling and disassembly of electronic devices 239

With regard to these requirements it seems to make sense to use the option of destructive or partial destructive processes for flexible disassembly. Some attempts for the development of tools using this strategy for the transmission of torques and forces will be shown in the following.

Regarding destructive processes there are many possibilities to fulfil the requirements. In this contribution only the most promising opportunity, the removing of material by rotating tools (i.e. drilling) will be investigated.

4.1 Transmission of torque in combination with drilling

The transmission of torque is especially needed for loosening and unscrewing screws. State of the art is the use of tools which are designed for assembly, i.e. different screwdrivers and spanners.

Taking the frame conditions of disassembly and destructive processes into account, a concept for a new tool for unscrewing, the so called drilldriver (patented) was found (Figure 6).

The drilldriver is a tool, similar to a left-turning twist drill with two cutting edges, which are shaped in a special geometry.

Drilldriver

Transmission of torques
- Using existing working points (cross recessed head, slot etc.)
- Using self created starting points (drill chips)

Drilling
- Removing material for the destruction of joining elements
- Creating drill chips as working points for the transmission of torques

Operations

Partial- or non-destructive disassembly
- Cross recessed screws
- Slotted screws
- Hexagonal socket screws
- Hexagonal head screws
- Torx
- ....

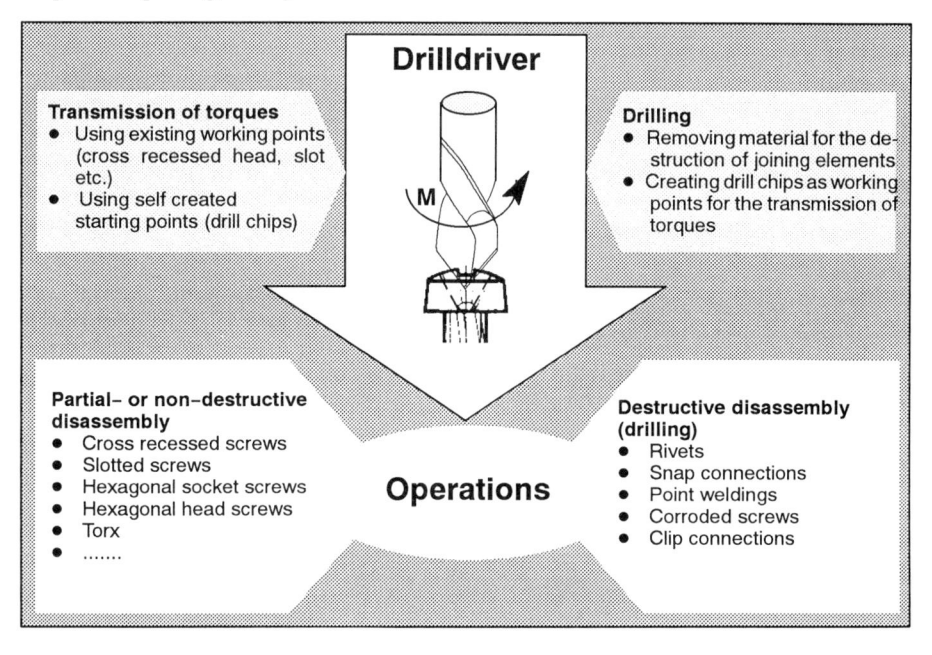

Destructive disassembly (drilling)
- Rivets
- Snap connections
- Point weldings
- Corroded screws
- Clip connections

Figure 6 Profile of the drilldriver (patented)

Using a drilldriver the following important disassembly functions can be fulfilled:

- **Transmission of torque using existing working points**
  The special geometry of the drilldriver allows to generate a form closure between the cutting edges and existing working points of many screw types. This form closure is used to transmit

the torque for unscrewing. For example cross recessed screws, slotted screws, hexagon sockets, and other screw types can be unscrewed.

- **Creating working points for the transmission of torque**
  In case, that there are no working points for form closures available, a short drill process is started, to create drill chips, which are used afterwards to transmit the torque. Both processes, drilling and the transmission of the torque are done in a left turning modus. With this option also hexagon head screws and hexagon socket screws can be disassembled.

- **Destruction of joining elements**
  In case, that the disassembly of the joining by transmission of torque is not possible (i.e. corrosion of screws), or not useful (i.e. disassembly of rivets), the tool can be used similar to a drill bit to remove material with high rotational speed until the joining is disassembled. In this way it is possible to destroy joinings as for example corroded screws, rivets, point weldings.

The advantages of the drilldriver are mainly in the extremely high flexibility, concerning different joining types and sizes as well as concerning different disassembly processes (non destructive, partial destructive, or destructive). In Figure 7 some examples of screws, which have been disassembled with the drilldriver are presented. The existing or generated working point s are marked by pointers.

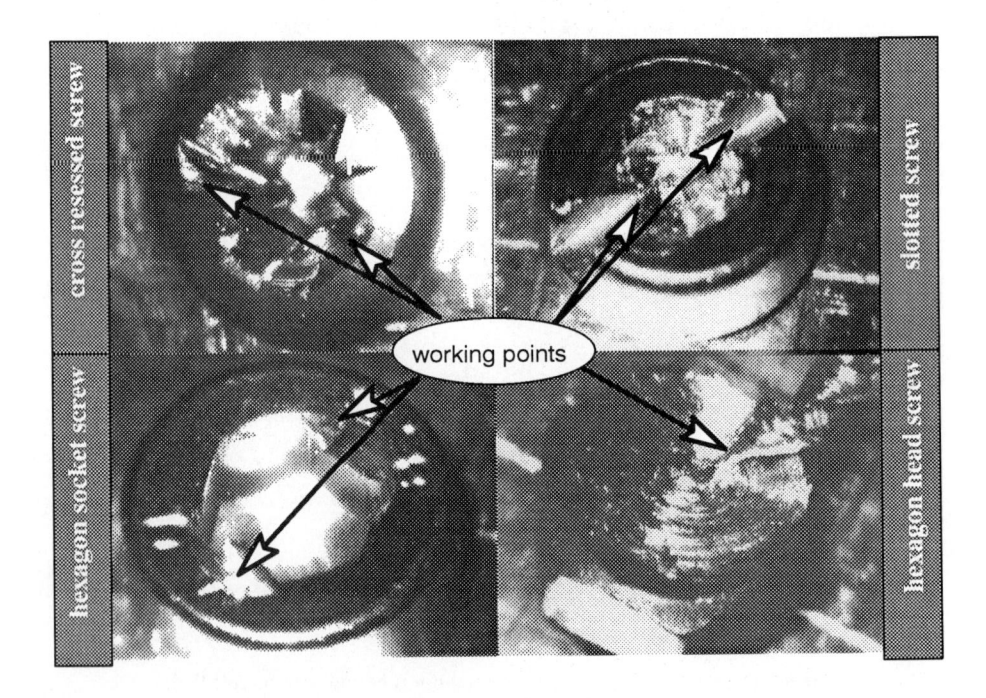

Figure 7 Examples for screws, which have been disassembled with the drilldriver. The existing or generated working points are marked by pointers.

Recycling and disassembly of electronic devices 241

There have been several investigations to optimize the geometry and the properties of the drill-driver–tool in order to fulfil the requirements. In this paper a short overview over the results will be given.

Geometry of the drilldriver

Investigations with different geometries show that depending on the screw head there exist several optimum geometries. So with regard to the distribution of screw types to be disassembled, these geometries can be combined to one overall-optimum. One example for a special shape of the top of the drilldriver is shown in Figure 8.

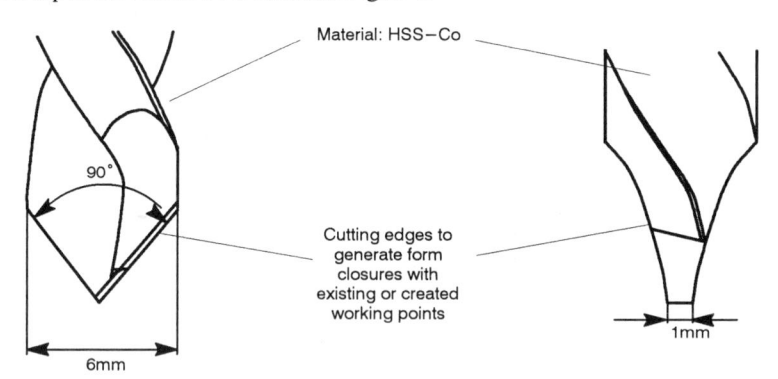

Figure 8 Example for an optimized shape of a drilldirver (screwsize M3–M5)

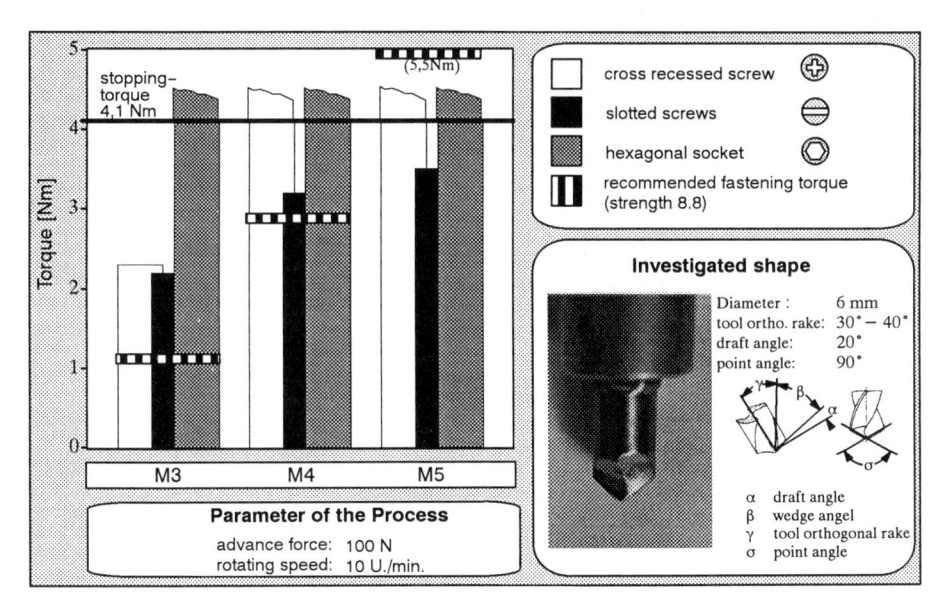

Figure 9 Maximum torque transmission with the drilldriver depending on different screws

The properties of a drilldriver shaped as in the drawing above, regarding the maximum transmission of torque on different screw types and sizes, are shown in Figure 9. As basis for comparison the recommended fastening torque for different screws (strength type 8.8), too, is shown in the figure.

In most cases the torque of the drilldriver is higher than the recommended fastening torques. In this context it has to be pointed out that the transmittable torque can be extremely increased by pressing the tool with higher advance force into the screw head.

4.2 Transmission of forces in combination with drilling

Regarding disassembly of products, the transmission of forces plays an important role, especially concerning gripping and fixing processes as well as processes for the destruction of components. Using tools, as they are state of the art, special working points are needed for the transmission of forces. Because of this extremely flexible gripper devices, as needed for disassembly processes, are not available.

Based on this frame conditions a flexible gripper, the so called drillgripper was developed and patented. The drillgripper is a tool that integrates a drill bit and a grip device. The function of the drillgripper is shown in figure 10. First a working point is created by drilling a hole in the product or component. Then the grip device is inserted into the hole. After that grip process is started by activating the grip device.

The grip device is an element which has in the inactive status a smaller diameter than the drill top. In the active status the grip device is shortened by pressure and the diameter increases so that a form closure between the tool and the component is created. An example for such a grip device could be an elastomer, as shown in Figure 10.

The grip process can be finished by elongating the grip device

By the use of this tool the following operations are possible:

- Grip and clamp processes for handling and fixing complex products without existing usable working points
- Transmission of high forces in order to achieve a destruction of components (i.e. housings)
- Many drillgrippers can be combined in order to increase the transmittable forces and to transmit torques
- With little changes the tool can be used for the continued transportation and handling of workpieces. In this case the tool is fixed at the workpiece and is used as an defined working point for handling and transportation systems.

Recycling and disassembly of electronic devices 243

Figure 10 Prototype of the drillgripper ; left: creating a working point; right: activating the grip device by shortening an elastomer

4.3 Transmission of forces and torques in combination with drilling

In the chapters above separate flexible solutions for the elementary functions "transmission of forces" and "transmission of torques" in combination with drilling have been presented. Both tools can create their working points themselves by drilling. So both concepts can be integrated without problems in one tool (Figure 11).

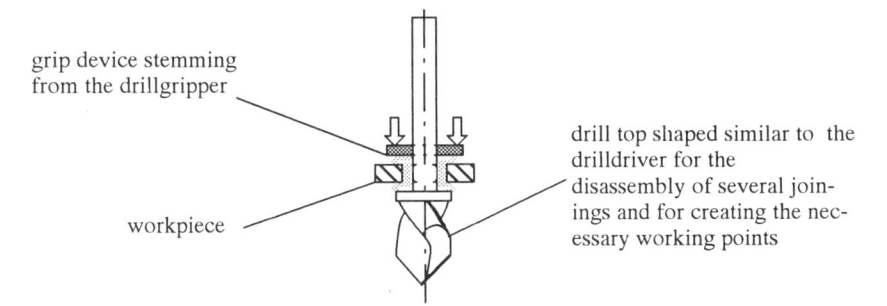

Figure 11 Integrated tool for the flexible transmission of forces and torques

The resulting tool can be used as drilldriver for the disassembly of joinings (see chapter 4.1) as well as for handling and transportation of workpieces like a drillgripper (4.2).

So using the integrated flexible tool shown in Figure 11 it is possible to fulfil the requirements of many disassembly processes without tool change.

5 CONCLUSIONS AND FURTHER WORK

In this contribution attempts and solutions have been shown to support the disassembly of electronic devices in order to make them more efficient.

The software-modules DisPlay and ReGrEd allow a fast simulation and planing of disassembly under different frame conditions, and the calculation of the optimal disassembly depth.

On the other hand tools and systems for the improvement of actual and future disassembly processes have been presented. This tools which are specially designed for disassembly are highly flexible and can be a major improvement for manual disassembly as well as a good starting point for further research on automated solutions.

In the further work the disassembly planning system will be completely implemented and improved. In the area of flexible tools for disassembly further partial destructive tools will be developed by investigating different remaining methods of creating working points during the disassembly process.

6 REFERENCES

/1/ Feldmann, K.; Meedt, O.; Scheller, H.: Life Cycle Engineering - Challenge in the Scope of Technology, Economy and General Regulations, RECY '94, Proceedings of the 2nd International Seminar on Life Cycle Engineering, Hrsg. v. K. Feldmann, Meisenbach Verlag Bamberg

/2/ Feldmann, K.; Meedt, O.; Meerkamm, H.;Weber, J. Entwicklung einer CAD-CADis Verfahrenskette auf der Basis recyclingrelevanter Produktkennzeichnung Proceedings of the Workshop "Umweltinformationssysteme in der Produktion" 9/1995, Berlin

/3/ Meedt, O.; Scheller, H.; Weber, J.:Entwicklung eines Werkzeuges zur recyclinggerechten Produktentwicklung und zur wirtschaftlichen Demontageplanung, VDI-Seminar "Kreislaufwirtschaftsgesetz: Entwicklung und Demontage recyclingfähiger Elektronikgeräte und Bauteile - Technik und Werkzeuge", Erlangen, 30.3.-31.3.1995

/4/ Krause, D.; Scheller H.: Recyclinggerechte Produktgestaltung und wirtschaftliche Demontageplanung feinwerktechnischer Geräte. Proceedings " Serienfertigung feinwerktechnischer Produkte" October 1994, Braunschweig.

/5/ Hesselbach, H.; Werder, K.; Elektronikschrott-Recycling: Engpass ist die Demontage, me Bd. 9 (1995) Heft 1

/[6/ Schröder, H.: Entwicklung eines rechnergestützten Tools Bestimmung der wirtschaftlichen Demontagetiefe, Master-Thesis at FAPS, University of Erlangen- Nuremberg

/7/ Kahmayer, M. Flexible Demontage mit dem Industrieroboter am Beispiel von Fernsprech-Endgeräten, Springer-Verlag 1995

/8/ Kriwet, A.: Bewertungsmethodik für die recyclinggerechte Produktgestaltung. (Produktionstechnik - Berlin, 163) München, Wien: Hanser- Verlag, 1995; TU Berlin, 1994

7 BIOGRAPHY

Prof. Dr.-Ing. K. Feldmann

Born 1943, Klaus Feldmann graduated Dipl.-Ing. and Dr.-Ing. in productional engineering from the Technical University of Berlin. From 1975 to 1982 he has had different leading functions at Siemens AG in the field of manufacturing automation and assembly. Since 1982 Feldmann is professor in manufacturing automation and production systems and member of the board of the Manufacturing Institute at the University of Erlangen–Nuremberg. Professor Feldmann is a member of the International Institute for Production Engineering Research (CIRP), of the Scientific Society of Production Technology (WGP), and of the Society of German Engineers (VDI).

Dipl. -Ing. O. Meedt

Otto Meedt, born in 1968, reached his Master of Production Engineering at the University of Erlangen–Nuremberg in 1993. Since 1994 he is scientific collaborator at the Institute for Manufacturing Automation and Production Systems (head: Prof. Dr.-Ing. K. Feldmann) in Erlangen. His research area is disassembly of Electronic Devices. Currently he is working on projects for development of tools and methods for manual and automatic disassembly as well as software development for disassembly planning, sponsored by the German and Bavarian government and from industry.

21
INFORMATION MANAGEMENT TO SUPPORT ECONOMICAL DISASSEMBLY OF TECHNICAL PRODUCTS

Prof. Dr.-Ing. D. Spath, Dipl.-Ing. C. Tritsch
Institute for Machine Tools and Production Science, University of Karlsruhe, Kaiserstraße 12, D-76128 Karlsruhe, Germany
Tel.: ++49 721 608 4010, Fax: ++49 721 699153

ABSTRACT

In view of overburdened landfills, of an increased environmental awareness and of ever stricter legislation, an environmentally adequate disposal of technical consumer goods is gaining more and more importance. Due to the fact that manufacturer's data of products which have to be recycled today and in the near future are relatively incomplete within the framework of this contribution an aproach is presented where the necessary data of product structure required for flexible planning and support of the disassembly and recycling processes are obtained by systematic manual product analyses. Based on this, a procedure to determine the economically best disassembly depht is described as a basis for the disassembly processes to be carried out and afterwards, possibilities for the flexible integration of automated subprocesses are shown.

Keywords
Information management, disassembly planning and optimization, automated disassembly

1 MOTIVATION

Over the last few years, the mass distribution of technical consumer goods and ever shortening product life cycles have led to a continuous increase in the amount of discarded electric appliances requiring disposal. Moreover, news about exhaustable raw material resources and an imminent ecological collaps due to overburdened landfills and growing waste heaps reach more and more into the awareness of the people as well as into that of the manufacturing companies.

This change in awareness is also reflected in current legislation as f.e. the law on circular economy and waste disposal which was passed in Germany and which is supposed to mark the beginning of a circular economy in order to lower the amount of waste (Hansen, 1994). In particular, the manufacturers will in future be obliged to show product responsibility which means on the one hand to design their products as environmentally friendly as possible and on the other hand to take care of their disposal at the end of the product life. This increasingly presents the challenge to recycle discarded products in the best possible economically and environmentally safe way i.e. to create cycles through the different phases of the product life.

2 DISASSEMBLY AND RECYCLING

According to today's perception thereof, disassembly, carried out at least partially, is indispensable as the first step in the recycling procedure to reduce the environmental impact by the separation of harmful substances or to close single material or part cycles (Jovane, 1993). A mostly non-destroying disassembly procedure is the basis of product recycling to allow the individual treatment and processing of single parts or components for the purpose of reuse. Additionally disassembly is necessary to produce valuable secondary raw materials for the purpose of material recycling because of the limited operating efficiency of available material separation technologies. Therefore, destroying disassembly technologies can be applied as well (Spath, 1995).

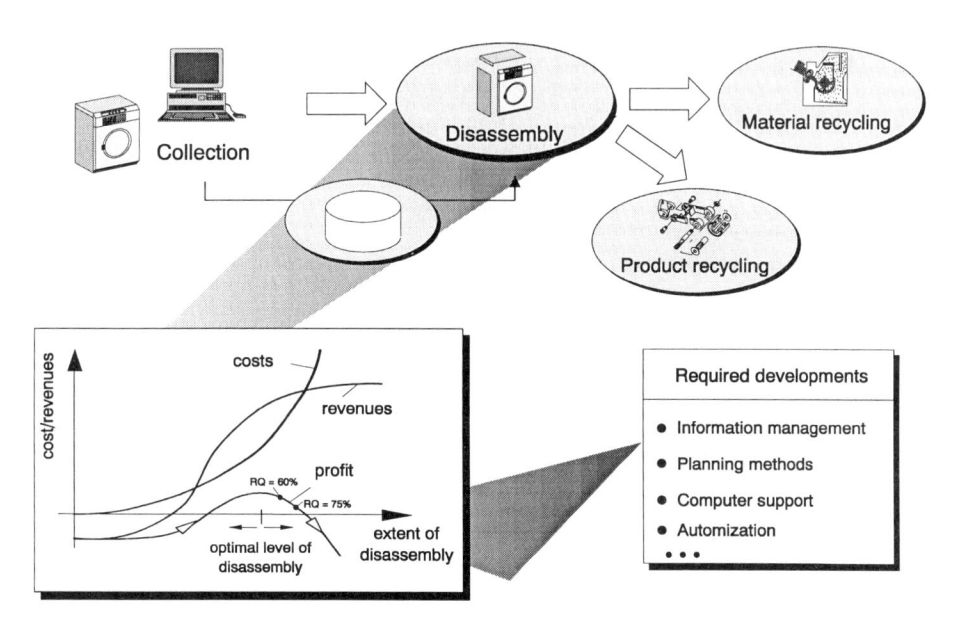

Figure 1 Economics of disassembly.

In practice one often meets with a mixture of both kinds of cycles and here the disassembly process only rarely means the direct reversal of the former assembly processes; instead there are some major differences. When assembling products, the shape, the materials and the quality of single parts or components are known and the assembly processes are always completely run through since only this guarantees the functionality of the products and leads to the highest possible value recovery. However, when disassembling products within recycling processes the highest possible value recovery may already have been reached when only a part of the materials, parts or components are separated. Typically the product development took place between 5 and 15 years ago so information on valuable and harmful materials and on the product structure are difficult to obtain or no more available.

Apart from influences in the phase of usage, such as repairs and corrosion which require great flexibility of the processes to be implemented, planning and execution of disassembly processes are determined by such influencing factors as the current legislation, the changing sales market for used parts and secondary raw materials or the recycling technologies available at the moment (Spath, 1995). Great flexibility is therefor required and particularly the disassembly depth is variable and for economic processes it has to be adjusted regularly to the current marginal conditions (Figure 1).

Considering the rising numbers expected for the future as well as the high personnel costs compared to the relatively low potential of value recovery, strictly manual disassembling will not be profitable in the long run. Automation is therefore the aim in this area, too, particularly if in future products of disassembly-adequate structure are also about to be recycled.

3   FLEXIBLE DISASSEMBLY CONCEPT

In order to make the disassembly and recycling process as efficient as possible in future, an approach is presented with which it is as well possible to compensate the above-mentioned information deficit as to guarantee the required flexibility of adjustment to external marginal conditions. Figure 2 shows the proceeding for the disassembly of technical consumer goods, subdivided into three steps.

First, a disassembly-oriented structuring of the products to be recycled is required on the basis of systematic test disassemblies. The analysis is used to determine missing product-specific information necessary for disassembly planning. It describes the product components obtainable via disassembly and thus serves to calculate possible proceeds as well as accompanying disassembly process chains necessary to calculate possible costs. First, only non-destructive or partly destructive disassembly steps are considered here since the aim is to recycle at a high value level.

Starting from this disassembly structure which describes the maximum degree of disassembly of the relevant products, the optimum extent of disassembly is determined at which the difference between obtainable proceeds and variable costs reaches the greatest value, taking into consideration current external and internal marginal conditions and provided that a concrete disassembly order was placed. To determine the proceeds a profitability assessment of the product components as specified in the analysis has to be carried out on the basis of the current market value of secondary materials and used parts or components.

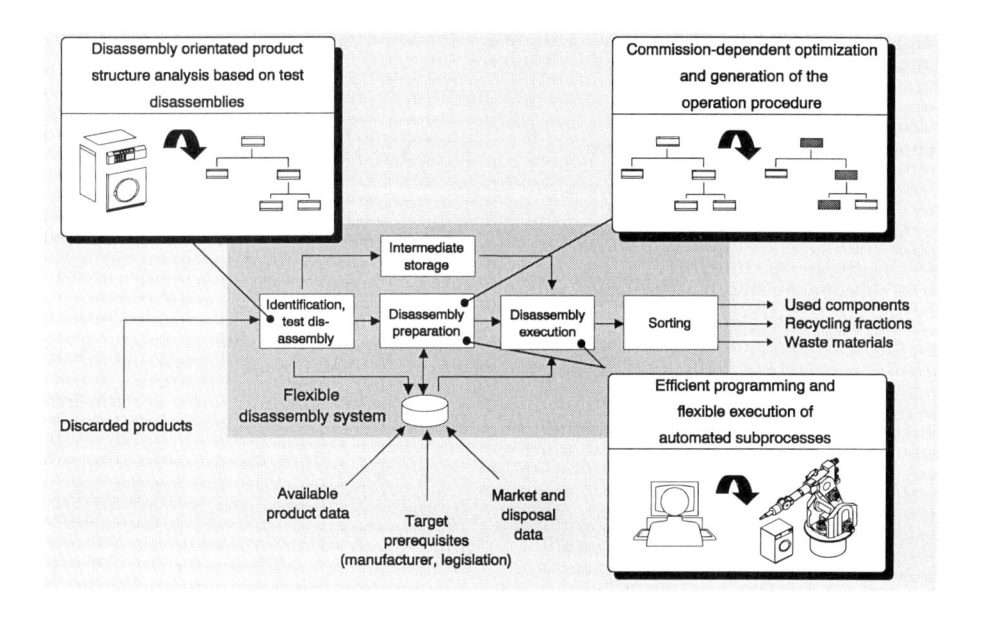

Figure 2   Disassembly concept.

To increase profitability, methods for flexible automation of essential subprocesses are required and also their integration into the overall procedure. Since there are typically small lot sizes and a great diversity of variants which require frequent changes in the disassembly process, computer aided tools for efficient program development and adjustment are particularly needed. As exact geometrical data on the products to be disassembled are not available, the exact geometrical data have to be generated via adequate sensor systems on-line based on a task-oriented description of the processes and a simple product modelling.

4   INFORMATION MANAGEMENT AND DISASSEMBLY PLANNING

To carry out disassembly of technical consumer goods economically and ecologically, information on their disassembly structure is required as well as on valuable or harmful materials contained or on those impeding the subsequent recycling processes. This leads to the task to create a corresponding information flow starting at the single product manufacturers. Apart from the option to enclose the required data in future directly in or at the product (Gerlach, 1994), standardized disassembly data and special recycling requirements depending on separate product manufacturers may also be stored in special databases and may be transferred to single disassembly companies when required. Thus, the manufacturering companies have the chance to directly influence the recycling path of their products and to adapt them to the latest developments of technology or to changing laws, regulations and strategies.

It is certainly imaginable in future to generate disassembly-specific data and precedence relations directly via analysis and evaluation of product-specific restrictions from CAD systems which are based on known procedures from the field of assembly planning (Zussmann, 1994). However, due to long product life cycles this will come into effect only in the medium term so in the period between it is necessary to generate the required informations using systematic product analyses. To achieve this, a test disassembler dismantles a new type of discarded appliance in a standardized test disassembly by the aid of a tabular record form and the currently available manufacturer's data (such as parts lists and exploded views from construction, after-sales service).

Since the test disassembler may easily observe the recording of disassembled parts and of parts-related disassembly targets, i.e. of those components directly accessible or disassemblable after a certain disassembly step is executed, the precedence relations between the different steps may subsequently be determined automatically (Figure 3).

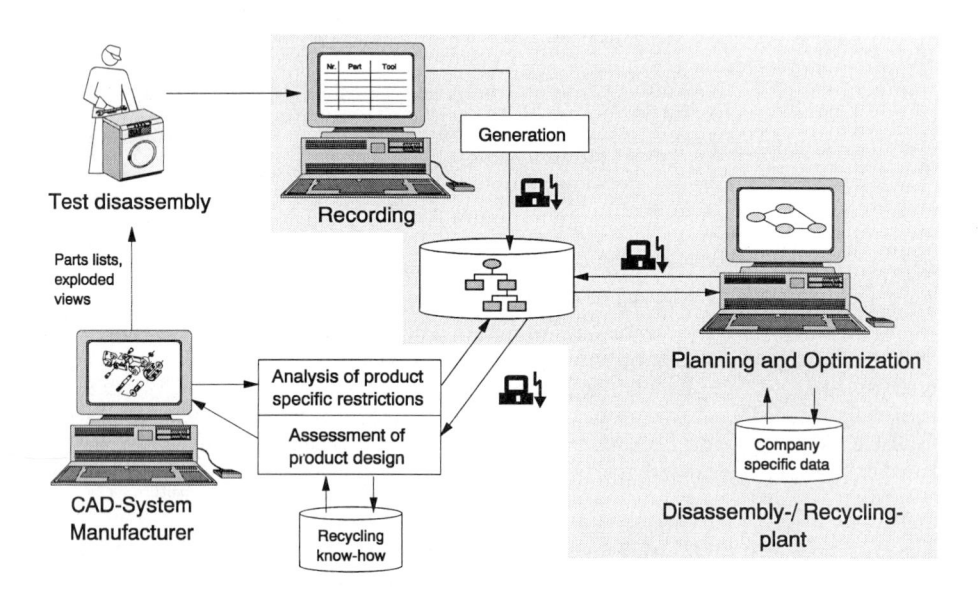

Figure 3 Communication and data exchange.

A formal description and presentation of these correlations may be given in the form of a disassembly graph. In a second step, by providing this information, the single disassembly companies may carry out order-specific planning when supplied with discarded products. Based on the manufacturer's requirements (e.g. minimum scope of disassembly, removal of single parts and components for reutilization) or on those of legislation (e.g. removal of contaminants) as well as adjustment to the current market situation (costs of landfill deposits and disposal, obtainable proceeds for secondary raw material and used parts) they can each time determine the profitable disassembly depth anew and thus decide on the operation sequence to be carried out.

Corresponding to the materials contained in the disassembled components or to the parts which may be reutilized, it is possible to assign them to a recycle fraction contained in the company specific database including corresponding information on purchasers and prices.

Here, the recycle fractions are organized hierarchically so it is possible to determine which mixed fraction components or assembly groups belong to by evaluating the min/max shares of basic fractions assigned to the materials within the mixed fractions (Figure 4). Manual planning intervention based on special analysis and evaluation functions offers the possibility of determining optimization prerequisites and parameters. Thus, the planner retains for himself the option to support the optimization process manually and accelerate it considerably.

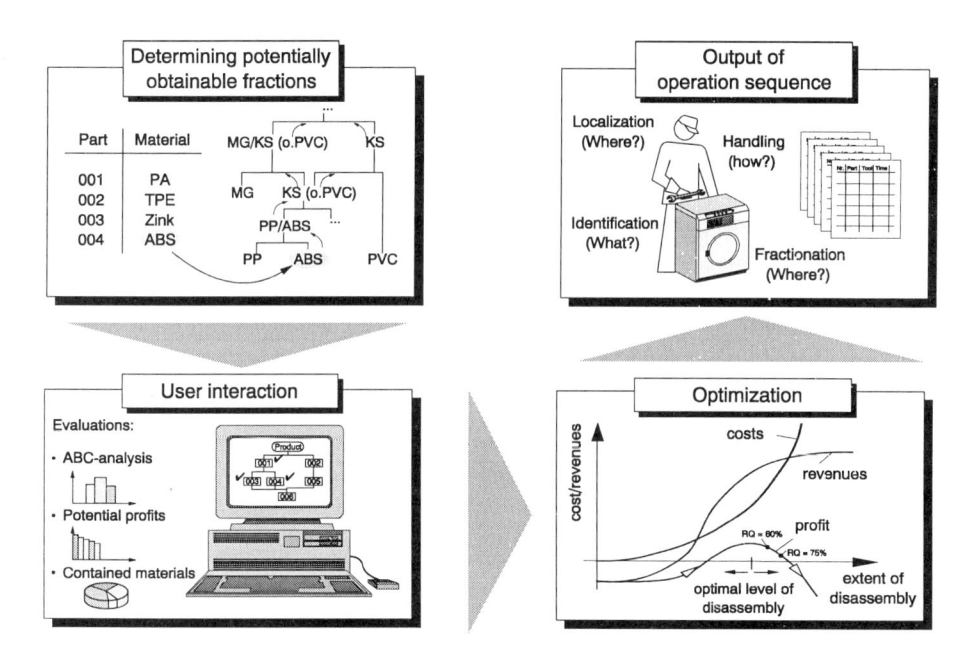

Figure 4   Disassembly planning procedure.

The target of optimization is to maximize the coverage which is received from the obtainable proceeds for material fractions or used parts minus the disassembly costs:

$$E = P_Z - P_A = \sum_{i=1}^{m} p(B_i) \cdot m(B_i) - \tau \sum_{i=1}^{p} t_i = \max. \tag{1}$$

The costs of a disassembly path Λ, i.e. of a certain sequence of disassembly steps, consist of the required disassembly time of the separate work steps t_i and of a cost rate τ. The corresponding proceeds P_Z are calculated by assigning the newly obtained parts or components B_i to one of the current recycle fractions and their market price as well as each mass.

To determine the optimum disassembly depth, all possible disassembly options have to be considered because even the execution of one single disassembly step may result in major differences in the total proceeds. As the number of disassembly possibilities for a product consisting of n parts may be a maximum of

$$|\xi| \leq \sum_{k=0}^{n-2} \binom{n}{k} + 1 = \sum_{k=0}^{n} \binom{n}{k} - \binom{n}{n-1} = 2^n - n, \qquad (2)$$

the appliance has to be structured into as many (dis-)assembly groups and subgroups as possible within the framework of the test disassembly so that the optimization algorithm may be applied recursively. Due to this procedure and by considering the product-specific restrictions, the maximum number of possibilities for disassembly may be drastically reduced. This renders possible a computer-aided evaluation by successively creation a graph of possible disassembly states.

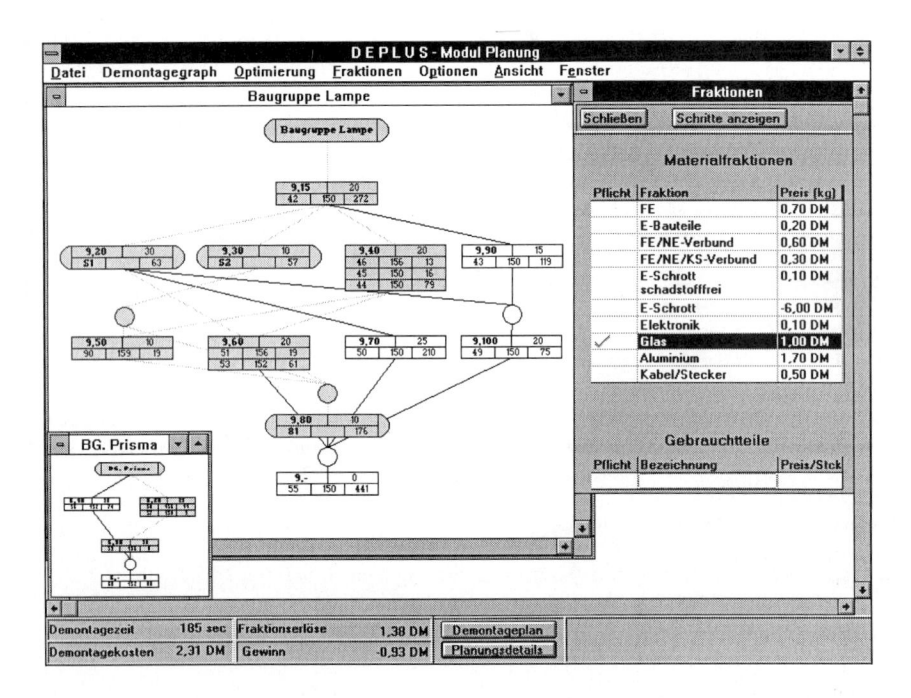

Figure 5 User interface of the disassembly planning module.

As the material proceeds rise with increasing disassembly depth it is possible to start the optimization procedure with a complete disassembly of the product or disassembly group and to stop when all disassembly steps of one disassembly level supply lower material proceeds than the maximum profit already known of the subordinate one do.

The order-dependent disassembly plan produced within the framework of planning is used as basis for the supply of information at the single disassembly workplaces together with further assignable data such as tools required, connecting techniques to be undone or fractionation hints. In the simplest case this can be done by printing the work instructions but it may also be carried out by a demand-adequate inquiry at a computer terminal (Weule, 1994).

Based on the described methods, a software prototype named "DEPLUS" (**De**montage**pl**anung und -**u**nterstützungssystem) was already implemented. Figure 5 shows the interactive use interface of the attached planning module.

In the right window the potentially obtainable recycle fractions of an example product are shown by evaluating the marketable used parts and secondary raw materials and the parts data from the subjacent database. Based on the disassembly graph, possible disassembly paths and the necessary disassembly steps for certain disassembly objectives (e.g. removal of contaminants) may be visualized and analyzed by calculating obtainable costs/ proceeds and further evaluation functions. This functionality gives the user the chance to gain a detailed overview over the potentials of value recovery contained in the product in order to systematically specify requirements and parameters for subsequent optimization procedure. Correspondingly, the results of the optimization process are finally visualized by the disassembly graph.

5   INTEGRATION OF AUTOMATED DISASSEMBLY PROCESSES

The planning tools described were developed first by considering purely manual disassembly processes and starting from parts lists, provided by the product manufacturer. In future this shall be used also as a basis for flexible integration of automated sub-processes into the overall procedure.

Recycling complex products places high requirements on the ability of an automated system to react to changes of products and variants as well as to product changes due to usage (e.g. repair, damage, soiling). Because of the complexity of the task and as the products are often not yet designed in a disassembly-friendly way, an automation step-by-step is advisable where first of all automated stations are realized for single processes which are ever recurring even in different variants (e.g. removal of cathode ray tube when disassembling TV sets). To identify such processes, humanitarian aspects (work safety, ergonomics of workplace) have to be considered alongside technical and economic ones. When designing such a hybrid disassembly system, the adaptability and flexibility of single work areas with regard to changed work contents are very important. Here a modular, clearly arranged program structure is needed as well as a simple, efficient programming aids for the automated stations. In particular, a high degree of autonomy has to be reached in order to avoid unnecessary standstill by independent, fault-tolerant execution of the required disassembly functions.

Figure 6 shows an approach for a macro-oriented robot program development which is based on a combination of special functions of the robot, disassembly tools and sensor systems to parameterizable functional macros. They are at the user's availability in the form of program modules with which he can describe the disassembly process at a high level of abstraction supported by a preceeding rough geometrical product modelling. By the integration of sensor devices, the program flow within such a module may be changed using data obtained on-line.

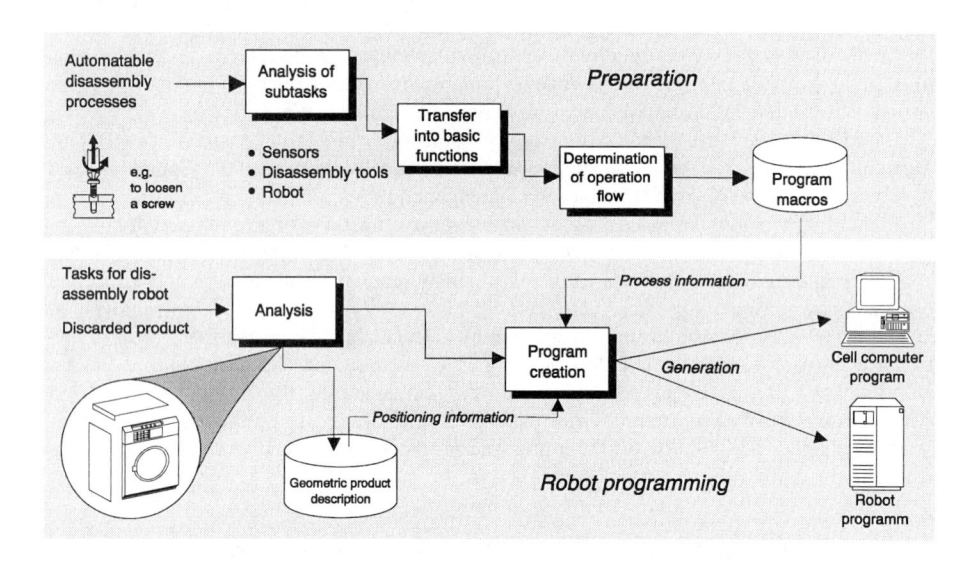

Figure 6 Robot programming procedure.

This provides the possibility on the one hand to carry out the precise robot positioning at the program running time and thus to simplify the geometrical product description required additionally or to be able to react to possible changes due to the usage phase, on the other hand it is possible also to present the detailed process knowledge directly on the prephrased program macros and to hide it from the user. Correspondingly, the result is a simple and fault-tolerant creation and execution of the disassembly programmes. Based on the concept described, a software prototype is currently implemented and in co-operation with an industrial partner, a disassembly system for (partially) automated disassembly of TV sets has already been realized as a first application.

6 CONCLUSION

In future a joint data base which contains this uniform representation of disassembly structures can be used both for comprehensive assessment of the products and for disassembly planning. That will be an important basis for the necessary information exchange between the product design and the recycling process in the context of life-cycle engineering.

Therefore, a computer-aided tool for product development with disassembly and recycle adequacy, with an emphasis on the life cycle oriented assessment of product alternatives is developed at the Institute for Machine Tools and Production Science. The assessment procedure of the *"Eco"-Portfolio Method* aims at a quantitative appraisal of life cycle oriented environmental effects of technical products. This is made possible by retracing environmental effects to so-called environmental drivers (Weule, 1994 and Spath, 1995). Therefore, a computer-aided partly automated generation of the disassembly graph is developed based on existing CAD data.

The presented disassembly planning methods were outlined and implemented in an initial implementation step, taking into consideration purely manual work steps. In the future the flexible integration of automated partial-processes into the complete procedure shall be facilitated. The prototype of a flexibly automated disassembly cell based on the presented methods is already being installed at the Institute for Machine Tools and Production Science.

6 REFERENCES

Gerlach, G., Sorber, J., Daniel, D. and Kaden, S. (1994) The problem of data storage and handling for selective disassembly processes of electronic scrap, in *Proceedings of the 2nd International Seminar on Life Cycle Engineering (RECY'94)*, Erlangen.

Hansen, U. (1994) Redistribution and disassembly in an industrial circular economy: *Umwelt Wirtschafts Forum,* **4**, 39-46.

Jovane F. e.al. (1993) A Key Issue in Product Life Cycle: Disassembly, in *Annals of the CIRP 42/2*, Edinburgh.

Spath, D. and Tritsch, C. (1995) Flexible disassembly of technical consumer goods - a prerequisite for future recycling concepts: *Technische Rundschau Transfer,* **18**, 12-15.

Spath, D., Hartel, M. and Tritsch, C. (1995) Life cycle assessment - Tools to support environmental product design and economical disassembly of technical consumer goods, in *Proceedings of 10th International Conference on Engineering Design (ICED 95),* Praha.

Weule, H. and Spath, D. (1994) The Utilization of Hypermedia based Information Systems for Developing Recyclable Products and for Disassembly Planning, in *Annals of the CIRP 43/1,* Singapore.

Zussmann, E., Kriwet, A. and Seliger, G. (1994) Disassembly-oriented assessment methodology to support design for recycling, in *Annals of the CIRP 43/1,*Singapore.

7 BIOGRAPHY

o.Prof. Dr.-Ing. Dieter Spath, born in 1952. Study of mechnical engineering at the Technical University of Munich. In 1981, a doctor's degree at the Institute for Machine Tools and Management Science at the TU Munich. Entering the company group KASTO in 1981, he has been managing director of the same since 1988. In 1992 appointment to the post of professor in ordinary at the University (TH) of Karlsruhe, Institute for Machine Tools and Production Science.

Dipl.-Ing. Christian Tritsch, born in 1964. Study of electrical engineering at the University (TH) of Karlsruhe. Since 1992 research assistant at the the University (TH) of Karlsruhe, Institute for Machine Tools and Production Science.

22

A proposal of CPR-Graph method for assembly sequences generation

Jian Wang, Kazuhiro Ohkura, and Kanji Ueda
Department of Mechanical Engineering, Kobe Univ.,
Rokkodai, Nada, Kobe 657, JAPAN
Phone: ++81-78-803-1123
Fax: ++81-78-803-1131
e-mail: ueda@mech.kobe-u.ac.jp

Abstract

This paper presents a Graph of Connection Precedence Relationship(CPR-Graph) for generating all the feasible assembly sequences for a set of mechanical parts. The CPR-Graph is defined as a special directed graph whose nodes and edges correspond to parts or subassemblies and the relations between them, respectively. Each edge in the graph has a label indicating a connection precedence relationship between the parts or subassemblies. Joining two parts or subassemblies corresponds to the following operation sequence in the CPR-Graph: deleting the two corresponding nodes and the edges between them, adding a new node corresponding to the assembled subassembly, and rewriting the labels of the other edges in the graph related to the change. Above operation sequence is called shortening operator. The central idea of the proposed method is that the process of assembling a product corresponds to the process of shortening the CPR-Graph of the product. Therefore, a complete assembly sequence is generated by using shortening operator repeatedly to change the CPR-graph from the initial state to the goal state. Where the CPR-Graph of initial state consists of the nodes only corresponding to the parts whereas the CPR-Graph of goal state consists of only one node corresponding to the whole assembly. The major advantages of the proposed method are its simplicity and less computational complexity compared to the other related methods.

Key Words

Assembly process, assembly sequences, assembly task, CPR-Graph, shortening operator, the initial state, the goal state

1   INTRODUCTION

The choice of the assembly sequence for a mechanical product affects the efficiency and flexibility of the assembly process, and characteristic of the finish assembly. The complexity of assembly steps, the needs of fixturing, the unit cost of assembly, and the occurrence of need for rework, are all affected by assembly sequence choice[1]. The choice of assembly sequence is an

important task for a production engineer. Exploring the choices of assembly sequence is difficult due to following two reasons.
(1) The number of feasible assembly sequences can be very large even at a small parts count and can rise strikingly with the increment of parts count.
(2) For some products, small changes in the design can have a large impact on the assembly alternatives.

For some complex products, the number of feasible assembly sequences may be so large that even skillful production engineers will overlook many feasible assembly sequences. There is growing need for systematizing and computerizing the generation of all feasible assembly sequences. Therefore, a computer representation of mechanical assemblies is necessary in order to automatize the generation of all feasible assembly sequences. A computer system for assembly planning must have a way to represent the assembly sequences. Several methodologies for generating assembly sequences have been proposed. The methodologies can be classified into following two approaches:
(1) The problem of generating the assembly sequences is transformed into that of generating disassembly sequences in which the disassembly sequences are the inverse of feasible assembly sequences.
(2) All feasible sequences are generated from a series of rules algorithmically. The rules are generated from the answers to a series of questions about the mating of parts and multiples of parts.

The former method includes the representation based on AND/OR graph suggested by Homem de Mello and Sanderson[2][3][4][5]. The latter method includes the representation based on liaison precedence relationships suggested by De Fazio and Whitney[6][7]. We will briefly explain these two methods before explaining our proposed method.

Homem de Mello et al. introduced an AND/OR graph representation of assembly sequences. The nodes in these AND/OR graphs correspond to parts or subassemblies, and hyperarcs correspond to assembly tasks in which two parts or subassemblies are joined to yield a more complex larger subassembly. The root node is associated with the set of parts that correspond to the whole assembly. The hyperarcs point from the node corresponding to the larger subassembly to node corresponding to the smaller subassemblies or parts. Figure 1.2 shows the AND/OR graph of feasible assembly sequences for a ball-point pen shown in Figure 1.1.

De Fazio et al. introduced an assembly sequence diagram representation of assembly sequences. The nodes in these assembly sequence diagrams correspond to the states of the assembly process, and the arcs correspond to the assembly move. The top node and the bottom node in these assembly sequence diagrams correspond to the initial state and the final state of assembly process, respectively. Figure 1.3 shows the assembly sequence diagram of the ball-point pen shown in Figure 1.1.

This paper proposed a Graph of Connection Precedence Relationship (CPR-Graph) as a new method to generate all the feasible assembly sequences for a mechanical product. The CPR-Graph is defined as a special directed graph whose nodes and edges correspond to parts or subassemblies and the relations between them, respectively. We can use the CPR-Graph to represent a state of the assembly process of an assembly. The initial state of the assembly process is the state that all parts are separated, whereas the final state is the state that all parts are joined together forming the whole assembly. The CPR-Graph in which each node only corresponds to a single part represents the initial state of the assembly process, and another CPR-Graph which only has one node corresponding to the whole assembly represents the final state. The former CPR-Graph is referred to as the initial state's CPR-Graph and the later one referred to as the goal state's

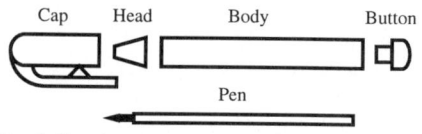

Figure 1.1 The ball-point pen.

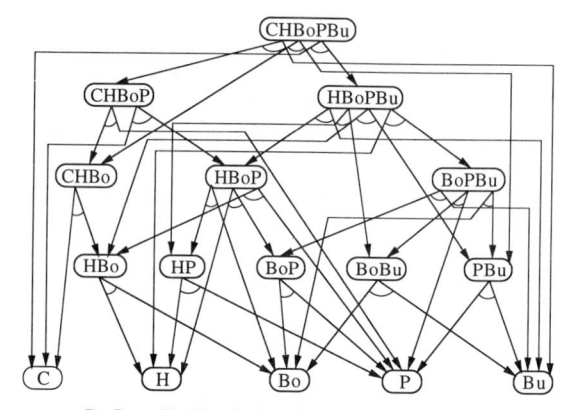

C=Cap, H=Head, Bo=Body, P=Pen, Bu=Button

Figure 1.2 AND/OR graph of feasible assembly sequences for the ball-point pen.

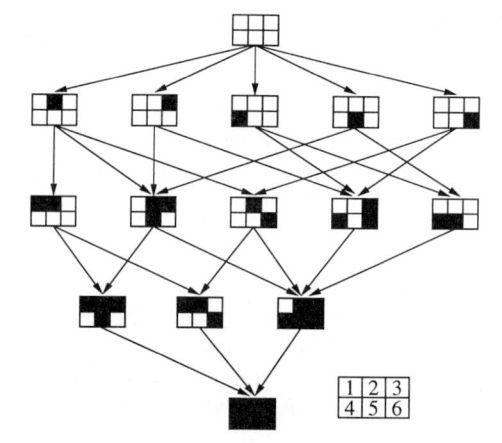

Figure 1.3 The assembly sequence diagram of feasible assembly sequences for the ball-point pen.

CPR-Graph. The CPR-Graphs whose nodes correspond to parts or subassemblies represent a state of the assembly process between the initial state and the final state. An operator, shortening operator(Its definition will be shown in section 3) is used to make the CPR-Graph change from the initial state to the goal state. A path for shortening the CPR-Graph whose initial node is the initial state's CPR-Graph and whose terminal node is the goal state's CPR-Graph corresponds to a feasible assembly sequence of the assembly. Therefore, a complete assembly sequence is generated by using shortening operator repeatedly to making the CPR-graph change from the initial state to the goal state. The major advantages of the proposed method are its simplicity and less computational complexity compared to the other related methods.

In this work, it is assumed that:
(1) Parts are interconnected if at least one contact surface exits between them.
(2) Two parts or subassemblies are joined at each time and after parts have been joined together, they remain together until the end of the assembly process.
(3) If parts or subassemblies are joined forming a large subassembly, all the contacts between the parts in that subassembly are established.
(4) The feasibility of joining two subassemblies is independent of how those subassemblies were built.

In the following, section 2 describes the CPR-Graph representation for an assembly. Section 3 describes the method for generating mechanical assembly sequences by using the CPR-Graph. Section 4 discusses our proposed method and finally, section 5 gives a brief summary of this work.

2   CPR-GRAPH REPRESENTATION FOR AN ASSEMBLY

2.1 CPR-Graph representation

As mentioned last section, the CPR-Graph of an assembly is defined as a special directed graph whose nodes and edges correspond to parts or subassemblies and the relations between them, respectively. If there is at least one contact between a pair of the parts or subassemblies, there will be two directed edges of the opposite direction between the two corresponding nodes in the CPR-Graph. Each edge in the graph has a label indicating a connection precedence relationship between the parts or subassemblies. Therefore, we can use the CPR-Graph to represent the connection relation and the precedence relation between parts or subassemblies in a state of the assembly process. The CPR-Graph representing a state of assembly process can be formally defined as follows:

Definition 1   *The CPR-Graph of an assembly whose set of parts or subassemblies is $P=\{p_1, p_2, \cdots, p_n\}$ is the special directed graph*

$$G= \langle V_G, E_G, (edg_{aG})_{a \in A} \rangle \quad (1\text{-}1)$$
where

$$V_G=\{v_1, v_2, \cdots, v_n\} \quad (1\text{-}2)$$
is the set of nodes,

$$E_G=\{(e_{ij}, e_{ji}) \mid Connection\ (p_i, p_j) \wedge (i>j)\} \quad (1\text{-}3)$$
is the set of edges, and

$$edg_{aG}(e_{ij}, v_i, v_j)=lab_{aG}(e_{ij}) \wedge edg_G(e_{ij}, v_i, v_j) \quad (1\text{-}4)$$
is the set of edge's labels.

$$A=\{Vtype1, Vtype2, Vtype3\} \quad (1\text{-}5)$$
is the set of label's types.

$$H(v_i)=p_i \quad (1\text{-}6)$$
where H is the bijection mapping.

In Eq.(1-3)

$$\text{Connection } (p_i, p_j) = \begin{cases} \text{true,} & \text{if there is at least one surface in contact between } p_i \text{ and } p_j; \ (p_i \cap p_j = \phi) \\ \text{false,} & \text{otherwise.} \end{cases}$$

In Eq.(1-4)

$lab_{aG}(e_{ij})$ means that e_{ij} is an edge labeled by a, and $edg_G(e_{ij}, v_i, v_j)$ means that e_{ij} is a directed edge which points from node v_i to node v_j.

In Eq.(1-5)

Vtype1, Vtype2, and Vtype3 indicate the types of connection precedence relationship between parts or subassemblies, and will be defined in section 2.2.

As an example, Figure 2.1 shows the CPR-Graph of the ball-point pen shown in Figure 1.1.

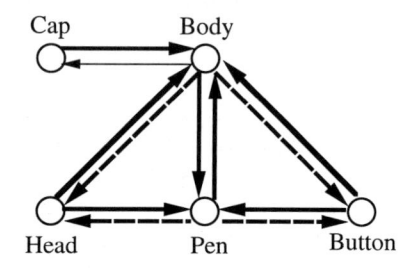

Figure 2.1 The CPR-Graph of the ball-point pen.

2.2 Connection precedence relationship representation

To facilitate the notation of connection precedence relationship, we will introduce the name of edges in the CPR-Graph and symbols as follows.

There are a pair of directed edges e_{ij} and e_{ji} with opposite direction between two nodes v_i and v_j in the CPR-Graph. The edges e_{ij} and e_{ji} are called the V-type edge. The edge composed of this pair of the edges e_{ij} and e_{ji} is called the G-type edge. The G-type edge is an undirected edge.

We will use a and b to represent a part or a subassembly. We will use the symbol P_{ab} to represent the priority of joining a and b. The symbol P_{ab} is different from symbol P_{ba} because P_{ab} is used for representing the connection precedence relationship belonging to a whereas P_{ba} is used for representing the connection precedence relationship belonging to b. We will use the symbol C_{ab} to represent joining a and b. C_{ab} is equal to C_{ba}. We will use the symbol S_{ab} to represent shortening nodes corresponding to a and b. S_{ab} is equal to S_{ba}.

It is assumed that a assembly is made of three parts a, b, and c, and that the part a contacts both part b and part c in the whole assembly. We will use the notation $P_{ab} > P_{ac}$ to indicate the fact that the establishment of C_{ab} must precede the establishment of C_{ac}. We will use the notation $P_{ab} = P_{ac}$ indicate the fact that there is no precedence order between the establishment of C_{ab} and the establishment of C_{ac}, that is, either C_{ab} or C_{ac} can be established first. A compact notation for logical combinations of connection precedence relationships will be used in the following; for example, $P_{ij} \land P_{ik} > P_{il}$ is used to represent $(P_{ij} > P_{il}) \land (P_{ik} > P_{il})$.

Definition 2 *If the connection precedence relationship of part I is represented as follows:*

$$P_{ij} = P_{ik} = P_{il} = \cdots \qquad (2\text{-}1)$$

$P_{ij}, P_{ik}, P_{il}, \cdots$ are referred to as the TYPE1 priority. Edges $e_{ij}, e_{ik}, e_{il}, \cdots$ in the CPR-Graph are labeled by Vtype1.

Definition 3 *If part I only contacts part J in an assembly, P_{ij} is referred to as the TYPE1 priority.*

The Vtype1 edge is represented by a thick solid line in the Figure. Figure 2.2(a) shows an assembly and its CPR-Graph.

Definition 4 *If the connection precedence relationship of part I is represented as follows:*

$$P_{ij} > P_{ik} > P_{il} > \cdots \qquad (2\text{-}2)$$

$P_{ik}, P_{il}, \cdots$ are referred to as the TYPE2 priority. Edges $e_{ik}, e_{il}, \cdots$ in the CPR-Graph are labeled by Vtype2. P_{ij} is the TYPE1 priority.

The Vtype2 edge is represented by a thin solid line in the Figure. Figure 2.2(b) shows an assembly and its CPR-Graph.

Definition 5 *If the connection precedence relationship of part I is represented as follows:*

$$C_{ab} \wedge C_{cd} \wedge \cdots \rightarrow P_{ij} > P_{ik} > P_{il} > \cdots \qquad (2\text{-}3)$$

If the precondition is not satisfied, $P_{ij}, P_{ik}, P_{il}, \cdots$ are the TYPE1 priority regardless of $P_{ij} > P_{ik} > P_{il} > \cdots$. If the precondition is satisfied, $P_{ij}, P_{ik}, P_{il}, \cdots$ are decided by $P_{ij} > P_{ik} > P_{il} > \cdots$ based on the Definition 4.

For example, the connection precedence relationship of part A shown in Figure 2.2(c) can be represented as follows: $(C_{BC} \rightarrow P_{AD} > P_{AB}) \& (C_{CD} \rightarrow P_{AB} > P_{AD})$.

The connection precedence relationship of part B shown in Figure 2.2(c) is represented as follows:

$$\begin{aligned}
C_{BA} &\rightarrow P_{BD} > P_{BC} \\
C_{BC} &\rightarrow P_{BD} > P_{BA} \\
C_{BD} &\rightarrow P_{BA} = P_{BC} \\
\neg(C_{BA} \vee C_{BC} \vee C_{BD}) &\rightarrow P_{BA} = P_{BC} = P_{BD}
\end{aligned}$$

A compact notation $P_{BD} > P_{BA \wedge BC}$ will be used to indicate the above connection precedence relationship of part B.

Definition 6 *If the connection precedence relationship of part I is represented as follows:*

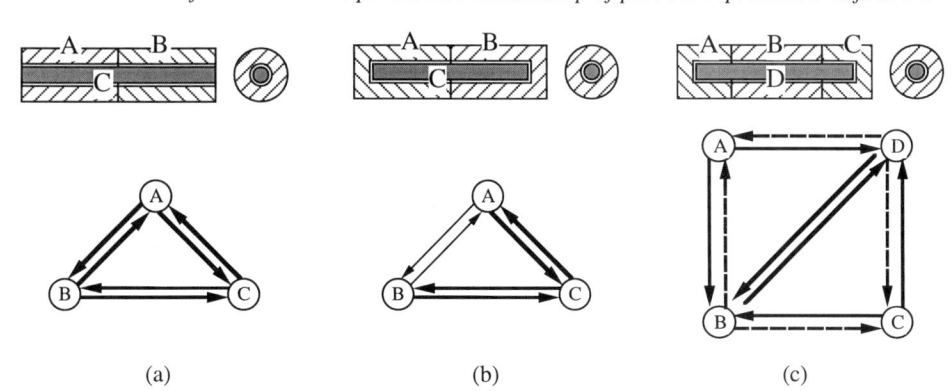

(a)   (b)   (c)

Figure 2.2 The assemblies and its CPR-Graphs.

$$P_{il} \wedge P_{im} \wedge \cdots > P_{ij \wedge ik} \qquad (2\text{-}4)$$

P_{ij} and P_{ik} are referred to as the TYPE3 priority. Edges e_{ij} and e_{ik} in the CPR-Graph are labeled by Vtype3. P_{il}, P_{im}, $\cdots$ are the TYPE1 priority.

We will indicate the Vtype3 edge by drawing a dotted line in the Figure. Figure 2.2(c) shows an assembly and its CPR-Graph.

Definition 7 *If the connection precedence relationship of part I is represented as follows:*

$$C_{ab} \wedge C_{cd} \wedge \cdots \rightarrow P_{il} \wedge P_{im} \wedge \cdots > P_{ij \wedge ik} \qquad (2\text{-}5)$$

If the precondition is not satisfied, P_{ij}, P_{ik}, P_{il}, P_{im}, $\cdots$ *are the TYPE1 priority regardless of* $P_{il} \wedge P_{im} \wedge \cdots > P_{ij \wedge ik}$. *If the precondition is satisfied*, P_{ij}, P_{ik}, P_{il}, P_{im}, $\cdots$ *are decided by* $P_{il} \wedge P_{im} \wedge \cdots > P_{ij \wedge ik}$ *based on the Definition 6.*

For example, the connection precedence relationship of part B shown in Figure 2.3 can be represented as follows: $C_{CD} \wedge C_{DE} \rightarrow P_{BF} > P_{B A \wedge B C}$.

Figure 2.3 An assembly.

2.3 G-type edges classification

Table 1 The classification of the G-type edges in the CPR-Graph

Gtype1	Gtype2
(Vtype1 & Vtype1)	(Vtype1 & Vtype2)
(Vtype1 & Vtype3)	(Vtype2 & Vtype2)
(Vtype3 & Vtype3)	(Vtype3 & Vtype2)

In above, V-type edges in the CPR-Graph have been classified into edges of three types, that is, Vtype1, Vtype2, and Vtype3 edges. G-type edges in the CPR-Graph will be classified into edges of two types based on classification of V-type edges as shown in Table 1. Three G-type edges (Vtype1 & Vtype1), (Vtype1 & Vtype3), and (Vtype3 & Vtype3) are referred to as Gtype1 edge. Three G-type edges (Vtype1 & Vtype2), (Vtype2 & Vtype2), and (Vtype2 & Vtype3) are referred to as Gtype2 edge. If one G-type edge between two nodes in a CPR-Graph is Gtype2 edge, joining the two parts or subassemblies corresponding these two nodes can not be established. For example, in the CPR-Graph shown in Figure 2.1, the edge between node *Body* and node *Cap* is Gtype2 edge, and actually, part *Body* can not be joined with part *Cap* in the state of assembly process represented by this CPR-Graph.

2.4 CPR-Graph generation

The connection precedence relationship of a part can be represented by several equations in some cases. In these cases, the label types of V-type edges in the CPR-graph are determined by the following rule.

Rule *If a V-type edge in the CPR-Graph is labeled by Vtype1, Vtype2, and Vtype3 based on the several equations at the same time, the label type of this edge is determined according*

to precedence order of Vtype2, Vtype3, and Vtype1.

In the case of the ball-point pen, before assembling the ball-point pen, its connection precedence relationships can be represented by the following equations:

C:

Bo: $P_{BoH} > P_{BoC}$ (a)

$P_{BoP} > P_{BoH \wedge BoBu}$ (b)

H: $C_{BuP} \rightarrow P_{HBo} > P_{HP}$ (c)

$C_{BuBo} \rightarrow P_{HP} > P_{HBo}$ (d)

P: $P_{PBo} > P_{PH \wedge PBu}$ (e)

Bu: $C_{HP} \rightarrow P_{BuBo} > P_{BuP}$ (f)

$C_{HBo} \rightarrow P_{BuP} > P_{BuBo}$ (g)

Equations (a) and (b) express the connection precedence relationship of part *Body*. The label type of edge e_{BoH} can be the Vtype1 or Vtype3 based on these two equations. Finally, this edge is labeled by Vtype3 according to above rule. The initial state's CPR-Graph of the ball-point pen can be generated based on above equations as shown in Figure 2.1.

3 GENERATION OF ASSEMBLY SEQUENCES BASED ON CPR-GRAPH

3.1 Procedure of generating assembly sequences

As mentioned in the introduction, it is assumed that once two parts or subassemblies are joined to form a subassembly, all contacts between the parts in the subassembly are established. It is assumed that only two parts or subassemblies are joined at each assembly task. The assembly process consists of a succession of the assembly tasks, each of which consists of joining two parts or subassemblies to form a larger subassembly. The assembly process starts from the state that all parts are separated each other ends to the state that all parts are properly joined to form the whole assembly. On the other hand, we can use the CPR-Graph to represent the states of the assembly process. The CPR-Graph of initial state consists of the nodes only corresponding to the parts whereas the CPR-Graph of goal state consists of only one node corresponding to the whole assembly. An operator, shortening operator, can be repeatedly used to make the CPR-graph change from the initial state to the goal state. The shortening operator is defined as follows: deleting the two nodes and the edges between them, adding a new node corresponding to the assembled subassembly, and rewriting the labels of the other edges in the graph related to the change. The central idea of the proposed method is that the process of assembling a product corresponds to the process of shortening the CPR-Graph of the product as shown in Table 2. As showing in Table 2, the initial state and the goal state of the CPR-Graph correspond to the

Table 2 The correspondence relationship between the assembly process of a product and the shorten process of the product's CPR-Graph

CPR-Graph		Product
Initial state	⇔	Unassembled state
Goal state	⇔	Assembled state
Shortening operator	⇔	Assembly task
Shorten sequences	⇔	Assembly sequences

unassembled state and the assembled state of the product, respectively. Each shortening operators corresponds to an assembly task. A path whose initial node is the initial state's CPR-Graph and whose terminal node is the goal state's CPR-Graph corresponds to a feasible assembly sequence of the assembly. On this path, the ordered sequence of shortening operators corresponds to the ordered sequence of assembly tasks, while the ordered sequence of the CPR-Graph corresponds to the ordered sequence of states of the assembly process. Therefore, a complete assembly sequence is generated by using shortening operator repeatedly to change the CPR-graph from the initial state to the goal state. The following precondition must be satisfied when the shortening operator is used in the CPR-Graph.

The precondition of shortening operator : *The shortened G-type edge must be Gtype1 edge.*

The procedure of generating an assembly sequence is as follows:
First, one Gtype1 edge in the initial state's the CPR-Graph of an assembly is shortened by the shortening operator, and the initial state's CPR-Graph is changed to the next state's CPR-Graph. The new node in the new state's CPR-Graph corresponds to an assembled subassembly. The shortened CPR-Graph is called the contracted CPR-Graph. Secondly, this contracted CPR-Graph is changed to its next state's CPR-Graph by the same operation. By this way, the new contracted CPR-Graphs are generated successively until the contracted CPR-Graph becomes the goal state's CPR-Graph. The ordered sequence of shortening operators making the CPR-Graph change from the initial state to the goal state corresponds to ordered sequence of assembly tasks assembling the assembly from the unassembled state to the assembled state. Therefore, the shorten sequences can be regarded as the assembly sequences, the assembly sequences is generated.

3.2 Procedure of shortening operations on CPR-Graph

Table 3 The changing operations of the shortening operator on the CPR-Graph

Connection relation representation	Precedence relation representation
Delete1(v_i, v_j, e_{ij}, e_{ji})	Delete2(P_{ij}, P_{ji}, C_{ij}, C_{ji})
Add(v_{ij})	---------
Rewrite($i, j \rightarrow ij$)	Rewrite($i, j \rightarrow ij$)

The representation of the CPR-Graph is composed of the representation of connection relation and the representation of precedence relation. When the CPR-Graph is changed from a state to its next state, the shortening operator including the operations of deleting, adding and rewriting operates the CPR-Graph as shown in Table 3. The contracted CPR-Graph can be generated from the operated result based on the definitions and the rule in section 2. The procedure of this operator is presented concretely as follows.

If Gtype edge (e_{ij} & e_{ji}) between nodes v_i and v_j in a CPR-Graph is Gtype1 edge, this Gtype edge will be shortened.

In the following, first, we will explain how the shortening operator operates the connection relation representation of the CPR-Graph.

First, the operator deletes node v_i, node v_j, edge e_{ij}, edge e_{ji}, adds a new node v_{ij} corresponding to assembled subassembly. Secondly, the operator rewrites i and j included in the edge symbol as ij. For example, edges e_{ia} and e_{bj} are rewrote as $e_{(ij)a}$ and $e_{b(ij)}$, respectively. It is assumed that there is a G-type edge between nodes v_k and v_i, and that there is a G-type edge between v_k and v_j. There will be two G-type edges ($e_{(ij)k}$ & $e_{k(ij)}$) between nodes v_k and v_{ij} after the rewriting operation. Because there is only one G-type edge between nodes in the CPR-Graph based on the definition 1, one of two G-type edges ($e_{(ij)k}$ & $e_{k(ij)}$) must be deleted.

Next, we will explain how the shortening operator operates the precedence relation representation of the CPR-Graph.

First, the operator deletes P_{ij}, P_{ji}, C_{ij}, C_{ji} from the equations representing precedence relation. Secondly, the operator rewrites i and j included in these equations as ij. For example, P_{ik} and C_{lj} are rewrote as $P_{(ij)k}$ and $C_{l(ij)}$, respectively.

The contracted CPR-Graph can be generated by above shortening operation.

3.3 The ball-point pen assembly

An assembly sequence of the ball-point pen shown in Figure 1.1 will be generated by the proposed method.

Figure 3.1(a) shows the equations of connection precedence relationship and the CPR-Graph of the initial state. It is assumed that parts come in an order of *Cap, Body, Head, Button,* and *Pen*. The CPR-Graph is shortened according to the order.

If S_{BoH} is established, the operator deletes node Bo, node H, edge e_{BoH}, edge e_{HBo}, adds a new node v_{BoH} corresponding to assembled subassembly BoH, and rewrites Bo and H included in the edges symbol as BoH. The operator deletes P_{BoH}, P_{HBo}, C_{BoH}, C_{HBo} from the equations shown in Figure 3.1(a), and rewrites Bo and H included in these equations as BoH. Figure 3.1(b) shows the equations of connection precedence relationship and the CPR-Graph after the initial state's CPR-Graph is shortened. In this way, the CPR-Graph is shortened according to an order of $S_{C(BoH)}$, $S_{(CBoH)P}$, and $S_{(CBoHP)Bu}$ as shown in Figure 3.1(c), (d), (e), and finally the contracted CPR-Graph becomes the goal state's CPR-Graph. Therefore, an assembly sequence of the ball-point pen is generated, that is, C_{BoH}, $C_{C(BoH)}$, $C_{(CBoH)P}$, and $C_{(CBoHP)Bu}$.

4 DISCUSSION

Although both the AND/OR graph and the assembly sequence diagram constitute a compact representation of all feasible assembly sequences, they may still be too large for assemblies with very large numbers of parts whereas the CPR-Graph is not very large. In the case of the eleven-parts assembly from industry introduced by De Fazio et al.[6], the number of nodes and the number of edges needed to store the representations of the assembly sequence diagram, the AND/OR graph, and the CPR-Graph are shown in Table 4.

It is assumed that only one or a few assemblies are assembled, and that parts may come in random order. In the case, the assembly sequence generated according to the coming sequence of parts is meaningful and practical than optimal assembly sequence. In both the AND/OR graph and the assembly sequence diagram, all feasible assembly sequences need to be calculated, whereas in our proposed method, only one assembly sequence need to be calculated according to the coming

Table 4 Storage requirements for 11-parts assembly from industry

	Assembly sequence diagram	AND/OR graph	CPR-Graph
Nodes	61	75	11
Edges	221	221	36

C:
Bo: $P_{BoH} > P_{BoC}$
   $P_{BoP} > P_{BoH \wedge BoBu}$
H: $C_{BuP} \rightarrow P_{HBo} > P_{HP}$
   $C_{BuBo} \rightarrow P_{HP} > P_{HBo}$
P: $P_{PBo} > P_{PH \wedge PBu}$
Bu: $C_{HP} \rightarrow P_{BuBo} > P_{BuP}$
    $C_{HBo} \rightarrow P_{BuP} > P_{BuBo}$

(a)

C:
BoH: $P_{(BoH)P} > P_{(BoH)Bu}$
P: $P_{P(BoH)} = P_{PBu}$
Bu: $P_{Bu(BoH)} > P_{BuP}$

(b)

CBoH: $P_{(CBoH)P} > P_{(CBoH)Bu}$
P: $P_{P(CBoH)} = P_{PBu}$
Bu: $P_{Bu(CBoH)} > P_{BuP}$

(c)

CBoHP:
Bu:

(d)

CBoHPBu.

(e)

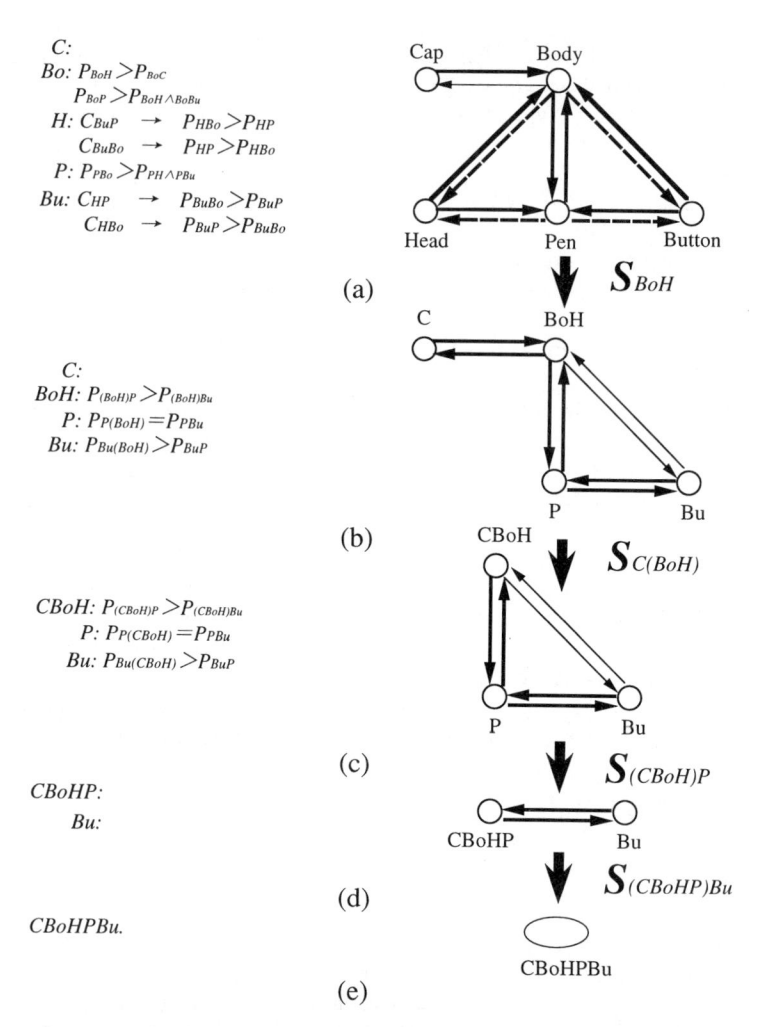

Figure 3.1 A process of shortening the CPR-Graph of the ball-point pen from the initial state to the goal state.

sequence of parts. Therefore, the computational complexity of our method is much less than that of the AND/OR graph and the assembly sequence diagram.

Whenever unexpected events cause the execution of an assembly to deviate from the preplanned course of action, it is preferred that the assembly proceeds from the unpredicted state toward the goal state in an efficient way. We can use the CPR-Graph for recovering from execution errors. For example, an error situation of which parts or subassemblies are joined according to a wrong sequence can be discovered based on the CPR-Graph. In this case, the execution error can be recognized by checking whether the G-type edge between the parts or subassemblies joined is Gtype2 or not. Since the execution error can be discovered immediately, the error can be recovered before the execution of the next assembly task. Therefore, the complete disassembly of the product for recovering from execution error can be avoided.

5 CONCLUSION

This paper proposed the CPR-Graph for representing the connection precedence relationship between parts or subassemblies of an assembly. The feasible assembly sequences can be generated by using shortening operator repeatedly to change the CPR-graph from the initial state to the goal state. The major advantages of the proposed method are its simplicity and less computational complexity compared to the other related methods. By using the proposed method, the feasible assembly sequence can be generated according to the arriving order of parts, and the errors or unusual situation can be discovered easily and immediately. A ball-point pen example has been used to illustrate the proposed algorithm for the generation of mechanical assembly sequences.

6 REFERENCES

(1) L.S.Homem de Mello and Sukhan Lee (1991) Computer-aided mechanical assembly planning. Kluwer Academic Publishers, Boston/Dorderecht/London.
(2) L.S.Homen de Mello and A. C. Sanderson (1990) And/or graph representation of assembly sequences. IEEE Trans. Robotics Automat., vol. 6, No. 2, pp. 188-199
(3) L.S.Homen de Mello and A. C. Sanderson (1991) Representations of mechanical assembly sequences. IEEE Trans. Robotics Automat., vol. 7, No. 2, pp. 211-227
(4) L.S.Homen de Mello and A. C. Sanderson (1991) A correct and complete algorithm for the generation of mechanical assembly sequences. IEEE Trans. Robotics Automat., vol. 7, No. 2, pp. 228-240
(5) A. C. Sanderson, L.S.Homen de Mello, and H. Zhang (1990) Assembly sequence planning. Artificial Intelligence Magazine, vol. 11, No. 1, pp. 62-81
(6) T. L. De Fazio and D. E. Whitney (1988) Simplified generation of all mechanical assembly sequences. IEEE J. Robotics Automat., vol. RA-3, No. 6, pp. 640-658 ; also, Corrections, vol. RA-4, No. 6, pp. 705-708
(7) D. F. Baldwin, T. E. Abell, M. C. M. Lui, T. L. De Fazio and D. E. Whitney (1991) An integrated computer-aid for generating and evaluating assembly sequences for mechanical products. IEEE Trans.Robotics Automat., vol. 7, No. 1, pp. 78-94

7 BIOGRAPHY

Jian Wang He received the BS degree in Mechanical Engineering from Nanjian Institute of Technology, People's Republic of China in 1986, and the MS degree in Mechanical Engineering from Kobe University of Mercantile Marine, Japan in 1991. He is currently a full-time doctoral candidate at Kobe University. His current research interests include assembly sequence planning and manufacturing systems. He is a member of JSME.

Kazuhiro Ohkura He received his BS in Precision Engineering in 1988 and his MS in Information Engineering in 1990 at Hokkaido University, Japan. He is currently a research associate at the Department of Mechanical Engineering, Kobe University. His research interests include theoretical and practical paradigms of genetic algorithms, autonomous agents and manufacturing systems.

Kanji Ueda He graduated the Department of Precision Engineering at Osaka University, Japan in 1970, and completed Ph.D. at the same university in 1977. Since joining Kobe University in 1972, he has been an associate professor, then a professor at kanazawa University. He is now a professor at the Department of Mechanical Engineering, Kobe University. His research field covers Manufacturing Systems, CAD/CAM/CAE, Intelligent Machines and Machining Theories. His current research activities are Biological Manufacturing Systems, Intelligent Artifacts, Genetic Algorithms, Artificial Life and Micro machining of advanced materials. He is a member of such societies and institutions as CIRP, JSPE, JSME, ASM/NAMRI and CAM-I.

23

A rule based system for design for manufacture and assembly

Dr. E. A. Warman
K Four Limited
Nene House, London Road, Peterborough, PE2 8AH, U.K.
Tel: + 44 1733 312019, Fax: + 44 1733 896046

Abstract
A system architecture, that can support proprietary CAD systems, has been developed and implemented for providing knowledge based support for design for manufacturing and assembly.
The rationale behind its development is presented together with an outline of products upon which it was tested. Future developments, for the next phase of this project are discussed.

Keywords
CAD design, manufacture, assembly, artificial intelligence, object oriented systems.

INTRODUCTION

It has been stated that up to 98% of manufacturing problems arise in the design office. This is in part due to the increased specialism of engineers who unlike their predecessors do not have a broadness of engineering experience. This specialism encouraged no doubt by the belief that it led to greater efficiency produced the serial approach in which a product is taken step by step through sequential processes from design to manufacture.

In order to produce more effective design, decisions and guidance concerning the ability to produce products and their component parts should be made during the actual design process and not after. Many design for 'x' systems need a complete design in order to subject it to an analysis process that might suggest changes or modifications thus adding time to a serial process. Sometimes these changes affect the functionality of a design thus compounding errors rather than removing them.

True concurrent engineering should result in a set of parallel (or nearly so) activities. It is of considerable value, if during the design process, the designer and manufacturing engineer can have their attention drawn to possible problems and advice given to enable corrective action.

It is implicit that in order to achieve a concurrent engineering environment, an expert system that has access to a wide range of knowledge and data; relating to manufacturing processes, resources, materials, constraints and the rules governing the application of this knowledge, is available to the designer.

The system developed to support this environment must be able to incorporate proprietary computer aided design systems. It must also be able to be changed as circumstances and applications change. The system must be configurable in order to suit different industries and organisations.

These aspects were some of the targets set by the DEFMAT project (Design for Manufacturing Architecture Tool Suit) a research project partially funded under the BRITE/EURAM research programme of the European Commission.

DESIGN

Whatever the viewpoint on what constitutes design and irrespective of industry all design activity starts from a conceptual phase through to the final production of 'manufacturing' data. Underlying all design activities some theories have been extracted (Warman and Yoshikawa, 1986). These theories can be related to computer implementations of design (Warman, 1990). To summarise this previous work it has led to the result that extensional data representation provides the best method of handling design data structures. Furthermore the extensional representations can be mapped directly onto the concepts of objects, as originally defined by (Goldberg and Robson, 1982). This leads directly to one of the key concepts utilised in DEFMAT - extensional data structures, features and objects are synonymous.

Objects and hence features are independent and possibly reusable components of a complex system. This means that a product and its individual components may be described using object structures. The object structure concept is thus central to the product model that constitutes a further key element of DEFMAT.

As the design proceeds in the DEFMAT system from concept through detail, the system can guide or suggest to the designer particular approaches related to design for manufacture and as design action occurs the product model is updated and grows at each stage of the design process.

The approach used is for the product model within the DEFMAT system to be documented as STEP/EXPRESS-G schemata. This approach is described in detail in a following section.

DEFMAT ARCHITECTURE

At the commencement of the project there was an elementary notion of an architecture that contained a CAD system, knowledge engine and data base. Such sample models are needed in documentation to justify the project for funding purposes and to provide a stimulus for the development of ideas. Using this simple model as a basis, four simple constructs were assembled. Each being an independent effort, differed from one another.

The experience gained from these tests and a logical analysis of the requirements of design, manufacturing and assembly processes led to an evolution of the present DEFMAT architecture.

This architecture in its present conceptual form is generic. It can relate to any product or process (in the sense of process engineering), manufacturing process and design regime. The drive towards genericity was influenced by the four simple test cases that addressed Printed Circuit Board (PCB) assembly for a leading computer manufacturer, the design of compressor rotors and drive shafts for a family of air compressors, the automatic assembly of miniature electromechanical units (hearing aids) and the automatic assembly of mobile telephones. It was possible from this diverse set of problems to move toward the architectural basis upon which the prototype system, described in this paper, was constructed.

IT became mandatory that the architecture must be able to support proprietary modules (CAD systems, data bases, etc.). This need imposed strict control over the interfaces.

The present realisation of the architecture is shown in figure 1.

The DEFMAT architecture is based upon the principles of modularity and - where possible - distributed functionality. Each module is designed to handle its own data and knowledge and interact with the other modules through the communication bus. As illustrated in figure 1 the architecture is also zoned into :- product specific, action specific, design regime and local world sectors. The data bus is the key link to all of the modules and the control module directs the functioning of the architecture. The practical implementation of the architecture may be in its entirety in one single computing platform or distributed between units. The function of each module is outlined as follows.

Figure 1.
DEFMAT Architecture

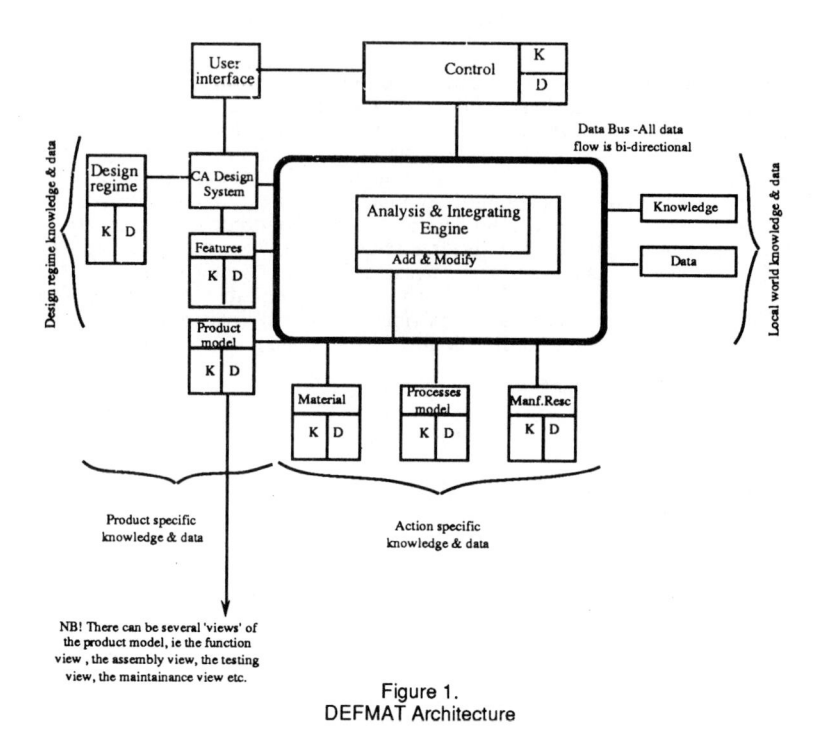

Control Module

The control module does not have a fixed plan of action with respect to the design process. In order to ensure and maintain genericity, any encapsulation of a sequence of design actions has been deliberately avoided. The control module manages the DEFMAT system state by reacting to the user interface or reacting to design changes signalled through the CAD interface or to the overall state of the system. The actual sequence of analysis is controlled by the analysis engine. All changes in the system state are captured by 'System Variables' that define which actions trigger analysis and which knowledge domain is to be applied. Take for example the system variable 'Analyse-Create-Joint', this can be set to ON or OFF. If set to ON, the control module will initiate the analysis engine and supporting processes whenever a joint is created.

The control module maintains it's own knowledge base in order that controls associated with system variables and the system variables themselves may be updated without recoding.

It is the control module that enables the variety of action, as shown in figure 2, to be undertaken by the DEFMAT system.

The control module thus initiates the loading, into the analysis engine, the relevant knowledge bases, and then starts the analysis engine on its respective analysis sequence.

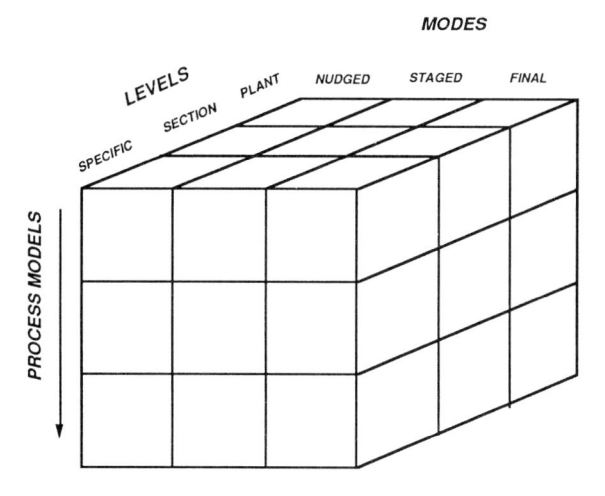

Process Model Complexity Matrix

Figure 2

User Interface

It was considered to be of paramount importance that the user interacts with the DEFMAT system through a single user interface. This interface presents the system as a coherent entity and not as a collection of individual module interfaces.

The present implementation uses OSF/Motif objects in a C programming environment. It is of importance to note that as well as the elements of each module being constructed from objects, each module may also be considered as an object. The interface module incorporates the window manager of the DEFMAT system and all functionality relating to windows. All user interaction with the DEFMAT system is thus through the User Interface (UI). The UI provides any of the windows necessary for user input and analysis output during the analysis processes. The overall target of the command module and user interface is to provide as generic as possible software solution. For example, whenever possible, the functions that call windows contain the required parameters that customise those windows.

Prototype tests in working design environments have found that the UI is totally acceptable to practising designers.

The Analysis Engine

The analysis engine together with its associated infrastructure, loads, under the direction of the control module, the specific rule sets related to the particular analysis that has been requested. Tools are available - restricted for use only by the knowledge engineer - for modifying rule sets. Rules may be prohibitive, directive or suggestive in nature. At this stage the rules sets that have been implemented are relatively simple in that multiple trade off are not handled. Such trade off rule sets are under development by one of the first commercial users. Nevertheless the simple rule set for PCB placement consists of 1500 individual rules. The rules themselves may be segmented and may reference particular system variables such that a degree of user customisation is permitted.

Suggestive rules may produce pictorial output to illustrate desirable alternative or graphical data to show better regions of performance. The approach gives the designer the confidence that he is in control and has - within bounds - freedom of choice.

The Product Model

In DEFMAT the product model contains all of the information required to realise the product. In a projected extension it will also contain pointers to the Computer Aided Engineering systems and associated results (functional, performance, stress analysis, etc.) obtained from these and other processes used by the designer in the respective design regime to arrive at geometric forms.

The present implementation is based upon STEP/EXPRESS-G schemata and code listing. EXPRESS-G is a standard syntactical representation for a product. The strict syntactical representation allow schemata to be directly transformed into EXPRESS code for product representation.

The product instances are filled as the design progresses on the CAD system. An example of a product tree according to EXPRESS-G is illustrated in figure 3 and the related EXPRESS code is shown in figure 4.

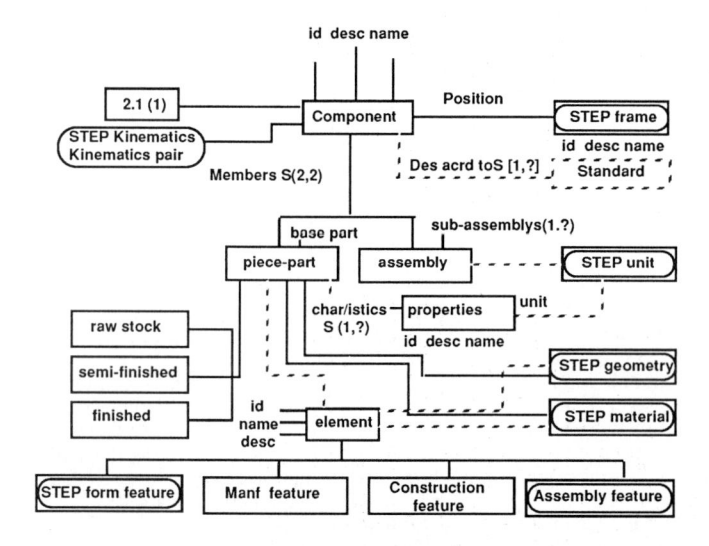

Figure 3 Example of Product Model in Express G Representation

In the present implementations of DEFMAT, features are obtained from the feature library or are specified during the design stage, by the designer, and subsequently entered into the library for future use. Feature recognition by the system is not a part of the current development for this needs to be the subject of further research (groups of design features may constitute a single manufacturing feature in an organisation or in another be seen as having a different correspondence in another organisation.).

In the present system geometrical representation are stored as pointers to CAD files. Long term storage of product data is achieved at present using the ONTOS data base, which is also read/write accessible at run time via the User Interface.

```
ENTITY
    SUBTYPE OF (Object)
    parentRef       :Reference;
    supplierRef     :Reference;
    orderRef        :Reference;
    sourceRef       :Reference;
    manf_docsRef    :Reference;
    standardRef     :Reference;
    locationRef     :Reference;
    id              :STRING
    description     :STRING
    name            :STRING
    (+functions to define entity)

END ENTITY
```

Figure 4 Express code for sample Product Model

The Process Model

In order to provide flexibility of application and maintain as much genericity as possible a distinction is made in the DEFMAT system between the process model and manufacturing resources.

The process model contains knowledge and data concerning the manufacturing processes to be considered by the system. The information is expressed as sets of information following a simple hierarchy. The processes are naturally related to the knowledge and data concerning material properties that is held in an associated data base. The information groups are linked by corresponding rule sets.

If required processes can be investigated that are not within the current manufacturing resources of an organisation.

Thus for example, if a designer selected tungsten as a material he would be advised against casting as a process and directed toward sintering.

Manufacturing Resources

The manufacturing resources relate to specific equipment and processes available at a specified manufacturing facility. A DEFMAT analysis may direct the designer to a drilling process for producing a set of holes to a specified tolerance but the equipment may then not be available at the specified plant. The choice then is to out source the component, purchase new equipment or consider further design alternatives.

It is at this type of conflict that design for manufacture may start interaction with the design functionality.

The actual equipment models that form the manufacturing resources under current reference by the user in a DEFMAT analysis, may be changed by the user. The system can thus reflect changes that occur and the DFM/A rules will reference the new equipment parameters.

The Design Regime

The DEFMAT architecture incorporates design regime knowledge and data. The term design regime refers to a specific field of the design such as machine tool, prime mover, etc. It may even be further reduced to consider a subset of these main design areas. The design regime contains the knowledge and tools to employ that knowledge such as algorithms, classical solutions, data tables, etc., and all the other tools used to derive and ascertain the functionality, kineasthetics and strenuosity of a design. Also tools for the simulation - where possible - of the functionality of a design. Thus the design regime for PCB's would require circuit simulation tools to be part of the design tool set.

Partitioned knowledge bases may be associated with each respective design regime and will be loaded by the control module when the user selects the design regime using the user interface.

THE FIRST DEFMAT PROTOTYPE

The first DEFMAT prototype was constructed using the tools available to each of the partners in the research consortium. The Computer Aided Design system used is PRO-ENGINEER. The knowledge engine used is NEXPERT and the data bases were constructed using ONTOS. The control module and user interface were designed by the team and written in C++. The data and knowledge reside in the ONTOS data bases and during systems operation the relevant data and knowledge sets are extracted and loaded into the NEXPERT knowledge base. It is the control module that ensures NEXPERT is loaded and the specific model of operation are established.

All this activity is transparent to the designer.

Pro-Engineer is a feature based CAD systems, used because it was available to the project team but any contemporary CAD system can be used and in later systems that are being built for other industrial users other CAD systems are utilised.

The interfacing strategy within DEFMAT allows for online and offline interfaces. The online interface allows the simultaneous updating of the product model as the CAD model is updated. The offline interface exports representation of the CAD model in ASCII file format. The ASCII file contains information to generate product model instance that may subsequently be translated into product model objects for subsequent analysis.

These strategies relate to the three possible modes of operating DEFMAT (figure 2). These modes are nudging, phase checking and complete design. In the nudging mode, the designer is nudged by the system as checks and advice are given on each new design feature as it is added. Phase checking is invoked by the designer at the completion of a component or other well defined steps. The complete design checking analyses the complete design or an assembly of components.

The design of a totally new product commences with conceptualisation. DEFMAT is able to provide some input at this stage by providing the designer with assembly and fixing options that result in partially filled objects in the product model schema. These objects can be utilised or discarded as the design progresses. Browsing tools are also provided that allow the designer to browse through the knowledge and data bases that are within the DEFMAT system.

A new design may constitute modification of existing products through to a blank sheet for a totally new product. This range of starting conditions poses no problems to the DEFMAT system.

In using the system the designer interacts with it through a series of menus. When the option 'new' is selected for a new design, followed by the selection of a specific design regime, the control module initiates all of the actions needed to load the knowledge bases and tools. This activity is transparent to the designer. The designer then selects the model of operations and then starts designing or detailing.

As the designer works, knowledge bases are loaded to meet specific intent such that the designer sees a continuum. Messages have to be responded to, values given or received in the two way dialogue between the designer and system.

THE TEST APPLICATION

Naturally during the development of DEFMAT test were conducted at each stage. The first full test required that all aspects of the system were exercised to the full, so an electromechanical product was selected. This product was a portable digital telephone. The tests were concerned with a wide range of manufacturing and manufacturing areas. The product contained a PCB with digital and analogue sections that posed interesting component placement problems. These problems had to be solved for two different assembly lines. The PCB was required also to be designed to facilitate robotic placement within a die cast housing where board to wall clearance was very small. The housing required that some machining and fixings had to be minimised and suitable for assembly by machine. In addition to these requirements was the production volume to be addressed a requirement that made the assembly a far from leisurely activity.

Information was obtained for the plant and processes and rules developed. Design rules were also extracted and the whole process was put to the test. The processes involved are shown as a process tree in figure 5.

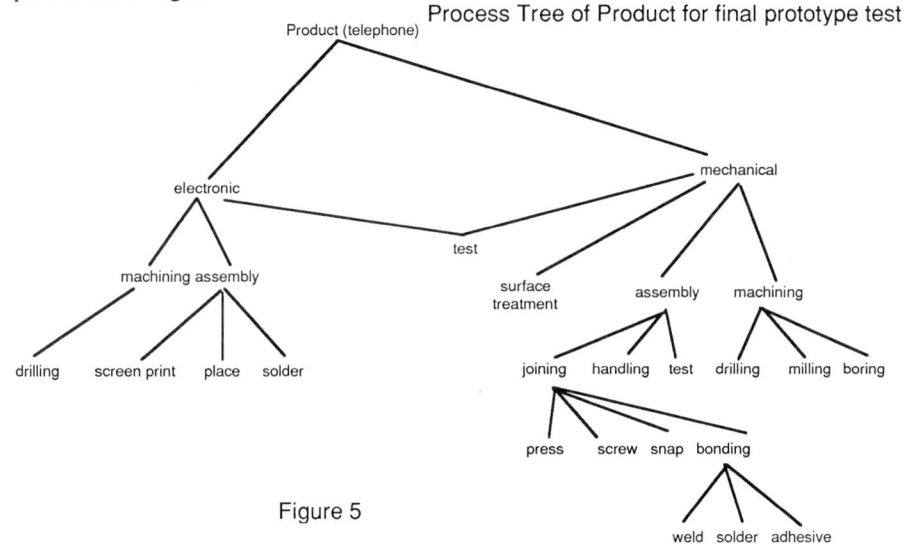

Process Tree of Product for final prototype test

Figure 5

Following a detailed demonstration, two designers were selected by one of the industrial partners to use and evaluate the system after a long period. After two weeks the response was very positive and a DEFMAT system is now under installation in the main design office.

SYSTEM MAINTENANCE

Rule based systems such as DEFMAT are not static but increase in richness with their continual use. System maintenance tools are provided to enable the knowledge engineer to edit the knowledge and data editor is utilised. The contents of the ONTOS data bases may also be edited. It is essential that in later versions of the system generic editing tool are provide in order to maintain system openness. The acquisition of knowledge and data for entry in a system such as DEFMAT is a nontrivial exercise. The transition from one rule set scenario to another often has unclear boundaries. The design and manufacturing engineer sometime make decision based upon context. Contexts must thus be established as rules sets.

The rules within DEFMAT are implemented in instances, thus the same rules may be used for different class attributes so that though a product under design, the rules are still relevant and need only to have the attribute values updated. Nevertheless the problem of rule boundary variation is still required to be addressed in future representations.

FUTURE DEVELOPMENTS

The current DEFMAT implementations do not capture fully the potential of the architecture to operate at the conceptual design level. To move toward this end of the design process will require further work in developing dynamic product model schemata that can support the inherent qualitativity of conceptual design work. To incorporate down stream design for 'x' considerations whilst operating at the conceptual design stage will involve a great deal of work in quantifying the design for 'x' constraints relative to conceptual design actions.

More attention must be given to the design regime and its relationships with the down stream processes. Consider a PCB and the possible effect upon rerouting of tracks if a component position were changed by 5 mm to suit an automatic assembly machine. Designer and manufacturing engineer often fail to appreciate that some manufacturing processes such as broaching can impose more load on a component that it experience in actual use.

Though DEFMAT is looking at problems as they occur and not after the event, more attention needs to be given to relating the early stages of design to the manufacturing consequences.

The multilevel tradeoffs that designers perform, dependent upon context, need also to be studied and it may be that knowledge engines need to be developed to work efficiently with increasingly complex rule structures.

The DEFMAT project and resultant system was a success and has resulted in systems being introduced into main stream design work in several industrial organisations.

The process of building up system characteristics from some of the fundamental aspects of design theory has produced a sound and robust architecture. Whether or not the architecture will become a definitive approach for design for 'x' is still open to speculation. The modular object oriented approach does not open up the practical implementation of the total product model where all activities related to a design and its realisation may be traced back interactively at a workstation.

The architecture allows for the 'plugging in' of features - that could be proprietary parts such as fixings - raises the subject of the suitability of standards for CAD. A subject that has again become unsettled by new proposals emerging from leading system vendors (Anon, 1995).

Though the system presents to the user an apparent single user interface it is evident that more attention needs to be given to user interface development in order to achieve greater user friendliness. Proprietary systems still produce too many cryptic responses.

A new generation of knowledge engines and associated system building tools needs to be developed so that complex rule sets can be extracted from the 'whys and wherefores' resulting from the knowledge elicitation process. Multiple level tradeoffs will certainly need to be addresses in other areas of design for 'x'.

It is strongly believed that DEFMAT is the starting point for a whole new series of developments of design systems.

ACKNOWLEDGEMENTS

The work presented was the result of a team effort between O Molly - CIMRU, S Neu and S Kruger - IWF, S Tilley - WTCM/CRIF, H Rothenberg - AEG and myself - K Four. Special thanks are given to Antonio Colacao of the European Commission for his support throughout the project.

REFERENCES

Anon (1995) If the world won't come to CAD. *Manufacturing Computer Solutions*, July/August 1995, 12-15
Goldberg, A. and Robson, D. (1982) The Smalltalk programming system in *Eurographics '82* (ed. R.J. Hubbold)
Warman, E.A. (1990) Object oriented programming and CAD. *Journal of Engineering Design* Vol. 1., No. 1
Warman, E.A. and Yoshikawa, H (1986) *Design Theory for CAD*. North Holland

BIOGRAPHY

E A Warman is the founder and a director of K Four Ltd. a consultancy and R & D company specialising in the application of computer techniques to all aspects of design and manufacturing. He is the author of over 60 publications devoted to product design, manufacturing and testing. He is a member of several professional societies, a chartered mechanical engineer, manufacturing engineer and computer scientist. He holds the degrees of Ph.D. and M.Tech. from Brunel University and is a past chairman of IFIP WG 5.2 (CAD) and WG 5.9 (Computer is agriculture and food production) and holds the IFIP Silver Core award.

PART SIX

Rapid and Virtual Prototyping

24

Choosing the Right Rapid Prototyping Technology for Each Phase of Product Development

H. Grabowski, J. Erb, K. Geiger
Institute for Computer Applications in Planning and Design,
University of Karlsruhe, Kaiserstr. 12, 76131 Karlsruhe,
Tel: +49-721-2129, Fax: +49-721-66113
e-mail: grabowski@rpk.mach.uni-karlsruhe.de

Abstract
To achieve fast development cycles, rapid prototyping technologies must be integrated into the product development process. Depending on the phase of product development the requirements (e.g. accuracy, material properties, cost, availability) to the prototype are varying very much. This paper describes a concept that integrates the use of prototypes into the design process supported by an integrated product modelling system. The system allows to store characteristics of prototyping technologies and criteria for the use of specific prototypes. A new prototyping problem can be solved using solution patterns that provide knowledge about similar problems. Based on the requirements specified to the new prototype, the optimum prototyping technology is than chosen on the basis of multicriterial optimisation theory.

Keywords
rapid product development, rapid prototyping, solution patterns, modelling of requirements, multicriterial optimisation

1 INTRODUCTION

The global markets of the upcoming 21st century appear to be an existential challenge for the traditional industrial countries from Europe, USA and Japan. High quality and client oriented products are not any longer competing against low cost and mass production. The competitors coming from Asia or Eastern Europe have learned their lessons well. Investigations in technology in combination with low wages attract the interest of international enterprises that want to reduce their cost of production and to get in touch with the clients of tomorrow. On the other hand products coming from those countries are not regarded as being of second choice any more.
    This trend implies the need to react for all established countries if they want to keep their top positions. They will of course be constrained to reduce production costs, to keep quality, and to increase their creativity. The lead times to develop new products - that must be as close as possible to the markets needs - have to be reduced as well.
    Rapid prototyping technologies such as stereolithography, laser sintering etc. are advertised as strategic tools in achieving short product development times. Even though these technologies play an important role it is often denied that they are one component of rapid product

development only. To achieve fast development cycles, rapid prototyping technologies must be integrated into the product development process and combined with other methods that are used to reduce development time.

2 METHODS TO REDUCE PRODUCT DEVELOPMENT TIME

In product development the design process is invoked after the decision is taken to develop a new product. During product design the designer is investigating to realise the functional requirements on the product defined in a list of requirements. The traditional design process consists of different states of solution which can be distinguished by the abstraction level they describe. The states of solution are related to one of the four modelling levels of the design process, that are the **requirement, function, principle and shape level**. In addition to that it has to be taken into account that process planning, production, the product's use and recycling should substantially influence the design process. The design process can be regarded as a procedure that develops from the incomplete to the complete, from the abstract to the concrete, and from alternative solutions to the optimal solutions (Grabowski and Huber, 1993) It must be taken into account that the design process is not straight forward but iterative. That means if a solution in any state is not meeting the requirements, the designer steps back onto a more abstract level to check decisions already taken, to modify them if necessary and to develop new solutions. The traditional design process is broad proceeding step by step towards the final product shape definition, reasoning for variations and possible solutions (see figure 1). It is obvious that an extensive support of the designer should be of high preference for each enterprise.

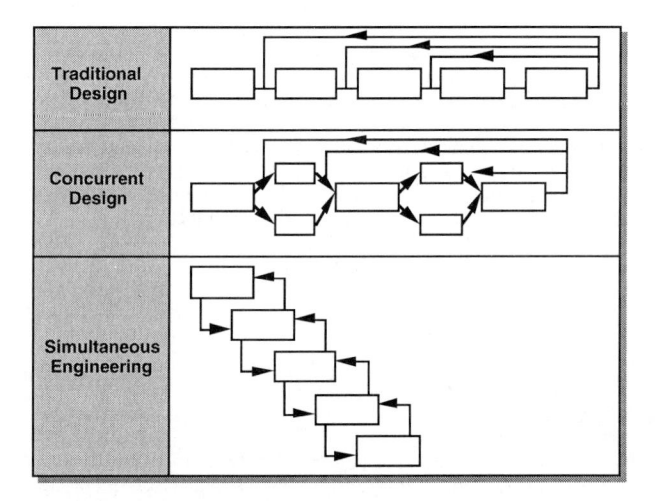

Figure 1 Methods to reduce development time

Within the last years methodologies such as Concurrent Design and Simultaneous Engineering had been developed that tend to overcome the traditional sequential processing in design by (see figure 1):
- splitting up work packages and parallelize work,

- building interdisciplinary teams for the development of a component or a product that allow the contemporaneous work on different levels of abstraction, i.e. on shape and process planning level (Anderl, Malle and Schmidt, 1992).

Both methodologies imply new organisation concepts with which work has to be split of, responsibilities have to be declared, processes have to be analysed and defined, and information flow between process units have to be organised. On the other hand communication and cooperation is an essential element for the realisation of short development cycles and high quality products.

The technical support of these concepts has been realised partially. Some of the commercial CAD-systems already offer communication functions that allow team members to have multi media sessions during which a design can be discussed and modified. In addition to that design spaces (corresponding to components of subassemblies of a product) can be defined by a team chef and responsibilities can be assigned to the team members involved. Information flow between processes can be defined and organised using the workflow components of so-called EDM (Engineering Data Management) systems. To overcome the problems of rigid structures and tayloristic processing, business process reengineering is a major keyword.

Analysing the techniques mentioned above at least one common aspect can be derived, that is the **integration aspect**. No matter if interdisciplinary cooperation is required or business processes are reorganised, the installation of small, autonomous, and flexible units is essential for the operativeness of the overall system. A product meeting perfectly the requirements of the customers, being produced under optimum conditions with respect to time, cost and quality will thus be a result of this effort.

Taking into account the possibilities offered through the use of rapid product technologies the shortening of development time and the reduction of iterations while keeping or improving product quality is within reach, if those technologies can be integrated into the product development process.

3 INTEGRATION OF RAPID PROTOTYPING TECHNOLOGIES

Within the product development process prototypes are used for different reasons and in different development stages. Prototypes produced with different technologies are very versatile tools for (Ulrich, 1994)

- **communication**, allowing visual and tactile communication between persons of various background such as top management, vendors, partners, customers, team members.
- **integration**, to ensure that components and subsystems coming from different experts and various domains fit and work together. The definition of responsibilities and interfaces is a main issue in integration.
- **learning**, to find errors or unintentional effects as early as possible in the development process.

Focusing on product development in mechanical engineering three different prototypes are classified (Bullinger 1995):

- **Design Prototypes**
  are built to verify optical, aesthetic and ergonomic requirements to the product. Accuracy of the prototype is not of importance. The material used is in general chosen under economical aspects taking into account its characteristics to form complex shapes (e.g. clay models). This type of prototypes is often used for market analysis purposes.
- **Geometrical Prototypes**
  are used for proving fit, form and accuracy of parts and assemblies. The material used to produce the prototype is not necessarily to correspond with the final product and therefore is chosen due to accuracy requirements.

- **Functional Prototypes**
  should cover all interesting features of the product's functionality, i.e. they behave in the same way as serial products will do. Depending on the functional tests that will be carried out, the material of the functional prototypes should have similar characteristics with respect to hardness, elasticity, strength etc.

Some applications of prototypes allow for the replacement of the physical prototype by a virtual prototype. This is the case if the prototype shall not be used for tests but to get an optical impression of the part or if the test may be replaced by simulations in the computer. In some applications only part of the prototype may be replaced by a computer internal model. Hybrid prototyping systems are under development making the advantages of both, physical and virtual prototypes, available.

Due to the information required the right prototype and herewith the right prototyping technology must be chosen with respect to time, cost and quality. It is not always necessary to use rapid prototyping technologies to produce the prototype. A fast production of prototypes might also be achieved with high-speed tools using traditional processes like milling. In the following, criteria for the classification of rapid prototyping technologies are given.

3.1 Specification of Requirements

The word "rapid" in rapid prototyping implies that the factor time is of great importance, but the significance of cost and quality cannot be denied. Similar to traditional product development processes the use of rapid prototyping technologies is ruled by three principal factors that are **time, cost and quality** (see figure 2).

The *time* from request to delivery is the only time period interesting for the client. This time period is strongly influenced by the following factors:
- in house rapid prototyping facilities can be used,
- prototype has to be purchased from a supplier service,
- availability and burden of rapid prototyping machines,
- lead time for the creation of a prototype with the chosen prototyping technology,
- CAD data quality and data processing requirements (model repair, slicing, triangulation).

Besides the factor time the prototype has to be of an appropriate *quality*. The different RP-technologies may be classified according to the **geometrical accuracy** of the prototype and the **material** used to create the prototype. The combination of these two key elements is appropriate to express nearly all relevant requirements to Design, Geometrical or Functional prototypes such as:
- dimensions of the prototypes geometry (Geometrical prototype)
- surface characteristics (Design, Geometrical, Functional prototype)
- weight, stability, or resistance to rupture (Functional prototype)

All these requirements together with the requirement to produce the prototype as cheap as possible have to be taken into account at the same time. It is impossible to fulfil all these requirements at the same time having available only one machine. In many cases it might be possible to produce the product with high cost but in other cases even this is impossible. The optimisation of the function is a task which has to be supported by a system taking into account time, quality and cost.

To get the best possible solution it is important to have as many machines based on different technologies available as possible. To afford this, a network has to be set up linking together end-users, service suppliers and support centres.

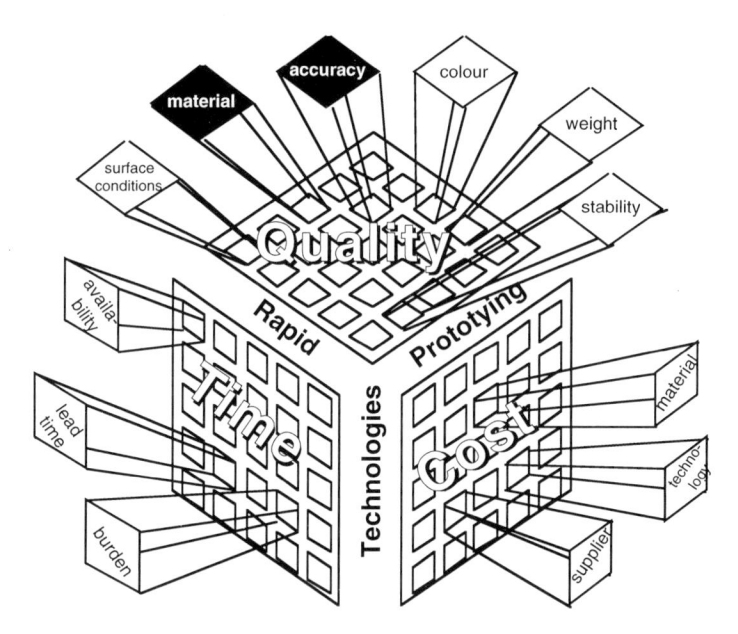

Figure 2 Classification criteria for the selection of (rapid) prototypes

3.2 Prototyping Service Junction between Suppliers and Users

The initial cost of a RP-machine is between US$ 75.000 and US$ 750.000. Besides this there is an investment in the maintenance between US$ 5.000 and US$ 90.000 and highly qualified personnel required to run such a machine. The training duration is between 5 and 13 days, additional fee for training is up to US$ 12.000 (Aubin, 1994). The installation of rapid prototyping technologies is thus a significant decision for each enterprise. On the other hand the availability of a RP machine at any time might be of such an importance that some companies buy their own machine. Of course this solution is only feasible for large companies, that can guarantee the minimum burden required for amortisation.

To offer rapid prototyping to those companies that cannot afford to buy this cost intensive technology might be a new service for specialised RP service providers. This RP service providers will be of interest not only for SMEs (small and medium size enterprises) but as well for large companies, because a network of RP service providers can offer the wide range of different prototyping technologies. This allows to have an optimum support of the development process using the best prototype available.

The proposed ESPRIT project PRONET is aiming to support SMEs to improve product innovation processes using rapid prototyping technologies. This goal shall be reached by the implementation of a network linking together end-users, service suppliers and support centres. The three main services offered as a result of this project will be:
- RP best practice (references and introduction guide)
  RP-references of best practice in different application areas will be documented.
- RP support and dissemination facilities
  This service will include awareness events, seminars and support for the RP-introduction in the individual companies' product development processes.

- Value added network (electronic purse and communication)
  The electronic purse will offer services like the generation of STL files from CAD models or the production of the entire prototype. For the communication standards like STEP and EDIFACT will be used and corresponding processors will be offered to read in or generate the files.

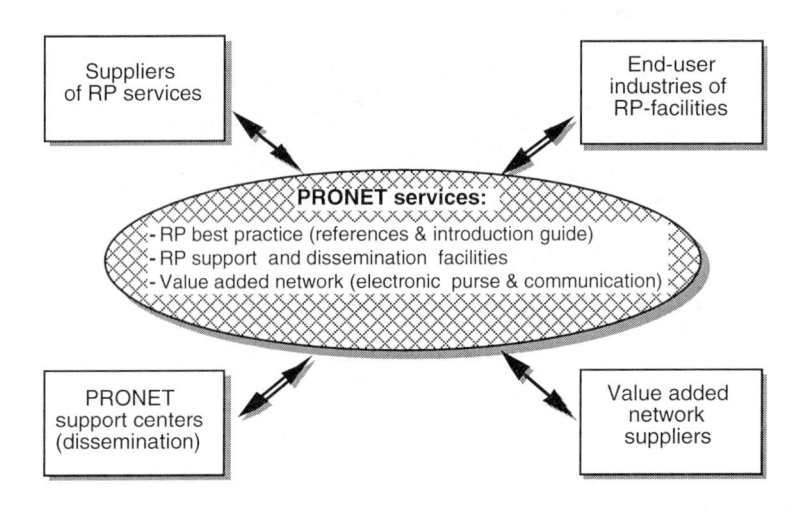

Figure 3 Services of a rapid prototyping network (PRONET - ESPRIT)

3.3 Standards

The CAD model designed by the client has to be exchanged to the supplier. Depending on the technology that shall be used for the production of the prototype a new file has to be generated triangulating or the slicing the original model. The resulting file contains the geometric data necessary as input for the processor of the RP machine to generate the information for the rapid prototyping process. Receiving the original CAD-file the RP supplier has to estimate time and cost for the prototyping process. Based on this information the client gives the order to produce the part or not. After the production of the part, the part will be shipped to the client. The overall effort of this RP procedure can be estimated between 2 to 4 days depending on the availability of the chosen RP machine. If a RP network will be used as a vehicle to connect client and supplier, data exchange becomes significant.

Data exchange between the user and RP-supplier should be done via standards that are offered by the application protocols of ISO 10303, known as STEP. It is currently under discussion in the ISO technical committee 10 how the standard shall support the necessary data exchange for rapid prototyping. Because of the fact, that a standard should not contain process specific data and each implementation of STEP shall be based on a specification given in an application protocol the option to specify the RP relevant data by restricting geometric data to certain elements should be preferred.

4 CONCEPT FOR AN INTEGRATED DESIGN SYSTEM SUPPORTING THE USE OF PROTOTYPES DURING PRODUCT DEVELOPMENT

As mentioned in chapter 2 working with "rapid" prototypes might be a suitable method to speed up the product development process, presumed that prototyping has been integrated into the design and development process from organisational and technical point of view.

Within this chapter a concept is described on how the creation of prototypes can be supported by a product modelling system. As a basis for concept development and implementation the integrated product modelling system DICAD (Dialogue oriented Integrated CAD system) from the institute RPK of the University of Karlsruhe was chosen. DICAD aims to support all stages of the product development process starting with the specification of requirements, the functional and preliminary design, allowing the detailed design followed by process planning stages like manufacturing or inspection planning.

4.1 Supporting the Design Process with Solution Patterns

Studies on the design process showed that, if a new product has to be designed usually 90% of the components have already been designed before and can be derived from former design solutions by varying parameters (Grabowski, Gittinger and Schmidt, 1994). Thus, in average only 10% of the designers work requires creative solution capability from the designer. Taking this fact into account, the design process can be improved in terms of time and quality by the use of solution patterns. Solutions of problems from the past are taken to solve the actual problem and, thus, to step forward from one state of solution to another (see figure 4). If the solution patterns are of high complexity, i.e. the solution is much more concrete than the problem, the design process is speeded up considerably. On the other hand the use of such complex solution pattern leads to the generation of less variants and less control in the design process by feedback to more abstract levels, e.g. by analysing calculations. The integration of calculations into the solution patterns and the necessary information about the generation of adequate prototypes in different states of product design can help to overcome these problems and to improve the design process. Erb (1995) describes a concept for the integration of calculations into solution patterns. The necessary equations are part of a constraint network which allows the use of the solution patterns independent from the direction of application of the solution pattern. Additionally to this concept information about prototypes that were generated for the considered part and the results of tests that were performed with these prototypes are stored. By the integration of the information about the process necessary to produce the prototype and the knowledge about the best kind of technology to produce the prototype the use of prototypes can be made more efficiently and the best possible results can be achieved.

Although calculations are integrated into the solution patterns some control cycles might be missing in the design process of the actual product. Some specific calculations that would have been applied in the traditional design process of a product might not be applied for a similar product. This makes the use of prototypes even more important in the down stream approach (see figure 4). The tests that are performed with the prototypes will help to analyse the product and to discover errors.

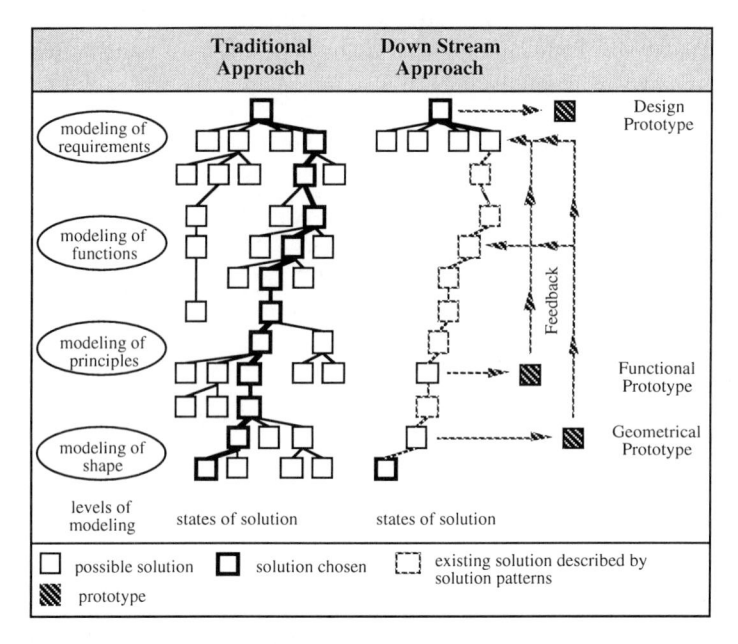

Figure 4 Integration of prototypes in the design process

4.2 Selecting a Prototyping Technology using Mulitcriterial Optimisation Theory

As soon as the design process has reached a status where a solution pattern exists that includes the necessary information to produce a prototype it has to be checked which prototyping technology is the best one. In addition to that a provider has to be selected who can deliver this kind of prototype within the appropriate time frame.

For that purpose a set of requirements has to be specified that describes all qualitative and quantitative demands the prototype should fulfil. Requirements can be of three different types:
- **fix requirement**: specifies a fixed value that has to be fulfilled by the prototype (example: colour of prototype is white)
- **range requirement**: specifies a range of values in which the prototypes properties should lie in (example: density of material between 3 and 6 g/mm^3)
- **trend requirement**: specifies a trend that should be fulfilled by the prototype (example: delivery as soon as possible).

After a prototyping problem has been specified with the help of the three requirement types (see figure 5), an inquiry in a rapid prototyping data base is done aiming to find an appropriate prototyping technology and/or service provider. This presumes that different prototyping technologies and different service providers have been assessed before and the significant data was stored in a rapid prototyping data base. The result of the inquiry is a sorted list presenting the prototypes meeting the requirements specified.

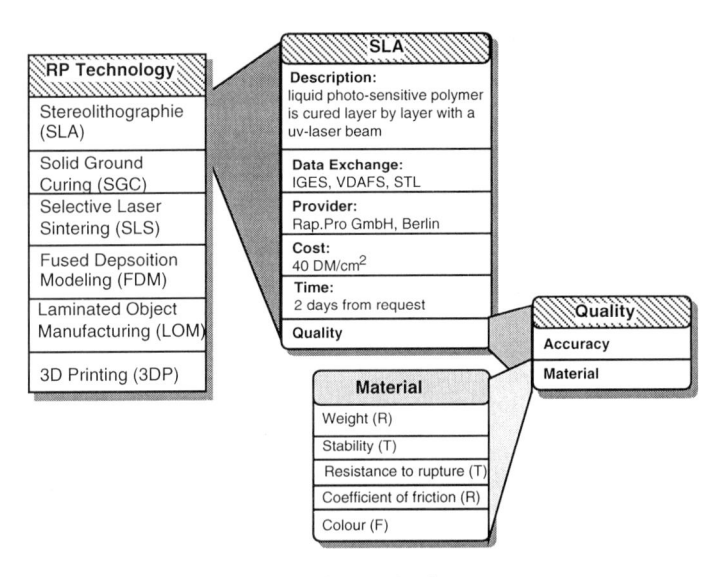

Figure 5 Specification of requirements for a technology request

In general a great number of requirements specified are contradictory, what means that if one requirement is fulfilled 100% another cannot be fulfilled at all. For example if the cost of the prototype should be as low as possible, but the accuracy should be as high as possible none of the available prototypes will be found to meet these requirements. For that reason a multicriterial optimisation algorithm will be used to find the optimum solution taking into account the competing demands.

For that purpose a demand must be formulated as a target function f(x) that has to be minimised or maximised. The prototype selection algorithm has 3 target functions, that are:

Quality: $\max f_1(x) = f(\text{material, process,}...)$
Cost: $\min f_2(x) = f(\text{material, process, volume }...)$ (1)
Time: $\min f_3(x) = f(\text{availability, burden}...).$

If f(x) is the set of all target functions $f_i(x)$:

$$f(x) = [\, f_1(x), f_2(x), f_3(x)\,], \tag{2}$$

than a value x* has to be found that f(x*) becomes an optimum:

$$f(x^*) = \text{opt } f(x). \tag{3}$$

Fix requirements, range requirements and trend requirements can be captured using auxiliary functions $h_i(x)$ formulating the limiting conditions. As an example the weight w of the prototype that has to be minimized can be defined as a function of the density d and the volume v:

$$\min h_1(x) = w = d \bullet v. \tag{4}$$

The resulting system of balanced and unbalanced equations is then evaluated determining the Pareto optimum solutions /Götze/. The Pareto optimum solutions are defined as the solutions that meet all specified requirements best at the *same time* (see figure 6). From the set of Pareto optimum solutions the designer is supported in choosing the best solution from his point of view by assigning weights to the three target functions quality, time and cost. The prototype solution derived with multicriterial optimisation is then compared to the prototype technologies described in the data base and the list of appropriate technologies is generated and displayed.

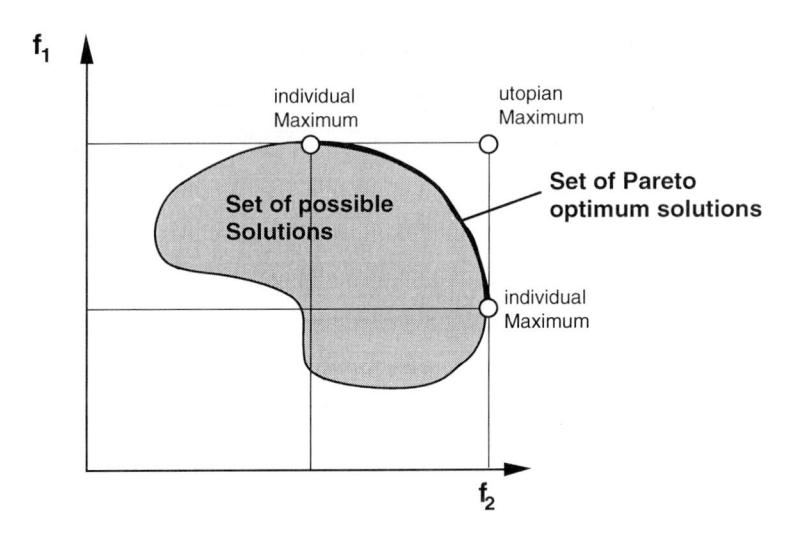

Figure 6 Multicriterial optimisation and Pareto optimum solutions

4.3 Scenario

After the definition of the physical principles that shall be realised in the design the effective geometry of the product is fixed. Based on this effective geometry a functional prototype may be generated. Those areas of the product that are not defined at that point of time have to be described in a temporary manner, e.g. using a skinning functionality. Depending on the design stage the adequate category of prototype is offered by the system. Figure 7 shows the result of an inquiry for a prototype of category functional prototype. The system is searching for a similar problem which means a prototype that was generated in the past for the same kind of product with similar effective geometry and in the same design phase. The result of this search is certain RP-technology that was chosen in the past to generate the prototype. The properties of this technology are given to the system user. Depending on the requirements of the current development process some of these parameters have to be modified and priorities may be determined. The change of these parameters initiates a new search for the best technology for the current problem. Another reason for the change of technology that shall be used might be the availability of the machines (see 3.2). This factor is important for the delivery time and is not only depending on the technology used.

In most cases it is not necessary to change the technology but to adopt the parameters that are relevant for the production of the prototype with the same technology that was chosen for a similar design task in the past. The availability of those solutions from the past will lead to a faster production of the prototypes because the preprocessing time will be minimised. Even more important than this is the fact that this process offers the possibility to learn from the experience made in the past.

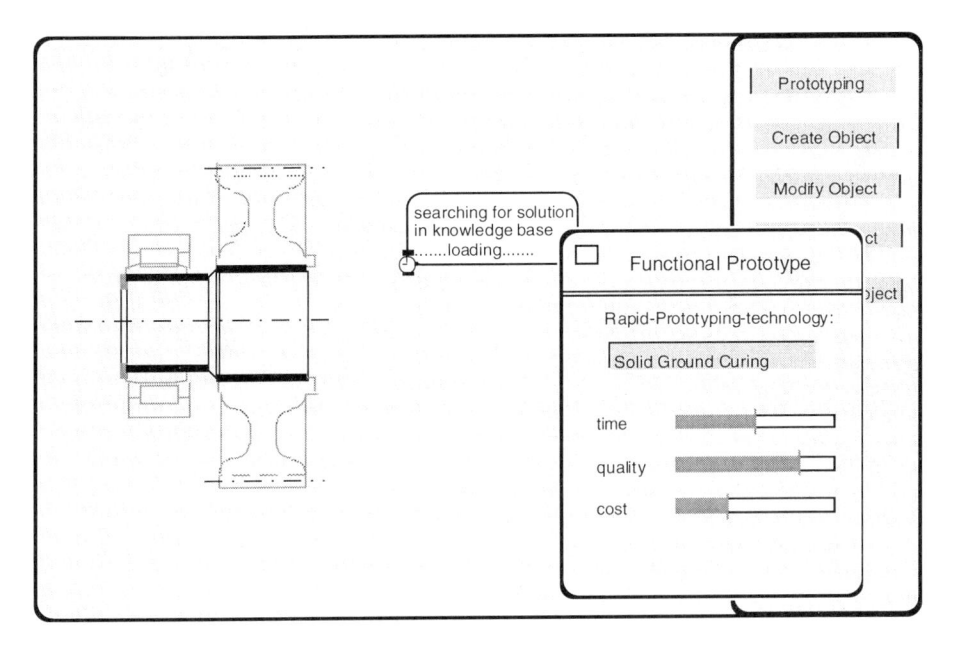

Figure 7 Generating a functional prototype from the effective geometry

5 CONLUSIONS

The aim of Rapid Product Development is to reduce development times by applying rapid prototyping technologies. Even though rapid prototyping technologies have advantages in process planning and fabrication time compared to traditional manufacturing techniques a remarkable reduction of lead times can only be achieved if rapid prototyping technologies are integrated into the Rapid Product Development process chain including design, fabrication and inspection.

Based on our work achieved in reverse engineering, aiming to reconstruct product models from geometric data of prototypes and the support of inspection and manufacturing planning (Grabowski, Erb, and Geiger, 1995) we did focus on the selection of prototyping technologies during the design process within this paper.

Taking into account that the rate of standard components within a new product design is remarkably high we did propose solution patterns to help the designer to find the former solutions that are appropriate to solve the actual design problems. Those solution patterns also contain information on the use of prototypes during the design process of former solutions.

If a prototype is regarded as helpful for the further development of the solution by either the designer or the system, the selection of an optimum rapid prototyping technology is invoked by specifying the requirements the prototype should fulfil and determining the technology required with the help of a mulitcriterial optimisation algorithm.

The selection of prototypes presumes that prototyping is not only done with in-house facilities but offered by different rapid prototyping service providers. The great variety of available technologies and services has to be assessed and documented in a rapid prototyping data base. Of course the work load could be dramatically reduced if neutral support centres would continuously survey the market and provide information on rapid prototyping benchmarks.

Considering the progress networking technology did within the last years the formation of a prototyping network linking together clients, service providers and support seems to be near.

6 REFERENCES

Anderl, R.; Malle B. and Schmidt M. (1992) Concurrent engineering based on a Product Data Model, *Proceedings of the 3rd International Conference on CALS*, CALS Europe 92

Aubin, R. (1994) A World Wide Assessment of Rapid Prototyping Technologies, United Technologies Research Center, *UTRC Report No. 94-13*, Initiated by the Intelligent Manufacturing Systems IMS, Test Case No. 6 on Rapid Product Development

Bullinger H.-J. (1995) What Makes Product Development Rapid? An Integration Approach of Organisational, Technical and Human Aspects; in: *International Conference on Rapid Product Development*, Stuttgarter Messe- und Kongressgesellschaft mbH, Stuttgart 1995

Erb, J. (1995) *Rekonstruktion von Informationen für die frühen Phasen der Konstruktion*; Dissertationsmanuskript; Universität Karlsruhe; 1995

Götze, S (1995) *Die multikriterielle Entscheidungsfindung als Modell für die Simultane Produktentwicklung*; Forschungsberichte aus dem Institut für Rechneranwendung in Planung und Konstruktione der Universität Karlsruhe, Shaker Verlag, Aachen, 1995

Grabowski, H.; Erb, J. and Geiger, K. (1995) Generating a Product Model from Surfaces Using Solution Patterns; Proceedings of the 28th ISATA International Symposium on Automotive Technology and Automation, Dedicated Conference *on Rapid Prototyping in the Automotive Industries*, Soliman J. (ed), Germany, September 1995,

Grabowski, H.; Erb, J. and Geiger, K. (1995) The role of CAD in Rapid Product Development. Proceedings of the 4th European *Confernce on Rapid Prototyping*, October 4 and 5, 1995, Paris

Grabowski, H.; Gittinger A.; Schmidt M. (1994) *Informationslogistik für die Konstruktions - Realisation von Informationsspeichern als wichtiger Schritt zu mehr Effizienz*, VDI-Zeitschrift, 136, Nr. 10, VDI-Verlag, Düsseldorf 1994

Grabowski, H.; Huber, R. (1993) DIICAD - Supporting Product Development Stages with an Integrated Knowledge based Design System; JSME/ASME-*Workshop Frontiers in Engineering Design*, Tokyo, 6/1993

Ulrich K. (1994) The Role of (Rapid) Prototyping in (Rapid) Product Development in: Proceeding of the International *Conference on Rapid Product Development* 31.01 - 02.02.1994 Stuttgart, FpF - Verein zur Förderung produktionstechnischer Forschung e.V., 1994

Wohlers T. (1995) Rapid Prototyping Markets, Applications and Trends in Asia, Europe and the USA; in: International *Conference on Rapid Product Development*, Stuttgarter Messe- und Kongressgesellschaft mbH, Stuttgart 1995

Innovative Product Development and advanced Processes by Solid Freeform Manufacturing

Kochan, D.
SFM GmbH
Bismarckstraße 57, D-01257 Dresden, Germany
Phone: (++49)(0)351/2806-261
Fax: (++49)(0)351/2806-262

Abstract
The advanced methods for product development and manufacturing by Solid Freeform Manufacturing, Rapid Prototyping, Layer Manufacturing or named otherwise offer new possibilities for time and cost savings in connection with quality improvements. Another important aspect is the overall impact on sustainable product development and manufacturing. The paper described the general principles on the new technologies, the typical application ranges and advantages related to sustainable product development and manufacturing processes.

Keywords
Solid Freeform Manufacturing, Rapid Prototyping, Layer Manufacturing, Clean Production, Near Net Shape Technologies

1 INTRODUCTION

The entire technological development is driven by different influence factors, especially time to market, cost and quality. For the general further development of the human society the environmental aspects becames more and more important. Therefore the evaluation of new production technologies from this point of view is most relevant.

Advanced production technologies have to be characterized by a best possible utilization of materials and energy resources.

If we consider, that for example in some cases - especially in connection with complicated parts - ninety percent and more of the material must be destroid by cutting procedures an essential requirement is clear. Any future oriented production must be in agreement with the possibilities of Near Net Shape technologies.

The following paper will demonstrate the good accordance with this general requirement.

Before the impact on SFM-technologies to the "Near Net Shape"-technology is characterized, some general explanation on the new principle will be necessary.

2 OVERVIEW ON THE NEW PRODUCTION PRINCIPLES AND USABLE MATERIALS

The industrial application of the new manufacturing principles began more than 7 years ago.

Based on sliced 3D-CAD-data it was possible to harden a photopolymerical material layer by layer by means of a laser beam. This so called Stereolithography-principle is currently the most applied procedure. But during the first years of industrial utilization the material allowed only the application of "show and tell" objects, in other words for design verification or presentation.

The further development can be characterized by some essential features:

- application of new and better materials for the stereolithography procedure
- invention and industrial application of new physical principles especially also with other materials.

An overview on the most frequently applied industrial procedures is given in Figure 1.

	Liquid		Powder		Solid	
physical effects	Point by point Laserbeam (single frequency)	Lamps			Melting+ Solidification	Gluing sheets
company	- Stereolithography (3D-Systems, USA) - Stereography (EOS, Germany) a.o.	- Solider (Cubital, Israel)	- Selective laser Sintering (SLS-DTM, USA; EOS, Germany) - 3D-Printing / Gluing (MIT, USA)	- Fused Deposition Modelling (Stratasys USA, Perception Syst.) - Ballistic Particle Mfg. (Perception System, USA)		- Laminated Object Manufacturing (Helisys, USA; KIRA, Japan)
usable materials	polymer epoxyd	polymer	Wax, PVC, Nylon, Ceramic, Alloys		Wax, Nylon	Paper, Plastic, Ceramic

Root: Solid Freeform Manufacturing

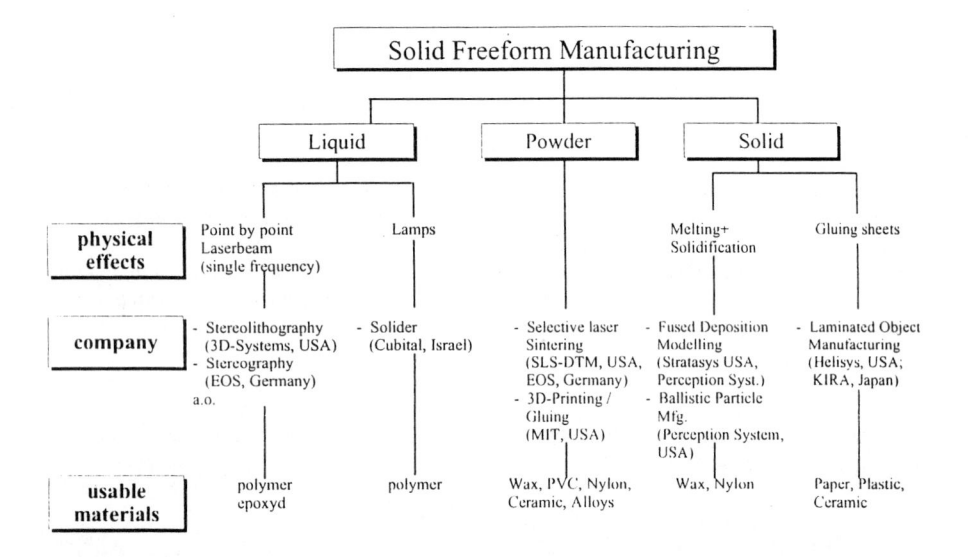

Fig. 1: Overview about industrial most applied procedures

The evaluation of the available procedures from the point of health and environmental aspects leads to the following characteristics:

Procedures which are using polymerical fluids (e.g. stereolithography; Solid Ground Curing) require an extensive postprocessing.

Generally it has to be considered that the photopolymers used are toxic materials. Independently of the different properties and the material-dependent recommendations, the following requirements are essential:
- The wearing of plastic gloves, laboratory coats, eye goggles, etc., is necessary for avoiding any skin contacts with resins or "green parts".
- The cleaning equipment (using chemical materials like isopropanol ore acetone) and the SFM equipment themselves require exhausters.
- The safe storage of new and old resins has to follow specific recommendations.

Whereas the Stereolithography resins can be used almost completely, the Solider principle (Cubital) does not allow a full utiliziation of the applied materials. The waste mixture, of resin and wax, cannot be recycled by the user and has to be returned to the vendor.

The materials for all other applied SFM procedures such as powder, wire and laminates are non-toxic. In particular the FDM procedure (using nylon wires) and the LOM procedure (using Laminates from paper, wood and plastic) do not require any specific expense. Also for the SLS procedure, one advantage of the materials used (Polycarbonate, Nylon, Wax, ABS) is the non-toxic behaviour.

For a general evaluation of the new technologies, the impact of the entire integrated product and process development is most important.

3 SFM-PROCESS CHAINS

The SFM-process chain can be characterized by the following essential components:
- 3D-modelling, as an absolute prerequisite for SFM-procedures
- SFM-building process
- Secondary (or follow up-) processes like vacuum casting, investment casting a. o.

Essential advantages and possibilities of 3D-modelling are:
- multiple possibilities for the description of complicated shape elements
- unification of surfaces to surface units by determination of different connecting conditions between the single surfaces
- different and independent description of outside and inside contours of objects
- simple and fast possibilities for the local and global changing of surface description.

All this and other advantages lead to reduced failures during the product developmental phase. In this context it can be understood as an essential contribution to the optimized product development from different points of view, especially also reduced waste.

4 THE IMPACT OF SFM-PROCEDURES TO „NEAR NET SHAPE-TECHNOLOGIES"

Near Net Shape technologies are the best possibility for sustainable manufacturing processes. The impact is given by the utilization of SFM-technologies along the product development and manufacturing cycle.

4.1 SFM-building procedures and typical secondary processes (Kochan, 1993)

Additionally to the 3D-CAD-modelling the availability of real objects of all technical objects and components is absolutely necessary. The real relationships and proportion in complex and complicated devices or assembly groups can not be simulated by computer support. Therefore the building of design or function pattern for many investigations is a compelling need. The rise of any product development can be minimized if test objects are available at an early stage.

For this reason the SFM-procedure can be applied to all mechanical parts. The manufacturing of the models can be realized with different SFM-procedures. For the produced prototypes the following demands are normally requested:

- checking of the design of outside shape
- checking of the construction of the single parts related to productibility and functionality
- suitability as basic models for vacuum casting parts, which can be used for assembling of applicable test products
- suitability as basic model for manufacturing of forms for injection molding by metal spray procedures. This parts must be usable for medium-sized production (up to 5000 pieces).

Because of the necessity for parallelization of product-development tasks and preparation of the manufacturing processes several experimental samples of the camera have to be put ready. For the multiplication of single stereolithographic models into 10 to 20 plastic parts the vacuum casting procedure can be successfully used.

The preparation of the serial production starts also simultaneously to the development and construction of new products. The selection of suitable manufaturing procedures must be in agreement with the complexity and number of the parts.

That means it is necessary to apply manufacturing procedures which allow a low price level also for small and medium-sized production.

An essential cost factor are the manufacturing costs for the injection molding tools. These manufaturing procedures of complicated freeform shapes require multiaxis milling and EDM processes for the quality justified production. In this case it must be possible to the NC-programming directly by base on the 3D-CAD-data.

This is also an opportunity for the reduced amount.

Some parts are suitable for the new metal-spray procedure. This procedure reduces the cost up to 20% in comparision with the traditional NC-manufacturing.

The tools, manufactured by the metall-spray procedure, can be used up to 5000 injection moulding parts.

Another new possibility is given by Rapid or Composite Tooling.

4.2 Economic effects and results

The application of the described innovative methods and procedures for the product development and manufacturing of new products leads to a lot of positive effects. Currently the exact qualification is not easy, because there don't exist analogues traditional processes for such complicated parts. Most important is the drastical reduction of the development time and costs from 50 to 80%. In this case the increased degree of complexity of the parts has to be considered.

We have realized several examples with such drastical advantages.

The consistent realization of the methods of Simultanious Engineering including extented and early tests at real models yields the following advantages and principles.

- investigation and evaluation of constructive and technical variants
- optimization of application features
- reduction of construction failures up to nearly "Zero"
- multiple usage of optimized CAD-data for follow up and parallel processes
- preparation of serial manufacturing and tooling in parallelization with the product testing.

The effects are summarized in Figure 2.

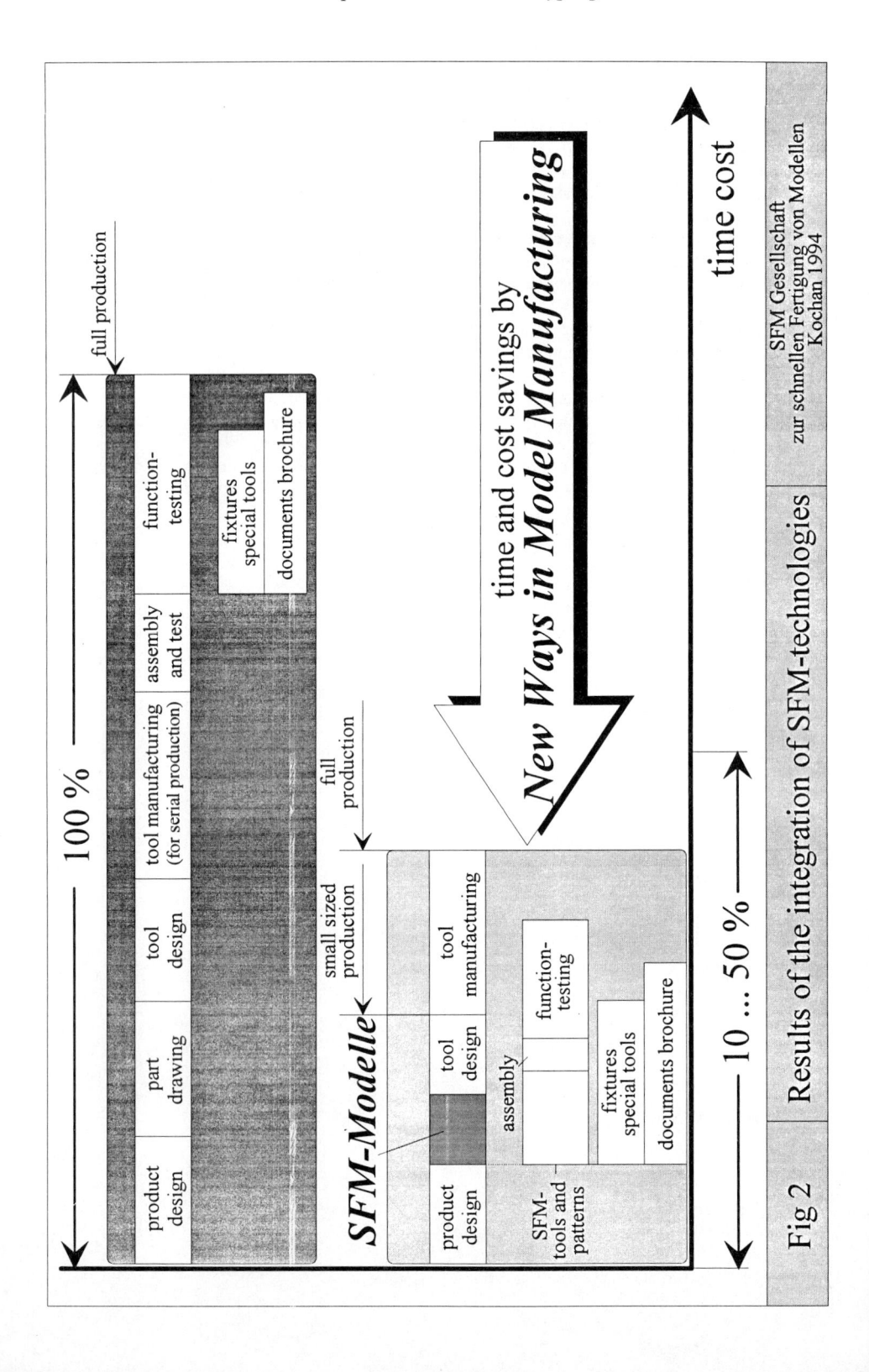

Fig 2 | Results of the integration of SFM-technologies

5 EXAMPLES AND CONCLUSION

The characterized effects will by demonstrated by different examples, f. e. camera housing, different gardening equipment, technical components a. o.
Related to this examples the following facts are most important

- for product design and manufacturing it is possible to realize more complex parts
- in one single part can be included some more functions
- that means new products can be designed nearly without traditional restrictions from the manufacturability point of view
- such complex parts-direct generated based on 3D-CAD-data can be manufactured by advanced casting, lost foam casting a. o. Near Net Shape Technologies
- especially lost foam casting is a typical example for more clean production.

The new technologies of Solid Freeform Manufacturing are still in the first phase of industrial application. But if this possibilities are not only applied as specific procedures for determined requirements, f. e. show and tell models, the entire advantages are most important. If the new technologies are understood and applied as a strategic tool along the entire process chain from product design till manufacturing the effects can be enormous. The broad utilization of all given advantages guarantees highest quality, shortest time and lowest cost.

The succesful development of the new NOBLEX Panorama camera (Sir Noble, 1995) is one evidence of the new development stage in direction of the new epoch of advanced production technology.

In general, by application of the new technologies, time and cost savings in the range of 50% up to 90% can be expected. The explanation for such tremendous advantages is given in Figure 3. This figure demonstrates also some essential trends for further developments.

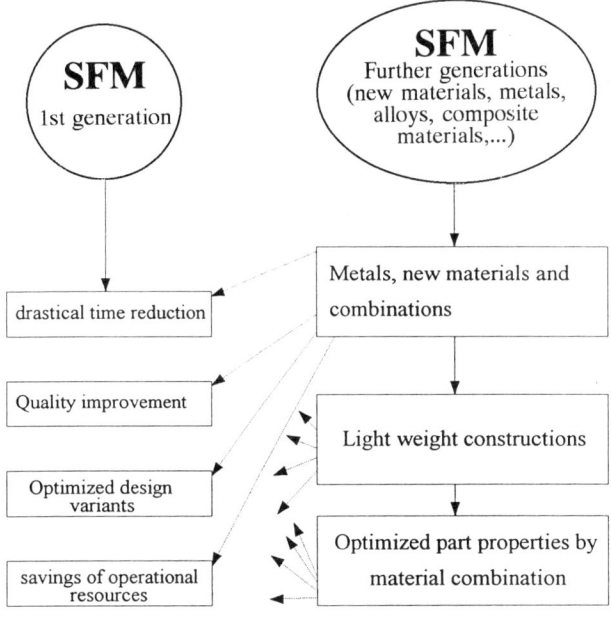

Fig 3
Essential effects of the new epoch of Production Technology

6 REFERENCES

Kochan, D. (1993) Solid Freeform Manufacturing - Advanced Rapid Prototyping. *ELSEVIER*, Amsterdam

Sir Noble, J. and Kochan, D. (1995) Contribution to renovation of a camera factory by intelligent production technology, in *Proceedings of CAPE '95*, Chapman & Hall

26

CAD data preparation in a virtual prototyping environment

H. Kress, J. Rix
Fraunhofer-Institut für Graphische Datenverarbeitung
Wilhelminenstraße 7, 64283 Darmstadt, Germany
Tel. : +49 6151 155 (212 | 220), fax: 155 299
E-mail : (kress | rix)@igd.fhg.de

I. Kiolein
Tallinn Technical University
Ehitajate tee 5, EE0026 Tallinn, Estonia
Tel. : +372 6 56 29 48, fax: 56 29 32
E-mail : kiolein@edu.ttu.ee

Abstract

The ability to rapidly prototype a proposed design is becoming a key contributor towards fulfilling the business requirements embodied in a short time-to-market, in cost-effective and high quality manufacturing. Reorganizing the design and development process along the lines implied by the concept, **Concurrent Engineering**, means that advanced information technologies must be taken advantage of. The integration of existing technologies in CAD modelling, design, analysis, simulation, networked cooperation, and virtual reality (VR) leads to a distributed desktop-environment which provides the basis to take the greatest advantage of the **Virtual Prototyping** technique.

Some of the basic difficulties in the virtual prototyping process are the contrary requirements of the design and presentation applications concerning the representation of the product data. Usually, 3D CAD models are extremely detailed, containing all categories of chamfers, bore holes, small pockets or protrusions. Contrary to this, VR data structures focus on visualization and real-time interaction aspects. They have to be less detailed, due to the restrictions of the visualization hardware. Therefore this paper describes strategies for the preparation of CAD data and the reduction of the model's complexity.

Keywords

Virtual Prototyping, CAD, tessellation, reduction of complexity

1 INTRODUCTION

The use of physical prototypes is a widely adopted technique to verify the design and the functional behaviour of complex products. Especially in the automotive and aircraft industry physical prototypes, also called mock-ups, are built to check aesthetical details and overall appearance of a product, to confirm that products are easy to assemble and disassemble or to prove the ergonomical design of cars and aircraft. In the development cycle of these products a huge series of different prototypes is produced. Normally these prototypes are made by hand in specialised departments, which makes this technique especially expensive and time-consuming. Consequently the industry tries to establish new processes which reduce both costs and production time.

In the past several years, the development of powerful computing hardware led to an increasing number of sophisticated engineering applications. Especially in the area of mechanical systems design an immense progress in the development of modelling, analysis, and simulation software has been achieved. 3D CAD systems provide high-level modelling operations in combination with kinematics, NC simulation or FEM analysis. Real-time analysis and simulation of multi-body systems has been realized in specialized applications. The need for a neutral representation of the different data sets related to one product has been identified. Therefore concepts for the covering of all these information arising during a product's life cycle have been defined in the STEP project of the International Organization for Standardization (ISO 10303-1, 1994). The information is represented by a conceptual information model, called product model. By means of the STEP product model and rules for the handling of these models it will be possible to describe product data in a unified way and to share product data between different systems without any loss of information.

These developments led to a vision of a computer based prototype. While the occurrence of a physical prototype represents a realistic first model of a part, a computer based prototype realizes the functional behaviour of the underlying product model by means of computer based simulations. In this context the term **Virtual Prototype** has been formed. It is defined e.g. by Haug et al. (1993) as „a computer based simulation of a prototype system or subsystem with a degree of functional realism that is comparable to that of a physical prototype". This phrase emphasises as one of the most important characteristics of a Virtual Prototype the realistic, dynamic simulation of the functional behaviour. The simulation can either be an off-line simulation without human action involved or a real-time operator-in-the-loop simulation including human interaction.

The Virtual Prototyping Process can be regarded as an important method to shorten the product development time and to reduce or eliminate the use of real physical mock-ups. The greatest advantage of this process can only be taken if a full integration in the whole engineering process is achieved (Kress, 1995). The realization goes along the lines implied by the concept, **Concurrent Engineering**, which can be regarded as a systematic approach to an integrated product development which embodies parallel work, cooperation and information sharing among designers, mechanical engineers, manufacturing engineers and project managers. Consequently, the paper focuses on one of the most important aspects of the process, the integration on the data level. Concepts for the preparation of CAD data for Virtual Prototyping are given. An algorithm for the tessellation of CAD models is described and different possibilities for the reduction of the complexity are discussed. Finally, the results of the first prototypical implementation and a realized demonstrator are explained.

2 VIRTUAL PROTOTYPING USAGE SCENARIO

The improvement of the product development process is a major goal for companies and enterprises to remain competitive. Organizational changes and the use of innovative software tools are inevitable to reduce product development time and costs. A key concept used to achieve this goal is Concurrent Engineering. Focusing on the design phase of a product, the notion of Concurrent Engineering has to be mapped on the requirements in this phase. The project management requires a continuous coordination of the teams and the involved persons. Designers, mechanical engineers, and analysts have to be assigned to their work, as well as tasks and deadlines have to be communicated and checked. Hence a computer-based environment provides an optimized base for the execution of these tasks. Only if the team members and their leaders have the possibility to keep track of the work-in-progress of their colleagues and check their own work against the efforts of the others, the concept of parallelised work in a group can be realized.

The variety of the information handled in the design process and the complexity of the decision-making require graphical-interactive applications based on user-friendly concepts for the visualization of the information. Characteristical data handled and created in the design process are e.g. information about person and organization structures, the design progress, the approval status of specific parts, the product structure or the product's shape. According to the structure of the information units different modes of presentation are chosen. A product structure can be visualized as a tree structure, administrative information are schemed in panels, and 3D CAD data is presented by interactive 3D viewing applications. To realize the idea of a virtual prototype, additional visualization capabilities have to be provided. Virtual reality (VR) technology offers the possibilities to create such interactive environments. They are defined as real-time interactive graphics with three-dimensional models combined with a display technology that gives the user immersion in the model world and direct manipulation (Bishop et al., 1992). This technology is especially suited for design activities, because of the iterative analyse-refine cycle. Any method which aids the designer during this process will improve the entire activity. Because a strength of VR is the capability for direct manipulation of objects within the virtual space, design activities should benefit greatly from this technology (Bishop et al., 1992). A system incorporating real-time simulation of the functional behaviour, direct manipulations, and VR display technology realizes the interactive visualization of the virtual prototype.

Several implementations have been realized in the area of the aircraft and automotive industry. The Advanced Technology Center at Boeing Aircraft in Seattle investigates full immersive VR with the goal of integrating it into the design, test and mock-up process (Pimentel, 1993). In October 1991 a first virtual prototype of the tilt-rotor aircraft V-22 has been shown (Aukstakalnis, 1992). Broad use of VR technology has been accomplished in the development of the commercial aircraft Boeing 777. The number of physical mock-ups for assembly verification has been reduced to one item for the extremely complex cockpit section. Automotive manufacturers like Chrysler (Beier, 1994), Ford (Deitz, 1995), Mercedes-Benz (Grebner and Fay, 1995), VW, or BMW are examining the VR technology for the virtual prototyping of cars. In Europe the research in this area is driven by the AIT Consortium, an alliance of the European automotive and aircraft industry (AIT Consortium, 1994). The goal is to establish research projects within the fourth research framework of the European Community. The research will focus on applications for styling, ergonomics, and assembly simulation.

3 DATA INTEGRATION

In the past years several approaches to define VR data exchange formats have been accomplished.

The Fraunhofer standard (FHS) for VR data exchange was developed by four Fraunhofer institutes (IAO, IBP, IGD, IPA) within the Fraunhofer demonstration centre for VR (Fraunhofer, 1995). The format allows the specification of hierarchical scene descriptions, polygonal geometry, material, texture and lighting definitions.

The Virtual Reality Modelling Language (VRML) is a language for describing multi-participant interactive simulations (Bell et al., 1994). The language is under development for the definition of virtual worlds networked via the global internet and hyperlinked with the World Wide Web. The syntax is based on the Open Inventor ASCII file format from Silicon Graphics, Inc. (Strauss, 1994 and Wernecke, 1994). The language supports descriptions of 3D scenes with polygonally rendered objects, lighting, materials, and ambient properties.

The Naval Postgraduate School Monterey developed the object description language NPSOFF (Zyda et al., 1993). The language allows the storage and retrieval of 3D objects. Graphics primitives, lighting models materials, textures and colours can be defined.

In Beier (1995) the object-oriented graphics kernel YART is characterized, which can be used in a heterogeneous, network-distributed virtual reality environment. The YART graphics kernel supports classes of analytical primitives like sphere, torus, and cylinder. YART provides abstract attributes like resolution, fillstyle or mapping, which do not belong to a given class of a primitive. A format for the distribution of the scene description via the network has been defined.

These data exchange formats are specialised for the use within VR environments. They are extremely tailored to the needs of the visualization within a VR system. This makes it rather complicated to integrate these formats in an environment, which handles primarily product data of the design process. The description of the product data in the design systems differs highly from the description in the VR systems. As a solution the extension of the existing product models in the design process with the needed VR data structures is proposed. Consequently, the models defined in the STEP project should be used as the initial input for this effort. First approaches are been proposed by Kress (1995) and Henson et al. (1995).

4 DATA CONVERSION AND ENRICHMENT

Some of the basic difficulties in the virtual prototyping process are the contrary requirements of the design and presentation applications concerning the representation of the product data. 3D CAD models are usually extremely detailed, containing e.g. all categories of chamfers, bore holes, small pockets or protrusions. The geometrical representations include e.g. canonical, Bezier or B-spline techniques and the topological representation can vary from CSG (Constructive Solid Geometry) to boundary-representation. Contrary to this, VR data structures focus on visualization and real-time interaction aspects. They have to be less detailed, due to the restrictions of the visualization hardware. The preferred representation is a polygonal description of the product's shape. Sequences of planar polygons approximate the shape and serve as the input for the VR systems.

CAD data preparation in a virtual prototyping environment

Today's high-end 3D graphics workstation can visualize scenes of approximately 20 to 30 000 polygons with 10 to 20 frames per second in real-time. Unfortunately, a complex CAD model, e.g. an engine of a car, consists of up to 200 000 polygons after tessellation (with an accuracy of less than half a millimetre). A whole engine compartment with all aggregates and sheet metal parts can reach 1 000 000 polygons. Obviously, these models are 10 to 50 times larger than the models existing hardware can handle. Therefore a reduction of the complexity of these models is necessary to enable real-time visualization and interaction.

CAD data preparation

A series of operations has to be performed to realize a continuous and self-regulating conversion procedure from the CAD to the VR data structures. The following operations can be distinguished: a geometrical and topological correction of the CAD data, a reduction of the complexity of the CAD model, the tessellation of the CAD model, a reduction of the complexity of the facetted model, and the enrichment with visualization and simulation information

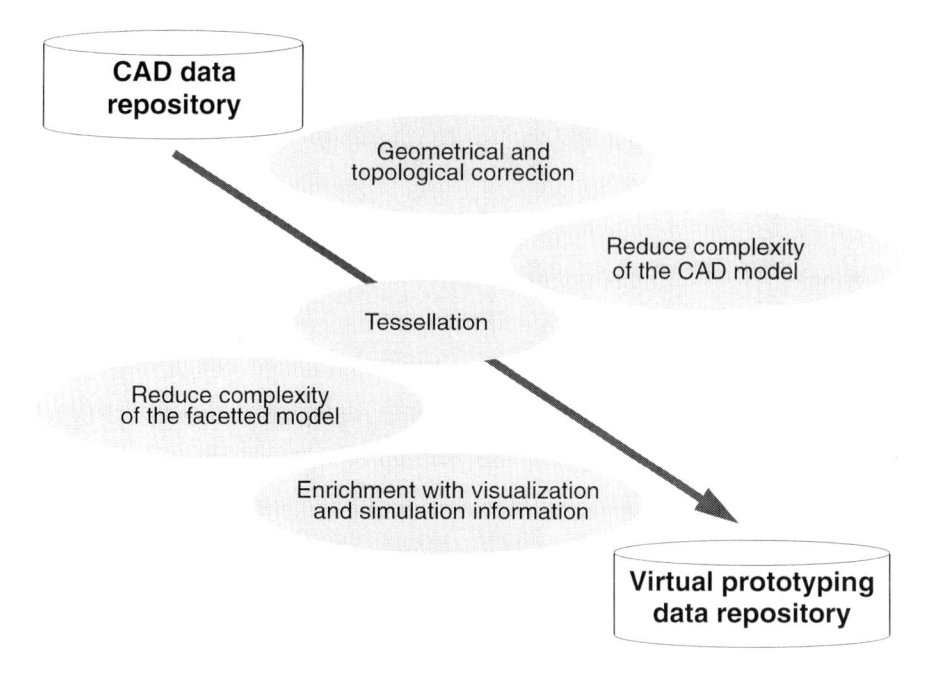

Figure 1 CAD data preparation for Virtual Prototyping applications.

Geometrical and topological correction of the CAD data

The first step in the process is a consistency check of the CAD model. This step ensures that the geometrical and topological representation of the model is correct, and the following steps can be performed without interruption. The structure of the model is scanned for inconsistencies which can appear during the design process due to defects defined by the user or numerical problems. Examples for these defects are missing patches in a multi-patch surface, gaps in a multi-patch surface, or multiple defined objects. As far as possible the defects should be eliminated automatically. If this is not possible, the defect is brought to attention of a user and will be eliminated manually.

Reduction of the complexity of the CAD model

As mentioned before, the reduction of the complexity of the CAD models should start with the initial CAD data. Only in this stage geometrical and topological features of the CAD data can be accessed and used for the reduction. Later this information will be lost during the tessellation process. CAD models contain numerous detailed objects like small bore holes, chamfers, or blendings. E.g. for a clash and clearance check of a virtual prototype these information are not relevant in this complexity and can therefore be eliminated. This gives the opportunity to develop efficient reduction algorithms.

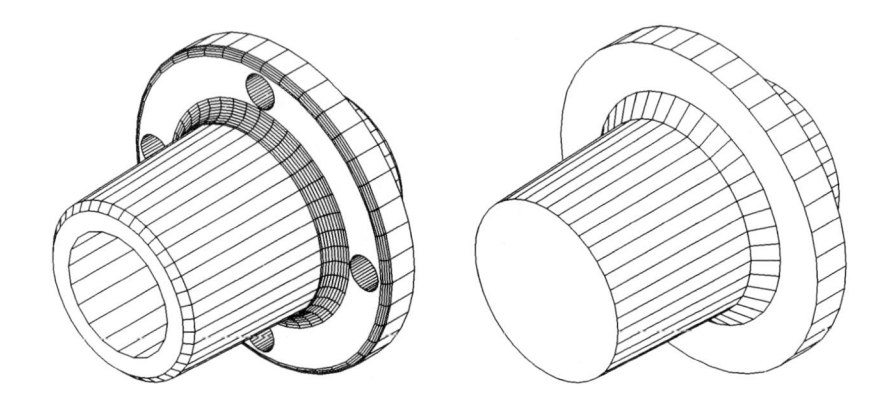

Figure 2 CAD model with different levels of complexity.

Figure 2 gives an example for an extreme reduction of complexity. A facetted CSG model of a hub is reduced by editing the CSG tree and eliminating through holes, chamfers, and blendings. A reduction of 90% (from 1711 facets to 164 facets) can be achieved. This example illustrates in principle the concept for the reduction. These strategies can be used in a similar way for the reduction of freeform models, both for open shells or closed b-rep. Feature-based CAD systems promise an even easier way for this approach, since these systems have more knowledge about the characteristics of the CAD model. Accessing form features instead of pure geometry and topology allows for application dependent reduction of the model.

Tessellation of the CAD model

For real-time visualization and interaction the CAD models are converted into a polygonalized representation, which can be loaded by the VR system. Depending on the geometrical and topological representation of the CAD data, different conversion algorithms have to be used.

The following aspects have to be considered during the conversion procedure: the accuracy of the tessellation, the object structure of the CAD model, and the orientation of the generated facets. In chapter 5 a procedure for the tessellation of trimmed surfaces is described.

Reduction of the complexity of the facetted model

After the generation of the facetted model, a further reduction of the complexity can be performed. For this tasks algorithms are used which reduce the original model either depending on a given minimum of roughness, which has to be preserved, or try to generate an optimized model with a given number of facets.

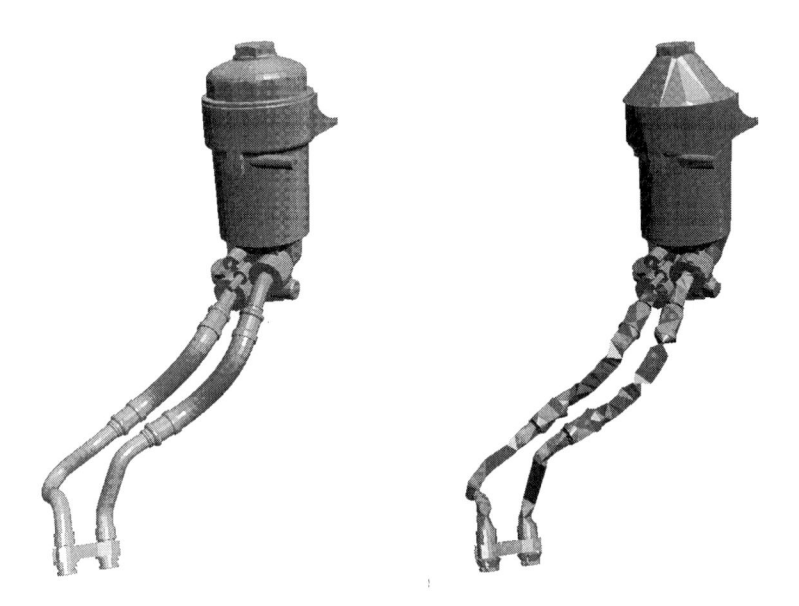

Figure 3 Facetted model of an oil filter (a) in high resolution and (b) in reduced complexity (model provided by AIT project).

Figure 3a shows an oil filter of a car. This model is part of the data set which has been prepared for the demonstration of **Digital Mock-Up** in the AIT project (see chapter 6). The original CATIA CAD model was tessellated with the VOX utility of TECOPLAN INFORMATIK GmbH (Ottobrunn, Germany). This was followed by a reduction of complexity using a tool developed by Schröder and Roßbach (1994). Different levels of reduction were tested. The results show, that a reduction of 64% (17 936 to 6456 facets) shows severe destruction of the model's shape (Figure 3b). Accordingly, reductions based on the facetted model can only be applied up to a reduction of 10 to 30% in order to prevent loss of details and of important overall shape aspects.

Enrichment with visualization and simulation information

In a final step the model has to be enriched with information for the visualization and simulation. This includes the definition of colours, textures, lighting definitions, behaviour of the objects, kinematic relations, or collision detection parameters.

These different procedures should be based on a common information model, which suites

5 IMPLEMENTATION

A first prototypical implementation of a converter for the CAD system *CATIA Solutions 4* (Dassault, 1993) has been realized. The functionality of the software module embodies the correction and the tessellation of CAD models represented by trimmed surfaces. The supported output formats are the Fraunhofer Virtual Reality format (FHS) and the Open Inventor ASCII file format of Silicon Graphics, Inc.. The system behaviour has been tested using complex models of the automotive industry. Both the CAD system and the converter are integrated in the open engineering framework *CoConut* (Jasnoch et al. 1994).

Trimmed surfaces have a fundamental role in CAD technique. Most of the complex CAD objects are generated by some sort of trimming process. Unnecessary parts of the objects are trimmed away. Even the result of Boolean operations on solid objects can be represented as b-reps based on trimmed NURBS surfaces.

There is a need to visualise these surfaces or patches in the CIM pipeline (rendering for visualisation, virtual prototyping etc.). The easiest way for this is to approximate the trimmed patch by triangular facets within a user given tolerance. This method has a number of advantages:
- It is not sensitive to the complexity of the trimmed patch.
- The same triangular, irregular network can be used for different purposes.
- Algorithms that operate on triangles are far easier to implement and numerically more stable than those dealing with freeform geometry.
- The stepwise triangular approximation is independent of the parametrization of the trimmed surface, e.g. very general algorithms can be written to process this geometry.

The two major disadvantages are:
- Adequate representation of a trimmed patch with high curvature areas requires large numbers of triangles.
- The triangulation, if not done properly, can result in triangles of different sizes and in particular in long and skinny triangles which in turn can cause numerical problems.

The implemented algorithm triangulates trimmed surfaces with a predefined triangle size or tolerance. The algorithm has some prerequisites for the model. The model has to be correct and precise. It is important that all surfaces in the shell are trimmed correctly, gaps and overlapping surfaces are not tolerated. In the case of a correct model the program can collect all surfaces that belong to the shell. Surfaces constructing the shell should have only one patch. The algorithm performs a patchwise triangulation of the surfaces. Patches may have an arbitrary number of holes. Patches containing holes are divided by the pre-program.

A generated mesh must satisfy different requirements. The mesh should be topologically and geometrically correct. The elements of the mesh should not intersect and have to be topologically correct. The nodes of the mesh should be positioned exactly on the edges and faces of the model. The mesh should be boundary conforming. No elements may intersect the boundary of the object, and there should be no gaps in the mesh. Regarding the given limit of accuracy, the mesh should match the CAD model exactly. Within one shell the normal vectors of the elements should point to the same side of the shell.

Program structure

The workflow of the program is written in C programming language and uses geometric functions of the CATIA system (CATGEO and CATMSP). The program is divided in two parts: the pre-program and the main program. The pre-program identifies and marks all incorrect surfaces so that the user can check these manually before starting the main program. Pre-discretisation and the search for modelling errors are the main tasks of the program. The functional behaviour includes the splitting of multipatch surfaces into single patches, dividing patches containing holes, and detecting free edges on the shell boundary.

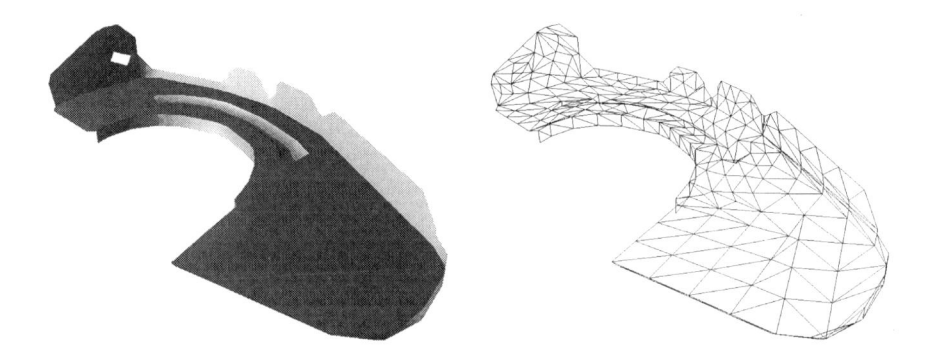

Figure 4 Sheet metal part in shaded (a) and wireframe (b) presentation.

The task of the main program is to triangulate the prepared model. The workflow of the program is divided in different operations:
- Reading surface information.
- Calculating boundary curves.
- Breaking boundary curves at the corners of neighbouring surfaces (if necessary).
- Deletion of double boundary curves. The previous and the actual action are necessary to collect separate surfaces into one shell and to get neighbouring information. Abreast located surfaces have common boundary edges.
- Generate points on surface boundaries. The points can be placed at equal distance or to satisfy the approximation tolerance.
- Delete double points on the corners. Intersecting curves have common points.
- Checking loop directions (to ensure the right direction of normal vectors). Every surface has a normal vector, which will be used to ensure right point order and direction of the point normals in the triangular stripes.
- Dividing complex areas to areas with three or four edges. The decomposition of the input geometry results in a number of simple shaped mesh areas.
- Triangulating areas with three or four edges, where the opposite edges have the same number of points.

The triangulation is performed in parametric space. Considering that the surfaces consist only of three or four edges, we used a very simple method for the triangulation. Co-ordinates for other points on the surface are calculated using linear interpolation, where four corresponding points on the boundary are used. This method takes advantage of the distribution of the boundary points. The surfaces with three edges are processed by the same function as the surfaces with four edges. The only difference is, that the points on one edge have to be used twice. The points are calculated row by row and stripes are formed right away. This allows to create larger stripes. This is important, because the larger the stripes are the faster the visualization and interaction can be performed.

Figure 4 shows an example sheet metal part, which has been processed by our software module. In the wireframe presentation the triangles of the stripes are clearly visible.

6 DEMONSTRATION

For the final presentation of the results of the ESPRIT project No. 7704 **Advanced Information Technology in Design and Manufacturing** (AIT Consortium, 1994) the Fraunhofer-IGD in close cooperation with BMW realized a Virtual Prototyping demonstration. A Digital Mock-Up of a 7 series BMW (Figure 5) has been prepared for interactive assembly and disassembly of different generators in the engine compartment. The demonstration included real-time collision detection, kinematic simulations, and the simulation of flexible parts. For the demonstration IGD's Virtual Reality software Virtual Design (Astheimer et al., 1995) has been used for visualization and interaction with the virtual prototype. The data preparation was performed as described in this paper.

Figure 5 Digital Mock-Up of a 7 series BMW (AIT project).

7   CONCLUSION

The paper surveys the concepts and a user scenario of the Virtual Prototyping process. The data preparation has been identified as one of the major problems in this process. Therefore the fundamental problems of the CAD data preparation for Virtual Prototyping applications are described and an overview over the concepts of the first prototypical implementation is given.

Future developments should concentrate on a common data model for the Virtual Prototyping process and automatic conversion procedures for design and presentation applications, so that real-time data exchange will be achieved.

8   REFERENCES

Astheimer, P., Dai, F., Felger, W., Göbel, M., Haase, H., Müller, S. and Ziegler, R. (1995) Virtual Design II - An Advanced VR System for Industrial Applications, in proceedings of Virtual Reality World '95, Stuttgart, Germany, February 1995.

AIT Consortium (1994) Management Overview of the First Project Phase. ESPRIT Project 7704 Advanced Information Technology in Design and Manufacturing.

Aukstakalnis, S. and Blatner, D. (1992) Silicon Mirage. The Art and Science of Virtual Reality. Peachpit Press, Berkeley, CA.

Beier, E. (1995) An Object-Oriented Graphics System on the Way to Virtual Reality, in proceedings of the Second EuroGraphics Workshop on Virtual Environments, January 1995, Monte Carlo, Monaco.

Beier, K.-P. (1994) Virtual Reality in Automotive Design and Manufacturing. University of Michigan, SAE-Congress 1994.

Bell, G., Parisi, A. and Pesce, M. (1994) The Virtual Reality Modeling Language. Version 1.0 Specification (Draft), 2 November 1994, World-Wide Web: http://vrml.wired.com/vrml.tech/vrmlspec.html.

Bishop, G., Fuchs H. et al. (1992) Research Directions in Virtual Environments. *Computer Graphics*, Vol. 26, No. 3, pp. 153-177.

Dassault Systemes (1993) CATIA Solutions Version 4: General Information Manual. IBM Deutschland Informationssysteme GmbH, Stuttgart, Germany, Doc. No. SH52-0610-00.

Deitz, Dan (1995) Real engineering in a virtual world. *Mechanical engineering*, Vol. 117, No. 7, pp. 78-85.

Fraunhofer Demonstration Centres for Virtual Reality (1995) Syntax Description of the Fraunhofer VR-data-exchange Format (FHS). Fraunhofer-IAO/IBP/IGD/IPA, Stuttgart/Darmstadt, January 1995.

Grebner, K. and May, F. (1995) Applications of Virtual Reality Techniques in the Industry - Selected Examples, in proceedings of Virtual Reality World '95, Stuttgart, February 1995, pp. 451-468.

Haug, E.J., Kuhl, J.G. and Tsai, F.F. (1993) Virtual Prototyping for Mechanical System Concurrent Engineering, in *Concurrent Engineering: Tools and Technologies for Mechanical System Design* (ed. E. J. Haug), Springer.

Henson, B.W., Juster, N.P. and de Pennington, A. (1995) Towards a Product Model for Virtual Prototyping, in proceedings of the Computers in Engineering Conference ASME 1995, September 17-20, 1995, Boston, USA, pp. 941-950.

ISO 10303-1 (1994) Industrial automation systems and integration - Product data representation and exchange - Part 1: Overview and fundamental principles. International Organization for Standardization, Geneve (Switzerland).

Jasnoch, U., Kress, H., Schroeder, K. and Ungerer, M. (1994) CoConut: Computer Support for Concurrent Design using STEP, in proceedings of the Third IEEE Workshop on Enabling Technologies: Infrastructure for Collaborative Enterprises, April 17-19, 1994, Morgantown, West Virginia, USA, IEEE Computer Society Press.

Kress, H. (1995) Integration Aspects within a Virtual Prototyping Environment, in proceedings of the 5th International Conference on Flexible Automation and Intelligent Manufacturing, FAIM '95, Stuttgart, Germany, June 28-30, 1995.

Pimentel, K. and Teixeira, K. (1993) Virtual Reality: through the new looking glass. Windcrest Books, McGraw-Hill.

Schröder, F. and Roßbach, P. (1994) Managing the complexity of digital terrain models. *Computers & Graphics*, Vol. 18, No. 6, pp. 775-783.

Strauss, P.S. (1994) A BNF Specification for the Open Inventor ASCII File Format. Silicon Graphics Inc., November 1994.

Wernecke, J. (1994) The Inventor Mentor. Open Inventor Architecture Group, Silicon Graphics Inc., Addison-Wesley Publishing Company, 1994.

Zyda, M.J. et al. (1993) NPSOFF: an Object Description Language for Supporting Virtual World Construction. *Computers & Graphics*, Vol. 17, No. 4, pp. 457-464.

9   BIOGRAPHY

Holger Kress is a researcher in the Industrial Applications Department of the Fraunhofer Institute for Computer Graphics since 1991. He is currently involved in research projects in the area of product modeling, groupware, and CAD frameworks. He received a masters degree in mechanical engineering from the Darmstadt Technical University in 1991. His research interests include concurrent engineering, product modeling, and computer supported cooperative work.

Indrek Kiolein is a researcher in the Institute of Machinery of the Tallinn Technical University since 1989. He is currently involved in research projects in the area of product modeling. He received a masters degree in mechanical engineering from the Tallinn Technical University in 1994. His research interests include concurrent engineering and product modeling.

Dr. Joachim Rix is head of the department for Industrial Applications of the Fraunhofer Institute for Computer Graphics (IGD) in Darmstadt, Germany. From 1991 to 1992 he was Associate Manager of the Fraunhofer Computer Graphics Research Group (today; Fraunhofer Center for Research in Computer Graphics, Inc. (CRCG) in Providence, RI). Mr. Rix received his Diploma and Ph.D. in Computer Science from the University of Darmstadt. His topics of interest are in Computer Graphics, CSCW, CAD, and Product Modelling. This includes the integration and use of computer graphics with its presentation and interaction techniques in industrial applications, like CAD, CAM, Concurrent Engineering , and Groupwork Computing. Since 1985 Mr. Rix is member of the national and international committees (DIN NAM 96.4, ISO TC184/SC4) developing STEP (Standard for the Exchange and Representation of Product Model Data). Since 1994 he holds the position of a deputy convenor of its WG 3 „Product Modeling". Since 1981 Joachim Rix is member of the Eurographics Association.

27

Identification of process relevant form structures for stereolithography within solid models

F. Mandorli and U. Cugini
KAEMaRT Group, Dep. of Industrial Eng., The University of Parma
V.le delle Scienze, Parma, ITALY
tel. +39-521-905706 - fax. +39-521-905705
e-mail: [ferro, cugini]@labcad1.eng.unipr.it

Abstract

The need to reduce the time to market of new products has drastically increased the importance of rapid prototyping. Stereolithography is an emerging, commercially available, rapid prototyping technology based on a constructive methodology.

Manufacturing processes based on a constructive methodology simplify process planning and reduce the complexity of tooling and fixturing by decomposing a 3D model into sliced 2D representation. Those advantages also help them to integrate within CAD data more easily than classical milling and forming technologies.

However, the production of a part within unsolidified material, imposes several new problems, not being present in classical manufacturing technologies.

In the present work, we address the problems of inconsistent slice contour, drifting layer portions, and trapping material regions.

The paper overviews several procedures to identify and control the correctness of process relevant form structures of a sliced model generated from a 3D CAD solid model.

Keywords

Stereolithography, direct slicing, solid model

1 INTRODUCTION

Despite the remarkable progress of CAD systems in producing virtual models, today the realisation of a physical prototype model is still required to support either assessment of design and detection of functional faults. Due to life cycles of artefacts, which become shorter, rapid prototyping is gaining increasing importance, and new technologies for a cheaper and faster mock-up production are appearing on the market. Those new technologies are based on a constructive methodology, where sets of laminates are incrementally fabricated and pilled-up (Burrell 1992, Hartmann 1994, Kobayashi 1991, Michaels 1993, Weiss 1992).

Stereolithograpy is a constructive methodology where the layer-based material increasing manufacturing is obtained by the solidification of photo polymer liquid (Bjorke 1991, Jacobs 1992, Medler 1990).

Instead of traditional milling processes, that have to deal with the non trivial problem of tooling and fixturing, stereolithography simplifies process planning and production and reduces the complexity of fixturing by decomposing a 3D model into sliced 2D representation.

The model to be manufactured is sliced by a set of parallel planes. The entire part is produced on the table of a piston-driven elevator, which is integrated in a container filled with a photocurable liquid and positioned under a laser beam (see figure 1).

The solidification of a layer takes place just beneath the liquid polymer's surface, through photo-polymerisation. The thickness of a solidified layer is in the range 0.1 - 0.5 mm and represent a physical constraint of the technology.

The sliced model is used to drive the laser beam over the liquid surface. Two main strategies can be used to drive the laser:

- the laser beam scans the liquid surface following the slice contours and hatching the internal part of the slice;
- the laser performs a raster scanning and the intersections between the scanning path and the slice's contour drives the laser power on/off.

After solidification of one layer, the table is lowered by a distance equal to the layer thickness and the process is repeated until the entire part has been solidified. Finally the part is removed from the liquid and cured in a special oven.

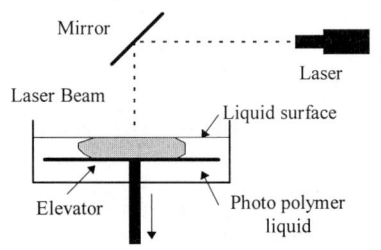

Figure 1 - Stereolithography apparatus.

The 3D model can be created starting from a solid or surface modeler. Next a faceted or sliced version of the model can be used as input for the stereolithography apparatus. This realises a CAD/CAM integration more easily than classical material removal technologies ever can.

However, the fabrication of a part within unsolidified material, imposes several new problems, not being present in classical manufacturing technologies, mainly regarding the correctness of layer contours and contiguity of successive layers.

The intent of the present work is to overview several procedures to control the correctness of slices shape generated from a 3D CAD model. The proposed approach is based on the identification of process relevant form structures that must satisfy particular conditions to ensure the data correctness. Experiment results as well as components of the implemented prototype slicing module will be introduced and discussed.

2   STEREOLITHOGRAPHY INPUT DATA

Current stereolithography apparatus can accept as input both a faceted model or a set of slice contours (see fig 2). A standard format for representing triangle-based faceted model used for stereolithography is STL (3D System 1988). A standard format to represent set of slice contours is CLI (Common Layer Interface) (BRITE-EURAM 1994). An alternative possibility is to represent slice contours using plotter driver standard as HPGL (Hewlett-Packard Graphics Language).

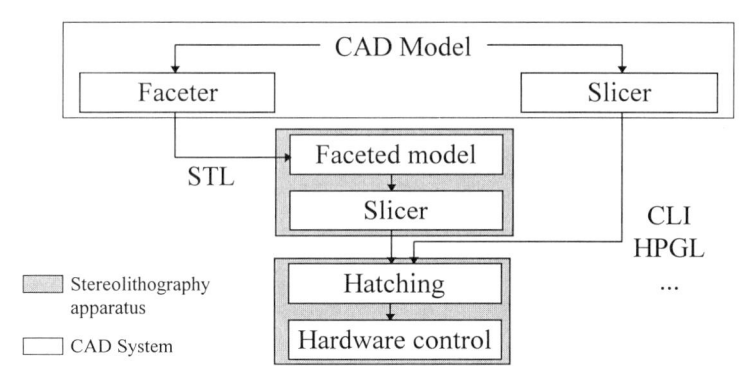

Figure 2 - Different data flow from CAD system to stereolithography apparatus.

An STL file can be easily obtained from a two-manifold solid model or a surface model bounding a closed volume. There are several advantages in using a faceted model as data input for a stereolithography apparatus:

1. Traditional CAD systems already include a faceter module for graphical purposes.
2. A standard data-exchange format to represent faceted models already exists (STL).
3. The algorithm charged to produce slices from the faceted model has to compute simply plane-plane intersections.
4. The integration with CAD systems easily follows from (1) and (2).

However, a faceted model produced by a faceting algorithm used for graphic purposes must be carefully checked before it can be used for stereolithography. Traditional faceting algorithms perform a local triangulation, iterating over all model surfaces, not taking in account information about surfaces adjacency. As a result, the faceted model can have some discontinuities along the model's edges.

These discontinuities are not significant for graphics, but they can affect the correct fabrication of the object. If a discontinuity is present on the faceted model, an open contour can be produced during slicing, affecting the correct layer solidification.

To overcome such a problem global faceting algorithms have been realised. These algorithms split the edges of the model by adding new points on the edge curves; for each edge, the added points are constrained to be vertices of the triangles adjacent to the edge and belonging to both faces adjacent to the edge (see fig 3).

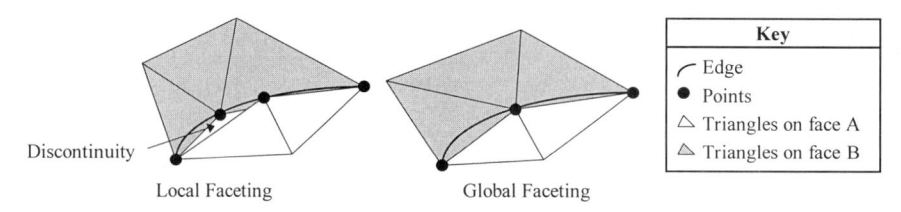

Figure 3 - Result of local and global faceting algorithms.

As a consequence of (3), the slicing software can be integrated within the front-end of the stereolithography apparatus; however, the low level information contained in the faceted model makes difficult an automatic control of the sliced model correctness.

Direct slicing is a different approach to produce input data for stereolithography (Jamieson 1995); following this approach, layers are directly produced by slicing a 3D solid model. Direct slicing requires more sophisticated algorithms to produce slices, but it can benefit from the geometrical and topological information present in the solid model to control the slicing procedure and ensure the sliced model correctness. The main advantages of this approach are:

- Elimination of the need of an intermediate faceted model.
- Volumetric correctness of the 3D model.
- Possibility to compute and check slices using the CAD system intersection algorithms.

In the next chapters we will describe some of the possible control that can be performed during slicing, taking advantage from traditional algorithms that we can expect to be present in a traditional CAD system.

3 DEFINITION OF PROCESS RELEVANT FORM STRUCTURES

Layered manufacturing techniques within liquid material imposes some physical constraints to be taken in account while generating data to drive the stereolithography apparatus.

Typical errors that may occur during fabrication derive from inconsistent slice contours, drifting layer portion and trapping material region. In this chapter we define form structures that are characteristic of the above mentioned situations, while in next chapter we define functions to identify such structures within the sliced model.

3.1 Inconsistent slice contour

The sliced model is used to drive the solidification process. For each slice of the sliced model, the laser beam has to solidify a layer having the same shape.

D_1: We define inconsistent slice contour (ISC) each contour that drive the solidification of a layer with a shape different from the slice's shape.

During solidification, hatching or raster scanning algorithms are used to drive the laser beam over the liquid surface. If a slice contour contains dangling/extra/open (poly)lines, then the driving algorithms can fail and solidify a layer with an unexpected shape.

Dangling/extra/open (poly)lines (see fig 4) may occur when the slicing plane is tangent to a surface of the geometric model (for further details about slicing procedures refer to Dolenc 1994).

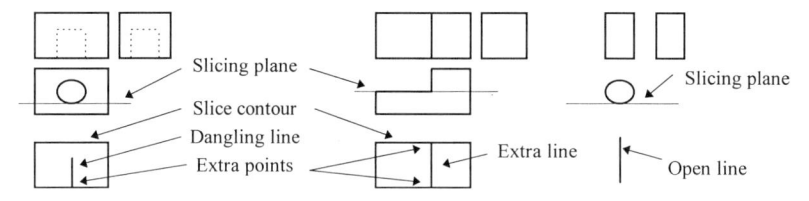

Figure 4 - Examples of inconsistent slice contours.

3.2 Drifting layer portion

A significant part of the rapid prototyping process, using traditional manufacturing technologies, is fixturing. When using stereolithography technology, fixturing can be seen as the problem of fixing the model that is created within the liquid material. We can split this problem in two sub-problems:

1. To fix the first created layer to the elevator's table.
2. To fix each new solidified layer to some already solidified part of the model.

The first layer glues to the elevator's table during solidification. If the layer has an extended surface, then it should glue too strongly to the table, making difficult to remove the object from the elevator at the end of the solidification process. In this case appropriated supports must be designed and added to the object model before starting the solidification process (for more details refer to to Otto 1995).

Each new solidified layer remains fixed to the model by gluing to the previous layer.

D_2: We define *drifting layer portion (DLP)* each layer portion that it is not fixed to already solidified portions of the model.

Drifting elements usually develop if overhangs are created (see fig 5). In those cases, layers are made by disconnected portions; while solidifying the new layer, if a layer portion has not contact points to the previously solidified layer portions, then it will float on the liquid.

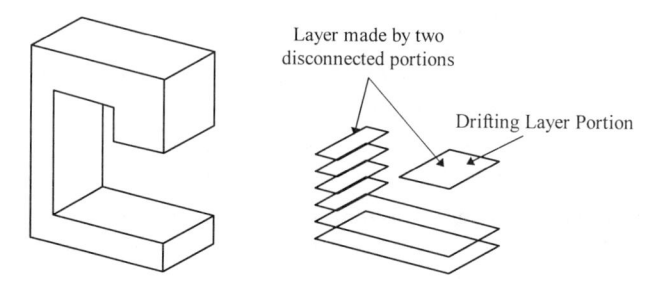

Figure 5 - Example of drifting layer portion.

3.3 Trapping material region

Constructive layered manufacturing can be used to create models with hollows (see fig 6). However, when using stereolithography technology, hollows will trap liquid that must be drained before post-curing.

D$_3$: We define *trapping material region (TMR)* a region in the model that will trap unsolidified material.

While designing, the identification of trapping material regions is fundamental to automatically suggest to the user the positioning of draining holes.

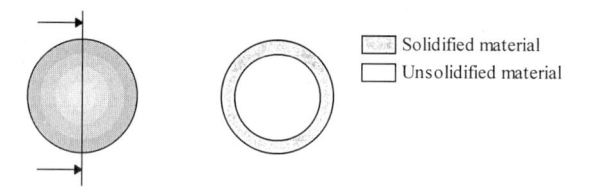

Figure 6 - The internal part of an hollow sphere is an example of trapping material region.

4   IDENTIFICATION OF PROCESS RELEVANT FORM STRUCTURES

The automatic identification of the previously defined process relevant form structures is needed to improve the automatic control of model correctness.

In most of the available systems the model control is left to the user, who, looking at the model graphical representation, verifies the model correctness before starting the fabrication process. As far as the model become complex, such a visual check is a non trivial problem and the risk to find a model error during the solidification process, with a significant lost of time and expensive photocurable material, becomes higher.

4.1 Identification of inconsistent slice contour

A slice contour cannot drive the solidification of a layer with a correct shape if it contains dangling/extra/open (poly)lines. The ISC condition can be stated as follows:

C_1: if a slice contour contains dangling/extra/open (poly)lines, then it is an inconsistent slice contour.

The identification within a slice contour of dangling/extra/open (poly)lines can be performed on the basis of the following considerations: each polyline is made by a set of lines and each line is bounded by a starting point and an end point. If a point exists with only one incident line, and this line has an open curve geometry, then the point bounds a dangling (poly)line or an open (poly)line. If a point exists with more than two incident lines, then this point is an extra point and it belong to an extra (poly)line.

When slices are produced by direct slicing of a solid model, we can expect they have a wire frame representation of its boundary. In this case each slice contour is made by one or more wires; each wire is made by a set of edges connected by vertices. Using appropriated functions to navigate the slice's data structure, dangling/extra/open (poly)lines can be identified looking at wires topological and geometrical properties.

4.2 Identification of drifting layer portion

A layer portion cannot be fixed to the model if it has not superimposition area with portions belonging to the previous layer. The DLP condition can then be stated as follows:

C_2: if a slice portion has not superimposition area with one of the slice portions belonging to the previous slice, then the considered portion corresponds to a drifting layer portion.

If slice portions are represented within the sliced model as surfaces, then superimposition areas can be easily founded by performing a 2D Boolean intersection among surfaces belonging to different slices; if no intersection is found, then the checked surfaces correspond to a drifting layer portion.

We can expect 2D Boolean intersection to be one of the provided functions of a traditional CAD system. On the contrary, not all systems return a surface as a result of a slicing operation. To overcome such a problem, the user can build a slicing procedure based on 3D Boolean operations returning a set of solid slices having a minimal thickness. In this case, area superimposition can be found by using traditional 3D Boolean operation among slices.

4.3 Identification of trapping material region

The identification of trapping material region corresponds to the identification of hollows. The TMR condition can be stated as follows:

C_3: if a set of connected faces is completely surrounded by another disjoint set of connected faces, then the space bound by the surrounded inner set is a trapping material region.

When more than one set of connected faces is present within the model, the problem of identifying a set of connected faces surrounded by another disjoint set of connected faces can be simplified by considering that, given two set A and B of disjoint connected faces, a point $\alpha \in A$ and a point $\beta \in B$, the set A is surrounded by B if and only if the point α is internal to the set B.

The identification of the surrounded set can then be solved by solving the 3D point-location problem of discriminating if a point if internal or external to a set of connected faces (Preparata 1985).

Computation can be performed firing a ray from the considered point and checking the number of intersections between the ray and the set of faces: if there is an odd number of intersections, then the point is internal to the set, if there is an even number of intersections, then the point is external to the set.

2 intersections
1 intersection
•P1
P2 •

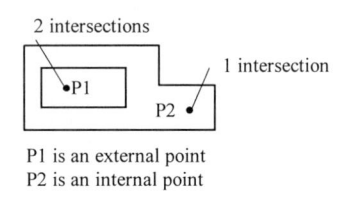

P1 is an external point
P2 is an internal point

Figure 7 -Example of point location solution.

The identification of TMR can benefit from the boundary representation of solid models. Most of the boundary representations implement data structures to store set of connected faces (usually called shells) and structures to store disjoint set of connected faces (usually called bodies). Looking at the topological relationship among shells belonging to a body, TMR can be easily identified.

5   EXAMPLES

Previous sections contain general identification procedures, but when implementing real applications we have to take into account that structure identification depends on solid/faceted model and sliced model representation.

A prototype version of a slicing module including the above mentioned controls has been implemented within the ACIS 3D Toolkit (ACIS 1994). The sliced model is obtained by direct slicing of solid's boundary representation.

Data automatically generated by the slicing module have been used to drive Sony JSC-2000 Solid Creator with a Sony NEWS workstation as a main control and processing unit.

In appendix A you may find three of the most significant functions implemented within the slicing module: slice-solid, consistent-contour? and drifting-layer-portion?.

The slice-solid function takes a 3D solid model as input and returns a list of slices described using a wire frame representation. The consistent-contour? function takes a wire frame representation of a slice as input and returns true if the slice has a consistent contour and false if it has an inconsistent contour. The drifting-layer-portion? function creates a solid slice from the solid model, and checks if the slice has some drifting layer portion. All

Process relevant form structures for stereolithography 321

functions are described by using the Scheme language extension adopted by the ACIS 3D Toolkit.

A selection of achieved results during our experiments with the implemented testbed and a prototype part is reported in the following. Shown test data are taken from two of our implemented design object descriptions, a designed tape drum holder from the mechanical high precision section of a video tape recorder and a wing rib from an aircraft (see Mandorli 1993, Otto 1995 for further details). The design object has been selected to test and demonstrate, in a first approach, the applicability of our introduced solutions together with their implementations within our test environment.

In figure 8-9, the investigated part's exact geometry is shown as a rendered model of our integrated viewer module. The following figures (10-11) show screen dump sections from our testbed during model evaluation, computation of inconsistent slice contours. In figure 12, a detected drifting region (indicated in true black) of the drum holder sliced model is shown within investigated slices where it occurs. In fig 13, the whole wire frame representation of the drum holder sliced model is shown.

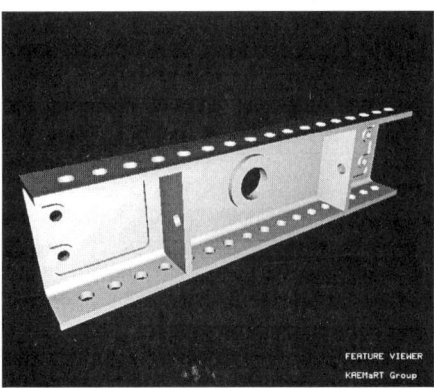

Figure 8 - wing rib.

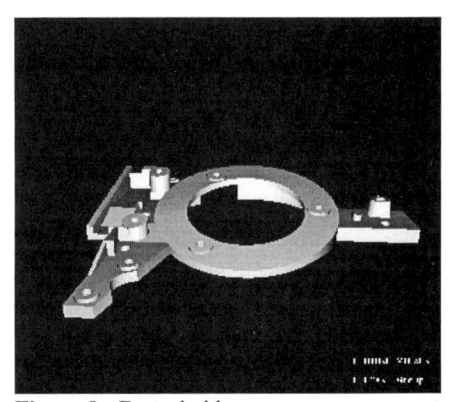

Figure 9 - Drum holder.

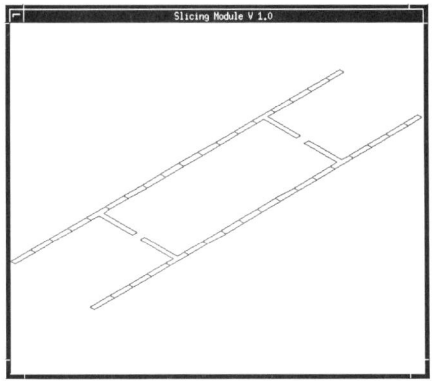

Figure 10 - wing rib inconsistent slice.

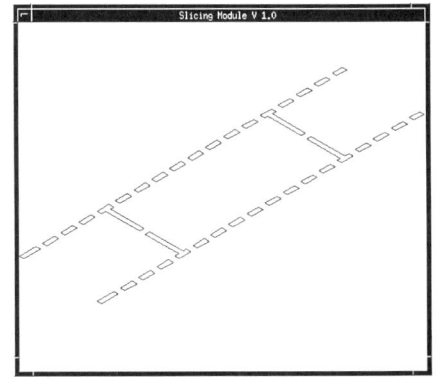

Figure 11 - wing rib consistent slice.

Figure 12 - drum holder drifting layer. 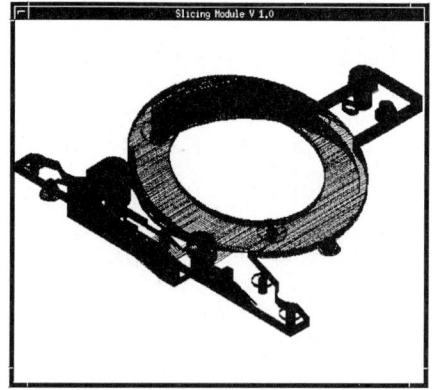 **Figure 13** - drum holder sliced model.

6   CONCLUSION

Stereolithography is a constructive methodology that simplifies process planning and production and reduces the complexity of tooling and fixturing. However, the fabrication of a part within unsolidified material, imposes several new problems, not being present in classical manufacturing technologies.

Stereolithography process is driven by a sliced model that can be obtained either from a solid model or a faceted (STL) model; in both cases, the sliced model must be checked to identify and control process relevant form structures before starting the solidification.

In the present work, we define three different types of process relevant form structures: inconsistent slice contours, drifting layer portions and trapping material regions.

The automatic identification of process relevant form structures reduces the risk to drive the production process with inconsistent data, and allows to save time and expensive material. For this purpose, conditions to identify the previously mentioned structures have been discussed and formalised.

If the stereolithography apparatus takes an STL file as input, then the sliced model is produced within the machine by the build-in software, and only a visual control on the front-end graphical device is left to the user before starting the production process.

If the stereolithography apparatus takes CLI data as input, direct slicing procedures can be implemented and completed with appropriated process relevant form structures identification functions, depending on the 3D model representation.

During experiments, we have verified that the process of identification of process relevant form structures can benefit from an homogeneous and integrated management of solid and surface objects. A prototype module to slice and control a 3D solid model with a boundary representation has been implemented to verify our taken approach.

AKNOWLEDGEMENTS

We would like to thank H. E. Otto for stimulating discussions on theorethical aspects of our current work and for giving technical support to carry out described experimental work.

APPENDIX A

```
(define  drifting-layer-portion?    ;;; returns the list of drifting layer portion
  (lambda (solid  z1  thick)        ;;; inputs: solid = solid entity, z1 = z value of layer, thick = layer thickness
    (let* ( (epsilon   0.01)
            (z2 (-z1  thick))
            (bbox  (entity:box  solid))
            (plane1 (solid:block  (position (position:x (car bbox)) (position:y (car bbox)) (- z1 epsilon))
                                  (position (position:x (cdr bbox)) (position:y (cdr bbox)) (+ z1 epsilon))))
            (plane2  (solid:block (position (position:x (car bbox)) (position:y (car bbox)) (- z2 epsilon))
                                  (position (position:x (cdr bbox)) (position:y (cdr bbox)) (+ z2 epsilon))))
            (slice1 (solid:intersect  plane1 (entity:copy solid)))
            (slice2 (solid:intersect  plane2 (entity:copy solid)))
            (shells1 (entity:shells  slice1))
            (body-shells'())
            (drifting-portions '())
            (res '())
          )
       (set! slice2 (entity:transform slice2 (transform:axes (position 0 0 thick) (gvector 1 0 0) (gvector 0 1 0))))
       (cond ( (= (lengthshells1)  1)  (set! drifting-portions '()))
             (#t (set! body-shells (body:separate slice1))
                 (do  ( (i   0 (+  i  1)))
                      ( (= i (lengthbody-shells)))
                    (set! res (solid:intersect (entity:copy  (nth i body-shells)) (entity:copy slice2)))
                    (if   (not (solid?  res))
                       (set! drifting-portions (append drifting-portions (list (nth i body-shells)))))
                 )
             )
       )
       drifting-portions )))

(define  slice-solid        ;;; returns a list of slice entities
  (lambda (solid  thick)    ;;; inputs: solid = solid entity, thick = layer thickness
    (let*  ( (box (entity:box  solid))
             (bottom (+ (position:z  (car  box)) epsilon))
             (top (- (position:z  (cdr box)) epsilon))
             (slices '())
             (slice   '())
           )
      (do  ( (i  bottom  (+ i thick)))
           ( (> i top)  slices)
        (set! slice (solid:slice  solid   (position  0 0 i)(gvector 0 0 1)))
        (set! slices  (append slices  (list  slice)))
      )
      slices )))
```

```
(define consistent-contour?   ;;; returns #t if the contour is consistent, #f if the contour is inconsistent
  (lambda (slice)             ;;; input: slice = slice to be checked
    (let ( (consistent? #t)
           (wires (entity:wires slice)))
      (do ( (i 0 (+ i 1)))                        ;;; for all slice's wire
          ( (= i (lengthwires)))
        (let ( (edges (entity:edges (nth i wires))))
          (if (and (= (lengthedges) 1)
                   (> (length(entity:vertices (car edges))) 1))
              (set! consistent? #f))
          (do ( (j 0 (+ j 1)))                    ;;; for all wire's edge
              ( (= j (lengthedges)))
            (do ( (k 0 (+ k 1)))                  ;;; for all edge's vertex
                ( (= k (length(entity:vertices (nth j edges)))))
              (if (> (length(entity:edges (nth k (entity:vertices (nth j edges))))) 2)
                  (set! consistent? #f))
            ))))
      consistent?
)))
```

REFERENCES

3D Systems, (1988) Stereolithography Interface Specification, USA.

ACIS, (1994) ACIS 3D Toolkit: Extension Reference, Spatial Technology, Inc., Boulder, Colorado.

BRITE-EURAM, (1994) Common Layer Interface (CLI), Version 1.31, *BRITE-EURAM Rapid Prototyping Techniques project*, Project n. BE5278.

Bjorke, Ø. (1991) How to Make Stereolithography into a Practical Tool for Tool Production, Annals of the CIRP, 40(1), pp. 175-177.

Bourell, D.L., Marcus, H.L., Barlow, J.W. and Beaman J.J. (1992) Selective Laser Sintering of Metals and Ceramics, *International Journal Powder Met.*, 28, 4.

Dolenc, A., Mäkelä, I. (1994) Slicing procedures for layered manufacturing techniques, *Computer-Aided Design*, 26, 2, pp. 119-126.

Hartmann, K., Krishnan, R., Merz, R. Neplotnik, G., Prinz, F.B., Schultz, L. and Weiss, L.E. (1994) Shape Deposition Manufacturing, in: *Proceedings of International Conference on Rapid Product Development*, pp. 69-78.

Jacobs, P.F. (1992) Fundamentals of Stereolithography, in: *Proceedings of Solid Freeform Fabrication Symposium, J.J. Beaman, H.L. Marcus, D.L. Bourell, J.W. Barlow and R.H. Crawford (eds.)*.

Jamieson, R., Hacker, H. (1995) Direct slicing of CAD models for rapid prototyping, *Rapid Prototyping Journal*, V. 1, N. 2, pp. 4 - 12.

Kobayashi, A. (1991) Application of Solid Creator System of Engine Components Design, in: *Proceedings of 2nd International Conference on Rapid Prototyping*, pp. 334-341.

Mandorli, F., Otto, H.E. and Kimura, F. (1993) A Reference Kernel Model for Feature-Based CAD Systems Supported by Conditional Attributed Rewrite Systems, *Second Symposium on Solid Modeling and Applications, J. Rossignac, J. Turner, and G. Allen (eds.)*, Montreal, Canada, pp. 343-354.

Medler, D.K. (1990) Stereolithography: A Primer, *Automotive Engineering*, 98/12, pp. 39-41.

Michaels, S., Sachs, E. and Cima, M. (1993) Metal Parts Generation by Three Dimensional Printing, in: *Proceedings of the Fourth International Conference on Rapid Prototyping*, pp. 25-42.

Otto, H.E., Kimura, F., Mandorli, F., Cugini, U. (1995) Extension of feature-based CAD systems using TAE structures to support integrated rapid prototyping, *Proceedings of the Computers in Engineering Conference and the Engineering Database Symposium ASME '95*, pp. 779 - 793.

Preparata, F., P. and Shamos, M., I., Computational Geometry: An introduction, Springer - Verlag, New York, 1985.

Weiss, L.E., Prinz, F.B., Adams, D.A. and Siewiorek, D.P. (1992) Thermal Spray Shape Deposition, *ASM Journal of Thermal Spray Technology*, 114, 4, pp. 481-488.

28

Rapid prototyping - new manufacturing tools for improving design and prototype cycle

Prof. Dr.-Ing. F. Klocke, Dipl.-Ing. S. Nöken, Dipl.-Ing. H. Wirtz
Fraunhofer Institute of Production Technology - IPT
Steinbachstraße 17, 52074 Aachen, Germany
tel.: +49-(0)241-8904-0, fax: +49-(0)241-8904-198,
e-mail: wiz@ipt.rwth-aachen.de

Abstract

Companies are increasingly forced to transform their technical innovations into fully developed products rapidly. In addition to strategic approaches in order to reduce the time of product development (concurrent engineering), rapid prototyping has become a very effective means to speed up and improve product design and process planning. The introduction of CAD/CAM technologies offers the possibility to produce models and samples directly on the basis of CAD data, not in weeks or months as with conventional prototyping methods, but in only a matter of hours. Combined with rapid prototyping methods, subsequent techniques, particularly conventional casting methods, plastic vacuum casting and metal spray tooling hold potential for the rapid manufacturing of plastic prototypes in greater quantities or of metal prototypes.

According to different demands resulting from the envisaged application of the prototypes, possibilities and limits of commercially available rapid prototyping techniques will be pointed out. The use of prototypes in product development and the conditions under which rapid prototyping technologies can lead to rapid product development will be presented. The Fraunhofer Rapid Prototyping Network, an alliance of seven Fraunhofer Institutes aiming at the technical, informational and organizational improvement of rapid prototyping will be introduced.

Keywords
Rapid prototyping, product development, CAD/CAM technologies, tooling

1 INTRODUCTION

New products are generally more demanding in functional terms and may therefore be more complex. Decreasing lot sizes are accompanied by the demand for greater variety and at the same time, product life cycles are steadily becoming shorter. Consequently, product engineers must strive to reduce the development phase to ensure that the scheduled date for launching the product to the market can be met. Time in the sense of time to market is becoming an increasingly important competitive factor besides product quality and production costs. However, the situation of product development today is characterized by long development times due to time consuming prototype manufacturing and frequent product modifications, see Figure 1 (König, 1993).

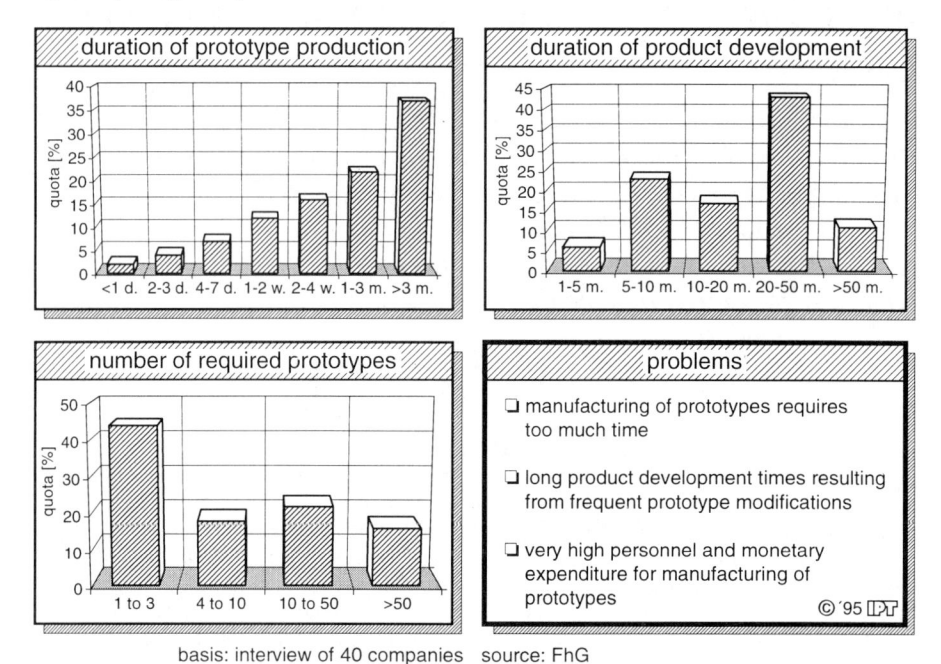

Figure 1 Situation of product development today.

Rapid prototyping is opening up new potential for decreasing product development times and enhancing product quality. Deploying rapid prototyping techniques, complex parts may be manufactured not within weeks or months as is the case for the conventional approaches, but in a matter of hours. The rapid availability of models suitable for design studies, function and assembly tests as well as for documentation and acquisition purposes contributes to an early improvement of product design and provides sound criteria on which decisions concerning the final product layout may be made more easily. During the development process, models and prototypes are useful for setting up serial production and management thereof. They represent the link permitting product and process design sequences to be synchronized. Their rapid availability determines the speed at which information may flow between various departments involved in product and process planning and thus exerts considerable influence on product development times.

2 NEED FOR PROTOTYPES IN PRODUCT DEVELOPMENT

The terms used to describe the different kinds of prototypes, models and samples required during product development vary widely in individual industries and companies. In the following, universally valid distinctions are drawn between individual product development phases and different sorts of prototypes are identified (Pfeifer, 1994).

An analysis of the product development phases leads to the conclusion that prototypes are required in all stages of development - from the product idea through to the launching of the new product to the market. In the consumer and capital goods industries, the product development cycle may be divided into five phases (Figure 2). The prototypes required in the individual development phases differ both in terms of quantity and material properties and in geometrical, optical, haptical and functional terms. The envisaged applications of these prototypes correspondingly vary concerning product and process planning.

product development phases	prototypes and characteristics	
concept phase	design model	virtual model (photo/CAD) physical model • number of parts: 1 • model making material • primarily optical and haptical requirements
pre-development phase	geometrical prototype	• number of parts: 1 • model making material • primarily geometrical requirements
function pattern phase	functional prototype	• number of parts: 2 to 5 • close to production material • primarily functional requirements
pre-production phase	technical prototype	• number of parts: 3 to 20 • close to production material • close to production process • pre-production tools
prototype phase	pre-production part	• number of parts: up to 500 • production material • production process • production tools
market introduction		

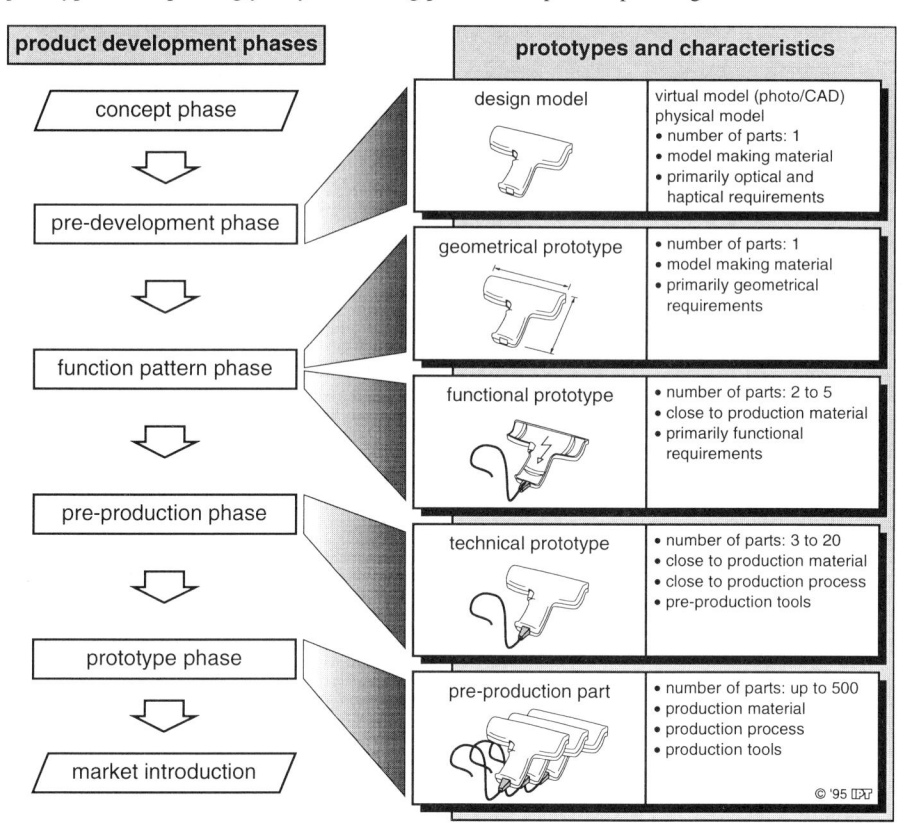

Figure 2 Requirement for prototypes in product development.

In the pre-development phase, design models are frequently required in a lot size of one. Generally speaking, the accuracy of the design models only has to meet relational requirements. All features must be as distinct as in the serial product, especially concerning their position and size, however a low overall accuracy of the model may be tolerated. As the functional requirements are not important, such prototypes are often manufactured using typical model materials. Besides being suitable for design studies, these models can also be used for ergonomical test and initial market surveys.

In contrast, geometrical prototypes have to closely meet the serial standards in terms of dimensional and form accuracy. Optical and haptical part characteristics are less important. A geometrical prototype is usually made of a material which is not necessarily identical to the one used for serial production. This type of prototype is required mainly for process planning operations such as the drafting of plans for manufacture and assembly and the development of production concepts. At this stage, prototypes can prove an invaluable means of communication.

Functional prototypes, produced in quantities between 2 and 5, are used to test and optimize the product's principle of operation and functionality. Functional prototypes are to assist the engineer during the planning of machinery, processes, manufacturing sequences, assembly and resource scheduling. The outer appearance and dimensional tolerances are of minor importance as long as they do not impair the function itself. The standards which the prototypes are required to meet in terms of mechanical, thermal and chemical stability under load are limited to the demands imposed by performance testing.

During the following development phase, technical prototypes are used in larger numbers of 3 to 20 for a detailed analysis of the function, the mechanical, thermal and permanent rating of the product as well as the acceptance of the clients and the producibility. These technical prototypes are to correspond as closely as possible to the final product, with regard to the material used and the final production process. Thus, pre-production tools are often deployed for deep-drawn, injection moulded and die-cast components.

Before a product is launched on the market, up to 500 pre-production parts depending on the branch and product involved are produced in pilot runs. These parts are made of the same materials as those to be used for serial production, using the same tools and manufacturing processes. Pre-production parts are required by the product planning department for comprehensive product and marketing tests. The process parameters required for serial production are determined and optimized. Any significant change in design to any of the product components at this stage of product development will result in very high subsequent costs.

3 METHODS OF PROTOTYPE PRODUCTION

Models and prototypes for product development are currently manufactured using conventional manufacturing methods combined, if necessary, with subsequent casting, NC milling, copy milling, turning and grinding as well as manual jointing and laminating techniques. The methods used in making prototypes and models are characterized by time-consuming production operations and make a significant contribution to the lavish amounts of time and money spent on prototype building due to small lot sizes and frequent changes to the product, see Figure 1. This is one of the main reasons for dispensing with models and prototypes altogether.

The introduction of CAD/CAM technology has provided companies with the possibility to manufacture models and prototypes directly on the basis of CAD data. New manufacturing methods known as 'rapid prototyping', 'desktop manufacturing', 'solid freeform manufacturing', 'layer manufacturing' etc. make consistent use of this option.

These methods permit parts to be manufactured rapidly without the use of moulds or tools. They are characterized by the fact that the workpiece is formed not by material removal as is the case in conventional cutting operations but by adding material or by the transition of a material from the liquid or powder phase to the solid state. An additional common characteristic is the fact that the NC data is generated directly and layerwise from the 3D part geometry. On the basis of the contours, the workpiece is subsequently built up layer by layer in the manufacturing process itself (layer technique).

About seven years ago, the first method of this type, stereolithography (SL) was introduced to the market. In the mean time, additional technologies such as solid ground curing (SGC), selective laser sintering (SLS), fused deposition modeling (FDM), laminated object manufacturing (LOM) and three-dimensional printing (TDP) have become available, see Figure 3. Recently, a new technique called ballistic particle manufacturing (BPM) was developed.

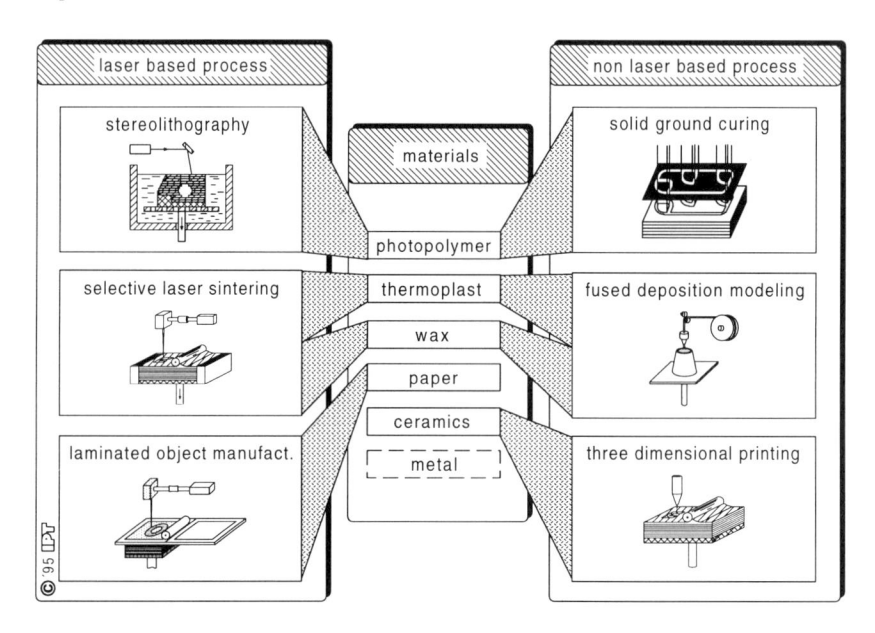

Figure 3 Commercialized rapid prototyping techniques.

Combined with rapid prototyping methods, subsequent techniques, particularly conventional casting methods, plastic vacuum casting operations and spray metal tooling techniques, hold out potential of rapid manufacturing of plastic prototypes in greater quantities or of metal prototypes.

3.1 Rapid prototyping methods and subsequent techniques

On the basis of the demands to be met by the categories of prototypes listed, a variety of possible manufacturing methods and sequences will be described in the following. New approaches to rapid prototyping methods permitting the direct manufacture of metallic prototypes will also be examined.

Geometrical, design and, to some extent, haptical prototypes may easily be manufactured deploying stereolithography, solid ground curing or laminated object manufacturing, see Figure 4.

In stereolithography and solid ground curing processes, the part geometry is formed in layer by layer hardening of a fluid photopolymer (photopolymerization) using an UV laser or an UV lamp respectively. The drilling machine housing shown in Figure 4, which was manufactured in only about 8 hours using stereolithography, demonstrates the impressive potential of this method.

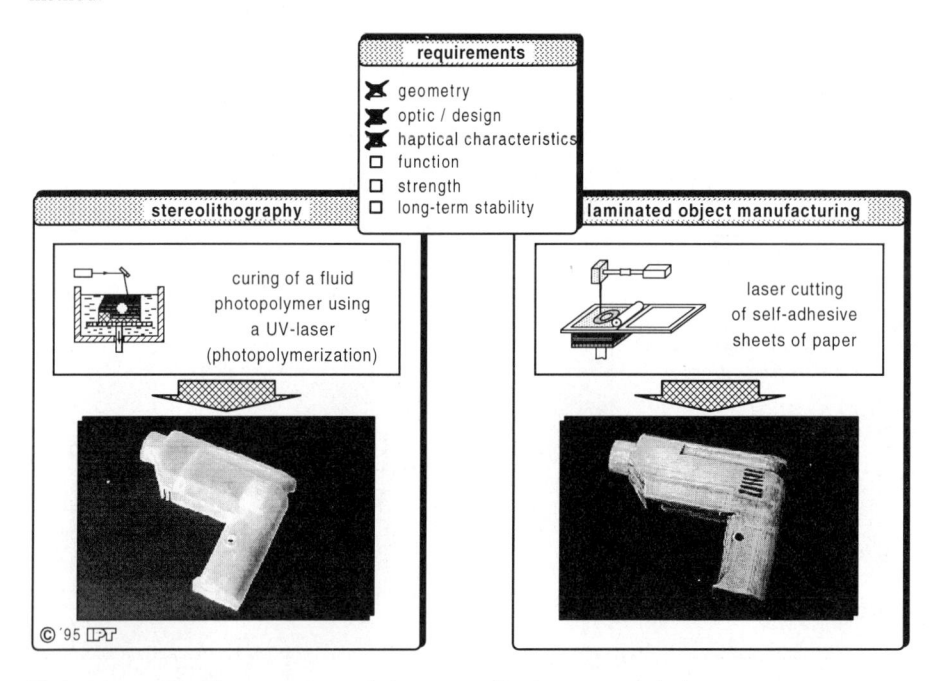

Figure 4   Rapid prototyping techniques - application: geometrical prototype.

Geometrical prototypes are generally sufficient to conduct limited function tests in which no particular requirements have to be met by the material. This applies to tests which were carried out on ease of assembly during the development of the above mentioned drilling machine. Conventional prototype manufacture of the complex internal structures would have taken several days to complete and would have restricted the number of optimization steps possible from a temporal and financial point of view. By virtue of it's extremely short manufacturing time per part, laminated object manufacturing, permitted all the tests to be carried out on all variants (Figure 4). The laminated part geometry is created by glueing individual sheets of paper onto one another and subsequently cutting along the contours using a laser beam.

In contrast to the previous examples, functional prototypes made of materials as similar as possible to those to be used in serial production are required at an early stage of development of components which are subjected to thermal and mechanical stress in order to test performance in service. The higher demands in terms of stability under load preclude the use of model or substitute materials in such cases. The selective laser sintering process shown in Figure 5 permitted a polyamide model of the drilling machine to be manufactured which is suitable for functional testing within the time frame of the production of the stereolithography part.

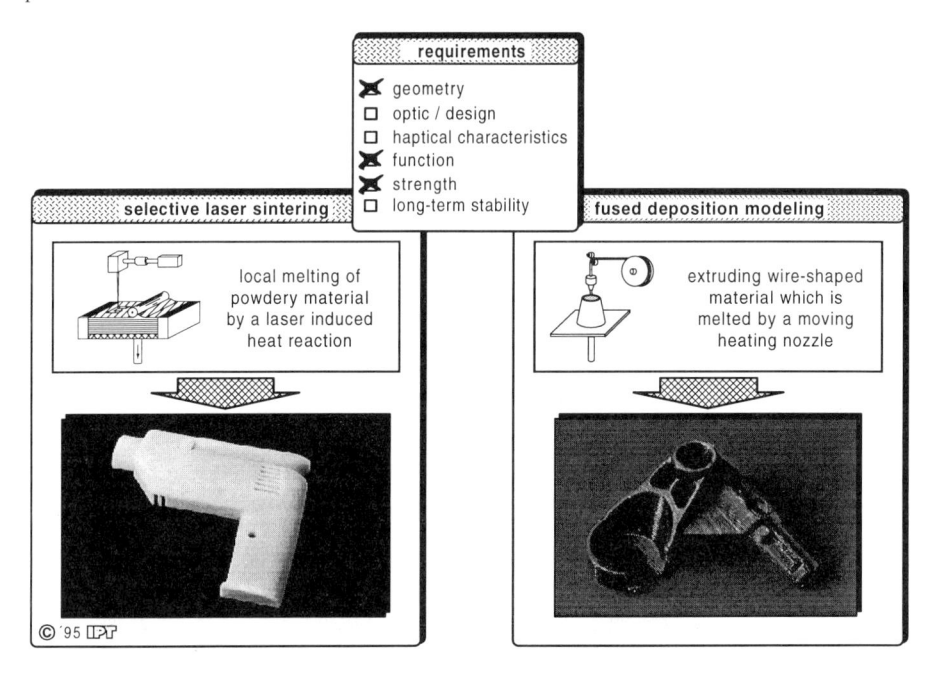

Figure 5   Rapid prototyping techniques - application: functional prototype.

The principle of the selective laser sintering process is based on the local bonding or fusion of powder materials by a laser induced heat reaction. The selective laser sintering process can be used to process virtually any powdery material. Polystyrene, polycarbonate and nylon are all currently in use.

Rapid prototyping techniques may be combined with subsequent casting operations in order to manufacture functional metallic parts. The articulated lever shown in Figure 5 made from forming wax was built up using fused deposition modeling and used as a lost wax model for investment casting. By extruding and layering a wire-shaped material which is melted by a moving heating nozzle, the part geometry is generated. Other materials which can be processed using this method are thermoplastics.

Technical prototypes must satisfy higher requirements in terms of stability under load and long-term stability. In order to obtain reliable information about the stability under load of the serially manufactured product, technical prototypes must be as similar as possible to the end product in terms of the material and manufacturing process used. In industry, hollow steel casting dies, taking an average of 4 to 8 weeks to manufacture, are currently used for the production of technical plastic prototypes and for small lot sizes (between 20 and 50 parts). The application of rapid prototyping methods in combination with follow-up techniques offers potential for a significant reduction of product development time. Plastic vacuum casting is one of the ways to quickly manufacture technical prototypes in small lot sizes, Figure 6.

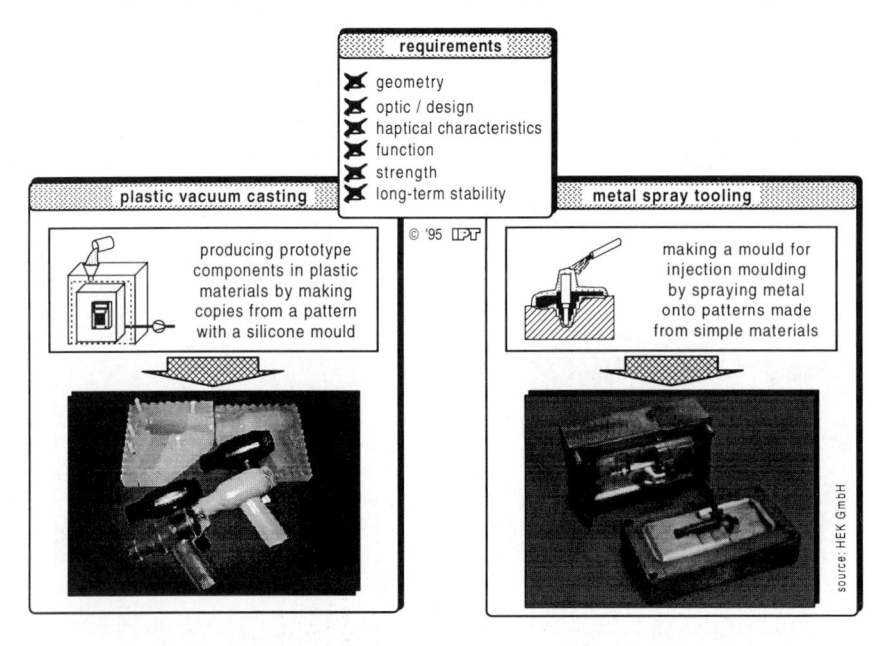

Figure 6 Rapid prototyping techniques - application: technical prototypes, pre-production parts and small lot sizes.

A stereolithographic model of a hair dryer was used as the premaster for the plastic vacuum casting method with which the original was duplicated producing polyurethane parts. This method is characterized by a high level of reproducible accuracy. Even areas with filigree undercuts can be manufactured without problems due to the elastic silicone moulds. The overall time frame from CAD data to stereolithography part to mould and 20 polyurethane parts with serial characteristics may be as short as 3 days.

When technical prototypes, pre-production parts or small lot sizes are required in larger quantities (50 to 1.000), the metal spray tooling method can be used to manufacture experimental or production tools, Figure 6. The principle behind the method and the characteristics of this technology will be explained using a tool for the manufacture of plastic spring hooks as an example. Like plastic vacuum casting, the metal spray tooling process requires the use of a premaster, which is coated with a layer of metal alloy which has a low melting point using a spray gun similar to an air brush. Once the metal coating is applied, the mould half is fed inside a moulding box and an alloy with a low melting temperature or artificial resin enriched with aluminium chips is fed in behind. The levels of dimensional and form accuracy are identical to those of the premaster. The use of this technology is limited in some cases by the geometry of the model. Slots with a width to depth ratio of less than 1 to 5 cannot be manufactured using this technique because of the restricted access and the danger of drop formation. The thermal properties of the tools manufactured in metal spray tooling operations are different from those of conventional injection moulded tool materials, resulting in differences in the geometrical and mechanical part characteristics. The manufacturing costs of the spray tooling tool with the metal injection facility shown in Figure 6 amounted to DM 1.500. The manufacturing time was 6 hours.

3.2 New approaches to the direct manufacturing of metallic parts

Due to the limited mechanical, thermal and chemical characteristics of the materials (polymers, wax, nylon, paper, etc.) which can currently be processed using industrially available rapid prototyping methods, parts manufactured may only suit visual, haptical, geometrical and to some extent functional purposes. The scope for conducting the comprehensive functional tests frequently demanded by users is limited.

In order to ensure greater congruence between prototype and serial part, the materials used to manufacture the prototype should, whenever possible, be identical to those intended for use in serial production. The manufacture of prototypes made of metallic materials similar to those used in serial production is still limited to a casting process chain. One of the goals of the ongoing development of rapid prototyping technologies is the manufacture of prototypes and prototype tools with enhanced material properties. Efforts are focused on metallic and ceramic workpieces, since these can currently only be manufactured using conventional manufacturing methods. For this reason, intensive research is being conducted concerning ways of optimizing existing rapid prototyping methods and of developing new ones to permit the manufacture of metallic parts or tools directly.

Selective laser sintering of metallic and ceramic powders is a promising approach towards the generation of prototypes with enhanced material properties, Figure 7. Problems arise from the high temperature gradients throughout the sintering area, resulting in thermal stresses and strains which will often lead to a delamination of single slices. Also, the porosity of the parts makes an infiltration necessary if high strengths and a good surface finish are to be achieved. There is currently only one machine commercially available utilizing a special kind of bronze-nickel alloy which requires no pre-heating or binder material and exhibits virtually no shrinkage.

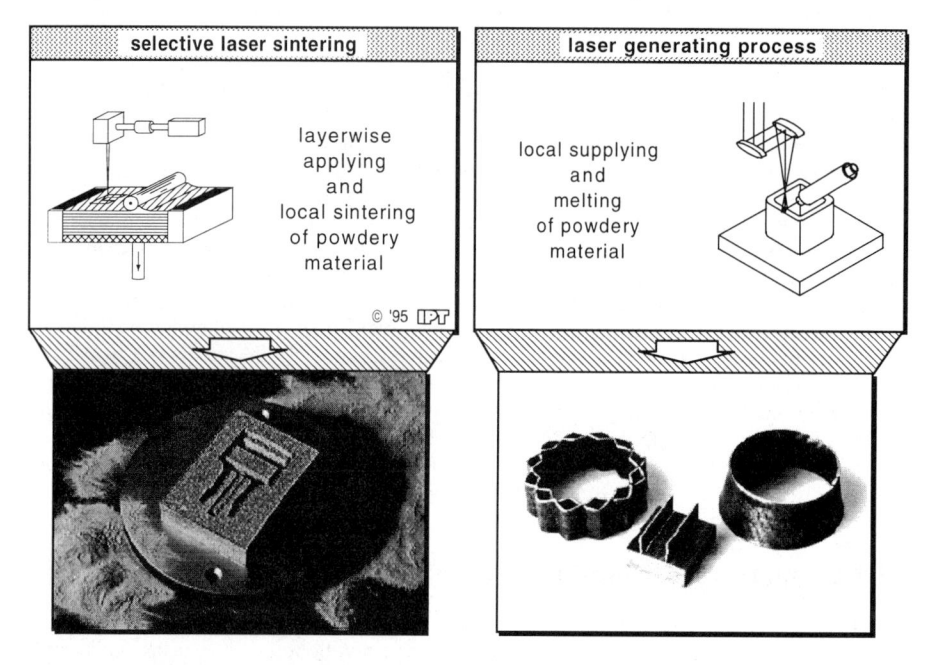

Figure 7 Selective laser sintering and laser generating for the production of metallic prototypes and prototype tools.

The principle of laser generating is shown in Figure 7. The principle involved is equivalent to a coating process in which a powdery material is melted using a laser beam and bonded with the substrate. Three dimensional structures are created by laying individual lines alongside and on top of one another. By virtue of the accuracy with which the dimensions of the focal point can be adjusted and the ease with which its intensity can be controlled, laser generating holds out considerable potential for the reproducible production of filigree structures.

4 PREREQUISITES FOR THE IMPLEMENTATION OF RAPID PROTOTYPING

The prerequisites for industrial application and efficient use of rapid prototyping are shown in Figure 8. Due to the high levels of capital cost and operation expenditure, the deployment of rapid prototyping methods is at present industrially viable only in cases in which time is a critical factor or for products with complex geometries. This technology can also only be profitably introduced when it gets integrated into the company's flow of data and information and when 3-D CAD is being used. Concurrent engineering structures along with the associated adapted planning and decision-making processes are additional prerequisites for a successful introduction of rapid prototyping.

organization
- synchronization of planning processes
- integration into the structural and operational organization

process technology
- selection of process or process combinations
- process generation

information processing
- 3D-CAD modeling
- reverse engineering

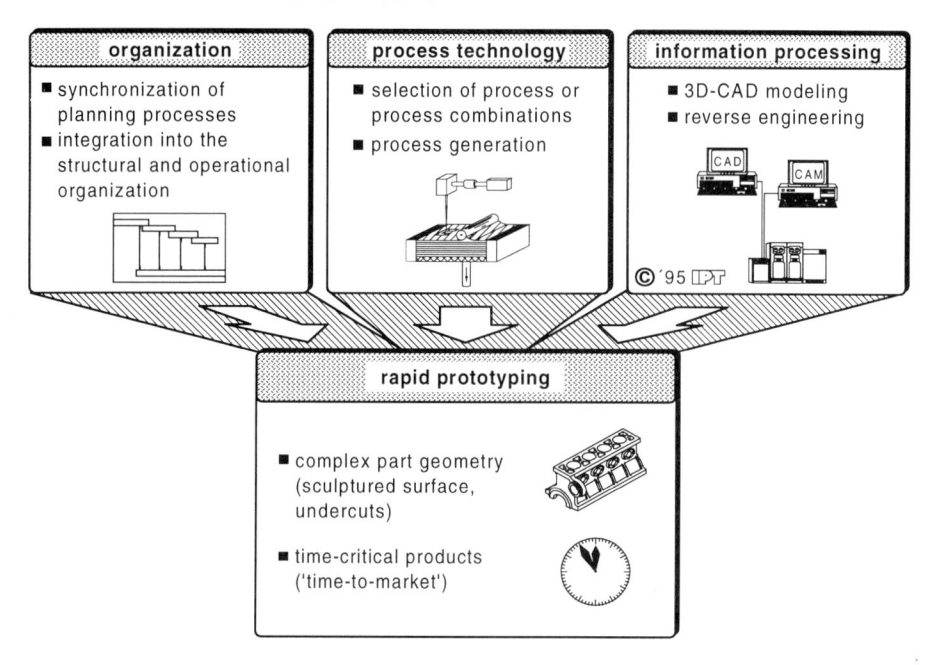

rapid prototyping
- complex part geometry (sculptured surface, undercuts)
- time-critical products ('time-to-market')

Figure 8   Requirements for the efficient application of rapid prototyping.

The analysis of the spectrum of products, of the development processes and of the demands which prototypes are required to meet during the development phase shows what requirements the technologies will have to satisfy. Rapid development of the rapid prototyping technologies themselves and the lack of knowledge about their potential and limits is hampering their selection and evaluation. The individual methods must not be treated as single solutions during the selection process, possible combinations of methods and subsequent techniques must also be included in the selection decision.

Once suitable technologies have been chosen, a make-or-buy decision has to be made within the scope of machinery and equipment investment planning. The difficulty here lies in estimating or evaluating the amount of time likely to be saved and the quality advantage to be gained from dropping conventional production methods in favour of rapid prototyping techniques since experience in this field is still very patchy.

5 CONCLUSIONS

The application of rapid prototyping techniques contributes to a reduction in the time required to manufacture a prototype. It also has a far-reaching strategic impact: Changing customer wishes can be reflected more quickly in product development, thus permitting companies to react more rapidly and therefore with greater agility to the requirements of the market. The availability of models and prototypes at an early stage results in a high level of product sophistication and in a greater availability of planning data for production. The concentration of resources in the early stages of development pays off later, as alteration costs are lower and the products are mature when they are launched to the market.

In view of the potentials rapid prototyping may open up in product development, the Fraunhofer Society has called a Fraunhofer Rapid Prototyping Network into being, which is devoted to the technical, informational and organizational improvement of rapid prototyping. The main objectives are
- the development of rapid prototyping methods for the manufacturing of metallic parts (laser generating, selective laser sintering),
- the processing of integrated information for rapid prototyping process chains and
- the geometric modelling and digitising for an easy generation and retrieval of 3D data.

Thus, the Fraunhofer Society is significantly contributing to the changes rapid prototyping is continually introducing to product development, leading to short development times, high product quality and thus competitive companies and satisfied customers.

6 REFERENCES

König, W., Eversheim, W., Celi, I., Nöken, S. and Ullmann, C. (1993) Rapid Prototyping - Bedarf und Potentiale. *VDI-Z*, **135**, 92-7

Pfeifer, T., Eversheim, W., König, W., and Weck, M. (1994) Rapid prototyping: the way ahead. *Manufacturing excellence - the competitive edge*. Chapman & Hall, London.

7 BIOGRAPHY

Prof. Dr.-Ing. Fritz Klocke, born 1950, is holding the Chair of Manufacturing Technology of the Laboratory for Machine Tools and Production Engineering (WZL) of the Technical University of Aachen since 1995. He is also director of the Fraunhofer Institute of Production Technology (IPT) in Aachen.

Dipl.-Ing. Stefan Nöken, born 1965, studied mechanical engineering at the Technical University of Aachen, since 1992 scientist at the Fraunhofer Institute of Production Technology in the Process Technology department.

Dipl.-Ing. Holger Wirtz, born 1969, studied mechanical engineering at the Technical University of Aachen, since 1994 scientist at the Fraunhofer Institute of Production Technology in the Process Technology department.

29

Rapid Prototyping of Engineering Methodologies

Nanxin Wang and Jie Cheng

Ford Research Laboratory
P.O. Box 2053/MD2122
Dearborn, MI 48121
EMAIL: nanxin@venus.srl.ford.com
Tel: (313) 323-9841
Fax: (313) 248-4602

Abstract

To reduce the development cycle time and increase the productivity, today's automotive industry relies heavily on the analytical solutions conducted by deploying complex multi-functional, multi-disciplinary design/analysis using all available analytical tools. This calls for the integration of existing product design/analysis programs into packages to perform a higher level of system functionality, such as total engine analysis, and powertrain system or vehicle optimization. The functional and procedural specifications for these integrations are often referred to as engineering methodologies. To enable the rapid prototyping of these engineering methodologies, we need to have a generic tool for the integration of rich, dynamically extensible process descriptions of the engineering methodologies, and a high level environment facilitating assembly and invocation of the engineering methodologies. A software integration framework called EMAT (Engineering Methodology Application Tool) has been developed for this purpose. It has been used to the integration of multiple design/analysis programs to prototype complicated engineering methodologies for a wide range of applications within FORD.

1. Introduction

In the automotive industry, knowledge regarding vehicle design, analysis and testing is frequently encoded in various procedural programming languages and used in a variety of program packages. Each of these program packages typically performs one particular computational function, such as: component modeling, vehicle simulation, finite element analysis, dynamic analysis, and design optimization. To meet the demand for more comprehensive design/analysis and optimization, these programs need to be integrated together to perform a higher level of system functionality, such as total engine analysis, and powertrain system or vehicle optimization. To be successful, these methodologies need to be rapidly prototyped and easily extendable for concept demonstration and production use. Current integration approaches, to a great extent, still rely on manual data exchange and developing application specific control programs for each analysis conducted. This has proven to be very time consuming and error-prone. Consequently, the ability to integrate complex design-analysis procedures becomes limited.

1.1. Engineering Methodologies

An engineering design/analysis process usually involves multiple design/analysis programs for modeling, simulation, and analysis. An engineering methodology specifies the procedure by which these programs are to be combined in terms of their input, output and order of execution, and a set of satisfaction criteria for the overall process. The sequence of the design/analysis is the determining factor in producing the design/analysis. Analysis programs run in different sequences will produce different results. Usually, many different engineering methodologies may be tried before a final design/analysis is produced.

Shown in Figure 1 is a greatly simplified methodology for crankshaft safety analysis. Three different programs are used in this example: (1) a design program for parametric design of crankshafts, (2) a classical stress analysis program for screening preliminary designs, and (3) an FEA analysis program for evaluating the final design. The example crankshaft safety analysis starts with the parameter design. At different design stages, different analysis programs are used. If the safety criteria are satisfied, the analysis process completes. Otherwise, design parameters are altered and the analysis is repeated until the desired safety is achieved. More complicated engineering methodologies often involve more design/analysis programs and require more complex sequencing and synchronization.

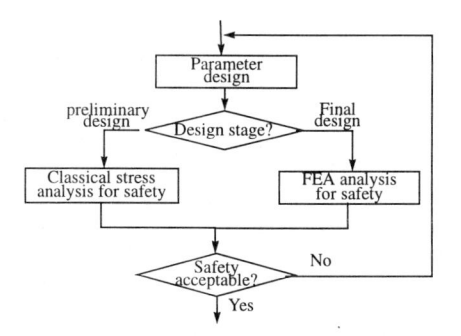

Figure 1. Crankshaft safety analysis methodology

1.2. Software Integration Issues

There are two issues in software integration that are of concern: (1) data exchange and (2) process automation. The first one concentrates on the data modeling, storage and communication. It's the basic element of software integration and it's the first problem one faces for any kind of software integration attempt. The second one emphasizes design/analysis process management. It addresses the problem of sequencing and synchronizing software programs that must be run under a particular methodology. Both data exchange and process automation are absolutely necessary in a true software integration environment.

For rapid prototyping of engineering methodologies, we address both issues. An integration toolkit collectively referred to as SPLICES (Cheng, 92) is used for data exchange. The main contributions of SPLICES software are (1) standardization of the IO interface of all models by making analogy to standardizing chip modules in VLSI circuit designs, (2) configurable dynamic protocols for inter-model data communication, and (3) a database management system for regulating the access to application data. A software integration framework called EMAT (Engineering Methodology Application Tool) (Wang 95) has been developed for the process automation. With the support of SPLICES, EMAT integrates the design/analysis programs as well as the lower level enabling tools to accomplish the rapid prototyping of complicated engineering methodologies. In this paper, we are going to focus on the discussion of EMAT and how it is used for rapid prototyping of engineering methodologies.

2. Blackboard Technology and Its Role in the Integration

Blackboard technology refers to a problem-solving method consisting of (1) knowledge sources - individual modules containing the knowledge required to solve a problem, (2) the blackboard - a global database recording all the general and domain specific problem data at various problem-solving stages, and (3) the control - a run-time decision maker determining the course of problem solving and the expenditure of problem-solving resources. The blackboard architecture is depicted in Figure 2.

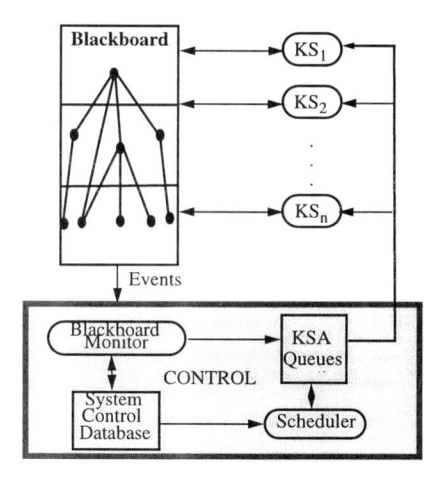

Figure 2. Blackboard architecture

where, KS stands for a knowledge source which can be a program of any size and in any language. KSA stands for a knowledge source activation. During the course of a typical problem-solving session, the blackboard is changed by the problem solving activity of KSs which may result in changes to the information stored on the blackboard database and causing predefined events to occur. These changes to the blackboard database are monitored by the control component which may trigger additional knowledge sources (KSs) according to the events. For each triggered KS, a KSA will be created and put into a KSA queue, waiting for the execution. The blackboard control will choose a KSA with the highest priority rate and execute it. The execution of the KSA may change the blackboard which can cause other events to occur. The whole process continues until either the problem is successfully solved or the KSA queue is empty and no solution can be found.

The blackboard technique is ideal for integrating design/analysis programs because (1) it provides a well-designed blackboard data structure as a common media to support inter-program communication, (2) with the blackboard as a media, programs can be loosely coupled to achieve a common task, disregarding their implementation language or style, and the system can be readily extended at any time, (3) with its event-driven and agenda-based control scheme, one can implement many kinds of control procedures, including deterministic, opportunistic, and even heuristic or rule-based process control.

3. Methodology Description Language (MDL)

MDL is an application-independent language developed as an abstraction, or high level representation, of engineering methodologies as a combination of procedural logic and data exchange protocol.

MDL describes an engineering methodology as a recursive structure of "processes", each process represents the dynamic execution behavior or state (starting, suspending, terminating, aborting, etc.) of a logical or program entity (one logical entity may correspond to one or many program entities). A process can be a compound process or an atomic process. A compound process can be decomposed into simpler constructs of compound or atomic processes, whereas an atomic process cannot be decomposed any further within a given methodology. A compound process can be classified as either a modular process, a looping process, a branching process, or a strategy process. A modular process consists of one atomic process, referred to as "Model" for carrying out some key computations, and two IO (Input/Output) interface processes, called "Acquisition" and "Disposition" processes. The Acquisition and Disposition processes themselves are compound processes. A looping process involves the repeated invocation of a compound process until the looping criterion fails to be satisfied. A branching process switches between two compound processes based on the satisfaction of a branching condition. A strategy process is used to represent a set of compound processes that can be executed either sequentially or parallel. A strategy process at the top-most level corresponds to a given methodology.

MDL provides a simple representation for the dynamic data exchange mechanism in a generic integration environment. Data exchanges are assumed to take place only among and within the modular processes. A data connection is established when an output data channel of an Disposition process is connected with an input data channel of an Acquisition process. The sequencing of the data flow and the formatting of the data are done at a lower level via a set of enabling tools and will be described in the next section.

When a methodology is to be developed, its execution control flow can be described by recursively applying the "process" definitions, while its data flow can be modeled partially at a high level by defining and connecting the channels of different modular processes. The execution control of the methodology is done by an MDL interpreter whose functions include program invocation, data channel connection, process state monitoring, and process execution sequencing control.

The basic control shell provided by the blackboard framework are modified to monitor the methodology execution. All activated knowledge sources are inserted into an execution queue ordered by their assigned priority rating. The blackboard control then executes the first available knowledge source activation (KSA) on the queue. The activation of a knowledge source is triggered by the occurrence of any predefined event. The control under the MDL interpreter is monitored by a strategy knowledge source (SKS). When an analysis is initialized, the SKS is activated first. The execution of the SKS processes the strategy objects and triggers the events which activate appropriate process knowledge sources (PKSs). When the execution queue is empty, the SKS is triggered again searching for next actions defined in the methodology until the whole process terminates.

Figure 3 illustrates the MDL specification for the crankshaft safety analysis methodology that was shown in Figure 1.

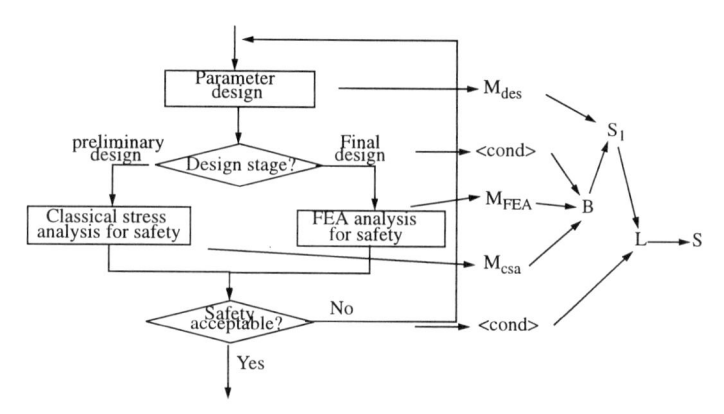

Figure 3. MDL representation for crankshaft safety analysis methodology

4. EMAT

EMAT is a generic software integration framework. It consists of a high-level language environment MDL (Methodology Description Language) (Wang, 1994) and a program for process execution scheduling and monitoring based on an artificial intelligence technology called Blackboard (Nii, 1986, Wang, 1993). Under the EMAT framework, a user can easily specify the control flow and data flow for any methodology in a declarative manner. Such a specification only needs to contain logical orders in which individual component programs will be executed (such as sequence, branching, or looping), and the input/output connections between the programs. EMAT will then dynamically interpret this specification into procedures that actually carry out the execution. In contrast to the conventional integration practices such as developing application specific scripts, EMAT provides a generic tool for integration of rich, dynamically extensible process descriptions of the engineering methodologies, and a high level environment facilitating assembly and invocation of the engineering methodologies.

The EMAT architecture is Shown in Figure 4. The left box contains the system level of components: an MDL language interface, a Methodology Library storing all developed methodologies for browsing, execution, or building new methodologies, a special blackboard for all globally shared information among processes of any given methodology, a MDL interpreter for carrying out methodology execution control, and a user interface. Through the user interface, an application developer can specify a methodology using MDL, modify or extend an already defined methodology, or test and debug a methodology by executing it. The same user interface can let an end-user easily find an appropriate methodology available and execute it.

The box on the right contains a collection of programs, design/analysis programs and integration enabling programs. This set of programs is often specific to an application domain.

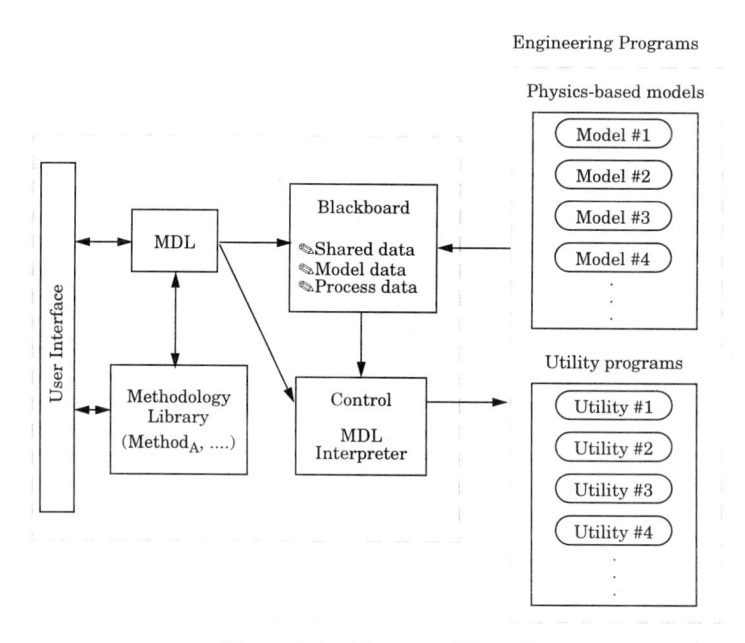

Figure 4. Architecture of EMAT

5. A Case Study

The development of EMAT has, to a large degree, been driven by the needs of vehicle engine design/simulation applications. Therefore its first application is in the integration of engine simulation programs comprising a general engine simulation environment (ENGSIM) (ENGSIM, 1993), for rapid prototyping of engine modeling/simulation methodologies.

Total engine analysis (TEA) is one of the engineering methodologies we developed using EMAT in the domain of engine modeling and simulation. It requires the integration of two major engine simulation programs: a thermodynamic simulation program (TSP) and a manifold dynamics simulation program (MSP). TSP is an in-cylinder heat transfer and combustion program for predicting the engine performance, fuel consumption, and emissions. MSP is capable of simulating the pressure waves inside the ports and manifolds for the prediction of mass flow across intake and exhaust manifolds. The combustion process of TSP is modeled as a set of time dependent ordinary differential equations. Since TSP does not simulate manifold, its usage is limited to the part-load and low-to-medium speed regions where the effect of manifold dynamics on combustion can be ignored. MSP does include an in-cylinder heat transfer and combustion program. That sub-program is, however, far too simple to accurately represent the combustion process inside an internal combustion chamber. Therefore, the two programs need to be integrated for the best prediction of engine performance over the whole space of engine operating conditions. In addition, integration of the two programs into one single program requires total restructuring of the programs and therefore is not feasible.

The TEA methodology is therefore developed by integrating the two programs as well as a consistency check function (CCF) under the EMAT framework (see Figure 5 flowchart). Since TSP is a single cylinder simulation program, when run with MSP together, it has to run Ncyl times, where Ncyl is the number of cylinders of an engine, for every MSP run in any iteration. TEA methodology starts with MSP execution as a bootstrap. When MSP achieves its convergence, it passes boundary conditions at valve to TSP, one per cylinder. TSPs are executed parallel (TSPs can be run sequentially if not enough CPU and memory resource is available). The program CCF checks the results from TSPs and MSP to see whether they are consistent. If the results are inconsistent and the number of total iterations is less than a threshold, CCF will provide information for adjusting MSP's heat transfer program to align it with that of TSP. The iteration repeats until either consistency is achieved or the iteration limit is exceeded.

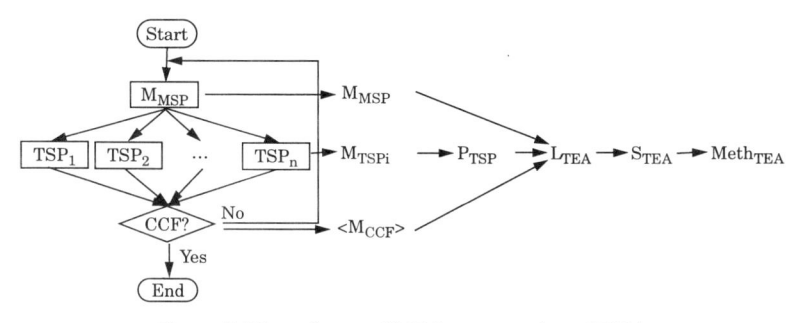

Figure 5. Flow chart and MDL construction of TEA.

Figure 5 also illustrates graphically the process of constructing an MDL specification for the TEA methodology. The three programs enhanced by their pre- and post-processing programs are represented by three modular processes: M_{MSP}, M_{TSP}, and M_{CCF}. The parallel and the iterative sub-structures of the methodology are abstracted into a parallel and a looping process: P_{TSP} and L_{TEA}, respectively. The strategy process S_{TEA} specifies the execution order of the methodology components, which is trivial in this case because only one compound process exists at that level.

After the completion of the TEA methodology, a need arises for the calibration of the combined programs because computer programs are not perfect images of the reality. Calibration is a process of adjusting certain program parameters to effectively shift or scale a program's outputs to match engine test data. The calibration of TEA methodology requires additional programs and iterative structures. It turned out to be very easy to develop an MDL specification for the new methodology, because MDL allows for reusing the specification of a methodology as an integral part of a new methodology.

The Total Engine Analysis Calibration (TEAC) methodology flowchart together with its MDL specification structure is shown in Figure 6. It can be seen that with the encapsulation of the details of the TEA sub-methodology, the structure of TEAC becomes very easy to describe. A new program CALIB is introduced into TEAC for the calibration of TSP. TEAC starts with the initial iteration between the MSP and TSP, preparing data for the first calibration. After the calibration, the calibration constants are fed into the TEA sub-methodology for the total engine analysis. The calibration is considered completed if TEA terminates successfully within one iteration. Otherwise, TEAC starts the loop again by branching back to TSP calibration. The elements included in TEAC are one sub-methodology, three modular processes, one logical function, one looping process, and one strategy process.

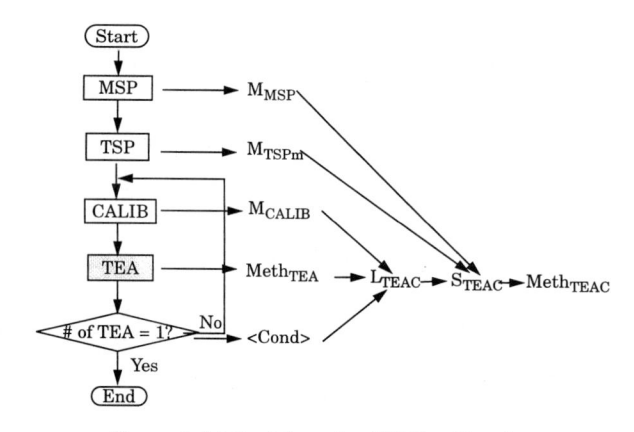

Figure 6. Methodology for CECS calibration

6. Conclusion

In this paper, we have presented a strategy for rapid prototyping of engineering methodologies using EMAT. EMAT has been developed as a generic framework for software integration and has been applied and tested in the vehicle component design/analysis application domain. Using EMAT, engineers can significantly reduce the time spent on prototyping engineering methodologies involving multiple programs. Based on the Blackboard architecture, EMAT can easily incorporate heterogeneous platforms and programs in a single methodology package. Most importantly, it enables the integration of complex and large-scale design/analysis processes, which would otherwise be very difficult or impossible to achieve. It also has the potential to incorporate more sophisticated engineering knowledge for process diagnostics and higher-level goal-driven design/analysis tasks. EMAT is currently being used and tested for prototyping more engineering methodologies by application engineers within Ford.

7. References

Cheng J., "SPLICES -- A Support Package for Layered Integration of Computational Engineering Software", CAE Technical Exchange Conference, Dearborn MI, December 1992.

Corkill, D.D., Gallagher, K.Q. and Murray, K.E., "GBB: A Generic Blackboard Development System", Proceedings of the Fifth National Conference on Artificial Intelligence (AAAI-86), 1986, pp1008-1014.

"ENGSIM User's Guide", Ford Research Laboratory, 1993.

"GBB reference Manual", Version 2.1, Blackboard Technology Group, Inc., August 1992.

Nii, H.P., "The Blackboard Model of Problem Solving and the Evaluation of Blackboard Architectures", Part one, The AI magazine, Summer, 1986, pp38-53.

Nii, H.P., "Blackboard Application Systems and a Knowledge Engineering Perspective", Part two, The AI magazine, August, 1986, pp82-107.

Wang, N., and Cheng, J., "MDL: A Methodology Description Language in a CAE Integration Environment", Proceedings of ASME Design Automation Conference, Sept. 1994.

Wang, N., and Cheng, J., "EMAT: An Engineering Methodology Application Tool", Proceedings of ASME Computers in Engineering Conference, Sept. 1995.

Wang, N. Fagan, D.J., and Staley, S.M., "XBB: A Blackboard System for Vehicle Suspension Tuning", Proceedings of ASME Design Automation Conference, Sept. 1993.

PART SEVEN

Sustainable Manufacturing

30

Architecture consideration for sustainable manufacturing processes re-engineering

Z. Deng
Narvik Institute of Technology
Teknologiveien 10, 8501 Narvik, Norway
Tel: 47-769-22181   Fax: 47-769-44866
E-mail: Ziqiong.Deng@hin.no

Abstract
This paper discusses how to link the requirements of sustainable manufacturing into product life cycle and the system life cycle of manufacturing system. To realize sustainable manufacturing, re-engineering cycles may be carried out over and over again in an enterprise. The architecture and methodology, therefore, should be developed for aiding these re-engineering. In this paper, a consideration of the architecture and the methodology is given.

Keywords
Sustainable manufacturing, product life cycle, system life cycle, architecture, methodology

1 INTRODUCTION

From the 1987 World Commission on the Environment and Development's report on sustainable development (United Nations World Commission, 1987), sustainability is defined as *"meeting the needs of the current generation without compromising the needs of future generations."* Therefore, from industry's point of view, the study of sustainable manufacturing involves in a study of the relationship between the environment and industrial production and seeks to meet consumer demand for products without compromising the resource and energy supply of future generations.

The primary task of sustainable manufacturing is to incorporate thinking environmentally into every step of the industrial operations. Several facets of industrial operations such as *design and materials selection, production, market use,* and *after-market disposal* are remarkable facets in developing a sustainable manufacturing strategy. The following are points to be considered for sustainable manufacturing (Kennedy M. L., 1993):

Materials selection

- Track material feedstocks for all products. Clarify their environmental effects and consider other options.
- Identify chemicals and materials used in the production. What reductions are available? Can packaging of incoming parts be reduced? Do the incoming parts require additional cleaning? etc.
- Find optimum sizes and shapes for all products to reduce materials use, machining time, and production costs.

Process changes

- Look for environmentally-friendlier chemical substitutes. These include cleaners, dyes, and finishes.
- Identify all wastes and by-products generated from production and how they are being recycled.
- Examine and improve worker health and safety conditions in the production area.
- Increase the facility's energy efficiency. Brainstorm on heating, cooling, motor usage, space heating, energy exchange, etc..
- Maximize renewable energy potential. For example, use partially renewable energy sources such as solar, wind, water, or an energy exchange with a neighboring industry.

Market and after-market

- Summarize the life-cycle of the products produced. Where do products go at the end of their life? Can products be brought back for upgrading, disassembly, or recycling?
- Reduce or eliminate excessive packaging.

All sustainable manufacturing requirements above can be considered and implemented in the product life cycle and the system life cycle of manufacturing processes.

2 PRODUCT LIFE CYCLE AND LIFE CYCLE OF MANUFACTURING SYSTEMS

2.1 Product life cycle

The *product life cycle* is well known notation used to structure the different phases of a product's life. Under the consideration of sustainable manufacturing, the product life cycle can be expressed as shown in Figure 1.

Architecture consideration for sustainable manufacturing processes 351

A great number of today's existing products and their manufacturing processes are less sustainable and therefore the closed cycle for the purpose of environmentally-friendly re-engineering is definitely necessary. During re-engineering, the sustainable manufacturing requirements for each phase should be considered. For example, in *'design'* phase, the

```
                           Product re-engineering
         ┌──────────────────────────────────────────────────────────┐
         │                                                          │
   ┌──────────┐   ┌──────────┐   ┌──────────┐   ┌──────────┐   ┌──────────┐
   │  Market  │──▶│ Product  │──▶│ Produce  │──▶│  Market  │──▶│After-market│
   │requirement│   │  design  │   │ products │   │   use    │   │    use    │
   └──────────┘   └──────────┘   └──────────┘   └──────────┘   └──────────┘
```

| Sustainable manufacturing requirements: | . Environmental effects of materials and other options
. Reduce materials use
. etc. | . Environmentally-friendlier chemical substitute
. Recycled wastes and by-products
. Increase facility's energy efficiency
. Maximize renewable energy potential
. etc. | . Reduce or eliminate excessive packaging
. etc. | . Bring-back for upgrading, disassembly, or recycling
. etc. |

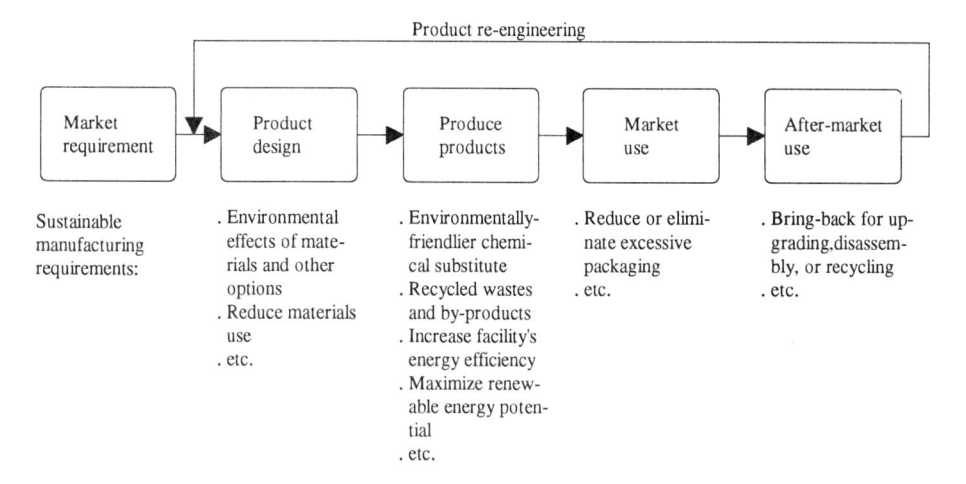

Figure 1 Product life cycle and re-engineering for sustainable manufacturing

environmental effects of materials will be identified and other options would be determined. Also, finding optimum sizes and shapes for products to reduce materials use may be carried out. In *'produce'* phase, the environmentally-friendlier chemical substitutes can be conceived (for example, using purificatory or less-toxic cleaner substitutes); recycling of wastes and by-products should be taken into account; increasing facility's energy efficiency would be considered and implemented. In *'market use'* phase, the reduction or elimination of excessive packaging must be studied. In *'after-market use'* phase, the bring-back of obsolete products would be re-engineered or re-produced by recycling, disassembly, or upgrading.

The product re-engineering and re-cycling based on the sustainable manufacturing requirements involve in the re-design of products, as well as the re-design and re-implementation of existing manufacturing systems (processes).

2.2 System life cycle of manufacturing system and its relationship with product life cycle

The procedure of the design and the implementation, or the re-design and the re-implementation of manufacturing systems (processes) can be explained with *system life cycle* as shown in Figure 2 (AMICE, 1991 and Williams T. J., 1992). The system mission, vision and values, management philosophies, and sustainable manufacturing requirements (see Figure 1) are defined at upper two phases in Figure 2. Based on these, through the system design (or re-design) specification phase, system implementation description phase, system build (or retrofit) and release phase, the new (or retrofitted) system will then work in system operation phase (or produce products phase in product life cycle) where the system life cycle is intersected with the product life cycle as shown in Figure 2. It means that a new re-engineering of products will cause a new re-engineering and retrofitting of the existing manufacturing systems (processes).

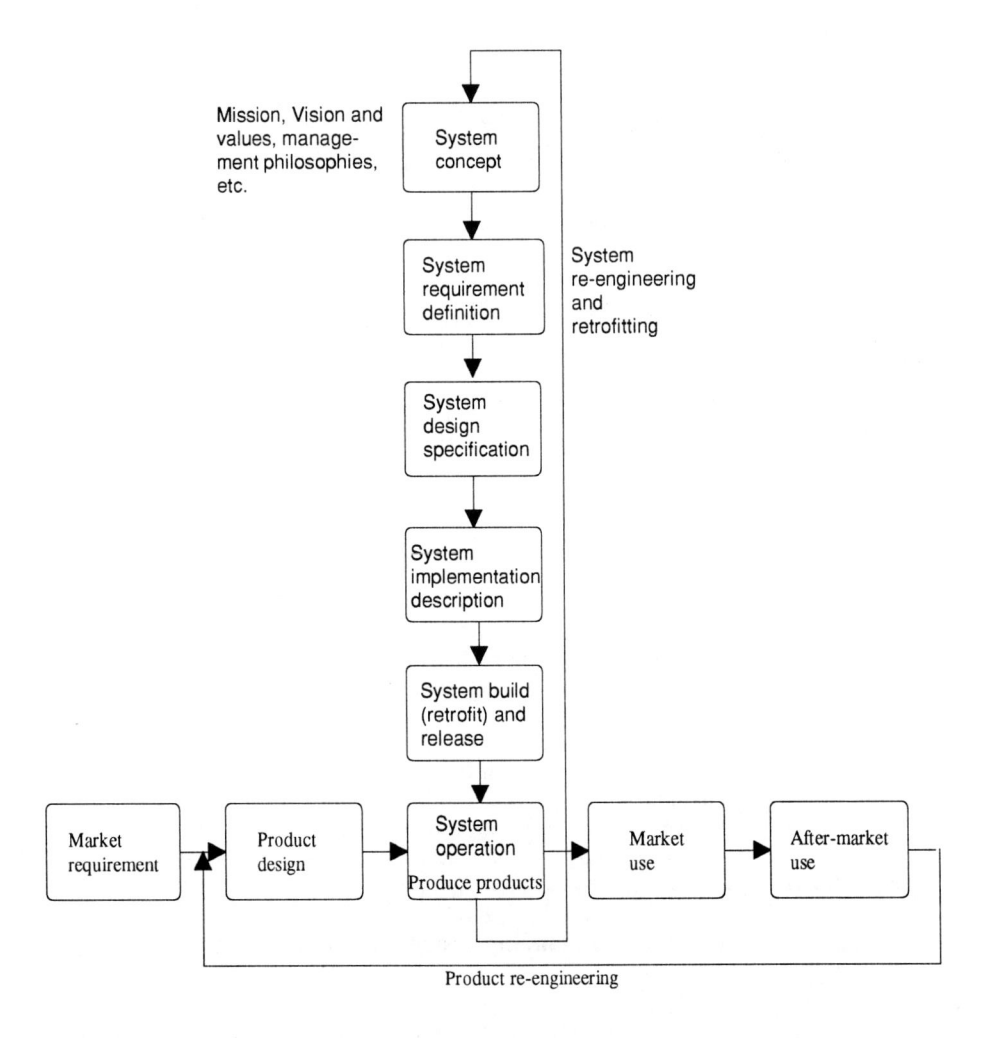

Figure 2 System life cycle of manufacturing systems, system re-engineering and retrofitting, relationship between product life cycle and system life cycle

3 ARCHITECTURE CONSIDERATION FOR SUSTAINABLE MANUFACTURING SYSTEMS

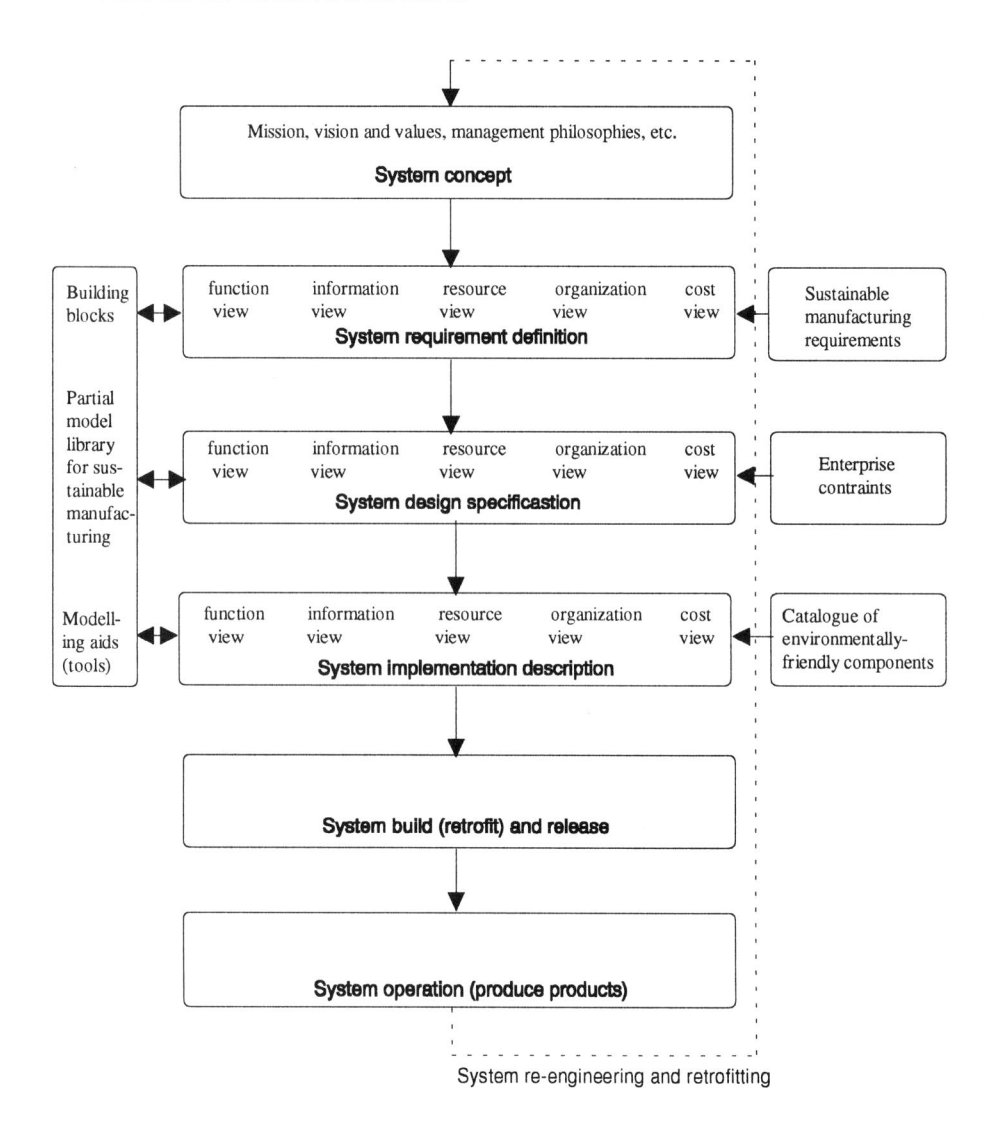

Figure 3 Architecture consideration for manufacturing system re-engineering

As mentioned above, when the products are re-engineered, the manufacturing systems (processes) follow to be re-engineered and retrofitted. To ease the re-engineering of manufacturing systems, an architecture should be suggested. Figure 3 shows an architecture consideration for this purpose.

In general, CIMOSA modelling methodology (AMICE, 1991) can be used in the three modelling phases shown in Figure 3: *system requirement definition* phase, *system design specification* phase, and *system implementation description* phase.

The sustainable manufacturing requirements are the input to the system requirement definition as shown in the upper right portion of Figure 3. We may use the language of building blocks suggested by CIMOSA (AMICE, 1991) to create the set of partial models. (See the upper left portion of Figure 3.) The partial models, in fact, are reference models which can be referenced when ones want to re-engineer a manufacturing system.

A partial model is created consisting of five sub-models called *function view, information view, resource view, organization view,* and *cost view.* (See central portion of Figure 3.) The first four views were suggested by CIMOSA (AMICE, 1991). We suggest an additional view, cost view, for the necessity of evaluating the re-engineering cost and ROI (return of investment).

In particular, the only specific features in the architecture consideration are that:

- the input to the system requirement definition modelling is the sustainable manufacturing requirements
- the creation of the set of partial models is for sustainable manufacturing
- the input to the system implementation description modelling is the catalogue of environmentally-friendly components.

The partial models can also be collected from sustainable practice of the pilot enterprises. Both the models created from CIMOSA methodology or other methodologies, and the models collected from pilot enterprises' practice can be summarized to form a reference models library. This library may help the users to speed up and to structurize their sustainable manufacturing re-engineering.

4 CONCLUSION

The re-engineering for sustainable manufacturing is an open end task for enterprises. According to a selective set of sustainable manufacturing requirements, one can model the re-engineering, do a cost and ROI analysis. Based on these analysis and considering the constraints of the enterprise, a full set of, or a portion of the set of sustainable manufacturing requirements may be selected as the prerequisite of one re-engineering and re-implementation of the manufacturing system. With the time going, new challenges (e.g. new market demands, appearance of new environmental technology, new environment prevention policies, etc.) may stimulate a new procedure of system re-engineering and re-implementation. Therefore, to settle a standard architecture, to create a complete reference models library, to develop a perfect methodology, and to develop a complete set of modelling tools are significant tasks for the industry of future to re-engineer their enterprises gradually approaching to the fully sustainable manufacturing.

5 REFERENCES

AMICE Consortium (1991) *CIMOSA: Open System Architecture for CIM*. Springer-Verlag

Kennedy M. L. (1993) Sustainable Manufacturing: Staying Competitive and Protecting the Environment. *Pollution Prevention Review*, Spring, 149-159

United Nations World Commission on the Environment and Development (1987) *Our Common Future*, Oxford University Press, Oxford, England

Williams T. J. (1992) *The Purdue Enterprise Reference Architecture*. Instrument Society of America

31

Manufacturing Objects as Uniform NC-Interface for Machining and Measuring

A. Storr, D. Handel, C. Itterheim, B. Rommel, H. Ströhle
Universität Stuttgart, Institut für Steuerungstechnik der Werkzeugmaschinen und Fertigungseinrichtungen
Seidenstr. 36, D-70174 Stuttgart
Tel: +49/711/121-2420   Fax: +49/711/121-2413
E-mail: claus.itterheim@isw.uni-stuttgart.de

Abstract:
Adapting existing NC programs to changed requirements, such as different machine tools or cutting tools, is time-consuming and expensive. The machine-oriented generation of NC programs prevents a flexible distribution of partial machining tasks to the suitable machine tools currently available.

A newly developed machining model based on machining and measuring objects, which can be applied to several manufacturing technologies, is presented by a flexible user-oriented NC programming method. All the objects are described by using the STEP-Toolkit currently available.

By using such objects for programming more information is provided at the control system and the machining process is made transparent. Furthermore the modification of the generated NC programs becomes more comfortable by interacting directly with the corresponding graphical model based on the ACIS-kernel. The user is supported by an interactive graphical user interface.

Keywords: NC programming using machining and measuring objects, interactive graphical user interface, CAM, STEP

1 Introduction

The process sequences used so far for the definition of machining a workpiece by means of a numerically controlled machine tool are shown in fig. 1.1. Essential characteristics are on one hand the uni-directional arrangement of the process sequences and on the other hand high loss of information due to the given interfaces. Especially the DIN 66025 being the first determination in regard to data exchange between the NC programming and the numerical controls used for manufacturing should be examined carefully and re-evaluated as well as re-defined, if the new open and user-oriented concepts for control systems are to be a success.

The following article describes the deficiencies of the existing DIN 66025 as found in various companies /1/. Then the requirements and first solutions for an innovative definition of NC-interface by making a connection between planning (CAP) and manufacturing (CAM) which

A uniform NC-interface for machining and measuring 357

is based on object-oriented structures are shown. Here it is basically irrelevant whether the NC programming is done on the shop-floor or in the operations planning department.

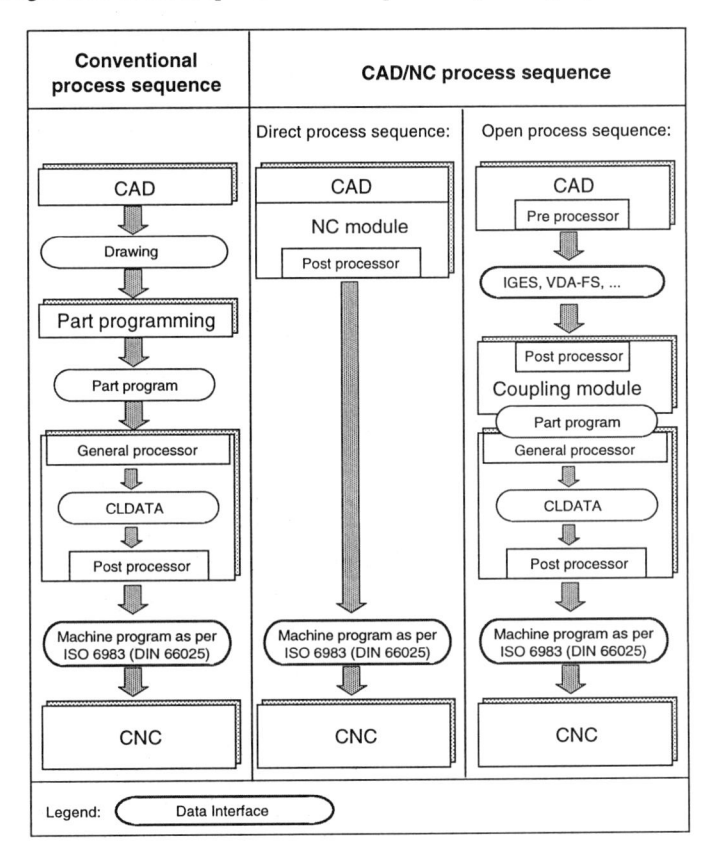

Conventional process sequence	CAD/NC process sequence	
	Direct process sequence:	Open process sequence:
CAD → Drawing → Part programming → Part program → General processor → CLDATA → Post processor → Machine program as per ISO 6983 (DIN 66025) → CNC	CAD → NC module → Post processor → Machine program as per ISO 6983 (DIN 66025) → CNC	CAD → Pre processor → IGES, VDA-FS, ... → Post processor → Coupling module → Part program → General processor → CLDATA → Post processor → Machine program as per ISO 6983 (DIN 66025) → CNC
Legend: Data Interface		

Fig. 1.1: Present CAD/NC process sequences

2 Present Deficiencies and Characteristics of the New Process Sequence

Present deficiencies of the DIN 66025 are:
- the extremely time consuming expenditure on testing and improving NC programs on the machine because of insufficient information and a lack of transparency concerning the NC programs;
- little possibility for the skilled worker to interfere in order to make modifications (i. e. for taking into account modified tools, modified mounting situations etc.);
- no backward documentation concerning optimizations of the machining process;
- inadequate transferability of NC programs to different machines/control systems.

Derived from the deficiencies mentioned above, the essential requirements to a new NC-interface for conventional and open NC machine controls are summarized in fig. 2.1. Especially important are:
- user-oriented supply of the information necessary for testing and optimizing programs on the machine;
- ensuring of the skilled worker's empirical knowledge by bi-directional information processing;
- use of the worker's know-how and increase of motivation by providing secure possibilities of user interaction.

According to the requirements as shown in fig. 2.1 a prototype of the new process sequence was developed. For this purpose design features and machining objects were defined as can be seen in fig. 2.2, which are connected by different types of relations. An object combines data concerning a partial area of the workpiece in a logical unit using a semantic component /2/.

For example, the workpiece model created by the technical designer is composed of several design features like the seat of a roller bearing, ribs or coolant drill-holes. These components determine and ensure as a whole the functioning of the workpiece. In order to guarantee this functionality quality-describing attributes are assigned to the design features. Consequently, a considerably higher level of information is achieved than in the former geometry-oriented interfaces like IGES or VDAFS (fig. 1.1).

Requirements with regard to manufacturing techniques:

More economical due to:	Prompt applicable due to:	Safe processes due to:	More flexible due to:
- omission of the post processor - reduction of the expenditure on testing and improving programs on the machine ("run-in")	- reducing the time for "run-in" - efficient use of CAD data - backward documentation	- user oriented increase of the qualitative information level at the machine tool - redundance free information - current information	- independence of machine tool and control system - "run-in" without programmer - absence of organizational determination

What is necessary:
User oriented increase of the qualitative information level at the machine tool

Bi-directional information processing i.e. for backward documentation

Dynamic possibilities of user interaction i.e. for replanning machining processes

Machining Model as a new interface

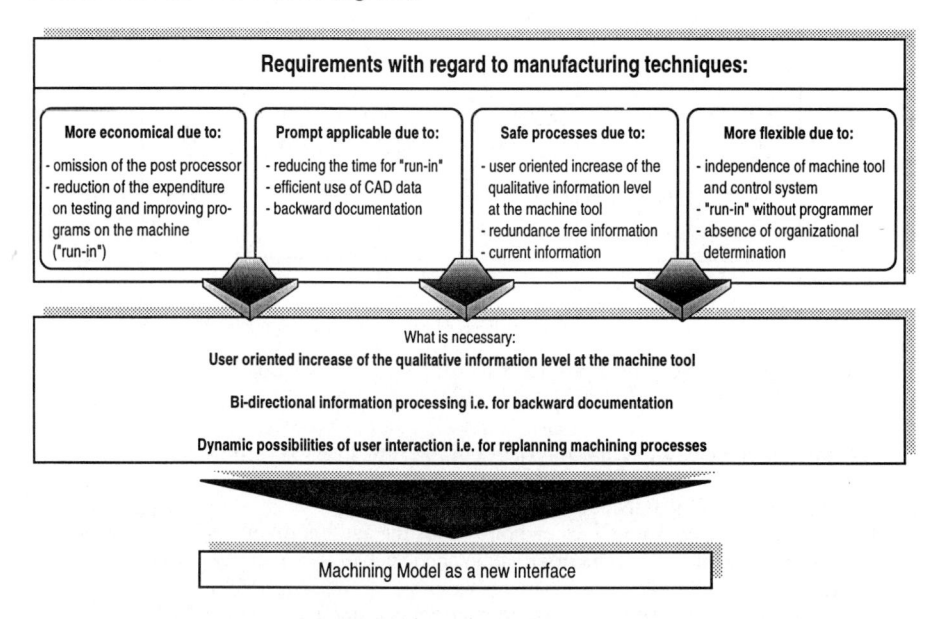

Fig. 2.1: Requirements on a new NC-interface with regard to manufacturing techniques

The design features generated during the design process do not contain any methods for defining the machining process. Therefore, they are converted into machining objects during the analysis and completion of the design features as is indicated in fig. 2.2.

If design features from CAD should not be available, then the machining objects have to be defined manually by identifying the machining tasks based on the design data. This can be done by using the CAD geometry or conventional technical drawings.

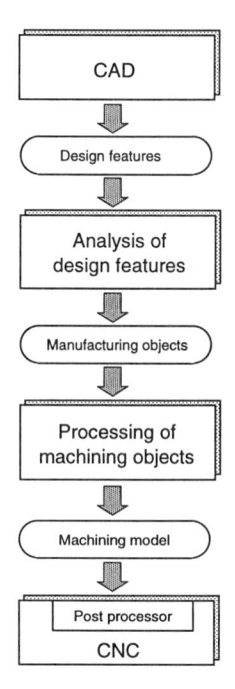

Fig. 2.2: New Process Sequence

The machining objects (MO) are the basis of the machining model. Its structure and processing in the numerical control system will be further explained in the following chapter. Because the machining of a surface can be normally done by using different technologies, a new solution was found. So-called macro and elementary machining objects are defined for describing the machining tasks and assigning of the machining technology /3/.

The macro machining objects (MMO) correspond to the description of the machining task and represent a higher form of organization as compared to the elementary machining objects. As a rule, several elementary machining objects (EMO) are derived from one MMO to represent the partial machining tasks such as "centre drill", "rough bore" etc.

3 Structure and Integration of the Machining Model into the New NC Process Sequence

Considering the mentioned requirements a machining model results which is structured as shown in fig. 3.1. As can be seen here, the machining model is subdivided in partial machining models. This subdivision is necessary to consider that a workpiece is often manufactured in more than one mounting situation.

A machining object consists of attributes and methods as for example methods used for determining the tools. From these methods, which are connected with the machining objects, object-oriented computerized machining instructions for the tool machine result. These instructions are transformed by post processor functionalities, which are integrated in the respective CNC, either directly into standard values for the individual axes and switching commands or, as a migrational concept, into conventionally machinable NC instructions as per DIN 66025.

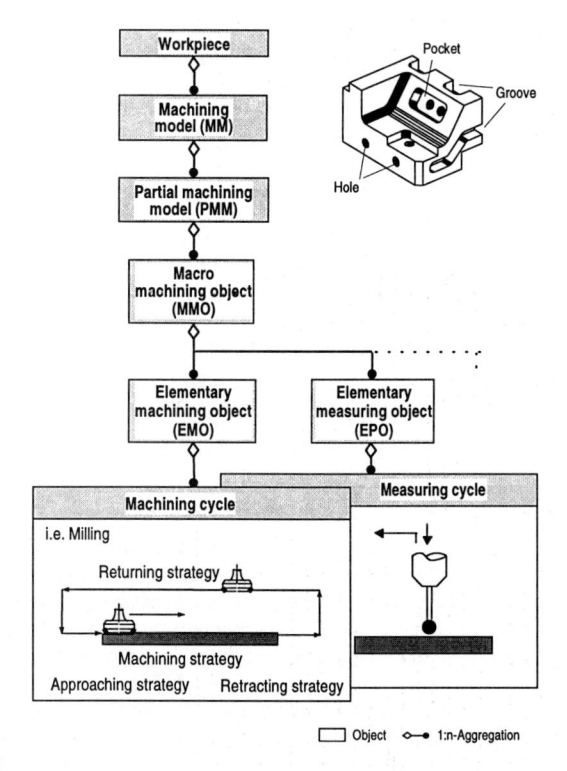

Fig. 3.1: Structure of the machining model

By the demonstrated subdivision into different interconnected machining object classes as well as by their window and pictograph oriented form of representation a user-oriented support of the skilled workers and their actions is ensured. At the same time with each object the information is deposited as to how it was derived and to which degree changes can be made without any risks involved. Such the required bidirectionality is guaranteed as well as the conditions are created for situation-oriented modifications shortly before manufacturing is started, as for example the changing of a tool.

On the right side of fig. 3.1 the so-called elementary measuring objects (EPO) are listed. They define measuring cycles with regard to measure workpieces on coordinate measuring machines (CMM). This is possible because macro machining objects have access to quality describing data like geometrical tolerances of form, position, orientation etc. Elementary measuring objects have methods for generating measuring and evaluation instructions in the DMIS format (DMIS Dimensional Measuring Interface Specification /6,7/) for the determination of the standard form features of the coordinate metrology like circles, cylinders, planes etc. Fig. 3.2 shows clearly the analogy between machining and measuring objects with the example "step drilling".

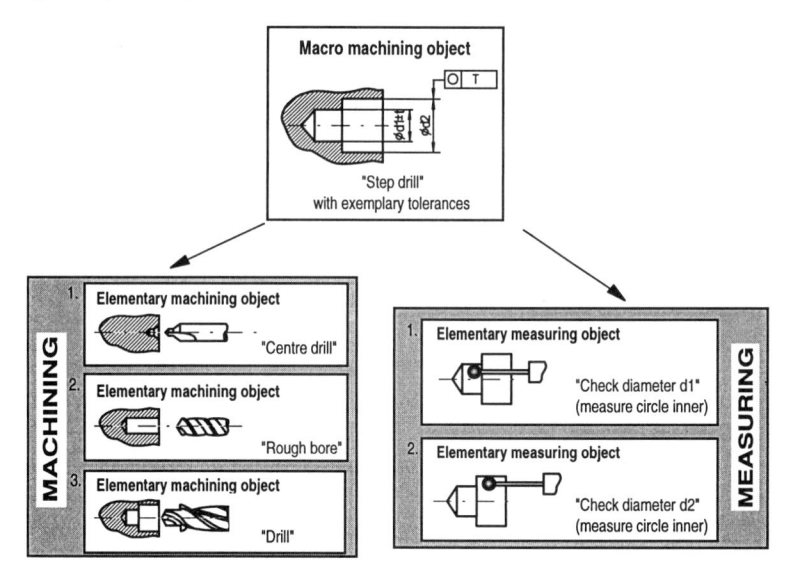

Fig. 3.2: Generation of elementary machining and measuring objects

4 Application Examples

For the realization of the new machining model two aspects push themselves to the forefront:

- integration of the workpiece graphics in order to make a connection between the abstract machining model and the real workpiece,
- the application of a window-oriented interface for data input and output.

An important aspect regarding concepts for the future programming of numerically controlled machine tools is the preparation and availability of machining data in a way that corresponds with the proceedings of skilled workers and their logical thinking and acting. It is therefore necessary to turn away from purely textual machining instructions and to turn to graphically interactive operation. The machining model consisting of machining objects offers here the opportunity to structure data according to the future requirements.

Handling of the Machining Model

Because a multitude of data levels would confront the skilled workers with a very confusing system, the machining objects are presented in two levels. In fig. 4.1 the information levels of a machining object are shown. Starting out from the first level "machining object" the information content is specified according to geometry, technology, tool and machining program.

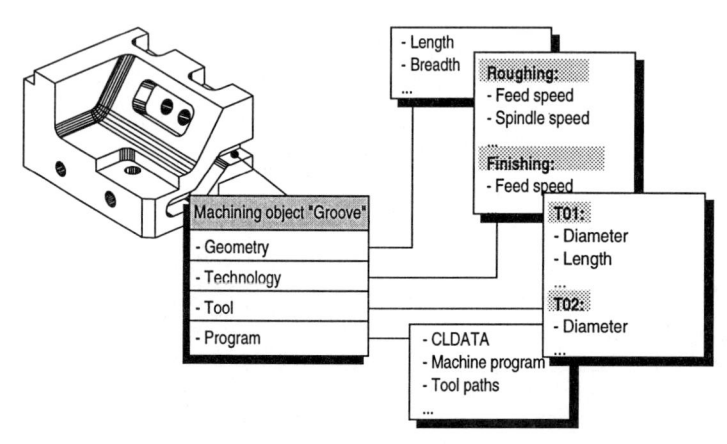

Fig. 4.1: Information structure of the machining model

The graphical interactive aspect of generating a machining object can be explained with a simple example. The geometry of a groove is defined either by using a contour (fig. 4.2) or by entering the parameters (fig. 4.1).

As a next step within the definition of the machining object the technology and the tool have to be determined (fig. 4.3). In order to support the skilled worker in the selection of the machining strategy and the necessary tool a connection to a tool and technology data base is conceivable.

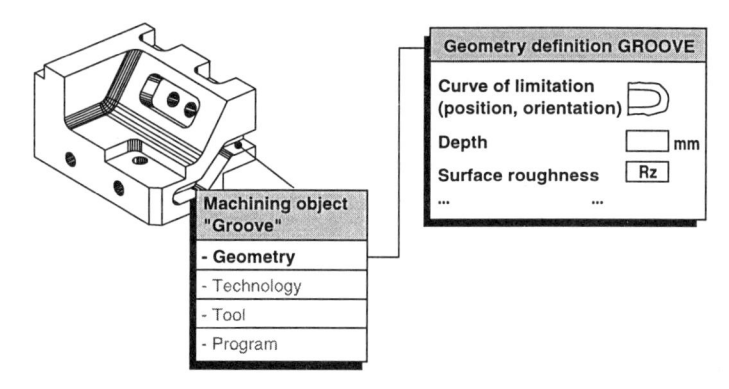

A uniform NC-interface for machining and measuring 363

Fig. 4.2: Geometry definition of a machining object

Data Interchange Format of the Machining Model

Using conventional interface formats as, for example IGES or VDAFS (fig. 1.1) often results in a loss of information. In the generation of part programs with machining-oriented NC programming languages the programming is often done in relation to the workpiece geometry. Transfering them according to DIN 66025 results in the machining instructions purely, without showing such a relation. This means that a backward connection between NC program and part program in case of modifications on the shopfloor cannot be made so easily. This deficit has to be removed with future interface developments.

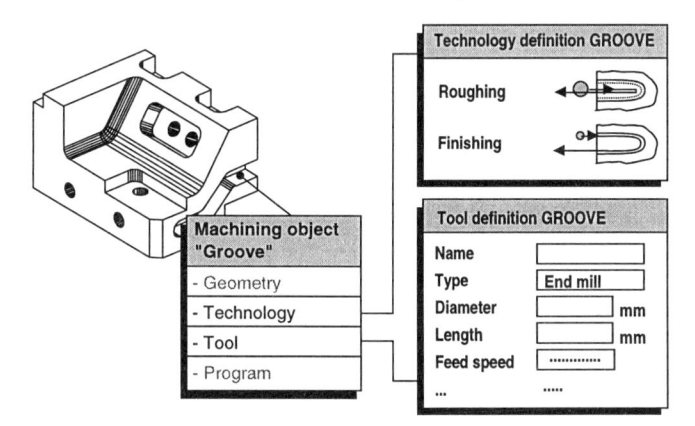

Fig. 4.3 Technology definition of a machining object

The Institute of Control Technology for Machine Tools and Manufacturing Units (ISW) intends to use this object-oriented machining model as a basis for future data exchange. This means: no loss of information during the transfer between different interface formats, because the machining model is always exchanged and so it is available all the time and at every point in the process sequence. Furthermore, a bidirectional data exchange between the manufacturing department and the other departments is possible.

With STEP there is a broad basis for the development of such a machining model. To be able to integrate further developments concerning STEP into the machining model, the corresponding tools like the modelling language EXPRESS as well as the necessary implementation methods in the area of data access und data exchange are applied. Fig. 4.4 shows the data interchange using the STEP format.

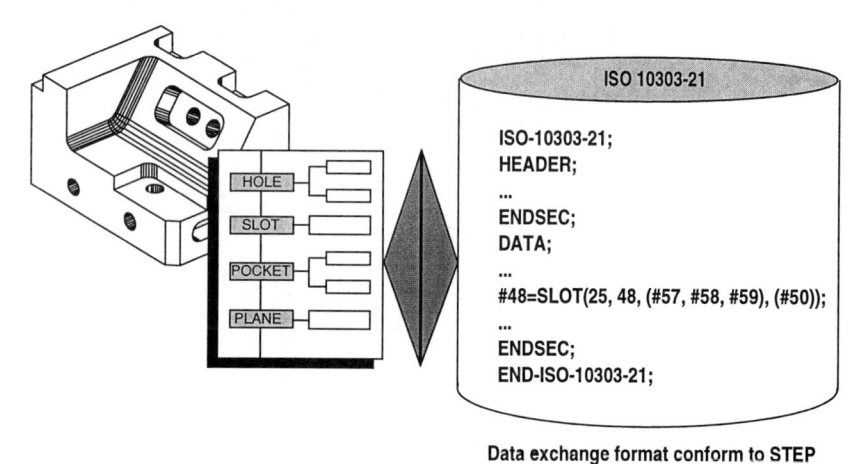

Fig. 4.4: Data interchange format of the machining model

Application Example "2 1/2-axes Milling and Drilling"

Machining objects can be assigned to machining tasks by the semantic component. For example, a skilled worker can, by using the name of the machining object "drill-hole", make an instant connection with the necessary machining task. Fig. 4.5 shows an exemplary workpiece.

The individual machining areas of the workpiece can usually not be machined in an arbitrary sequence. Therefore, it is necessary to define dependencies in order to avoid technological collisions during the machining process. Fig. 4.6 shows the technological interdependences in regard to the machining sequence for the exemplary workpiece, as they are represented in the computer. The vertical branches describe a compulsory order whereas the horizontal branches can be machined in any order whatever.

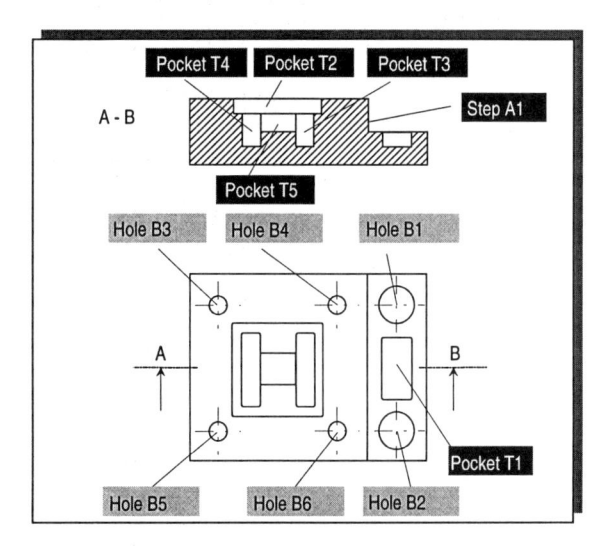

A uniform NC-interface for machining and measuring 365

Fig. 4.5: Exemplary workpiece for 2 1/2-axes milling and drilling

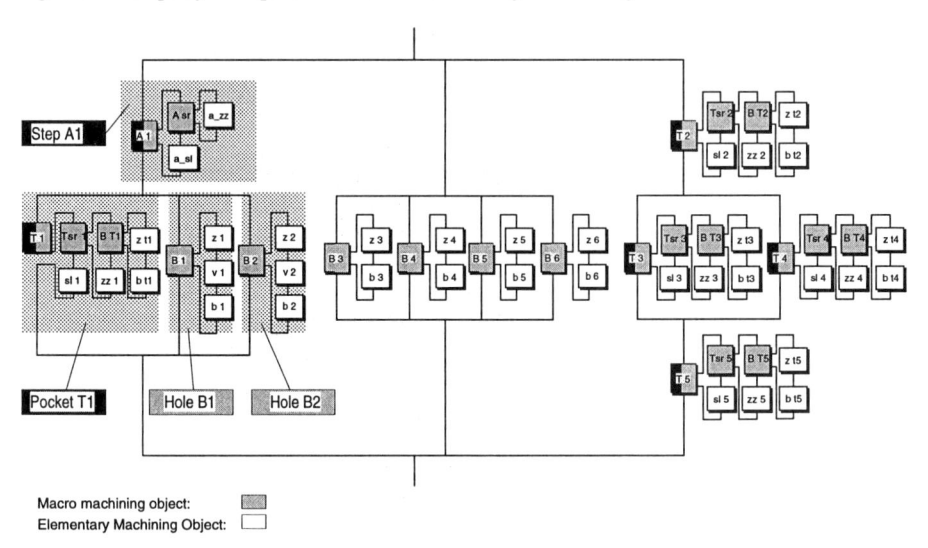

Fig. 4.6: Computer internal MMO/EMO structure of the exemplary workpiece

In the shaded part of picture 4.6 the workpiece area is illustrated with step A1, pocket T1 and holes B1 and B2. It was determined in the programming that step A1 should be machined before the pocket and the holes, because the start positions of the following operations are defined in a way that shorter machining paths result.

Alternative Machining Models

Present NC programs cannot be machined on a machine tool with different kinematic requirements in case of a breakdown, unless extensive adaptions are made. The object-oriented programming method allows during the generation of the machining model to assemble earlier defined machining objects according to the new kinematics without major expense. In fig. 4.7 we have two examples, a part of the machining program for a 3-axes machine and the modificated program for a 4-axes machine.

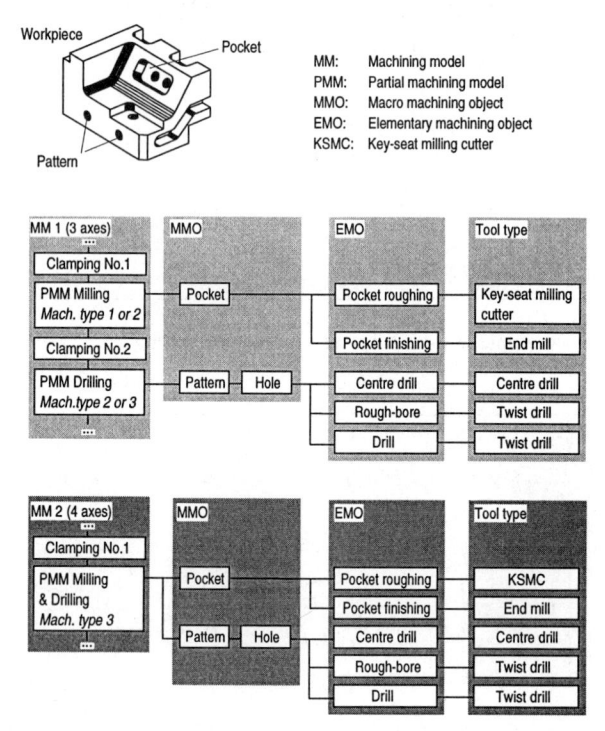

Fig. 4.7: Alternative machining models

Manufacturing Objects for the Machining of Sculptured Surfaces

The new process sequence seen in fig. 2.2 can not only be used for 2 1/2-axes drilling and milling machining but also to the same extent for 3..5 axes machining of sculptured surfaces. Fig. 4.8 illustrates, how multi-axes machining of sculptured surfaces is described by machining objects. However, for the process of NC programming it is necessary to replace presently used strategies like linear machining of the entire workpiece by object (or geometry)-oriented strategies. The analysis of the design features provides as a result macro machining objects which represent geometry-oriented individual machining of specific workpiece areas (e.g. pocket,

A uniform NC-interface for machining and measuring 367

groove, ...). These macro machining objects are related to strategies, which ensure a technologically useful and time optimal machining. In addition, such a classification makes it easier to compute the tool approach and collision free course, which is usually very time consuming. Especially in the case of 4- and 5-axes machining advantages in regard to collision-free computation of machining instructions are to be expected.

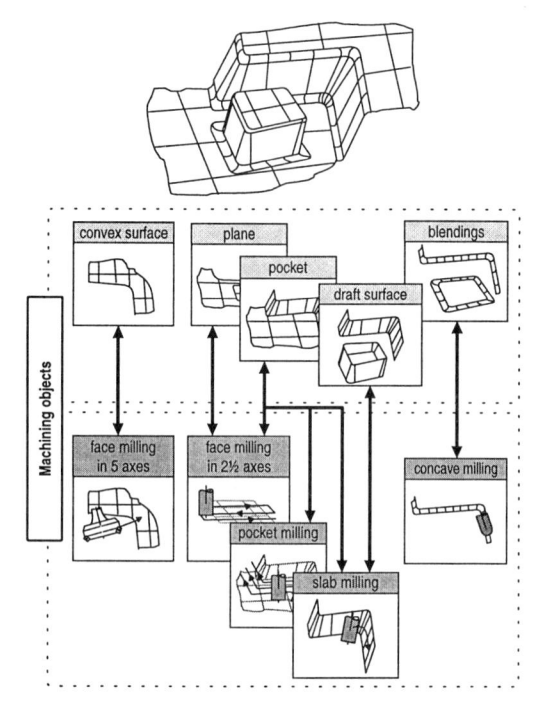

Fig. 4.8: Examples for multi-axis machining of sculptured surfaces based on machining objects

5 Summary

With the demonstrated solution the areas CAP and CAM are connected bi-directionally through a machining model which is based on object-oriented structures. Therefore the presently existing gap between the operations planning department and shop floor is closed. The comprehensive application of the machining objects contained in the machining model provides the user at the machine tool with highly informative machining and control data in a user-oriented way, so that the testing, modifying and optimizing of machining tasks becomes faster and more secure. Additionally, the experience and know-how of the skilled worker is now available to the planning department. Based on the parameterization connected to the machining objects the generation of variants and the reusability of solutions for individual machining tasks is improved considerably.

It is our goal for the future, to use the machining model as programming and NC-interface continuously in the process sequence and the control system. As migration concept it is planned to continue by means of machining objects with the generation of conventionally machinable NC blocks according to DIN 66025 in order to use the presently applied NC control systems without having to make technical modifications. However, thus decisive advantages of the represented solution are lost as, for example, the user-oriented presentation of machining and control data at the machine tool.

6. References

/1/ Fechter, Th.   Schnell und sicher zum ersten Werkstück.
    Reibetanz, Th.   mav - Maschinen Anlagen Verfahren 3 (1994), S. 46...47.
    Walter, W.

/2/ Storr, A.   Qualitätsgerechtes NC-Programmieren von Drehteilen.
    Reibetanz, Th.   wt Werkstattstechnik 83 (1993) 9, S. 120...125.

/3/ Reibetanz, Th.   Situationsorientierte Bearbeitungsmodellierung zur NC-Programmierung
                     Berlin, Heidelberg, New York: Springer, 1995.

/4/ Storr, A.   Werkstattgerechte NC-Programmierung von Frei-
    Itterheim, C.   formflächenbearbeitungen.
                     VDI-Z 136 (1994) 9, S. 72...75.

/5/ Storr, A.   NC-Verfahrenskette für werkstattgerechte Nutzerunter-stützung
    Itterheim, C.   von Freiformflächenbearbeitungen.
    Ströhle, H.   In: Vortragsband des 1. Workshops zum Verbundprojekt
                     "Werkstattgerechte Nutzerunterstützung bei der
                     Freiformflächenbearbeitung" am 15. März 1995 in Dresden.
                     Hrsg.: Inst. für Arbeitsingenieurwesen an der Technischen
                     Universität Dresden, 1995.

/6/ Hartmann, U.   DMIS - Dimensional Measuring Interface Specification.
    Hoppe, U.   Herstellerneutrale CNC-Programme für Meßmaschinen.
    Schmidt, W.   wt Werkstattstechnik 80 (1990) 5, S. 255...258.
    Steger, W.

/7/ Pietschmann, C. MICROTECNIC '92: Fertigungsintegrierte Meßtechnik.
                     tm - Technisches Messen 60 (1993) 5, S. 211...219.

32

The Post-Mass Production Paradigm, Knowledge Intensive Engineering, and Soft Machines

Tetsuo Tomiyama, Tomohiko Sakao, Yasushi Umeda
Department of Precision Machinery Engineering,
The Graduate School of Engineering, The University of Tokyo
Hongo 7-3-1, Bunkyo-ku, Tokyo 113, Japan.
Telephone: +81-3-3812-2111 (ext. 6454). Fax: +81-3-3812-8849
e-mail: {tomiyama,sakao,umeda}@zzz.pe.u-tokyo.ac.jp

Yasunori Baba
Research into Artifacts, Center for Engineering(RACE),
The University of Tokyo
Komaba 4-6-1, Meguro-ku, Tokyo 153, Japan.
Telephone: +81-3-5453-5887. Fax: +81-3-3467-0648
e-mail: baba@race.u-tokyo.ac.jp

Abstract

This paper first proposes a new production paradigm called the "Post-Mass Production Paradigm" (PMPP) which recognizes social, natural, and environmental limitations that conflict with the proliferation of artifacts produced with mass production technologies. PMPP aims at qualitative satisfaction rather than quantitative sufficiency and advocates sustainable production by decoupling economic growth from material and energy consumption. PMPP, therefore, requires innovative technologies and ideas that might revolutionize the current engineering activities over product life cycle. We will discuss not only technological issues as well as economical and social issues.

To achieve PMPP, we then briefly discuss knowledge intensive engineering that is a new way of engineering activities in various product life cycle stages conducted with more knowledge in a flexible manner to create more added value. The power of knowledge intensive engineering is demonstrated in knowledge intensive design, knowledge intensive products, knowledge intensive production, knowledge intensive operation and maintenance, and knowledge intensive recycling on an integrated computational framework.

Because PMPP aims at qualitative satisfaction with less material and energy consumption, it also requires to reconsider how we design artifacts. For instance, we need to design artifacts with such features as longer life, easy maintainability, and robustness against faults. We will describe a new design methodology for innovative products called "Soft Machines." As an example, we will illustrate "Cellular Machines" we have developed.

Keywords
Post-mass production paradigm, Knowledge intensive engineering, Knowledge systematization, Product life cycle, Soft machines, Cellular machines

1 INTRODUCTION

Modern capitalism and industrialism now dominate the global economy. The maturation of the modern capitalist system has depended significantly on mass production technology since Henry Ford started his automobile production. As a result of expanded production capacity, the competition to develop manufacturing technologies, and the technologies of manufactured products, cheaper products with higher quality have been offered, yielding tremendous social advantages such as high wage. However, just as there have been advantages, there have been disadvantages as well.

Limitations imposed by the natural environment, coupled with mass production's emphasis on volume-generated productivity and profitability, have led to the obvious problem of making, using and disposing of increasing numbers of artifacts despite constraints imposed by natural resource availability, energy supplies, and the ability of the biosphere to accept industrially generated waste. More than environmental limitations confront our reliance on contemporary patterns of mass production and mass consumption. Social limitations are also evident as are limitations of markets to accept increasing numbers of mass produced artifacts. Competition among those with increasing productive capacity for markets with decreasing absorptive capacity can lead to trade frictions and the difficulties they generate.

As a result, unless we develop revolutionary technologies, economic growth will be severely bounded by environmental and economic limits. To fundamentally attack and remove these problems, we need to reconsider the current mass production paradigm and to pursue a new manufacturing paradigm to reduce the production and consumption volume of artifacts to an adequate, manageable size and bring this volume into balance it with natural and social constraints. We call this idea the "Post-Mass Production Paradigm (PMPP)."

2 THE POST-MASS PRODUCTION PARADIGM

2.1 From Quantitative Sufficiency to Qualitative Satisfaction

PMPP is suggested as a way to overcome the modern problems described above by decoupling economic growth from resource/energy consumption and waste creation. It aims at transition from quantitative sufficiency to qualitative satisfaction, maximizing global productivity while maintaining individual corporate/regional/national productivity, and arriving at global sustainability. This stands in sharp contrast to the current forms of economic activity. The current industrial economy depends on the assumptions that certain resources (natural resources, waste disposal sites, available markets, etc.) are infinite in quantity and limitless in supply. Given these implicit biases of mass production and mass consumption, it is inevitable that solutions to the modern problems must explicitly consider the finite nature of environmental resources, market capacity, etc. In other words, we need to find out a methodology to reduce the production of artifacts in terms of volume, while maintaining and improving living standards. This is a paradigm shift from quantitative sufficiency of artifacts to qualitative satisfaction.

2.2 PMPP as a Paradigm Shift toward the Knowledge Intensive Society

PMPP should not be confused with on-going efforts to create production systems capable of generating many kinds of products in extremely small lot sizes. These "many-kind, variant-lot" or even "one-off" production systems are an incremental change from mass production, but are still based on the notion of competitive advantage through quantitative competition. In comparison, PMPP is a revolutionary change, since it rejects the dependence on quantitative measures of economic success and emphasizes qualitative measures of economic performance. It is meant to overcome the limits to growth imposed by the finiteness of the natural world by providing mechanisms for economic development based on the creation of high value products that depend on intellectual resources rather than natural resources.

In this sense, PMPP is not motivated just by environmental issues. Rather, it should be viewed as a survival strategy toward the 21st century in which knowledge plays increasingly crucial roles than nowadays.

Drucker predicts the arrival of the Post-Capitalist society in which the major role players are knowledge and management, while capital and labor are the primary sources of the current Capitalist society (Drucker, 1993). In fact, he points out the importance of management based on better use of knowledge. Business process reengineering (Champy, 1993), for instance, advocates removing of unnecessary processes; to do so, management based on better use of knowledge is a key issue.

In summary, the present society is directed toward a new one in which knowledge plays central, ever-increasing crucial roles. We call it a "knowledge intensive society" toward which PMPP can be regarded as a survival strategy.

2.3 Redefining Manufacturing Industry as Life Cycle Industry

This section defines the way manufacturing industries would exist under PMPP and be able to maintain economic growth without becoming dependent on volume-driven economic growth. We take the following three points into consideration.

First, we must consider life cycle issues of products, such as marketing, material acquisition, design, production, logistics, use and operation, maintenance, reclamation, reuse, recycling, and discarding. The present manufacturing industry provides services from marketing through maintenance, however reclamation, recycling and discarding of waste are normally outside the domain of manufacturing industries. In contrast, the boundaries and limitations imposed by the environment, will force the manufacturing industry to gradually provide services in these later stages of the product's life cycle. That is, the manufacturing industry will be forced to include and cover all life cycle stages. Examples of this movement include the ideas of environmental auditing and the recent German legislation for "forced take-back." Consequently, for most of the manufacturing industries new areas of value addition and market creation will be in the reclamation, recycling, and discarding phases of a product's life. In this sense, manufacturing industry should be redefined as life cycle industry.

Second, due to consideration of the life cycle, it is necessary to define the life cycle price of a product so that costs that were not formerly evaluated in the market can be included. For example, most of the reclamation and discarding costs of durable goods are paid nowadays by the local community or government. However, viewing possible legislative enforcement of the extended producer's liability, if these costs are correctly evaluated and included in the retailing price, it would mean an entirely new pricing system. It could be natural that in a relative comparison with production costs, operation and maintenance (e.g., inspection and repair) costs would also become more significant because of the environment protection costs.

Third, it is important to reevaluate the value of knowledge proliferation resulting from manufacturing activities and to reflect it to company's profit and income.

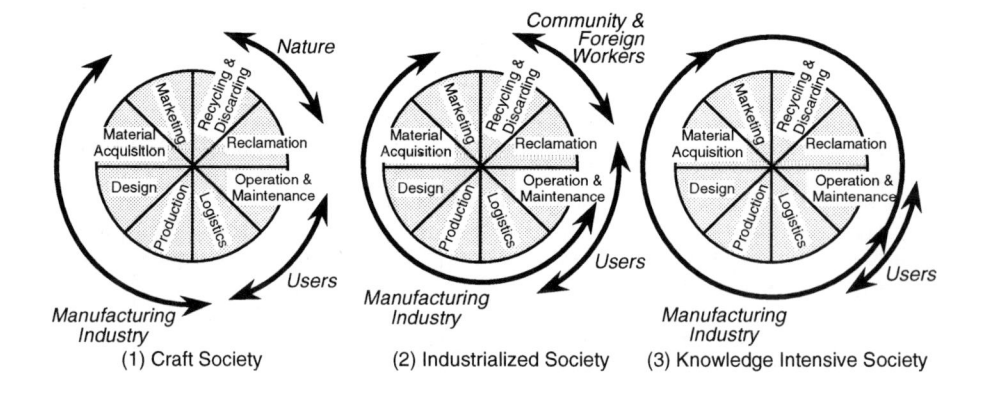

Figure 1 Craft, Industrialized, and Knowledge Intensive Societies

Based on the discussions above, Figure 1 compares the craft society, which existed before mass production, the current industrial society based on mass production technology, and the future "knowledge intensive" society which will realize PMPP.

First, under PMPP, manufacturing industries will clearly be recognized as and expected to serve as "life cycle industries." In the craft society, reclamation and recycling were carried out by nature, while in the industrial society local communities provide these services paid for by taxes or fees. In the knowledge intensive society, reclamation and recycling are considered material acquisition activities and are performed by manufacturing industries. Operations and maintenance will also become part of the services provided by manufacturing industries.

In previous times, end users were supposed to fix the purchased machine. Therefore, not only the machine but also drawings were also shipped out. Nowadays, it is very difficult to obtain technical information detailed enough to repair the machine. Most machines have become too complex for ordinary consumers to understand and to maintain because of advanced design, complex functionality, and the mechatronics principle. The machine has become nothing but a black box. Therefore, it is a very common practice now that most machines are, in fact, maintained or even operated by experts who are sent by the manufacturer. This is somewhat related to recent trends of outsourcing: we can even observe that many maintenance service companies (of, e.g., elevators) make more profits than the manufactures. Alternatively, some machines are not maintained at all. Instead they are disposed of when they malfunction, because the replacement cost is less than the repair cost.

3 KNOWLEDGE INTENSIVE ENGINEERING

PMPP pays substantial attention to product life cycle issues, including marketing, material acquisition, design, production, logistics, installation, operation, maintenance, reclamation, recycling, and discarding. The famous Japanese "design-in" activity is a supporting evidence of the fact that such intensively accumulated knowledge is more effective than a distributed, independent set of knowledge.

Thus, it is increasingly crucial to use and develop methods for facilitating mutual communications among those involved in every stage of the product life cycle. This idea resulted in the concept of "knowledge intensive engineering" which is a new way of engineering activities in various product life cycle stages conducted with more knowledge in a flexible manner to generate more added-value (Tomiyama, 1994). Technically speaking, in case of knowledge intensive design, this boils down to integration and management of various kinds of models to synthesize an artifact, to analyze its properties, and to evaluate its performance against requirements under certain circumstances, since design is a stage to generate knowledge and added-value. These models are not simply connected to each other but integrated , so that mutual data exchange and model sharing are possible.

Knowledge intensive engineering has another goal to increase added-value of engineering activities. For example, operation tasks require different models from design, such as a control model and sensory data, so do maintenance tasks. Knowledge intensive maintenance is a process in which various kinds of models for maintenance are used to diagnose a target machine more accurately and correctly than traditional association based diagnosis expert systems and to generate more useful information such as inspection instructions and repair plans. These should further result in feedback information to reliability design.

This can be further extended to the knowledge intensive enterprise concept which is based on a smooth flow of design knowledge from the design department to the production, sales, and other departments and even to products themselves as embedded intelligence. Figure 2 depicts such an organization in which various kinds of activities are performed on the framework that allows flexible knowledge transfer from CAD to manufacturing cells and robots, and to products.

Knowledge intensive engineering requests that various kinds of engineering knowledge must be systematically organized and installed in a computational framework called a

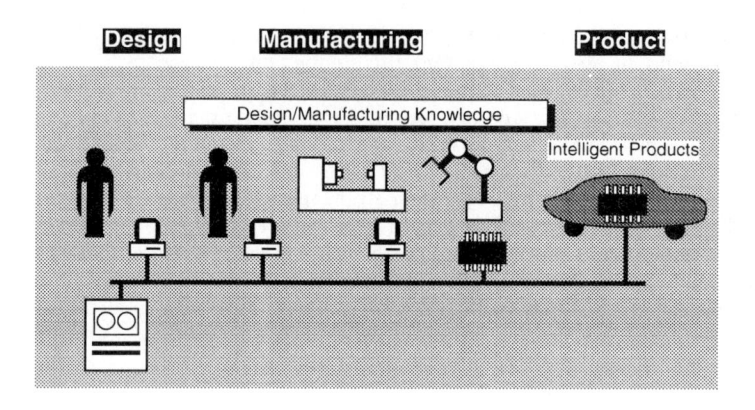

Figure 2 Knowledge Intensive Integrated Enterprise

"Knowledge Intensive Engineering Framework." We have been conducting a research on it (Tomiyama, 1994).

4 SOFT ARTIFACTS

PMPP requests technological advances including soft artifacts that can generate more added-value to compensate decrease of production volume. This can be achieved by the following three strategies.

4.1 Soft Artifacts with Longer Lives

The mass-production technology resulted in excessive productive capabilities that called for unnecessary, early replacement of artifacts. In other words, the mass-production technology contributed to shorter-life of products. Therefore, the first key strategy for making PMPP viable is to extend the lifetime of an artifact, and thereby using this quality to extend the value of that product. An artifact exhibits longer life in two ways; i.e., physically longer life and functionally longer life.

Physically extending the lifetime of an artifact might be possible with high reliability design technologies, which results in cutting down maintenance cost.

However, we must understand how consumers behave; they might buy a new product, when an old one is broken or worn out, but more likely, when it becomes obsolete or out of fashion. This suggests that an artifact simply with longer life does not suffice. Rather, soft artifacts should exhibit the features of "growth-sustaining" including easy upgrading and maintaining of functions. They might be physically aged but functionally updated through functional maintenance that neglects non-fatal, minor failures or deterioration. Such an artifact does not become functionally obsolete nor out-of-fashion; these features are necessary to avoid scrapping before the physical lifespan.

4.2 Artifacts that Generate More Added-Value

The second strategy aims at "more added-value generation" through a product's life cycle to obtain the maximum benefit of the integrated life cycle industry compensating the decrease in production volume. Possible ways to generate more added-value through a product's life cycle are to provide services in the operation and maintenance stage and to realize multiple use in a cascading chain.

We focus on the latter one here. In general, simple thermal recycling is not desirable, because only energy contained in the material is reclaimed neglecting the quality of the material. It is better to reuse the whole of the artifact, although we need to establish second-hand markets to do so. A technical approach to such highly reusable, cascadable artifacts is reconfiguration. If an artifact is not wanted any more, we might reconfigure it for a second use in some other place. Reconfiguration of artifacts also leads to functional maintenance and upgrading (see Section 5.3).

4.3 Re-Evaluation of Life Cycle Costs

Life cycle costs have to be fairly re-evaluated, so that manufacturing industry can redefine itself as life cycle industry to provide better services over a product's life cycle and to generate more added-value. In fact, manufacturers of "social-capitalized" artifacts (see Section 5.1) do not really sell artifacts as "material" but services provided by the artifacts. In an extreme case, for instance, selling how to use rather than material can be more profitable; for instance, a manufacturer of detergent can sell how to use it appropriately and this "knowledge intensive service" distinguishes this manufacturer from others.

Fair evaluation of remaining value after use is also needed to establish market for second-hand goods or recycled goods.

5 EXAMPLES OF SOFT ARTIFACTS

5.1 Social-Capitalized Artifacts and Growth-Sustaining Artifacts

First, we highlight social-capitalized artifacts and growth-sustaining artifacts. These two would make possible to produce economic growth by improving the quality of life, which would depend on qualitative expansion and recycling, not quantitative proliferation of artifacts, or shorter lifetimes.

An example of the social-capitalized artifact is a single-use camera which is returned to the factory and recycled or reused. A social-capitalized artifact still aims at quantitative sufficiency but is completely recyclable and reusable, forming an item of social capital (satisfying the low cost condition). This type of artifact is almost indistinguishable from a disposable product but is eventually replaced (whether or not it has broken down), while allowing multiple-use.

In contrast, a growth-sustaining artifact is a high-valued product and aims at qualitative satisfaction and will be individually owned. Its example is an expensive single-reflex camera. It sustains growth in functions and will survive indefinitely by successive modular maintenance. Also, when upgrading a module of such an artifact, the module that was replaced must be returned to the factory for repair and reuse.

Figure 3 compares a current car, a social-capitalized car, and a growth-sustaining car.

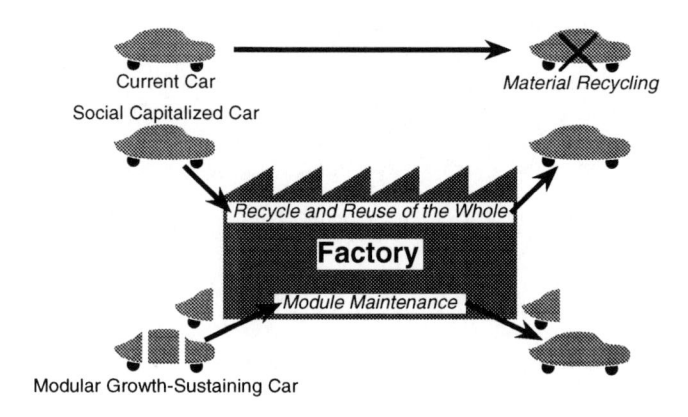

Figure 3 Comparison between Social-Capitalize Artifacts and Growth Sustaining Artifacts

5.2 Requirements of Soft Machines

Under PMPP, social-capitalized artifacts must be completely reusable and recyclable, while growth-sustaining artifacts must allow for modular upgrading by the user. In either case, a more modular design with simple disassembly and replacement of modules is essential. The difference will consist only in whether the modular assembly is fixed initially or whether development and innovative growth can be achieved by replacing and rearranging modules.

Therefore, modularity is critically important to both categories of artifact, with the ease by which a user can rearrange modules an important feature of those that can sustain growth. In this sense, growth-sustaining artifacts must be readily capable of reconfiguration, and this imposes a similar importance on the concept of self-organization capabilities.

Artifacts in the future must be the least possible trouble to use. Here, too, intelligence and autonomy are important, indeed vital, in the achievement of self-organization and reconfiguration capabilities. One result will be the need for a self-maintenance capability.

In summary, soft machines will be given the intelligence to provide a high level of service to mankind, and will have the autonomy to maintain and restore themselves without causing any trouble to their owners. This will be apparent in high modularity, with self-configuring abilities for the easy replacement and recycling of modules. Naturally, such modularity implies a high degree of standardization (which is also critical from the point of view of anti-trust legislation).

5.3 Self-maintenance Machines

As working examples of soft machines, we have developed two types of self-maintenance photocopiers (Umeda *et al.*, 1994, Umeda *et al.*, 1992). Self-maintenance is a concept that a machine continues to operate even when a fault happens by adjusting itself (the control type) or by reconfiguring itself (the function redundant type). These two types of self-maintenance machines are examples of artifacts with functionally longer life in that they perform functional maintenance as opposed to physical maintenance.

5.4 Cellular Machines

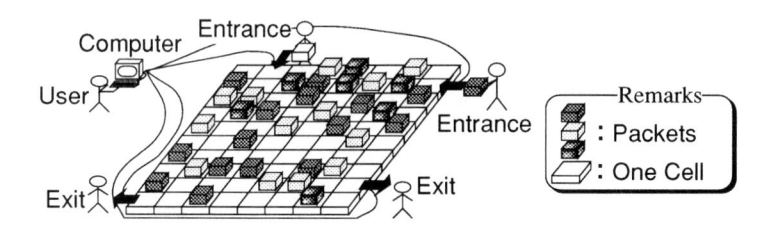

Figure 4 A Cellular Automatic Warehouse

We developed a cellular machine as an example of growth-sustaining artifacts. Figure 4 shows the overview of a "Cellular Automatic Warehouse" we designed and built as an example of a cellular machine that has extremely high modularity. This machine consists of homogenous units called *cells*. Each cell is equipped with a CPU board to exhibit embedded intelligence (see Figure 5), so that as a whole the machine is highly autonomous based on distributed intelligence. Each cell has the ability to receive a "packet" from and send it to any of four adjoining cells.

At the initialization stage, after each cell recognizes its adjoining cells by communication, the exit cells will send out a message so that every cell can recognize the direction of the exits. When a packet is placed on an entrance (there can be more than one entrance), the machine stores it somewhere. When a particular type of packet is called from one of the exits, a cell that carries it will send it to the direction of the exit. The exit cell cancels this call when the wanted number of packets arrived.

Since each cell is identical, it is highly reconfigurable; it even allows "hot reconfiguration" which is adding or removing cells during operation only by re-initializing (see Figure 6). Because it is modular and reconfigurable, modular maintenance and upgrading by the user are possible. When the machine is not needed any more, it can be disassembled to cells and reused again; in other words, it is a cascadable artifact.

Each cell can communicate with adjoining four cells. There is no global communication and each cell determines its behavior purely based on this local communication. In this sense, it is self-organizing. This machine also exhibits fault-tolerance by the ability of self-maintenance. It is achieved only by logically disconnecting normal cells from the faulty cell, which results in logical reconfiguration of the system. Another advantage of this machine is that since the behaviors of each cell is determined by itself based on local

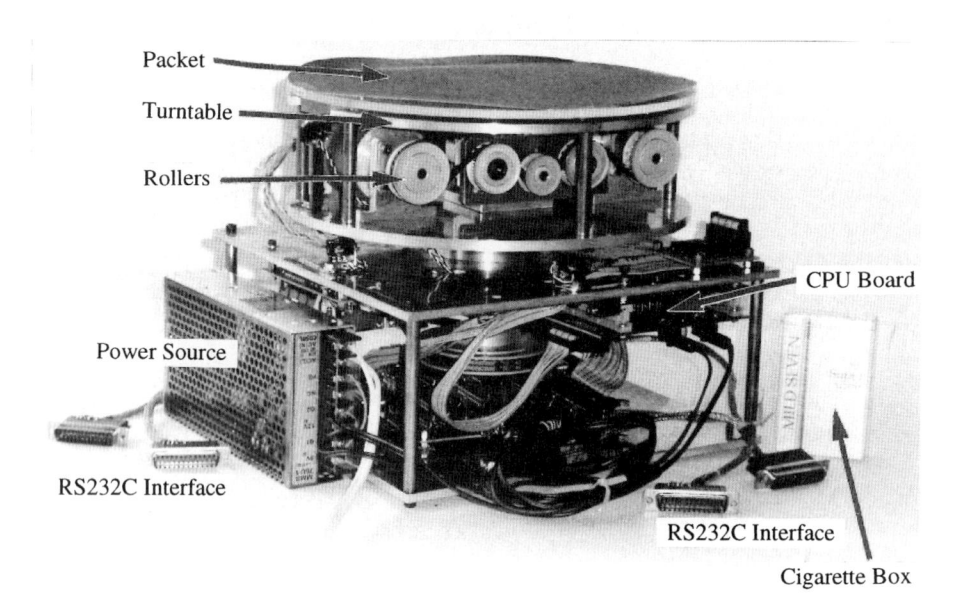

Figure 5 A Single Cell

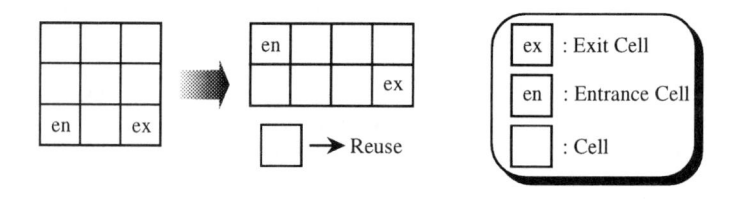

Figure 6 Reconfiguration of the Cellular Machine

communication, there is no programming for the whole system at all. These features make this machine the least possible trouble to use.

Of course, the cellular machine has disadvantages as well. For example, it has less efficiency in terms of both space and energy. It has also problems of dead locks, because each cell has only local information.

However, it demonstrates the feasibility of highly modular machines that are reconfigurable, autonomous, intelligent, and growth-sustaining.

6 CONCLUSIONS

This paper first proposed the Post-Mass Production Paradigm in order to solve the problems caused by technological advances that inevitably met social, natural, and environmental limitations. It primarily aims at bringing down manufacturing volume to a manageable size.

PMPP advocates the better use of more life cycle knowledge in a flexible manner, which resulted in the concept of knowledge intensive engineering. By doing so, we can generate more added-value to compensate decrease in volume under PMPP. To establish knowledge intensive engineering, systematization of knowledge about product life cycle issues is indispensable.

Technical issues such as soft artifacts must be investigated to economically compensate the volume decrease. It is also required to reconsider other changes including life cycle cost evaluation. Some concepts of soft machines were discussed in Chapter 4 and 5, including social-capitalized artifacts, growth-sustaining artifacts, and cellular machines.

Soft artifacts are not just innovative products but highly flexible and reconfigurable to meet ever-changing market demands. To manufacture soft artifacts, we also need innovative manufacturing systems. Chapter 5 demonstrated the feasibility of soft artifacts through examples of self-maintenance machines and cellular machines that exhibit the required features of soft machines.

7 ACKNOWLEDGMENTS

We would like to thank our colleagues who worked together for Gnosis Consortium of the International Collaborative Research Program, "Intelligent Manufacturing Systems" (IMS).

REFERENCES

M. Hammer & J. Champy. (1993). *Reengineering: A Manifesto for Business Revolution.* HarperBusiness, New York.

P. F. Drucker. (1993). *Post-Capitalist Society.* HarperBusiness, New York.

T. Tomiyama. (1994). From general design theory to knowledge-intensive engineering. *Artificial Intelligence for Engineering Design, Analysis and Manufacturing*, **8** (4), 319–333.

Y. Umeda, T. Tomiyama, and H. Yoshikawa. (1992). A design methodology for a self-maintenance machine based on functional redundancy. In L. A. Stauffer, editor, *Proceedings of Design Theory and Methodology – DTM'92–*, 317–324. ASME.

Y. Umeda, T. Tomiyama, H. Yoshikawa, and Y. Shimomura. (1994). Using functional maintenance to improve fault toleneance. *IEEE EXPERT*, **2** (3), 25 –31.

BIOGRAPHY

Dr. Tetsuo Tomiyama has been Associate Professor at the Department of Precision Machinery Engineering, the University of Tokyo, since 1987. From 1985 to 1987, he worked at the Centre for Mathematics and Computer Science in Amsterdam. He received his doctor's degree in precision machinery engineering from the Graduate School of the University of Tokyo in 1985. His research interest includes design theory and methodology, knowledge intensive engineering, applications of qualitative physics, large scale engineering knowledge bases, and soft machines (self-maintenance machines and cellular machines). He is a member of the IFIP Working Group 5.2.

Tomohiko Sakao is a Ph.D. student in the Department of Precision Machinery Engineering, the Graduate School of Engineering, the University of Tokyo. His research interest is soft artifacts, especially cellular machines. He received a BS and an MS in precision machinery engineering from the University of Tokyo in 1993 and 1995 respectively.

Dr. Yasushi Umeda has been Lecturer at the Inverse Manufacturing Laboratory, Faculty of Engineering, the University of Tokyo since 1995. He received his doctor's degree in precision machinery engineering from the Graduate School of the University of Tokyo in 1992. His research interests include soft machines (e.g., self-maintenance machines and cellular machines), green life cycle design, intelligent CAD for mechanical design, and functional reasoning.

Dr. Yasunori Baba has been Associate Professor of Manufacturing Science Division of Research into Artifacts, Center for Engineering (RACE) at the University of Tokyo since 1993. He received a BA in economics in 1977 from the Univ. of Tokyo and MPA in 1981 from International Christian University, Tokyo. After obtaining a Ph. D. from the Univ. of Sussex, U.K., he worked as a Research Fellow at the Science Policy Research Unit (SPRU) at the university from 1986 to 1988. He was a senior researcher at the National Institute of Science and Technology Policy (NISTEP) of Science and Technology Agency (STA) from 1989 to 1991 and had a joint appointment at the Saitama University at the Graduate School of Policy Science from 1991 to 1992. His research field is the economics of technical change and its application to science and technology policy. He is currently working on the information of "Artifactual Engineering" with a focus on global environmental issues.

PART EIGHT

Design for Environment

33
Multi-Lifecycle Design Strategies: Applications in Plastics for Durable Goods

Donald H. Sebastian, Marino Xanthos, Ezra Ehrenkrantz, Ming C. Leu, Kamalesh K. Sirkar, Reggie J. Caudill, and Richard Magee
Multi-Lifecycle Engineering Research Center
New Jersey Institute of Technology
University Heights
Newark, NJ 07102-1982
Tel: 201.642.4869
Fax: 201.596.6056
Net: sebastian@admin.njit.edu

Abstract

Envision an industrial economy where competitiveness is balanced with environmental responsibility: a production system in which the quality of the waste stream is engineered with the same concern as the product itself, where waste materials are re-engineered into valuable feedstocks, where next-generation lifecycle considerations are included in product design, and where "clean" manufacturing processes will not only minimize waste but maximize its quality and value. This concept–multi-lifecycle engineering–goes well beyond today's lifecycle considerations and green products into a new realm of re-engineered waste materials, process technologies and design methodologies. This paper describes the technical approach and research agenda of an engineering research center dedicated to the scientific principles and technologies of multi-lifecycle engineering can be a pathway across this new frontier and a guide to sustainability.

Keywords

Lifecycle, recycle, re-engineering, polymers, plastics, durables

1 INTRODUCTION

Over the last decade, America's manufacturing industry has struggled to achieve a balance between economic security and environmental responsibility. While individual efforts to reduce process wastes, avoid ozone-depleting chemicals, and design green products have been initiated, overall progress toward sustainability has been slow. The pace must now quicken as strong political and societal pressures have emerged, particularly in many Western European countries, leading to imposition of laws that will require manufacturers to recover and recycle their discarded products [Jovane, 1993]. These policies foretell a significant change to an entirely new industrial paradigm that will have profound impact on future U.S. economic competitiveness in the global marketplace.

Current practice leads to the design and manufacture of products for a single application, with the expectation that they will be discarded at the end of their useful life. A better approach is proposed, one that considers the potential of recycling or reprocessing material from one product to create another, not just once, but many times as illustrated in Figure 1. This new recursive approach, called multi-lifecycle engineering, will be used to design and manufacture products that are less expensive, higher in quality, and more friendly to the environment than can presently be achieved.

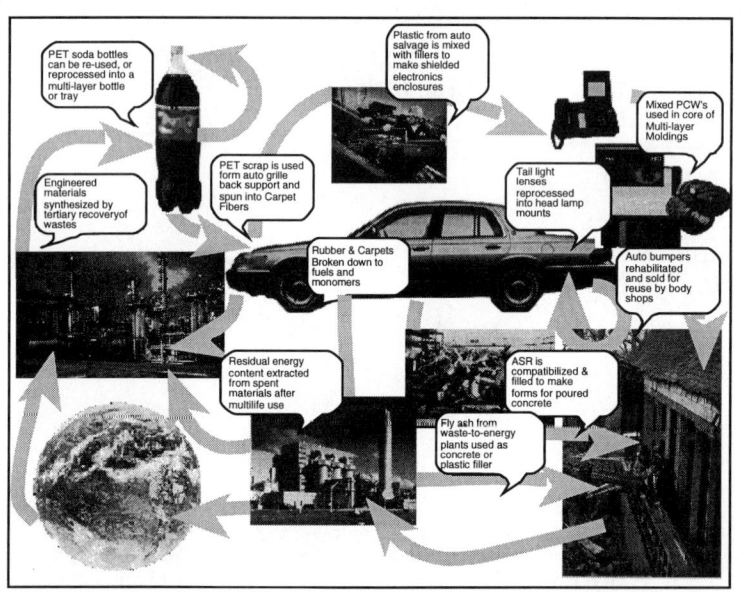

Figure 1 Material Flows in Multi-Lifecycle Scenario

There are some specific applications that demonstrate the feasibility of this approach. For example, in the automobile industry where 75 percent of every vehicle is recycled (mainly the steel components), bumpers from scrapped Chrysler minivans are being recycled and re-engineered into front closeout panels for the Jeep Renegade. Materials from post-consumer sterile-pack juice containers are being separated and used to make plastic parts and paper towels. And, a novel process developed in Wisconsin combines utility fly ash with municipal sewage and papermill sludge to produce lightweight aggregates for construction concrete. These examples indicate the potential of a multi-lifecycle engineering approach to identify products and materials suitable for use as feedstock in new generations of products that are both environmentally responsible and economically competitive.

Existing design tools, value analysis methodologies and process technologies are extremely limited in their ability to accommodate multi-lifecycle considerations. In fact, current practices have led to products that waste natural resources, harm the environment, and are at a competitive disadvantage in the world market. Numerous statistics point out the scope of the problem. The National

Academy of Sciences reported that 94 percent of all natural resources extracted from the earth enter the waste stream within months [Bylinski, 1995]. One indicator of the need for more efficient production is the enormous amount of energy that is expended to produce products and packaging that end up in the waste stream. In 1990, 2.4 quads of energy were used to make 45 billion pounds of plastic, of which 33 billion pounds ended up as solid waste. Only 2 percent of that waste is recycled. The amount of plastics in the waste stream is projected to reach 60 billion pounds by 2010 [Bisio, 1994]. If waste streams such as these can be recovered and re-engineered into new material feed streams, then we can break this trend and achieve sustainability.

A revolutionary change in the engineering of products and manufacturing processes is required. This new approach requires an understanding of the multi-lifecycle use of material and energy resources and fundamental advances throughout the entire production system from product design, processing and packaging, to demanufacture, recovery, and eventual processing into a next-generational use.

To lead a coordinated effort to realize this vision the **Multi-lifecycle Engineering Research Center (MERC)** has been created at New Jersey Institute of Technology (NJIT). The primary mission of the Center is to discover and disseminate basic scientific knowledge and to develop advanced technologies necessary for the successful implementation of multi-lifecycle engineering principles. MERC will work in close collaboration with industry partners to develop and demonstrate product design and process technologies giving special emphasis to cost-effective, re-engineered materials derived from industrial waste and discarded product streams. The Center will engage in research to overcome critical technical and policy barriers and will demonstrate and transfer multi-lifecycle technologies to the commercial sector. Technologies will be transferred through newly formed industry partnerships, and knowledge will be disseminated to students and society through educational programs and courseware materials emphasizing multi-lifecycle engineered products, processes, and methodologies.

New applications for reclaimed materials that provide a sufficiently large market to offset the costs of collection and refinement are necessary to accommodate the flow of industrial and post-consumer waste and to create an economic incentive for industry to develop the infrastructure for recovery and the culture of reuse. Major unexploited market opportunities in areas such as building components, construction materials and packaging may be technically feasible and commercially viable outlets for new applications of materials re-engineered from the waste stream. New materials will not only potentially provide better building products, with superior performance and cost advantages, but also conserve dwindling natural resources. Specific examples include organic or concrete and asphalt matrix composites and their fillers offer unexploited performance benefits; plastic-webbed steel studs which provide improved thermal-break properties can be economically viable if low-cost, recycled plastics can be processed efficiently; and, compatibilized waste stream plastics, both with and without reinforcement, can be used as internal structural cores in multilayer extruded sheets while virgin materials are reserved for surface dependent characteristics.

The manufacture of packaging and durable goods is one obvious area that is ripe for the application of multi-lifecycle engineering opportunities, but other industries have large unexploited waste streams as well. The semiconductor processing and pharmaceutical industries, for example, produce large solvent waste streams with trace contaminant levels that prevent reuse. If economical processing techniques were available and institutional issues resolved, these waste solvents could be reprocessed and used as process solvents in other industries instead of being incinerated. Also, exhausted thermosetting resins used by many industries and utilities in various separation and purification applications are now burnt or landfilled. This waste stream can be converted by pyrolysis to make superior adsorbent carbons or carbon molecular sieves or converted to specialty carbon blacks. Likewise, polymeric wastes in fiber form can be pyrolyzed under appropriate conditions to produce carbon fibers, glassy carbon or graphite. All these are possible, but research challenges remain before these concepts can be transformed to routine practice and commercialization.

A distinguishing feature of the Center is its application-driven focus on the multi-lifecycle use of waste stream materials for alternative applications in products, processes and packaging to meet marketplace requirements that cut across industry sectors. With this driving focus, the Center's vision is embodied in four program goals:

- To make a significant contribution to scientific and engineering knowledge in the form of production technologies, re-engineered materials, and computer-based design, engineering and economic analysis tools.
- To educate a new generation of engineers with a broad knowledge of multi-lifecycle engineering–strategies, methodologies and technologies–for careers as practitioners, educators, and researchers.
- To collaborate with industry to develop multi-lifecycle products and processes, to transfer these technologies from experimental testbeds into practice, and to form educational partnerships between industry and academia.
- To establish a national forum and become a source of technical information for public policy makers on issues critical to sustainability including the potential impact of proposed regulations.

2 FOCUS

The few materials that are commonly recycled today benefit from large source streams, ease of separation from the waste stream, and well-established reclamation channels (e.g., automobiles, beverage cans and bottles, and paper). These factors keep the recovery costs sufficiently low to be economically viable. Generally these materials–steel, glass, paper, and commodity plastics–are recovered as single separate components for blending with virgin feed streams of like material or for use in low-value applications with few performance demands, such as fence posts or park benches. The Center will explore a range of alternatives to the existing routes for materials recovery and reuse: (1) re-engineering material waste streams generated by one industry to form feed streams to other industries; (2) compatibilizing mixed waste stream elements, where separation is prohibitively expensive, to form new materials; (3) reprocessing waste streams to form structural materials for durable goods; and (4) incorporating waste materials as fillers, extenders, modifiers, and enhancers of conventional materials. As stated, the problems associated with this endeavor cut across too many technology areas to focus research on a single discipline or element of engineering science. Consequently, the Center will adopt a *vertically integrated, application-driven* research plan to focus the Center's resources to overcome critical research barriers and demonstrate technologies that will deliver realizable product and production process systems throughout the life of the program.

MERC has adopted an initial focus on classes of waste stream materials for which there is considerable resident expertise, a substantial investment in pilot-scale processing and production equipment, and well-established ties with industry: plastic packaging, mixed plastics (i.e. automotive shredder residue and computer housings), utility fly ash and selected streams from the chemical and semiconductor industry. Recovery mechanisms can be classified into one of five distinctly different approaches:

- Recycle and re-use of fabricated parts and components (disassembly)
- Segregation of waste streams (macro-separation)
- Purification and recycle of mixed waste streams (micro-separation)
- Utilization of mixed waste streams (compatibilization)
- Recovery of feedstock chemicals (tertiary recovery)

No comprehensive study exists that matches a material system to any of these five recovery approaches. Nor is there a quantitative basis for evaluating the economic viability of any suitable option for its cost/ performance capabilities. There is little understanding of the impact of multiple use and reprocessing on the fundamental performance characteristics of these materials or, in general, the relationship between processing technology and re-engineered materials. No general framework has been established for evaluating the characteristics of heterogeneous mixtures of waste stream materials with other material systems. There are no design tools that can deal with both the issues of current product life and future lives of the embodied materials at the early stages of conceptual design when such decisions have the greatest leverage in shaping product success. These research challenges will underlie and focus the Center's research objective to create, through carefully selected applications which exemplify the depth and scope of recovery approaches, a general knowledge base with supporting technical rigor and technology demonstrations necessary to advance multi-lifecycle engineering systems.

3 STRATEGIC PLAN

The Center's core research program is organized to respond to a range of critical scientific and technological barriers in four integrated, cross-disciplinary research thrust areas, as depicted in Figure 2, and described below

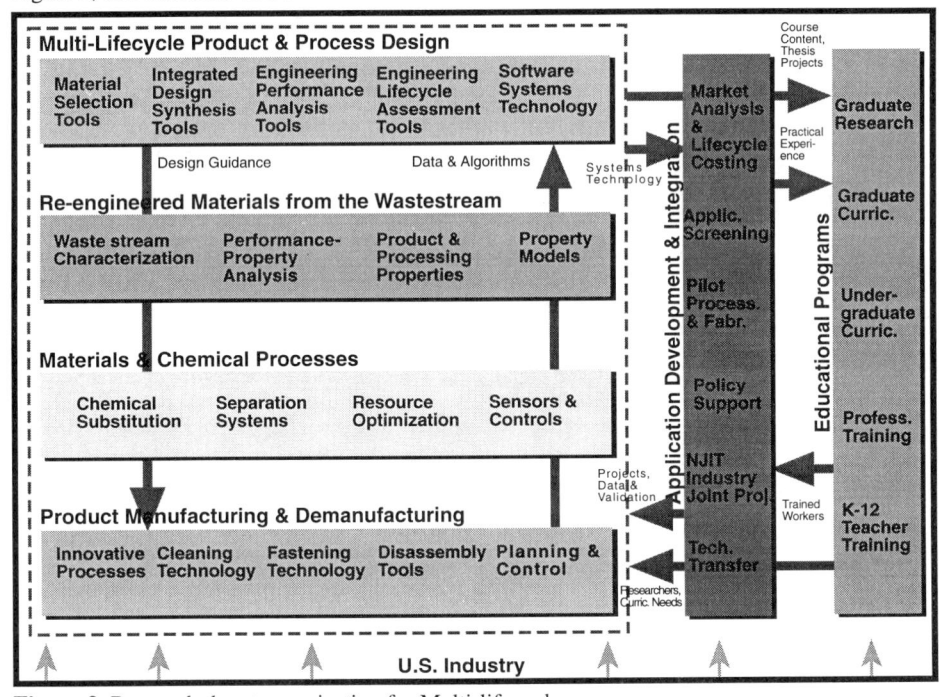

Figure 2 Research thrust organization for Multi-lifecycle program.

3.1 Multi-lifecycle Product and Process Design

The successful practice of multi-lifecycle engineering places a premium on innovation. The transformation of waste streams to feedstocks, whether by purification or compatibilization, demands process innovation. The migration of products from traditional raw materials of construction to alternative supplies demands product innovation. Product designs must be rethought from the front-end to allow material substitution with the concomitant form and fabrication process substitution that this will engender. In addition, designs must be rethought from the back-end concerns of life extension, product disassembly, and material recovery or disposal. This is a daunting task, that presses the individual designer and even most design teams beyond the normal limits of human background and experience. This thrust will address the development of formalized design methodologies that will elevate multi-lifecycle engineering above *ad-hoc* procedures and niche successes, and capture this process in computer-based technology that leverages decades of development in computer aided engineering, knowledge-based systems, object-oriented software design, and interactive computing.

The research conducted in this thrust will build on the existing methodology for integrated product and process design that represents each domain as a hierarchical decomposition of functionalities. To the process engineer, these are a familiar entity - unit operations or elementary steps: to the product designer, these may be known as features, although this research group has developed an organizational scheme that combines form and function in the feature entity [Sebastian, 1993 & 1994], while the common reference to "feature" is concerned with classification based on parametrically similar geometric form [Dixon, 1987]. As software objects, these functional elements encapsulate information on geometric form and associated constraints based on performance, material characteristics, economic and environmental factors. The other thrust areas of research will be generating heuristic and first-principle characterizations of materials, parts and processes. That information will be collected and embedded in these object elements. This scheme is a perfect mate for developing Lifecycle Assessments (LCA) from reusable component parts. The contributions to the LCA's overall material and energy balances are captured in the parameters and relations of the software object that corresponds to the unit process or product element. Complex systems are built from the sub-elements, and overall LCA information can be generated at any appropriate level of aggregation, masking the underlying complexity and eliminating the need to create every LCA from a bottom-up approach.

The resulting design system will grow in generic applicability with the addition of application-specific knowledge developed in each phase of the study. It is expected that the design system and its underlying databases will be implemented in a client-server mode suitable for access across the Internet. Specific areas of investigation will be:

- Material Selection Systems
- Design Synthesis Tools
- Engineering Simulation Tools
- Engineering Lifecycle Assessment Tools
- Computer and Software Systems

3.2 Re-engineered Materials from the Waste Stream

The characterization of waste streams and the development of a quantitative understanding of the performance and processing characteristics of materials derived from these waste streams are of paramount importance to achieving success in the Center's research program. This thrust area will focus on the characterization of waste streams with concurrent efforts in the search for applications with favorable cost/performance characteristics in comparison to virgin materials. The materials property database will provide the matching tie between end-use application requirements and technically feasible waste streams of supply and will engender designer confidence in the suitability of the materials for use in lieu of or in combination with virgin feedstocks. The collection of validated data will provide the foundation for both design synthesis and engineering analysis embodied in the computer-based system created in the *Multi-lifecycle Product and Process Design Thrust* and the basis for developing prototypes to be tested in the REMAPP testbed. The knowledge of material properties also becomes a key element of developing material-specific sensors and control algorithms.

Research efforts will be closely linked to the identification of cost/performance applications that may have the opportunities to utilize substantial volumes of materials from the waste streams. The construction industry, plastics packaging and the chemical process industry will be the primary targets for such efforts, initially. A property data base will be used to help identify the potential for a match with marketplace needs and specify material characteristics for a variety of alternative product applications with favorable cost/performance characteristics. Data will be systematized as product application, material-property threshold value sets for use in an advanced material selection system developed in the *Multi-lifecycle Product and Process Design Thrust*. Specific areas of investigation will be:

- Materials Characterization, Testing and Evaluation
- Performance/Properties/Structure
- Waste Materials Modification to Fit Needs/Performance
- Product and Process Development

3.3 Material and Chemical Processes

To produce chemicals and engineered materials, raw materials are subjected to various combinations of chemical reactions (including polymerizations and depolymerizations), separation processes and some elementary phase/thermal change operations. With the advent of multi-lifecycle engineering, these fundamental unit operations must be adapted to the transformation of waste materials into alternative feedstocks. Such production processes are the focus of this thrust area. To operate these production processes in an environmentally sound and efficient manner, reaction systems and separation technologies for traditional chemicals and materials as well as re-engineered materials have to be studied. For processes using newer benign or re-engineered chemicals and materials, the reaction kinetics and property-performance characteristics are needed for optimum design of the production processes. These production processes for polymeric materials require development of compatibilization and processing technologies based on the property-performance relations for mixed plastics waste streams that are difficult to separate. An alternate route will employ thermolysis (thermal decomposition) processes to recover high value carbons in a variety of structural forms and monomers and chemicals from mixed as well as pure plastics waste streams. For mixed gaseous, liquid, and solid waste streams that can be purified, this thrust area will develop more efficient separation techniques leading to re-engineered products and process feed streams in a clean production environment. All such processes will employ process control and pollution prevention. These research efforts are central to the Center Strategic Plan.

Research will be conducted to support the specific needs of the selected applications. The following is a brief discussion of potential research topics associated with processing waste stream materials. Specific areas of investigation will be:

- Chemical/Material Substitution
- Separation Technologies
- Resource-Optimized Reaction Systems
- Process Sensors and Controls

3.4 Product Manufacturing and Demanufacturing

The research in the Product Manufacturing and Demanufacturing thrust is integrated with the activities in the other thrusts as follows. The re-engineered materials from the waste stream are utilized in the products and manufacturing processes with recycled materials. The parts obtained from disassembly may be used as the feed stream to the material/chemical processes that concern separation and substitution of materials. The models of processes and systems generated from these thrust areas provide the knowledge base for the multi-lifecycle product and process design. The manufacturing and demanufacturing technologies developed are demonstrated in the testbed facilities.

Our existing technical expertise on mechanical design, manufacturing process design, discrete-event system analysis, process optimization, polymer science and technology, cleaning technology, sensing and control, modeling and simulation, and industrial robotics will allow us to tackle the critical technology barriers described above. Research activities have been identified and grouped under five topic areas described below. Specific areas of investigation will be:

- Products and Processes with Recycled Materials
- Cleaning Technology
- Fastening Technology
- Disassembly Process Planning
- Flexible Disassembly System

3.5. Applications Development and Integration

The Applications Development and Integration thrust performs three primary support functions: (1) to identify and screen potential new applications for re-engineered materials, perform feasibility studies specifying cost and performance objectives, and identify critical research and development barriers; (2) to develop research prototypes, demonstrate new product and process technologies and field test pilot applications in the REMAPP testbeds; and (3) to conduct policy and economic studies to overcome implementation barriers to selected applications and to develop costs models for the Multi-lifecycle Product and Process Design thrust.

The activities begin with the selection of appropriate applications for research and development and culminate with the implementation of the results into practice. The establishment of research targets based on marketplace needs requires the translation of those needs into cost and performance criteria. This definition enables feasibility studies to be undertaken to determine which candidate applications merit selection for research and development support. Final selection of applications will consider the views of our industrial partners, feasibility testing in the REMAPP facilities, preliminary analysis and modeling, and the economic and policy challenges. In this way, technology transfer begins with the determination of feasibility not just at the end. Furthermore, this process leads to projects with a high probability of success, a defined set of measurable targets, and increased likelihood of implementation.

Re-Engineered Materials, Products and Process (REMAPP) Testbeds. REMAPP is a network of laboratories, experimental facilities and equipment located at NJIT and at MERC affiliate institutions and industry partners. The objective is to utilize existing facilities, equipment and laboratories and not to duplicate equipment and capabilities, whenever possible. In addition to supporting the research and development program of the Center, REMAPP also serves as an educational laboratory providing "hands-on" learning experiences for undergraduate and graduate engineering students and working practitioners.

Multi-lifecycle Economic Analysis and Modeling. Economic analysis supports decision making on technological issues during three phases of the development of an application: (1) At the time of the initial review of project *feasibility;* (2) At various times during technological *development* to assist technology choices; and (3) At a final review when *commercialization* is decided. Because there are no standardized models for lifecycle analysis, specific methodologies will be adopted by economic analysis group in consultation with industry partners.

3.6. Program Integration

The Center proposes a vertically-integrated, application-driven approach to implementing its research plan. This means that fundamental research will target the unaddressed technology roadblocks that are associated with specific demonstrations of multi-lifecycle engineering. The selection of these demonstration applications is paramount in determining the breadth of applicability of Center research findings and the likelihood of commercialization of its product and process prototypes. A progression from relatively segregated waste streams, used in single component products to successively more complex streams and products is planned.

The selection process for applications, depicted in Figure 3, has three primary screening criteria: (1) market potential, need, and impact; (2) multi-lifecycle knowledge base expansion; and (3) capabilities match between MERC research teams and the research barriers identified. The process begins with the identification of marketplace opportunities, defined as a need by industry for disposal of a large, unutilized waste stream. Candidate waste stream materials will be classified and characterized to determine where there may be opportunities for a match with specific needs. The screening process for applications requires a market and economic analysis including information about potential market size and composition, product cost-performance criteria, code or other regulatory information, and potential barriers. In addition to market criteria, an application will also be judged by its potential to bridge gaps in the general knowledge base and practice of multi-lifecycle engineering and on the capability of MERC research teams to address the critical research barriers identified for the specific application.

The identification of a viable candidate application will initiate the following process: target applications will be analyzed to express performance requirements in a quantitative fashion. An analysis of product geometry decomposed around the functional elements that satisfy the performance criteria will be used to generate alternative product designs best suited to candidate replacement materials and fabrication processes. Suitable waste streams will be characterized as new material systems – alloys, blends, composites and multi-layer structures – to generate the parameters required to project their performance in the selected application. Potentially viable material systems will then determine the necessary processes for transformation of waste streams by sortation, separation and refinement, compounding and compatibilization, or tertiary recovery, that will, in turn drive the selection of pertinent investigations to overcome the barriers to process realization. The product selections will also drive the product fabrication technologies, and the ultimate recovery and disassembly operations that must be implemented to address multi-lifecycle issues.

Three examples of potential application areas and their associated critical research barriers are described in Table 1. Application projects will be chosen, in close consultation with industry partners, to capture the breadth of suitable applications that cut across American industry. By way of example, the incorporation in concrete of large quantities of fly ash from coal-fired power generation plants and the recovery and utilization of polyethylene terephthalate (PET) from beverage bottles both represent segregated waste streams that can be transformed for use in single-part fabrication steps. In addition, a mixed-plastics stream recovered from post-consumer sources, such as automotive shredder residue (ASR), computers and home appliances, has potential application as internal core material in multi-layered plastic sheets for building products and other plastic parts. The development of appropriate design and costing tools as well as the technical challenges of waste stream decontamination and characterization are topics ready for immediate study and integration with the testbed activity. As the program evolves, fundamental work relative to re-engineering mixed waste streams and complex product and process design will lead to broader industry applications such as computers, consumer electronics, home appliances, and automobiles.

	Plastic-webbed Steel Studs	Plastics Packaging	Fly Ash / Concrete
	Use of mixed plastics for thermal-break material in novel stud design as a potential low-cost, high-performance replacement for wood studs in affordable housing. *Waste Stream: Mixed plastics from ASR and computer housings*	Post-consumer polyethylene terephthlate (PET) for use as internal layers in co-extruded plastic sheet, then thermoformed into finished package container. USFDA recently approved the use of this material structure for food packaging applications. *Waste Stream: PET from beverage bottles*	Use of modified coal fly ash in cement creates concrete with special performance characteristics: pH-modified and acid resistant, amenable to glass, plastic, or mixed fiber reinforcement. (Feasibility study completed.) *Waste Stream: Coal fly ash from electric power utilities*
Multi-lifecycle Product & Process Design	Material selection algorithms for structural applications; Function-form taxonomy for building structural components; Part/mold/process design tool integration for reinforced plastics; Cost models for ASR compatibilization and web production	Material selection algorithms for food packaging w/multi-layer ; Function-form taxonomy for packaging/ container; Part/die/process design tool integration for multi-layer extrusions ; Cost models for packaging recyclate & multi-layer fabrication	Material selection algorithms for filled and reinforced concretes Function-form taxonomy for formed concrete shapes Part/process design tools for poured and formed concrete Cost models for fly ash recovery concrete formulation
Re-Engineered Materials From the Waste Stream	Waste stream characterization study; Rheological Characterization of Mix; Material modification; Stuctural Property Data	Waste stream characterization study; Material modification; Material characterization study & database	Material characterization Interfacial reactions with fiber-reinforced materials
Material & Chemical Processes	Compounding Studies; Separation and refinement; Reactive processing & polymer modification	Multi-layer Processing Studies; Processibility modeling; Optimal process design for extruding Re-engineered materials	Removal of carbon using separation technologies; Mixing Studies; Process Modeling
Product Manufacturing & Demanufacturing	Manufacturing process development; Disassembly of plastic and mechanical components; Disassembly process planning	Robust mold design for forming; Sensor recognition and mechanical sortation of mixed waste streams	Manufacturing process plan Develop tooling modeler
Applications Development and Integration	Feasibility & cost/performance specification--economic analysis including price-demand elasticity; Prototype design and testing; Code testing; Codes changes and public acceptance	Feasibility & cost/performance specification--economic analysis ; Public acceptance Collection & processing cost analysis; Demonstration of products and closed-loop process technology	Product feasibility study: expand into new applications ; Impact on statues & regulations(RCRA); Lifecycle costing & cost-benefit analysis; Demonstration of full-scale process and finished products

Table 1 Potential applications of re-engineered materials with critical research barriers

Specific objectives common to each application include:

- Ability to match the marketplace cost and performance requirements with characteristics of waste stream materials and associated processes.
- Identification of economic and policy factors pertinent to multi-lifecycle analysis of the application and development of economic models which accurately reflect the interaction between these cost factors and changes in product and process design variables.
- Detailed characterization of each component of the production process, including energy usage, environmental effects, material properties required for each product cycle, and waste stream generation.
- Formulation of computer-based software systems, utilizing existing design tools and architectures, that can accommodate multi-lifecycle design attributes and value analysis considerations in a methodology based upon the concurrent engineering approach to product and process design.
- Construction of a material properties data base that is capable of relating product specifications to waste stream materials.
- Design of products or system components to meet user requirements.
- Development of demanufacturing and recovery techniques to extract desired material constituents from target waste streams effectively and efficiently.
- Analysis of process robustness and operational performance and development of process technologies to accommodate re-engineered material and reduce environmental impact.

To relate the Center strategic plan to the organization of research by thrust area consider the roles of each component as follows: The Applications Development and Integration (ADI) thrust identifies, evaluates and selects the applications to be studied and defines the critical research and policy barriers that must be overcome. These barriers then drive the research agenda for each thrust area, as well as the additional ADI testbed, economics and policy support activities. The Multi-lifecycle Product and Process Design thrust (MPPD) is responsible for the abstraction of product performance specifications for the application selected to form material system requirements. These specifications will be used to match product needs with material application opportunities. The Re-engineered Materials from the Waste Stream thrust (RMW) must develop the waste stream characterization and property evaluations that feed the decision process to match the product performance requirements. The Product Manufacturing and Demanufacturing thrust (PMD) must develop methodologies and technologies for disassembly and cleaning of used products, including fasteners, tooling, sensors and control. The Materials & Chemical Processes thrust (MCP) is responsible for the material-level refinement or mixture compatibilization steps that transform the waste stream into a new, useful feedstock, and the development of appropriate material sensors and controls technology to achieve that end. The PMD thrust is then responsible for the fabrication steps that transform these materials into new products. All three, RMW, PMD, and MCP will develop fundamental material, product and process models that can be incorporated in the computer-based design system of the MPPD thrust, and all will channel information necessary for economic modeling studies that are coordinated by the MPPD task. The outcome of each application study is implementation and validation in the ADI testbeds, with continuous feedback of corrective data into the system models.

4 CONCLUSION

This paper presents a roadmap for developing tools and technologies that will empower a fundamentally different approach to product and process innovation. The comprehensive program uses a system approach to drive basic and applied research that will not only produce useful, single point products, but will develop generalizable knowledge that transcends the individual application.

The MERC actively solicits partner research groups from academedia and industry to work in building and extending is work to accelerate widespread adoption of multi-lifecycle engineering principles.

5 REFERENCES

Biso, A.L, and M. Xanthos, Eds. (1994) *"How to Manage Plastics Waste: Technology and Market Opportunities"*, Carl Hanser Verlag, Munich.

Bylinski, G., (1995) "Manufacturing for Reuse", *Fortune,* pp102-112, February 6.

Dixon, J.R., et. al., (1987) "Expert Systems for Mechanical Design: Examples of Symbolic Representation of Design Geometries", *Engineering with Computers,* Vol 2.

Jovane, F., et. al., (1993) "A Key Issue in Product Life Cycle: Dissassembly", *Annals of CIRP,* Vol 42(2), pp 651-58.

Sebastian, D.H., (1993) "Function Based Design for Injection Molding", *Proc. 51st ANTEC,* 39, 115.

Sebastian, D.H., (1994) "Concurrent Design Methodology for Injection Molded Parts*", Proc. 52nd ANTEC,* 39, 115.

34

Environmentally sound Computer Aided Design

Prof. Dr.-Ing. R. Anderl
Dipl.-Ing. J. Katzenmaier
FG Datenverarbeitung in der Konstruktion (DiK)
TH Darmstadt, Petersenstraße 30, D-64287 Darmstadt
Phone: +49 6151 16-6001, Fax: +49 6151 16-6854
E-mail: anderl@dik.maschinenbau.th-darmstadt.de
katmaier@dik.maschinenbau.th-darmstadt.de

Abstract

Changing laws and markets demand the development of environmentally sound products. This requires the consideration of ecological aspects originating from all stages of the product life cyle within the design process. In order to meet these changed requirements further developments in the area of information technology are necessary. This concerns the modeling of requirements, process characteristics, and environmental properties in an information model as well as the enhancement of corresponding applications like CAD, EDM, or simulation systems.

Keywords

Sustainable development, environmentally sound CAD, requirements modeling, management of ecological requirements, engineering data management

1 INTRODUCTION

The international industry faces an increasing need to design and to produce environmentally sound products. This demand applies more and more to all stages of the product life cycle and means that the products are expected to consider various requirements such as a good care for resources, low emissions, few byproducts of the production, environmentally sound installation, distribution, usage, re-usage and disposal by the way they are designed. The reasons for these changes are on the one hand side an increasing awareness of ecological dangers within the population and on the other hand side changing laws, regulations and directives for environmental protection which are being created and introduced by governments as a result of this development. Latest examples for these changes are the European Union directive for a voluntary participation of commercial enterprises in a community of environmental management and company audit (European Union 1993) and the new German law for recycling and disposal (Kreislaufwirtschafts- und Abfallgesetz, Hulpke 1995).

1.1 Current situation

Many of the larger companies have already reacted on the changing requirements in the environmental sector and are presently establishing an extensive environmental management (Fischer 1995, Becker 1995). This affects the organizational structure of the company as well as processes, responsibilities and regulations of the company's environmental protection. It is supposed to produce transparency and to reduce the risk of environmental harms by an improved organization (Zimmermann 1995).Therefore the environmental management is an essential organizational basis and important prerequisite for concrete measures to be taken by the companies in order to introduce environmentally sound products. Specific measures for environmental protection are being taken at different stages of the product life cycle (e.g. production, distribution, usage or recycling), which are meant to reduce the expected emissions and amounts of waste (see Förstner 1993). These compensating measures differ in the amount of influence the producing companies have on them and in the ratio of effect over investment (see figure 1).

Production

The production as one of the first stages of the product life cycle is usually completely under control of the company so that the producer can take direct measures to improve environmental protection in that area. Such measures may be the use of manufacturing processes with low emissions and harmless lubricants, the avoiding of harmful byproducts, or the arrangement and scheduling of processes in a way that minimizes negative effects on the environment. Presently optimization efforts in this area are mainly done by an integration of environmental aspects into the companies' planning and controlling systems (Haasis 1994, Kaiser 1995).

Distribution

Distribution is a stage of the product life cycle which can only partly be influenced by the producing company. Provided that the product is not distributed by the company itself possible measures to avoid environmental damages would be e.g. to choose haulage companies which are known to take ecological measures, to combine the order of distribution with ecological conditions or to distribute via railway (see Kühnert 1991, Pfohl 1992).

Usage

The stage of usage is of the lowest influence for the companies. Whether the usage of the product by the consumer is carried out in an environmentally sound way or not is usually completely beyond the influence and control of the producer. Therefore - provided that the product allows the (mis-) usage of a product in an environmentally harmful way - it is almost impossible for companies to prevent such a behavior. The only way to influence the users' behavior is to offer proper operation instructions and explanations or to appeal to the users. Nevertheless it remains difficult to evaluate such measures.

Recycling/Disposal

Today there are a lot of advanced measures of environmental protection for this stage of the product life cycle. Various methods and techniques have been developed which enable the recycling of products, parts or materials and which are documented in many publications (e.g. Lund 1993, Tiltmann 1993, Kohler 1995). Recycling or the environmentally friendly disposal of products has become mandatory for almost all classes of products and the implementation of such processes is strongly enforced. As a result of the increasing strictness of directives, regulations and laws which even include taking back obligations for some products (see Hulpke 1995) the product life cycle comes full to circle in the producing company at this point. This increases the company's amount of influence on the recycling and disposal processes.

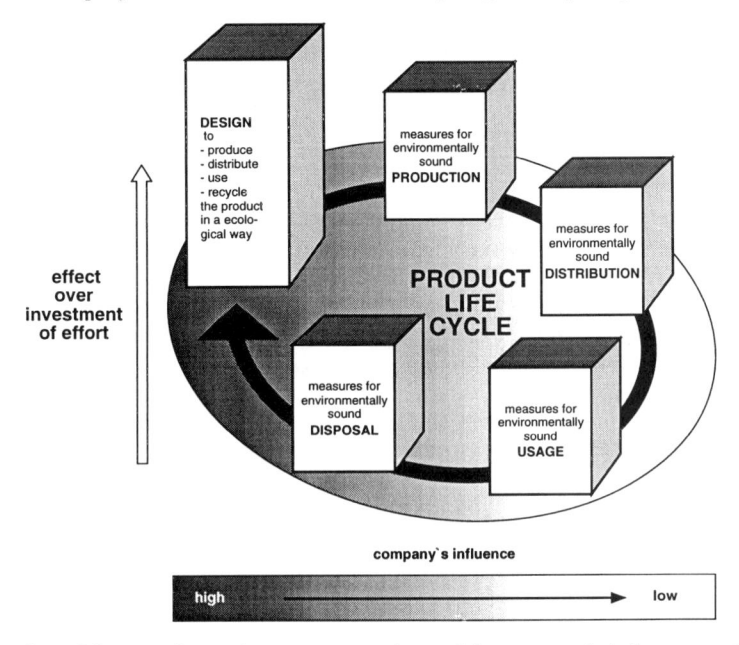

Figure 1 Measures for environment protection and the company's influence on them.

All of those measures applied during the different stages of the product life cycle aim at the ecological optimization of processes which are more or less environmentally harmful. Basically the kind and amount of environmental damage occuring during the product life is caused by the design of the product. Examples for such causalities are:

- the use of materials and geometrical details which can only be manufactured by using certain environmentally harmful processes,
- geometry, mass, or other properties of the product which affect the distribution in a negative way,
- the use of fossil combustion materials for engines which inevitably results in harmful emissions, or
- composite materials which are hardly recyclable and have to be disposed of in an environmentally harmful way.

Yet, since in the near future it will probably not be possible to design the ecologically perfect product the compensating measures aiming at the different stages of the product life cycle will stay of importance and have to be developed consequently to a more advanced level. Nevertheless it is necessary to develop methods which concentrate on the early stages of the product development and guarantee the efficiency of the above mentioned measures.

1.2 Approach

The approach of environmentally sound computer aided design, which is described in the following, aims at the consideration of ecological requirements originating from all stages of the product life cycle in early stages of the product development by applying modern means of information technology.

The great economical responsibility of development and design regarding the entire product life cycle is well known. Studies showed that the stage of design causes a large share of 70% of the future production costs whereas the cost for the design itself only has a share of 10% of the final costs for the product (Ehrlenspiel 1980). Regarding environmental aspects this ratio becomes even more drastic since the amount of ecological damages resulting from the design of the product are by far higher than the ecological damages caused by the process of design itself.

By the design of a product a considerable amount of future ecological harms is set and can only be reduced by environment protecting measures during succeeding stages of the product life cycle (see 1.1). Therefore the importance of an early consideration of environmental requirements at the stage of design in order to reduce this pre-set amount of damages becomes more than evident. Apart from this there are legislative approaches like the new German law for recycling and disposal which focus on the „product responsibility" and will in the near term future force companies to establish measures within the early stages of design in order to design environmentally sound products (Hulpke 1995).

Since design and development are in general entirely within the sphere of influence of the producing companies the producer has all possibilities to meet the conditions and demands of this product responsibility with regard to ecologically friendly production; yet, this task requires the development and application of new methods and tools of information technology in this functional area of the company.

2 ENVIRONMENTAL INFORMATION TECHNOLOGY

The consideration of succeeding stages of the product life cycle at the stage of product design is a key issue of current research projects in the area of design theory. Corresponding projects and activities cover all stages of the product life cycle and focus e.g. on the improvement of design for manufacturing (Meerkamm 1994), design for assembling (Schmitz 1994), design for maintenance (Eichinger 1994), design for disassembling and design for ease of recycling (VDI 1993, Wende 1994). All these approaches have in common that they examine the connections between the results of the design process and the influence they have on a specific stage of the product life cycle respectively the influence of requirements of one succeeding stage on the product design. Several or all stages of the product life cycle have so far been taken into account mainly under economic aspects (Ehrlenspiel 1994, Frech 1995, Koch 1994). These projects aim at the improvement of the design process by the determination and evaluation of the costs that are caused by the design.

2.1 Environmental requirements

Apart from technical and economical requirements ecological demands are a third category of requirements which cover all stages of the product life cycle. (see figure 2). Presently the consideration of environmental criteria during the process of design is applied only partly; methods and evaluation procedures for the destination stages disassembling and recycling have been developed by Spath (1995) and Wende (1994); for the stage of manufacturing this is mainly done within the IMS project (Kimura 1995, Yoshikawa 1995). The topic of modeling corresponding requirements is being researched into e.g. by Grabowski (1995).

The problem of a holistic coverage of ecological requirements along all stages of the product life cycle and of methods for appropriate compromises in the case of contradictory technical, economical or ecological requirements has not been solved yet and is subject of research.

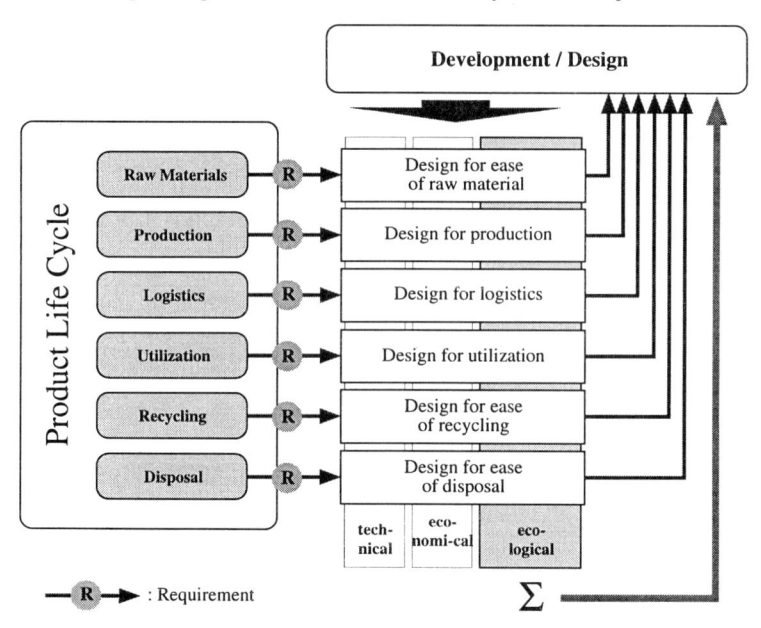

Figure 2 Categorization of requirements for the development and design of products.

An important step towards the integration of environmental requirements into the process of design is the provision of all relevant environmental information in the CAD environment. The creation of possibilities for the designer to get access to all needed ecological data in the same way as he already has access to technical information and - in an increasing way - to information concerning costs is an essential prerequisite for the design of environmentally sound products. By this way the designer has to be enabled to develop the product according to the ecological requirements, to receive a permanent overview of the ecological effects of his decisions and to assess his design solution comparing the results with the given requirements. This applies to the initial conception of a product as well as to the determination and quantification of physical effects and to the process of designing the product shape. Due to the variety of environmental information and the complexity of interrelations new requirements have to be met by the computer aided tools and by the information models they are based upon.

2.2 Information model for environmentally sound design

The computer aided administration and processing of environmental requirements using CAD and EDM technology makes it inevitable to represent all necessary information in a common and standardized basis. This requires an appropriate information model approach.

One basis for such an information model could be the integrated product model according to ISO 10303 STEP which already covers various product information regarding the life cycle besides product shape and product structure (Grabowski 1994). However under the objective of developing products which meet ecological requirements originating from all stages of the product life cycle the existing model is not sufficient. Therefore the development of a coherent information model is necessary which is able to describe all environmentally relevant requirements as well as environmental effects of a product and assessment information. Basically such an information model consists of four major partial models (see figure 3):

- A model for the product definition,
- a model for ecological requirements,
- a model for process data and functional dependencies, and
- a model for analysis and evaluation information.

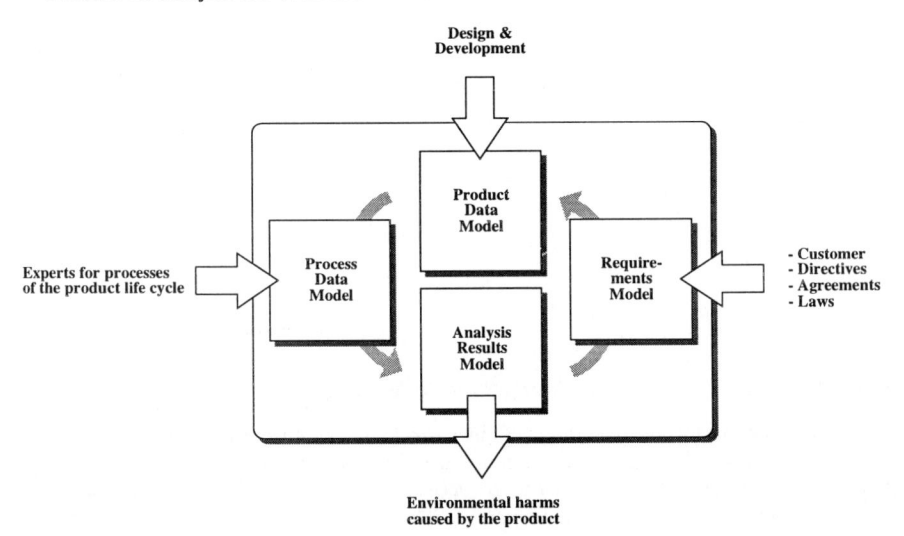

Figure 3 Parts of and relationships within an information model for the environmentally sound computer aided design.

Model for the product definition

The partial model for the product definition corresponds to the integrated product model according to ISO 10303 which is available today. It includes all information set up by designers and developers as a result of their designing process such as geometry, functional, structure, material, tolerances and surface conditions. In order to evaluate ecological damages evolving from the various processes additional information like on the production planning or on the manufacturing processes, which will be applied, have to be included in this part of the model (e.g. information like „This geometric detail can be realized using that manufacturing process" or „In order to disassemble this assembly that disassembling process can be used").

Model for ecological requirements

The partial model for ecological requirements is of central significance for a system that is meant to support the development of environmentally sound products. In this part of the model all requirements for the product referring to environmental protection have to be represented. This model serves two purposes: One is to enable the evaluation of design solutions and the other is to restrict possible solutions during the design process according to the existing requirements. In order to create the possibilities to analyze these requirements and to evaluate the product it is necessary to introduce various classifying attributes and cross-references. The following list gives some examples for possible classifications of ecological requirements and for relationships which may be necessary in order to handle this kind of requirements:

- Like in conventional requirement models the actual requirements have to be classified into fixed requirements, minimum requirements and desirable requirements. Regarding environmental aspects the class of fixed requirements is probably not of a very high significance since requirements such as „the emissions are supposed to have a specific value of xy" are unlikely to be applied to a product. It is more common to formulate minimum requirements to set specific limitations which may not be exceeded. For the class of desirable requirements further diversification according to the degree of priority seems sensible since in terms of environment it is likely that given the possibility customers will formulate a lot of wishes in order to build up a good environmental reputation which can make the realization of the product almost impossible.

- Another important attribute to be treated by a classification is the kind of requirement. Requirements can affect the product in two ways: First of all a requirement can be formulated explicitly in a way like „The product mustn't produce emissions higher than ...". Such requirements usually apply to the stages of usage or disposal. Yet, there are other requirements which affect the product in a more implicit way. These requirements apply for example to the stage of production and are not requirements for the product itself but for processes a product requires. This can be a requirement like „The process of manufacturing mustn't produce more waste than ..." which primarily affects the process but the process itself is dependent on product properties like shape which have to be considered during the design process.

- In order to evaluate the determined ecological damages it is important to know the origin of the corresponding requirements to be able to decide whether a non-fulfillment is critical or not. Examples for the origin of environmental requirements are e.g. customer requirements, company directives, internal agreements of associations or limitations set by law.

- A fourth classification attribute which is needed for a complete description of requirements is the assignment of requirements to a specific stage of the product life cycle for which they are of relevance. This can be necessary since the customer or the company assesses ecological damages at different stages of the product life cycle with different priorities.
- Especially in the area of design it also makes sense to introduce assignments between requirements and technical functions or principles for which these requirements apply (e.g. there are different laws limiting the emissions of spark-ignition engines and compression-ignition engines).

Additionally the model has to consider various cross-references which make it possible to build up relations between the different classes of environmental requirements. This can e.g. become necessary if a product is produced for a foreign customer. In this case the legislative limitations of the foreign country might be expressed as a requirement by the customer. In order to take this requirement into consideration with the right priority when assessing a designed solution such relationships have to be included into the model as well.

Model for process characteristics and functional dependencies

In order to determine the environmental effects of a product it is necessary to represent process characteristics in the model which describe the ecological properties respectively the environmental „behavior" of the various processes during the different stages of the product life cycle (e.g. amount of lubricants remaining to dispose per unit of volume of cut metal, energy necessary for recycling per unit of composite material).

Furthermore it is important to consider relationships within the model that can be used to assign the attributes of a product set by the designer to the process characteristics represented in this model. This is essential in order to determine the ecological damages originating from design and development. These process characteristics and functional dependencies have to be determined and fed into the model by experts for the various processes occurring in the course of the product life cycle.

Model for analysis and evaluation information

Information concerning ecological damages determined by the evaluation of product information represented in the product data model using process characteristics has to be represented in another partial model in order to re-use it during the assessment of the design solution. It is especially important to represent relationships between the determined ecological product properties and the functional or geometrical details that are primarily responsible for such a property. This is necessary in order to identify environmentally relevant weak points of the product faster and to be able to keep track of environmental effects in the case of changes of the product.

With respect to the abilities and properties of the information model described above it is not sufficient if the model, like most of the conventional information models, supports only pure data or facts. Very often environmental information and requirements are formulated not as exact requirements but as limitations or even as imprecise expressions. Furthermore the model is requested to describe complex functional dependencies which relate information originating from different partial models to each other. In order to develop such an information model it is necessary to expand the presently provided modeling techniques and languages. This affects e.g. dependencies and constraints for different classes of information which can be expressed explicitly or implicitly, inequations serving as comparison operators for such dependencies or for the processing of limited values, rules and deduction mechanisms for the evaluation of environmentally relevant correlations, and imprecise information to be processed e.g. by fuzzy logic methods.

An information model showing a structure and abilities as described above can be a suitable informational technology basis for the computer aided design of environmentally sound products.

3 SCENARIO OF AN APPLICATION IN A CAD/EDM ENVIRONMENT

The consideration of environmentally aspects at the stage of design of technical products requires both a suitable information model and the further development of software tools which are based on this model.

CAD and calculation systems are presently the most important computer aided tools used for design. Yet an increase in the number of information- and administration systems in the form of engineering data management systems and of simulation tools can be observed. These four types of tools are also the most important components of a system for the design of environmentally sound products which have to developed further in order to meet the new requirements. Each of these tools corresponds mainly to one of the described partial models (see figure 4). At the same time they form important bridges between these different parts of the model.

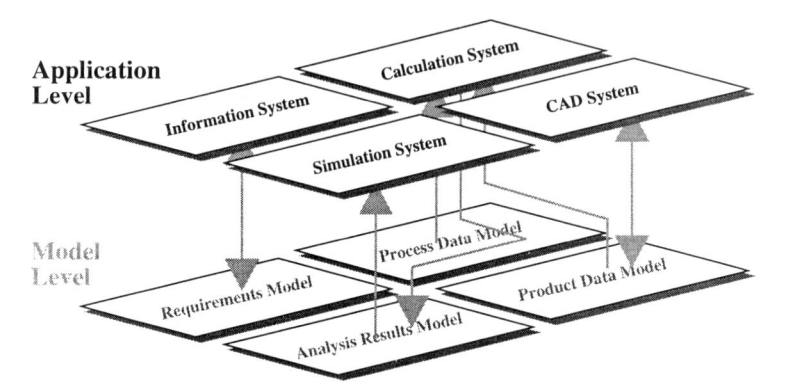

Figure 4 Applications forming a computer system that supports the design of environmentally sound products and corresponding parts of the information model.

Design System

Modern CAD systems not only support explicit shape design but in addition are able to relate geometric elements to each other by the use of constraints and parametric dependencies.

A CAD system as one part of a system for the design of products that take environmental requirements into consideration also requires methods of parametrics and constraints to a great extent. Yet, in contrast to traditional parametric CAD systems the new requirements make it necessary not only to relate geometric items to each other but also to environmental effects. Those functional relationships can be part of the model for process characteristics and functional dependencies or - in the case of not too complex dependencies - set up by the designer himself using methods of parametrics and constraints (e.g. water consumption of a cleaning device depending on pipe and nozzle shape).

By such means physical effects for example as part of the design solution respectively the geometric elements producing or representing these effects have to get assigned to those processes of the product life cycle for which experts or the designer himself have provided appropriate environment related evaluation methods.

Calculation System

Calculation systems are presently used to calculate various technological properties of products (e.g. mass, center of gravity, moments of inertia). They evaluate the shape information resulting of the design process using additional information e.g. on the material which has to be added. In many cases those systems are integrated in CAD systems. For the determination of more complex properties like economic or ecological aspects a complete integration in the CAD system is not appropriate since such calculation systems are very complex units themselves. Instead the calculation systems, which exist today mainly for the determination of cost properties, are built up as separate tools and connected to the design systems or using the same data model files.

A calculation system for the determination of ecological properties can also work according to the described principle: It evaluates shape or functional information fed into the product data model using additional pieces of information such as environment related process characteristics or functional dependencies represented in a process data model and combining them by the use of cross-references between these models which were set up by the designer with the CAD system. By this way the ecological damages expected to be caused by the product can be determined and assessed.

Information System

In order to enable designers to manage, use and process the ecological information represented in the requirements model an appropriate information and management tool offering various function for searching, selecting and analyzing the stored information is necessary. The main reason for this necessity is that in contrast to technical or cost related demands not the customer is the primary origin of requirements but other interest groups or government laws. For this reason the number of requirements increases to a great extent. Apart from this problem it becomes in many cases also very difficult to put the existing requirements into relation with distinct functions or properties of a product or to determine the validity of those requirements.

Not long ago design departments started to use Engineering Data Management systems in order to manage product data. The main task of these systems is, amongst others, to support the designer in structuring, configuring, classifying and retrieving products and product data. Apart from this they are also used to label and track different states of product data (e.g. in release procedures). Therefore it makes sense to use EDM tools to manage the ecological requirements as well. Such a solution offers many advantages since the management of requirements demands functions similar to the management of product data (e.g. structuring, configuration, classification, see 2.2).

The main task of an information system for ecological requirements is to provide the designer with appropriate and momentarily relevant pieces of environmental information at every stage of the design process according to a priority scheme. Apart from this such a system is an important aid for the designer during the assessment of those ecological properties of the product that are determined by the calculation system. Generally spoken designers have to be put in the position

- to retrieve certain requirements by the naming of search criteria (e.g. origin or affected stage of the product life cycle),
- to find requirements in the order of priority, which are of relevance at a certain stage of development,
- to determine if there are requirements including other requirements partly or completely such that those don't have to be taken into consideration separately,
- to summarize requirements that come from different origins or affect different stages of the product life cycle,
- to validate the requirements for different principal solutions (e.g. different emission criteria for Otto or Diesel engines),
- to relate requirements to different product aspects (e.g. functions, principles, shape) or
- to check for contradicting requirements (This applies for example if a certain requirement with origin in the stage of usage demands a certain shape that again calls for a certain way of manufacturing that contradicts an environmental requirement with origin in the stage of production).

The representation of ecological requirements in an EDM system and the application of classification criteria as described in 2.2 are a first step in direction to the solution of these problems. This enables the information system to apply existing management and retrieval functions to ecological requirements using the various classification methods like article characteristics tables or markers for state/origin recognition. For the great number of requirements which are valid independent from distinct orders or products (e.g. laws, directives, agreements) the introduction of hypermedia systems as retrieval or navigation tools, which are able to handle the various cross-references that exist between those requirements, can be additionally helpful.

Both kinds of tools, EDM systems and hypertext/-media tools, have to be further developed towards „intelligent" information systems in order to meet the extended requirements like context dependent views, the graphical presentation of requirement structures or the analysis and recognition of interdependencies.

Simulation System

New methods of representation and management of ecological requirements are a major precondition for an efficient consideration of such requirements during the stage of development and design. Another important aspect regarding the ecological assessment of a product is the simulation and presentation of ecological product properties, which are caused by the design and determined by calculation systems, together with the corresponding permissible limits. This enables the designer to evaluate his decisions and to assess the consequences using the given requirements.

One possible form of presentation is the visualization of the consumption of energy, emissions or pollution values in charts or graphs. This can be done in different forms (see figure 5). Examples are:

- Presentation of distinct values and permissible limits in bar graphs in order to show exceedings or clearance.
- Presentation of values together with additional comparisons that underline the effects on humans and nature (e.g. energy that is necessary to compensate for the pollution).
- Presentation of environmental effects in graphs showing the dependency of other (geometric) parameters together with the corresponding limits in order to find an optimum solution.

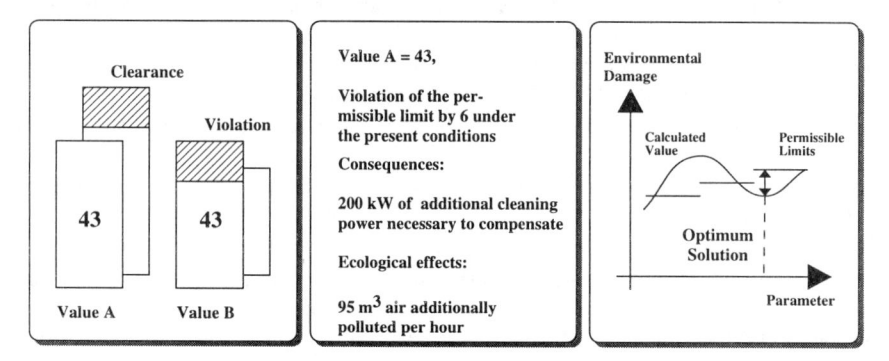

Figure 5 Different ways of presenting ecological product properties.

These forms of presentation can easily be realized using the Hardware presently available in a computer aided design environment. Future developments in this area will have to make use of more advanced presentation methods. This can comprise the presentation of environmental effects like noise using multimedia tools or even the use of virtual reality to simulate more complex effects like vibrations (of engines for example).

Since pure figures as delivered by calculation systems are often not very suitable to put ecological effects in context with actual strains on humans and nature, such methods of presentation and „live" demonstration of ecological effects caused by the design of a product can help to form a stronger environmental awareness in development and design departments.

4   REFERENCES

Anderl, R.; Kruse, P.; Polly, A.; Sabin, A.; Stephan, M.; Ungerer, M. (1995) Produktmodellierung - Die Basis für integriertes Qualitätsmanagement in der Konstruktion. *Zeitschrift für wirtschaftlichen Fabrikbetrieb*, 4, p.171-173.

Becker, J.; Winzer, P.; Martini, J. (1995) Integrierte umwelt- und qualitätsgerechte Gestaltung betrieblicher Prozesse. *Zeitschrift für wirtschaftlichen Fabrikbetrieb*, 6, p.283-287

Ehrlenspiel, K. (1980) Möglichkeiten zum Senken der Produktkosten - Erkenntnisse aus einer Analyse von Wertanlysen. *Konstruktion*, 5, p.173-178.

Ehrlenspiel, K. and Schaal, S. (1992) In CAD integrierte Kostenkalkulation. *Konstruktion*, 12, p.407-415.

Ehrlenspiel, K.; Seidenschwarz, W.; Kiewert, A. (1994) Target Costing - ein Rahmen für zielkostengesteuertes Konstruieren. *Konstruktion*, 7/8, p.245-254.

Eichinger, P. (1994) Servicegerechte Konstruktion. *Konstruktion*, 9, p.292-294.

European Union (1993) Verordnung Nr. 1836/93 über die freiwillige Beteiligung gewerblicher Unternehmen an einem Gemeinschaftssystem für das Umweltmanagement und die Umweltbetriebsprüfung.

Fischer, J. (1995) Umweltmanagement in dezentralen Unternehmensstrukturen. *Zeitschrift für wirtschaftlichen Fabrikbetrieb*, 6, p.278-282.

Förstner, U. (1993) Umweltschutz Technik. Springer, Berlin.

Frech, J. (1995) Kostengerechte Konstruktion. *CAD-CAM Report*, 3, p.131-148.

Grabowski, H.; Anderl, R.; Erb, J.; Polly, A. (1994) Integriertes Produktmodell. Beuth, Berlin.

Grabowski, H. and Rzehorz, C. (1995) Anforderungsmodellierung - Ein Weg zu treffsicheren Angeboten, in: Konferenzband zum gleichnamigen Workshop, Wissenschaftsverlag, Aachen.

Haasis, H.; Rentz, O. (1994) PPS-Systeme zur Unterstützung eines betrieblichen Umweltschutzes. *CIM Management*, 3, p.48-53.

Hulpe, H.; Mischer, G.; Schendel, A. (1995) Das Kreislaufwirtschaftsgesetz und seine Auswirkungen. *Umweltmagazin*, 1/2, p.34-35.

Kaiser, H. (1995) Effiziente Reststoffplanung und -steuerung mit Standard-PPS-Systemen. *CIM Management*, 4, p.59-62.

Kimura, F. (1995) Expectations for IMS research, in: *The future of intelligent manufacturing systems* (ed. I. Hoppner), Japanese-German Center, Berlin.

Koch, R.; Fischer, J.; Jakuschona, K.; Szu, K.; Hauschulte, K. (1994) Konstruktionsbegleitende Kalkulation auf Basis eines Prozeßkostenansatzes. *Konstruktion*, 12, p.427-433.

Kohler, H. and Kaniut, C. (1995) Umweltschutz und Recyclingtechnologien in der Automobilindustrie. *Zeitschrift für wirtschaftlichen Fabrikbetrieb*, 6, p.288-290.

Kühnert, H. and Trute, I. (1991) Problemkatalog zur Umweltwirksamkeit von Verkehrsprozessen. *Forschungsdienst ökologisch orientierte Betriebswirtschaftslehre*, 12, p.16-26.

Lund, H. (1993) The McGraw-Hill recycling handbook. McGraw-Hill, New York.

Meerkamm, H.; Rösch, S.; Storath, E. (1994) Wissensmodellierung und -verarbeitung für das fertigungsgerechte Konstruieren, in: *CAD '94 - Produktdatenmodellierung und Prozeßmodellierung als Grundlage neuer CAD-Systeme* (ed. J. Gausemeier), Hanser, München.

Pfohl, H.; Hoffmann, A.; Stölzle, W. (1992) Umweltschutz und Logistik. Eine Analyse der Wechselbeziehungen aus betriebswirtschaftlicher Sicht. *Journal für Betriebswirtschaft*, 2, p.86-103.

Schmitz, U. and Baier, C. (1994) Gezielte Unterstützung des montage- und recyclinggerechten Konstruierens, in: *VDI-Berichte Nr. 1171*, VDI-Verlag, Düsseldorf.

Spath, D.; Hartel, M. (1995) Entwicklungsbegleitende Beurteilung der ökologischen Eignung technischer Produkte als Bestandteil des ganzheitlichen Gestaltens. *Konstruktion*, , p.105-110.

Tiltmann, K. (1993) Handbuch Abfall-Wirtschaft und Recycling: Gesetze, Techniken, Verfahren. Vieweg, Braunschweig.

VDI-Richtlinie 2243 (1993) Konstruieren recyclinggerechter Produkte. Beuth, Berlin.

Wende, A. and Schierschke, V. (1994) Produktfolgenabschätzung als Bestandteil eines recyclingorientierten Produktmodells. *Konstruktion*, 3, p.92-98.

Yoshikawa, H. (1995) The manufacturing industry - From competition to cooperation, in: *The future of intelligent manufacturing systems* (ed. I. Hoppner), Japanese-German Center, Berlin.

Zimmermann, G. (1995) Prozeßorientiertes Umweltmanagement. *Zeitschrift für wirtschaftlichen Fabrikbetrieb*, 6, p.282.

35

Model based approach to life-cycle simulation of manufacturing facilities

H. Hiraoka, D. Saito
Chuo University
13-27 Kasuga 1-Chome, Bunkyo-ku, Tokyo 112, Japan
telephone: +81-3-3817-1841, fax: +81-3-3817-1820
e-mail: hiraoka@mech.chuo-u.ac.jp

S. Takata
Waseda University
3-4-1 Okubo, Shinjuku-ku, Tokyo 169, Japan
telephone: +81-3-5286-3299, fax: +81-3-3202-2543
e-mail: takata@cfi.waseda.ac.jp

H. Asama
The Institute of Physical and Chemical Research (RIKEN)
2-1 Hirosawa, Wako-shi, Saitama 351-01, Japan
telephone: +81-48-462-1111, fax: +81-48-462-4658
e-mail: asama@cel.riken.go.jp

Abstract

For realizing effective utilization of manufacturing facilities, evaluation of various point of views in each phase of facility life cycle is essential. For this purpose, we propose a computer support system based on a model of the facility. The system, which is called *the life cycle simulation* system, can simulate the degradation of facility as it works. As essential modules of the system, we have developed a deterioration evaluation sub-system which provides qualitative estimates of the component deterioration of the facility, and a movable parts identification sub-system which can identify the parts in motion of the facility assuming that the predicted deterioration has occurred. The output of the sub-system can be used for kinematic analysis of the facility in order to identify failure modes induced by deterioration. As an illustrative example, the evaluation of a grinding robot is demonstrated using an experimental system.

Keywords
Computer models, maintenance, life cycle, deterioration, kinematic motion

1 INTRODUCTION

Today's manufacturing becomes increasingly dependent upon facilities with the advances of automation and integration of manufacturing systems. This lead to a growing concern about effectiveness of the facility, which should be evaluated from various point of views, such as functionality, reliability and maintainability. For improving its effectiveness, we have to properly manage the various activities in each phase of its life cycle.

In this study, we direct our attention to a maintenance activities of the facility. In the past, maintenance was regarded as a reaction to the occurrence of failure. Even through condition-based maintenance is concerned, most efforts have been directed at diagnostic issues, that is, identifying what happened or what is happening. However, in order to carry out the effective maintenance over the facility life cycle, it is necessary to take a proactive approach. If you do not know what to expect, you can hardly prepare for it. Only by predicting potential problems, can you devise countermeasures, such as improving design or planning for preventive maintenance.

For implementing the proactive maintenance strategies, we need an effective computer support for predicting problems which would occur in the facility. For this purpose, we propose a concept of a life cycle simulation which is executed based on a model of the facility, called a facility model. As essential modules for the life cycle simulation, we have developed a deterioration evaluation system which predicts potential deterioration of the facility components, and a movable parts identification system, which is used to extract the parts participating in kinematic motion from the facility model and to construct a model for kinematic simulation so as to evaluate functional failure induced by expected deterioration.

In Chapter 2 and 3, we will describe the concept of the life cycle simulation and the facility model respectively. Then we will present the method of evaluating deterioration and identifying movable parts in Chapter 4 and 5. In Chapter 6, an experimental system is explained and illustrative examples are demonstrated.

2 CONCEPT OF LIFE CYCLE SIMULATION

To take the proactive maintenance approach, we need to be able to evaluate the state of the facility in any phase of its life cycle based on the best available information at each point in time. For this purpose, we need to have access to any evaluation tools and information provided in

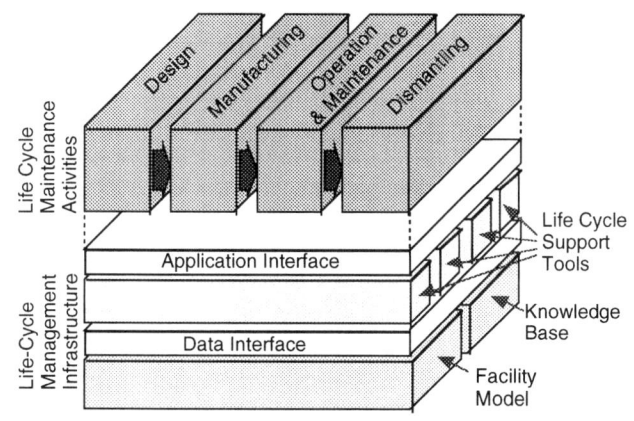

Figure 1 Life cycle facility management infrastructure

each phase of the life cycle. For example, it is essential to know the real operating situations and the problems experienced in the past in the design phase for the purpose of design for maintenance. On the other hand, it is necessary to have exact design information for the maintenance planning in the operation phase.

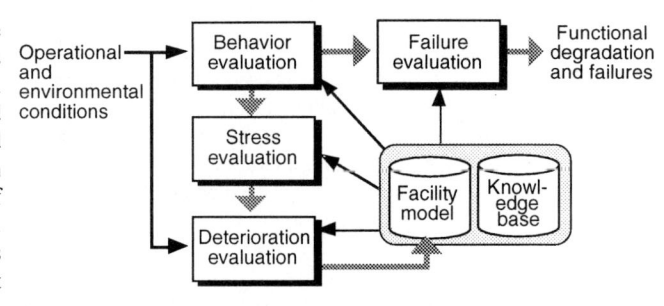

Figure 2 Life cycle simulation

For fulfilling these requirements, we propose an integrated information system which is called a life cycle facility management infrastructure. Figure 1 shows the architecture of the system. The system has two types of common data bases: a facility model and a knowledge base. The facility model maintains all information associated with the facility life cycle. In the knowledge base, generic knowledge, independent of an individual facility, is stored. Such generic knowledge includes deterioration mechanisms, theories and practices. Various tools which support maintenance activities based on the facility model are implemented in the system. They are accessible to any phase of the facility life cycle.

We have been developing a system for simulating the degradation of the facility as it continues to work. It is a essential part of the life cycle facility management system. We call it a life cycle simulation system. The process of the simulation is shown in Figure 2. First, the system evaluates behavior of the facility under given operational and environmental conditions in terms of various operational parameters, such as torque, speed and temperature. The results are used to evaluate the stress exerted on components of the facility during the operation. Then, the system evaluates deterioration which could be induced by the exerted stress and other factors existing in the component. Here, deterioration means a physical and/or chemical process occurring on a component of the facility, such as wear, fatigue, and corrosion. The change in the properties of the facility due to deterioration is reflected in the facility model. Based on 'the deteriorated facility model,' the behavior of the facility is evaluated again. The results are assessed in reference to required functions and potential failure modes are predicted. At the same time, the stress exerted on the components is estimated based on the newly evaluated behavior. By repeating the afore mentioned procedure, the change in the facility as the operation continues can be simulated.

3. FACILITY MODEL

The facility model for life cycle management system should represent the basic information of the facility such as parts, assemblies and its hierarchical structure. It should also have the capability of representing various additional information, such as design intention and information related to maintenance. As a base of the model structure, we have adopted the assembly structure of the facility (Sodhi, 1994, Sugimura, 1994). We used an object oriented data structure as shown in Figure 3 (Takata, 1995).

The model consists of *assembly items* and *assembly relations* between the *items*. Notations in the figure are based on EXPRESS-G (NN, 1992) where an attribute is shown as a thin line with its name on it and marked with a circle at the value end. A thick line shows a class hierarchical relation, with a circle attached to the sub-class side.

An *assembly item* represents a physical substance in the facility. Assembly items are classified into *part* and *form feature*. A *part* is an individual physical substance. We consider an assembly a kind of *part* that can be divided into multiple *part*s. This concept allows us to represent the hierarchical structure of the facility in a flexible manner.

Form feature is a group of geometric elements that carries some functions or behavior. Note that it does not necessarily belong to a single *part*. An assembly feature is a *form feature* that mates with another *form feature* of a different *part* or sub-assembly to make an *assembly*, such as holes/pins and grooves/extrusions. For allowing users to focus their attention on any level of hierarchy, neglecting the lower levels of assembly structure, we provide the access function for the model to get *form features* from *assembly items* at any level. If the *item* is an *assembly*,

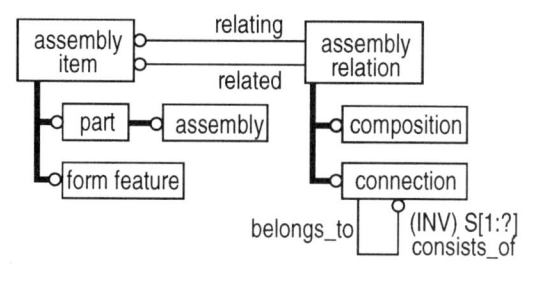

Figure 3 Structure of the facility model

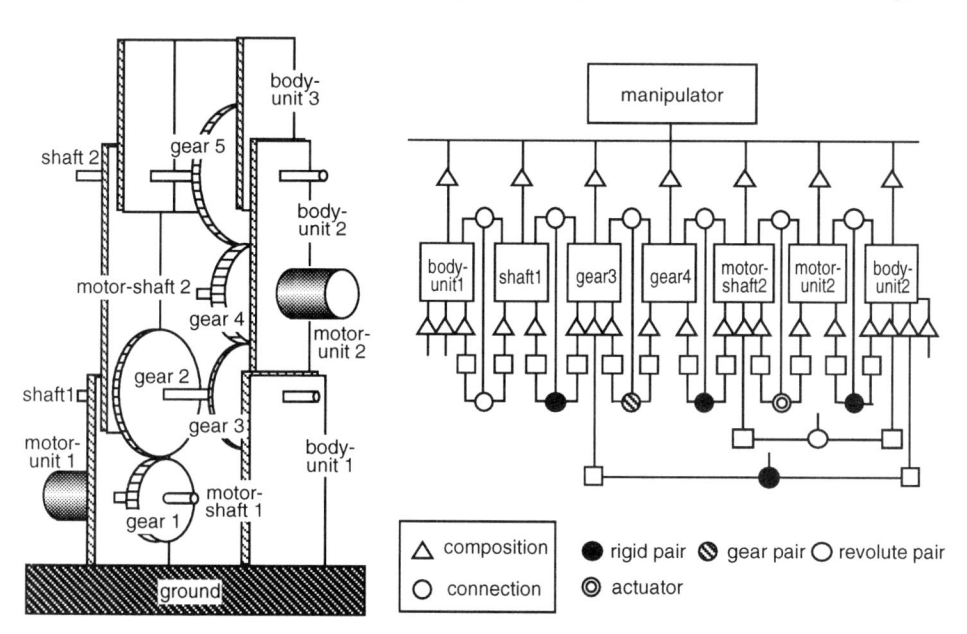

Figure 4 Robot manipulator

Figure 5 Model representation of robot manipulator

form features of the *part*s that are mated with those outside the *assembly,* are identified by the function.

Two types of *assembly relations* are identified to represent assembly structures. *Connection* is an *assembly relation* between two *item*s that have no inclusion relations with each other, e.g., a *part* to a *part*, an assembly feature to an assembly feature. As hierarchical information, *connection* has a pointer to the *connection*s in which it participates. For example, a *connection* between two parts may consist of several *connection*s between assembly features. They have a pointer to the parent *part*-to-*part connection*.

Composition is an *assembly relation* between an *assembly item* and another *assembly item* that consists of it. *Composition* has a transformation that represents the position and orientation of the child *item* in the coordinate space of the parent *item*.

Both *assembly item*s and *assembly relation*s can have technical information in addition to configuration structures. *Connection*, for example, contains technical information that represents the type of the *connection*, such as a sliding pair and a revolute pair. This is effective in performing various simulations and analysis based on the model.

As an example of the facility model, a model of a robot manipulator shown in Figure 4 is represented in Figure 5. Note that Figure 5 is an instance-level diagram. Attribute names have been omitted. *Connection*s and *composition*s are represented as circles and triangles. The entire model is not represented because of space limitations. The hierarchy of *assembly item*s are arranged vertically.

4. MODEL BASED DETERIORATION EVALUATION

4.1 Modeling of the deterioration processes (Takata, 1994)

Mechanisms, such as fatigue, wear, and corrosion, which induce deterioration at certain areas of parts or assemblies are called deterioration mechanisms. The resultant deteriorated states are distinguished by deterioration modes. The deterioration mechanism is caused by a certain set of conditions which we call causal factors. They are classified into four categories: 1) Inherent characteristics such as geometry, material and surface finish. 2) Exerted stress such as mechanical stress, thermal effects and electro-magnetic effects. 3) Relative motion. 4) Operating environment such as in a gas, in a liquid, or in particles.

Although there seems to be an infinite number of phenomena recognized as deterioration of the facility, we can identify a certain set of deterioration mechanisms which are basic and common for many types of facilities (Dasgupta, 1991). We call them fundamental deterioration mechanisms.

In many cases a chain of multiple fundamental deterioration mechanisms are related to failure. For example, fatigue failure could be initiated by a notch created by corrosion. In this way, one of the causal factors of a deterioration mechanism could be provided by other deterioration mechanisms. There is also a case where some of the causal factors are provided by mechanisms other than deterioration mechanisms, which we call causal factor formation mechanisms. An example of this type of mechanism can be seen when the rotation of a shaft with a radial load creates cyclic stresses which lead to fatigue at a stepped part of the shaft. The chain of deterioration mechanisms and causal factor formation mechanisms is termed a deterioration process. Figure 6, for example, schematically represents the deterioration process of fatigue for a spindle.

4.2 Method of deterioration evaluation

Figure 6 Example of the deterioration process

The qualitative deterioration evaluation is based on the facility model and a deterioration data base. The result is represented in terms of potential deterioration processes which may occur at particular parts of the facility (Takata, 1994, 1995).

Deterioration data base
The deterioration data base contains the fundamental deterioration mechanisms and the causal factor formation mechanisms. They are expressed in terms of a set of causal factors and the resultant deterioration modes or causal factors. This data base can be prepared independently from individual facilities.

Facility model
The structure and properties of a facility are retrieved from the facility model. The properties are represented in terms of values of attributes defined in assembly items and relations. The model has a mechanism to inherit the values of attributes from those of higher level items in the hierarchy, unless they are explicitly defined at their level.

Inference of deterioration process
Inference is conducted for every form feature of every part defined in the facility model, since different form features have different causal factors even in the same parts. Figure 7 illustrates the inference

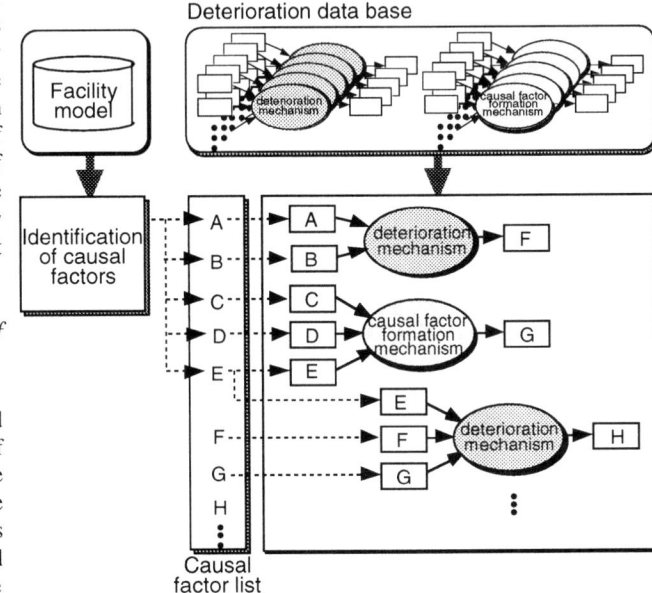

Figure 7 Inference mechanism of deterioration processes

mechanism. The causal factors associated with a particular form feature are identified from the facility model and placed in a causal factor list. They are selected from the attributes of the form feature itself, the related form feature connection and the parent parts. Then, mechanisms whose causal factors match with those in the casual factor list are searched in the deterioration data base. The output of the selected mechanisms are appended to the casual factor list and the other mechanisms are searched again taking the newly appended causal factors into account. Finally, the deterioration process(es) is (are) formed by the selected mechanisms.

Inference of range of effects of deterioration
In the above procedure, the effects of causal factors and deterioration modes are considered within the specified form feature. However, there are causal factors and deterioration modes which also have effects on other form features or parts. The following three cases were identified: 1) Some types of deterioration, such as loose bolt fastenings, have ranges of influence involving multiple parts or part connections. 2) Effects of causal factors, such as vibration and heat conduction, propagate from one part to another. 3) Effects of deterioration change environmental causal factors of other form features or parts. For example, leakage of coolant due to a damaged seal makes a bearing wet. In the system, the first and the second cases are dealt with. Inference algorithms were developed for the first case. For the second case, the propagation rules for the corresponding causal factors were defined.

5 IDENTIFICATION OF MOVABLE PARTS

To prepare appropriate methods to maintain a facility, it is necessary to predict what will happen to the behaviors of a facility when some of its parts fail. When we perform kinematic analysis of facility with failed parts for that purpose, a nominal kinematic model developed at the design phase may not be useful. For example, small parts such as bolts may not be explicitly represented in the model but those small parts could deteriorate and break, which may lead to the facility's malfunction. On the other hand, it is not economical nor realistic to analyze the whole facility model because it includes numbers of parts that may not contribute to the motion. So we should first identify the parts that can participate in the motion under the specific situation. In the following, we describe how to identify parts that are able to move using the facility model with information on failure. The method is based on kinematics and no friction is assumed.

The strategy is, starting from the actuator where the motion is generated, to repeat calculating the transmission of the motion by following the connection of parts in the facility model. If the motion is reached to the fixed end that does not allow the transmitted motion, the original motion is found unfeasible. Otherwise all the parts on the path from the actuator to the end will move and should participate in the kinematic model on which more detailed analysis (such as by Kramer (1992) or Haug (1989)) would be made later.

Transmission of motion
In our system, two types of motion are represented: translation and rotation. Translation is represented by a vector of its velocity. Rotation is represented by the center of rotation and a vector of the angular velocity.

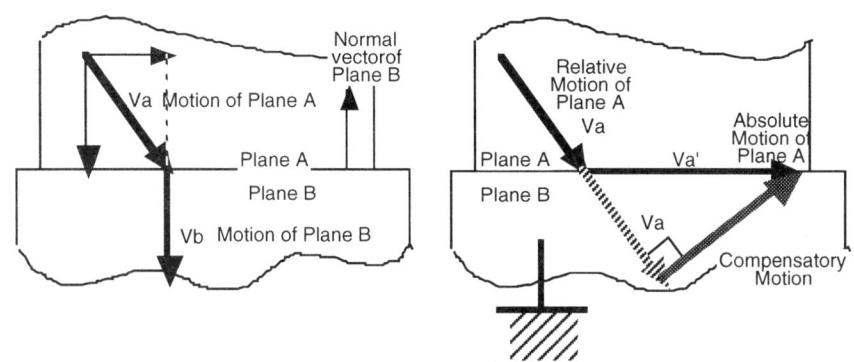

Figure 8 Transmission of translational motion for planar pairs

Figures 9 Compensatory motion for planar pair

Motion of an object is transmitted to another object through their physical contact. Transmission differs depending what type of connection is formed by the contact. In kinematics such connection of objects is called a kinematic pair (Duffy, 1980, NN, 1995). In our facility model, kinematic pair is represented by *connections* between a pair of features in contact. We currently have implemented three types of kinematic pairs that include a planar pair, a gear pair and a revolute pair.

For instance, in planar pairs, the translational motion of the input feature A with velocity Va is transmitted to the output feature B as shown in Figure 8. As no friction is assumed, the transmitted motion of feature B has the velocity Vb, that is the components of Va perpendicular to the contact plane. When the motion of feature A is rotational motion the transmitted motion of feature B is the same rotational motion. Similarly, we have analyzed and implemented the transmission of gear pairs and revolute pairs.

Propagation of motion
The motion is generated at the actuator, which is the single source of motion represented by a feature relationship, as relative motion between two features that compose the actuator. The motion of the feature is transmitted to the part it belongs to. It is transmitted from the part to all its features. Then, the motion of each feature is transmitted to its connecting feature as described above. Thus the motion is propagated to all the related parts by repeating this process.

A series of parts and features generated by this process is called a path of transmission. As a part has multiple features, paths have branches to make a tree with the actuator as its root. When the motion is not allowed at any point in the tree, the motion is unfeasible. The tree is called fixed in which all the parts are not movable. When the different motions from different trees coincide at a point, motions are also considered to be fixed.

The possibility of motion of the parts is checked by the following procedure. First, one feature of the actuator is assumed fixed and the motion is transmitted to the other feature. It is propagated as described above. Next, the second feature is assumed fixed and the motion is transmitted to the first feature. If one of the trees generated is found free and the other is found fixed, the motion is feasible. If both trees are found free, the motion is also feasible. However, because this means the mechanism is 'floating,' absolute motion against the world coordinate system cannot be decided. If both trees are found fixed, the motion is not feasible.

Compensation of motion

Even if the motion is found unfeasible using the above algorithm, there is a possibility that the actuator itself moves relative to the fixed part in the world coordinate system. The algorithm to find the possibility is as follows:

Where the tree of transmission reaches a fixed part, it calculates compensatory motion that complies with the fixed part and generates possible absolute motion from the transmitted motion relative to the actuator. For example, compensatory motion for a planar pair is calculated as shown in Figure 9. The relative motion of plane A is transmitted to plane B, but plane B is fixed to the ground. The only possible absolute motion of plane A is sliding along the contact surface. To make the absolute motion of plane A along the contact surface, the difference between the relative motion and its absolute motion should have a vector shown as compensatory motion in the figure. It would generate absolute motion Va' along the contact surface, which realizes both the relative input motion of A and fixed plane B. So if it is possible that plane A has the absolute motion that is the sum of the input relative motion and this compensatory motion, the motion becomes feasible. To ensure this, our algorithm transmits and propagates the compensatory motion backwards from the fixed point to the actuator and checks to see if it is feasible. For a gear pair, compensatory motion generates planetary motion around the fixed gear feature.

Though it has fairly good functions, our method has some limitations. First, it analyzes instantaneous motion and derives movable parts at a specific instance where all the information of the facility is known. To follow the behavior of the facility for some period of time, we need precise kinematics analysis with some method to handle changes in the situation.

Representation of motion with a single value brings some difficulties for handling a mechanism with loops, or generating compensatory motion. For resolving these problems, we consider that a method will be necessary to represent ranges of values and to make calculation on them. Although our method has these limitations, it derives rather easily and intuitively, a kinematic model that is sufficiently correct.

6 EXPERIMENTAL SYSTEM

The experimental system has been developed by using the object oriented expert shell G2. Elements of the facility model are defined as object classes. The deterioration mechanisms and the causal factor formation mechanism are also described as objects. The inference of deterioration is executed by production rules. The identification of movable parts is also executed by rules and procedures attached to objects.

As an example, a grinding robot whose structure is shown in Figure 4 was analyzed. First, deterioration evaluation was performed. Figure 10 shows the display of the system when potential deterioration of a bearing between body-unit 1 and shaft 1 was evaluated. The lower right part of the figure shows a corresponding part of the facility model of the robot. The upper left part of the figure indicates prediction of a potential deterioration process. The process

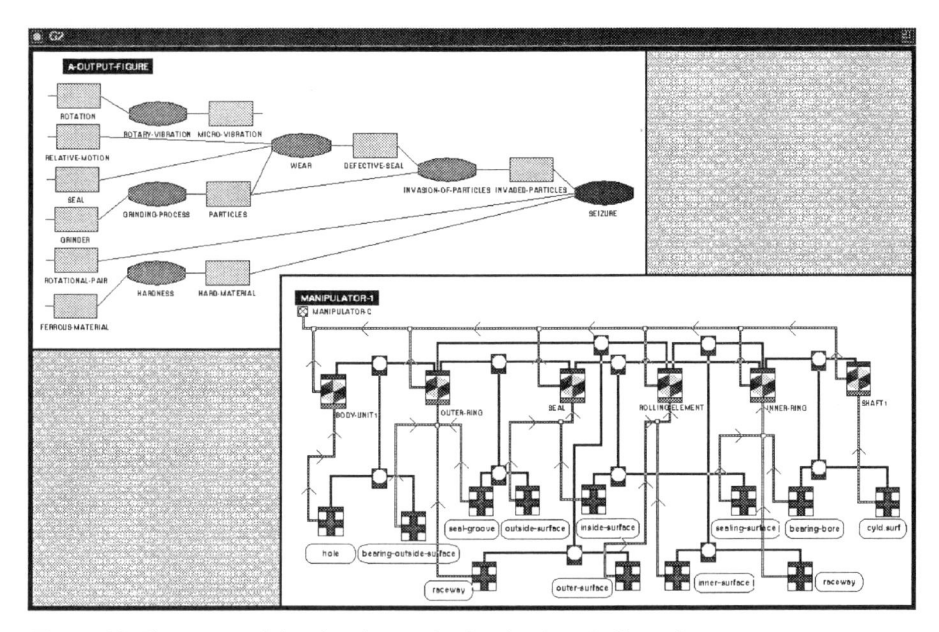

Figure 10 The output of deterioration evaluation for the grinding robot

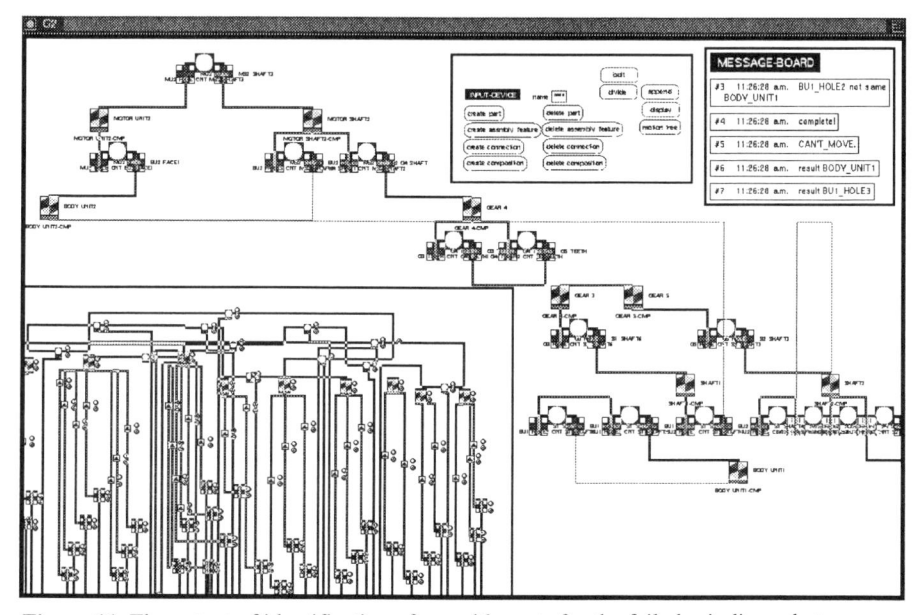

Figure 11 The output of identification of movable parts for the failed grinding robot.

shows that the seal of the bearing may be worn by grinding swarf. This leads to the invasion of grinding swarf into the bearing which may cause the seizure of the bearing.

Based on the result from deterioration evaluation, the connection of the features of body-unit 1 and shaft 1 was changed to be rigid to show the shaft got stuck. Identification of movable parts was performed based on the facility model with this information on failure. The upper part of Figure 11 shows the tree of identified movable parts. The mechanism is originally designed so that motor-unit 2 controls body-unit 3 independent of the motion of body-unit 2. With the failure of the shaft sticking, the motion of motor-unit 2 was found to generate compensatory motion and to bring the motion of body-unit 2 as well as that of body-unit 3.

7 CONCLUSION

For realizing proactive maintenance strategies, we propose the concept of life cycle simulation to evaluate the change in the facility as it works. We adopted model based approach, and proposed the model structure of the facility.

Based on the facility model, we have developed two modules of life cycle simulation system. One is the deterioration evaluation system that infers possible deterioration of facility components. The inference is performed by combining the information derived from the facility model with the deterioration mechanism and causal factor formation mechanisms which are stored in a deterioration data base.

The other module is the movable parts identification system which finds the parts participating in the motion when there is a failure in the facility. The system searches the facility model simulating the transmission of motion and generates a mechanism tree that will be used for precise kinematic analysis.

Currently we are trying to integrate these two modules with other modules necessary for completing the life cycle simulation such as the stress evaluation module and the failure evaluation module shown in Figure 2.

ACKNOWLEDGMENTS

This research is supported by the research program conducted in RIKEN (The Institute of Physical and Chemical Research) which is a part of Cross-over Research Program for Nuclear Base Technology promoted by Science and Technology Agency, Japan.

REFERENCES

Dasgupta, A., Pecht, M., (1991) Material Failure Mechanisms and Damage Models, *IEEE Trans. on Reliability*, 40, 5, 531-536

Duffy, J. (1980) *Analysis of Mechanism and Robot Manipulators*, Edward Arnold, London.

Haug, E.J. (1989) *Computer-Aided Kinematics and Dynamics of Mechanical Systems, Volume I: Basic Methods*, Allyn and Bacon, Mass.

Kramer, G.A. (1992) *Solving Geometric Constraint Systems - A Case Study in Kinematics -*, The MIT Press, Cambridge, Mass.

N.N.(1995) ISO 10303 Industrial automation systems and integration- Product data representation and exchange - Part 105: Integrated application resources: Kinematics.

N.N. (1992) ISO CD10303-11 Product Data Representation and Exchange - Part 11, The EXPRESS Language Reference Manual, ISO TC184/SC4/N151.

Sodhi, R., Turner, J.U. (1994) Towards Modelling of Assemblies for Product Design, *CAD*, 26, 2, 85-97.

Sugimura, N., et al. (1994) A Study on Product Model for Design and Analysis of Mechanical Assemblies Based on STEP, *Japan-USA Symp. of Flexible Automation*, 495-498

Taga, H. (1993) Extraction of Movable Parts for the Maintenance of Machines, Master Thesis, Chuo University (in Japanese).

Takata, S., et al.(1994) Model Based Evaluation of Component Deterioration, *RECY'94*, 168-180.

Takata, S., Hiraoka, H., Asama, H., Yamaoka, N., Saito, D. (1995) Facility Model for Life Cycle Maintenance System, *Annals of CIRP*, Vol. 44/1, 117-121.

36

Methods for Continual Improvement of Products and Processes

Prof. Dr. R. Züst
Dipl. Ing. ETH G. Caduff
Institute for Industrial Engineering and Management, ETH Zürich
Zürichbergstr. 18, CH-8028 Zürich, Switzerland
Phone: 0041-1-632 05 50, Fax 0041-1-632 10 45
e-mail: zuest@bwi.bepr.ethz.ch

Abstract
In businesses today, questions concerning the environment engross increasingly more employers and employees, in addition to stakeholders. The question arises: How could ecological as well as economical solutions become more effective and efficient?
The proposed paper will present methods to continual improvement to increasing ecological efficiency and ecological effectiveness.

Keywords
Eco-performance, environmental information system, environmental management system, environmental performance evaluation (EPE), environmental performance indicators (EPI), ISO 14000 ff, life cycle assessment (LCA), sustainable product and process design

1 INTRODUCTION

Activities for avoidance and reduction of damage to the environment are limited at the moment mainly to the operational sphere, where final-stage measures are generally taken in response to official restrictions. Acting only when the level of environmental damage becomes too great is as a rule costly and not very effective. Individual measures required by law in the process stage have, however, led to significant improvements in energy efficiency and in emission reduction (Meadows, 1992). Progress has also been made in minimizing damaging by-products.

Today the main emphasis in "integrated environmental safeguarding" is on "*eco-performance*". Eco-performance signifies effective and efficient avoidance and reducing of operational effects damaging to the environment with simultaneous regard for economic aspects. Here the following measures (among others) are required, and are dealt with in this article:

- the setting up of an "environmental management system" for the successful deployment of environmentally-oriented goals and strategies;
- deployment of "sustained product and process creation" integrating ecological aspects;
- realization of an "environmental information system" as the basis of an effective management instrument;
- quantification of eco-performance for the management process and for external communication.

2   SETTING UP AN ENVIRONMENTAL MANAGEMENT SYSTEM

Industrial enterprises are making increasing use of voluntary self-check mechanisms. For this reason a model for voluntary self-checking ("Environmental Auditing") was developed by the International Chamber of Commerce (ICC) about ten years ago. With this proposal for voluntary environmental self-checking by industry all areas of the enterprise, from management through to construction and production, are activated to safeguard the environment via clear regulatory guidelines.
The concept of the ICC "Voluntary Self-Regulating by the Economy" is based upon two assumptions (see ICC Environmental Auditing, document 2, p. 63):
- self-regulation is, when correctly employed, more effective than abidance by rules and regulations;
- an excessive increase in legislation and official regulations has a counterproductive effect. Legislation also goes out of date very quickly and does not always cover all cases.

The EC directive No. 1836/93 concerning the "voluntary participation of commercial and industrial businesses in a common system for environmental management and environmental audits" is based largely upon ICC principles. It calls additionally for checks on company audit management by external parties and the publication of the results.
Industrial firms are confronted with many diverse requirements in connection with "eco-performance", as seen for example in EC directive No. 1836/93 or the future Standard ISO 14000 ff. "Environmental Management". The significance for industry is on a par with that of Standard ISO 9000 ff. "Quality Management": fulfilment turns into competitive advantage for companies.

Standards in the area of environmental management take into consideration the principle of "Continual Improvement", whereby permanent improvement in eco-performance is to be reached. In ISO 14000 ff. this principle is articulated in the procedural steps "Environmental Policy", "Planning", Implementation and Operation" and "Checking and Corrective Action".
Examples in practice have shown that for the successful deployment of environmentally-oriented goals and strategies an *overall concept for all areas of the enterprise* is necessary. This concept should, among other things, include the following points (Bleicher, 1992):
- structure (model, building-up and running organisation, organisational processes);
- activities (environmental policy, environmental program, assignments);
- behavior (company culture, behavior in confrontation with problems, performance and cooperative behavior).

The real strengths of an integrated and holistic environmental management system lies, however, in the sum of the achievable progress overall. Many individual successes taken together bring many ecological and economic benefits.

3   SUSTAINED PRODUCT AND PROCESS CREATION

The product must be observed from conception through to disposal. It is useful to divide this period into product life phases. See Figure 1.

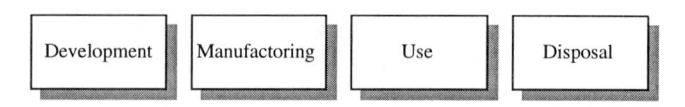

Figure 1   Product life phases

Product development must not be regarded in isolation from its sustainability. Recyclable packaging (eg. PET bottles) is only feasible if for example collection possibilities (recycling centers, collection cycles etc.) are taken into account in the concept phase and included in product design and the manufacturing process. A "product" becomes a "product system" (Züst, 1995a). Today too little attention is still paid to this fact.

In 1991/92 we investigated the disposal of small electric appliances in small/medium quantities, in the area of capital goods. The results indicate that expert disposal/recycling of existing product designs not adapted for recycling very quickly generates 30% - 200% of manufacturing costs. Dissassembly and handling of components and parts take time, and this generates high personnel outlay. The development phase represents the most effective area for pinpointing/steering. The share of costs fixed during development rises to over 90% over all phases in the product's life. See Figure 2.

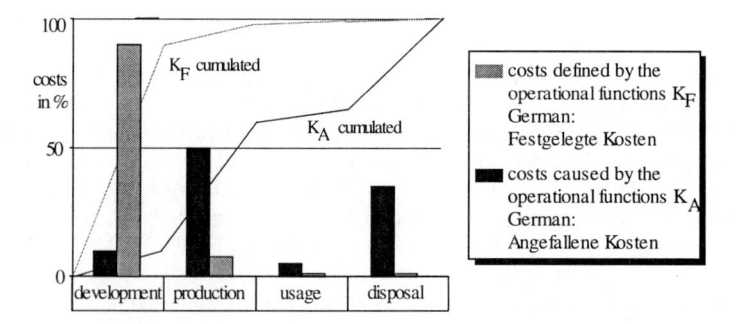

Figure 2   Fixed and accrued costs from the point of view of the manufacturer (Züst, 1992 and 1995a)

Because in practice the financial and environmental costs of manufacture and expert disposal of new products are very difficult to estimate, additional measures in the areas of organisation (process-oriented structures, teamwork etc.), planning and decision-making aids (environmental handbook, construction and disposal guidelines etc.) and company culture must be taken. In addition, the following questions must be answered:
- How can ecological aspects be taken into account in the development phase?
- What information is necessary in order to collate and evaluate ecological effects?
- How must this information be prepared in order that it be deployed by the respective user?

Figure 3 illustrates one possible answer to these questions using an example of hardware development from the electronics industry.

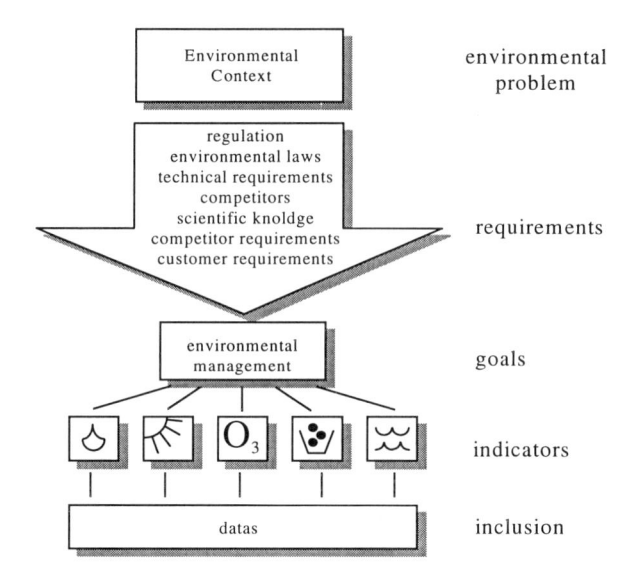

Figure 3   The integration of ecological aspects in hardware development

4 REALISATION OF AN ENVIRONMENTAL INFORMATION SYSTEM

Although various findings and aids are well-known, "eco-performance" has in practice only partially penetrated industry. Experiences in companies and in research indicate that, firstly, the area of management and organisation is very important and that, secondly, the product designer must be granted access to additional environmentally-relevant information via an environmental information system.

A great deal of ecologically-oriented data is already available in some firms. The purchasing department knows how many materials have been purchased and the environmental officer possesses information concerning the amount disposed of and concerning toxicity. Further, operational plans, piece lists, dispensing or dangerous materials lists exist which contain ecologically-relevant data. Nevertheless, ecologically-oriented information, as for example operational product and process balances, can at the moment only be collected with a great deal of effort. This is for two reasons:

- The dimensions of the materials flow: materials used (unfinished, parts, component groups, etc.) are often given in "number of pieces" or "lengths". A company balance sheet containing flow of mass and energy cannot be put together directly using information in this form.
- Procedures and aids: in order that ecologically-relevant aspects be taken into account in the operative decision-making process, information must be correspondingly collated, prepared and transmitted. Suitable procedures and aids, integrated into existing processes and planning tools, hardly exist as yet.

Not only the mere collection of but the adaptation of ecologically-relevant data for decision-making has an important function. Here, however, as with "cost-coverer's expense" or the "process-cost expense", methods need to be developed.

Via the networking of the environmental information system with the (existing) company information system, environmentally-relevant information can be made available for use in all planning and decision-making contexts. There are two basic possibilities:

- *Technical integration*: ecology-oriented information, as for example "recyclability", "damage to the environment up to now", etc., are made available. It is left up to employees to apply this additional information in the development of new products and manufacturing processes.
- *Organisational integration*: in a teamwork context the necessary information can be, for example, made available via consultation with various experts.

In practice a combination of both technical and organisational integration has proved useful. The advantage of organisational integration is that there development of instruments for adaptation and communication takes less time and trouble than in technical integration. With technical integration the availability of information is, however, much greater.

"Company modelling", ie. data and function modelling, has shown itself to be a suitable aid in the setting up of an environmental information system (Plötz, 1995).

How far environmental information systems should be computerized depends on the extent of computerization in a firm. It has been noted, however, that EDP-supported systems make enduring implementation of an environmental information system easier. The possibilities for rapid data retrieval are a great advantage.

5 QUANTIFICATION OF ECO-PERFORMANCE

5.1 Application of Eco-Performance

Eco-performance evaluation has two objectives (Züst, 1995b):
- Eco-performance should form the basis of a goal-oriented management aid. With it the process of "Continual Improvement" as called for as an environmental standard will be fostered.
- Eco-performance should also form the basis for an effective medium of communication to the public. It can be a means of influencing stakeholders (shareholders, banks, official bodies, employees etc.) positively.

Eco-performance in the management process:
"Integrated environmental safeguarding" requires long-term thinking. From the point of view of management, which comprises the activities "planning", "decision-making", "ordering" and "checking", this signifies above all intensive "planning" and differentiated "decision-making". The informational aspect is of even greater importance.

Eco-performance in communication with the public:
Economy - and the connected aspects profitability and liquidity - is central to all companies. It is and remains one of the most important criteria for evaluation. In addition a company must function in a legal manner, which involves abidance by environmental legislation. Figure 4.

Only believable and comprehensible publicizing of a company's environmental activities, plus continual improvement in eco-performance, can ensure the legitimacy and popularity of an enterprise and its products and services.

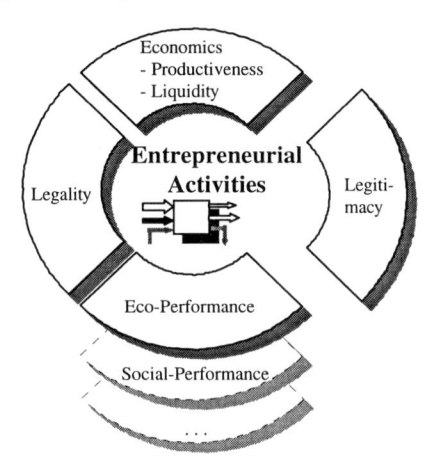

Eco-performance will, firstly, influence positively the legitimacy of a firm and its products and services. Secondly, its optimization of resource deployment will save money. Both of these aspects can have a direct effect upon economy.

Figure 4   Effects of company activities (Züst, 1995b), following (Frei, 1995)

5.2 Environmental Performance Evaluation

If a company commits itself to "Continual Improvement", as intended for example in Standard ISO 14000 ff., the quantifying of eco-performance takes on central significance. Standard ISO 14031 "Environmental Performance Evaluation" illustrates the targeted collection, analysis, evaluation and writing-up of damaging environmental effects of company activities.

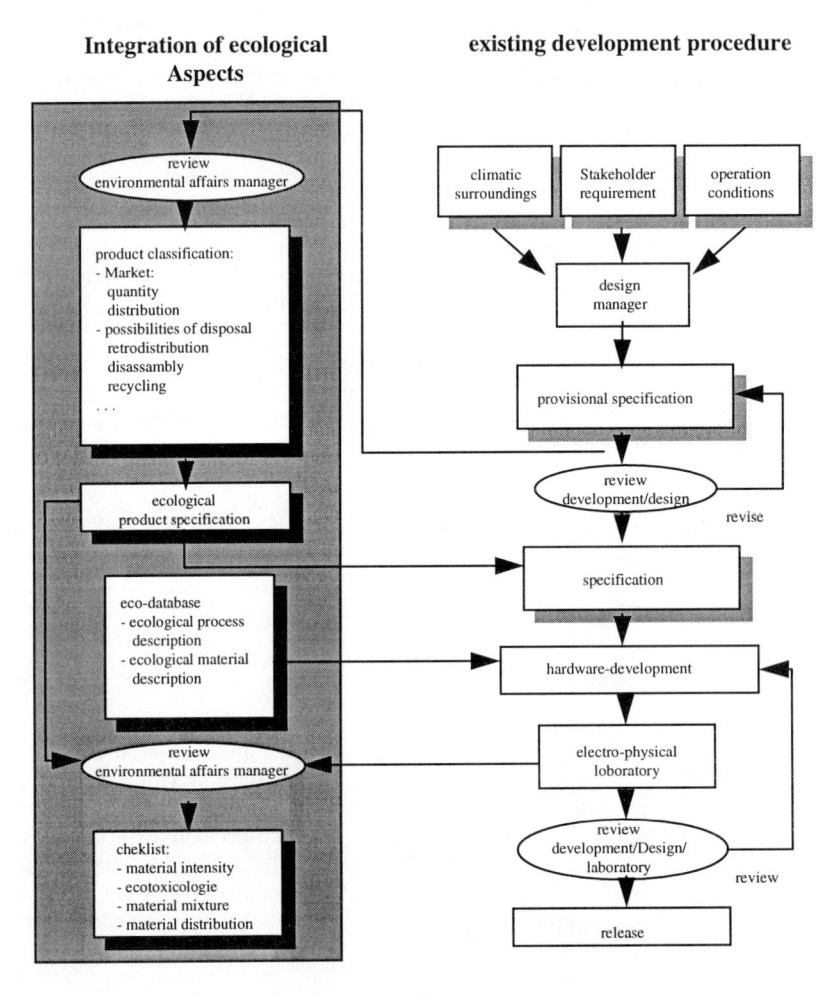

Figure 5 Process of determining relevant indicators

It is not always possible to determine a standard of measurement whereby an effect can be represented with sufficient precision. It is often possible, however, to identify standards in the sense of indicators which are suitable for a workable formulation (Büchel, 1995). For this reason eco-performance is often expressed in terms of "Environmental Performance Indicators".

The basis for the establishment of Environmental Performance Indicators is the recording of energy and materials flow. Figure 5 illustrates the process of determination of relevant Environmental Performance Indicators. Using the example of the environmental problem of smog, NOx emissions on the global level and yearly consumption of heating oil on the company level can, for example, be designated as indicators.

5.3 Life-Cycle Assessment

For the workable and successful application of an ecological evaluation process the following general requirements are necessary (Ahbe, 1995):
- clear ecological statements in the evaluation of products and manufacturing alternatives;
- quantitive comparability of products and processes;
- illustration of significant ecological effects as support to decision-making;
- workable integration of the process into existing decision-making channels;
- comprehensibility;
- the possibility of both company-internal and company-external application of results.

The most important *company-internal* requirements are as follows (Ahbe, 1995):
- user neutrality, ie. information and statements must not depend on the user of the evaluation process;
- full aggregation of individual environmental stresses into the total picture of environmental damage.

In the context of these requirements (Ahbe, 1995) the "Ecological Scarcity Method" (Ahbe, 1990) has proved itself to be the most suitable compromise.

The application of the "Ecological Scarcity Method" is therefore described below, using an example from practice: the comparison of two manufacturing processes (Ahbe, 1995).

A metals plant manufactures aluminium containers. Up until now these have been produced according to workshop principles. Because of increased sales capacity is no longer sufficient, and the alternative of implementing automated assembly-line production arises.

The first step involves delimitation of the study and its goals, the quantification of causes of damage and the evaluation of environmental damage.

The total environmental damage generated by the old process is illustrated in Figure 6, for the new in Figure 7. Figure 8 provides a comparison.

Standard-belastungs-träger	Ein-heit	Basisdaten (in UBP pro Einheit)				Menge	Umweltbelastung (in UBP pro 1000 Teile)			
		Luft	Wasser	Energie	Abfall		Luft	Wasser	Energie	Abfall
Polystyrol	Kg	1138,6	0,00	56,2	66,0	0,34	387,1	0,0	19,1	22,4
Papier "SKE"	Kg	423,0	10,70	25,2	49,9	0,43	181,9	4,6	10,8	21,5
Karton "GC"	Kg	532,9	66,60	31,0	58,2	0,73	389,0	48,6	22,6	42,5
El. Energie	KWh	155,9	0,01	9,5	10,9	554,64	86468,4	5,5	5269,1	6045,6
Per-Entfetten	m2	1188,8	0,00	0,0	0,0	87,00	103425,6	0,0	0,0	0,0
Lösemittel	Kg	14300,0	0,00	35,9	0,0	0,79	11297,0	0,0	28,4	0,0
Erdgas	MJ	5,5	0,00	1,0	0,0	25,76	141,7	0,0	25,8	0,0
						Summen	202290,7	58,8	5375,8	6132,0
						Total				213.857,19

Figure 6 Environmental damage generated by the old process (per 1000 pieces) (Ahbe, 1995)

Standard-belastungs-träger	Ein-heit	Basisdaten (in UBP pro Einheit)				Menge	Umweltbelastung (in UBP pro 1000 Teile)			
		Luft	Wasser	Energie	Abfall		Luft	Wasser	Energie	Abfall
Papier "KS"	Kg	631,4	80,23	38,80	66,38	0,30	189,4	24,1	11,6	19,9
El. Energie	KWh	155,9	0,01	9,50	10,90	159,44	24856,7	1,6	1514,7	1737,9
Wä. Entfetten	m2	1,8	3,09	-0,02	0,01	87,00	156,6	268,8	-1,7	1,0
Lösemittel	Kg	14300,0	0,00	35,90	0,00	0,20	2860,0	0,0	7,2	0,0
						Summen	28062,7	294,5	1531,8	1758,8
						Total				31.647,78

Figure 7 Environmental damage generated by the new process (per 1000 pieces) (Ahbe, 1995)

Because transport materials are no longer necessary and because of process improvements the new process is seen to generate fewer causes of damage.

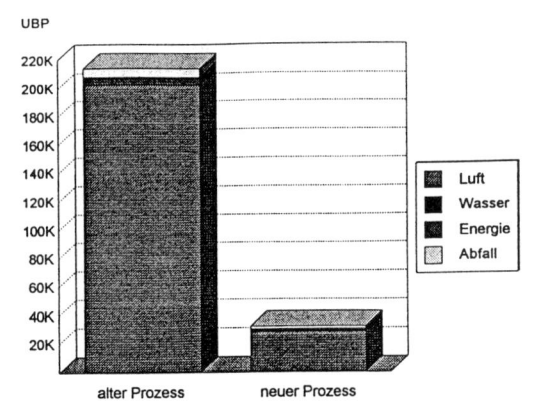

Figure 8   Process comparison (per 1000 pieces) (Ahbe, 1995)

The relatively large ecological difference between the two processes has three main causes:
- The new process requires considerably less electricity due to continuous running of production.
- Modern oven construction allows lower levels of heat loss.
- The new process creates much lower emissions of volatile organic compounds (VOC) due to use of water-soluble varnish and a metals de-greasing plant which uses water.

The final step of the ecological evaluation comprises the search for possibilities for improvement.

The example used shows that the aggregation of individual data into a total picture of environmental damage is possible and operationally feasible. For company decision-making the clarity of the results is important (Ahbe, 1995).

5.4   Distinguishing between "Environmental Performance Evaluation" and "Life Cycle Assessment"

Environmental effects are not only described via Environmental Performance Evaluation (ISO 14031: EPE) but also via Life Cycle Assessment (ISO 14040 ff.: LCA). The difference lies firstly in mode of observation. While EPE concentrates on the environmental effects of company activities, LCA describes the environmental effects generated by a product during its entire life cycle. Secondly, EPE and LCA have different objectives. While EPE concentrates on increasing eco-performance within the firm, LCA is a management tool for ecological evaluation of the product range. LCA can also become a basis for consumer purchasing decisions.

EPE and LCA are hierarchically linked. The "Oslo Paradigm", determined in June 1995 by Swiss ISO 14000 experts, illustrates this relationship and the corresponding conclusions:

- Basis: EPE and LCA describe environmental effects.
- Specifications: EPE describes environmental effects in the context of organization decision-making.
  LCA describes the environmental effects generated by a product during its entire life cycle.
- Conclusions: The LCA consists of an aggregation of the individual EPEs of the relevant organisation.
  EPE and LCA must therefore employ the same data structure. Data investigation also requires identical procedures.
  The same terminology must therefore be used for both EPE and LCA.

It is hoped that the above conclusions will be taken into account in the ongoing standardization process.

6. CONCLUSIONS

The conception and realization of an environmental management system as a component of integrated management will be of great importance in the future. The realm of management, ie. planning, decision-making, ordering and checking, plays an important role in its rendering.

Individual phases in product life with their various processes and environmental effects must be taken into account in the development phase. From an originally isolated product, therefore, a "product system" to be shaped emerges.

Ecologically-relevant information must be made available to the product designer. In order that the latter not be additionally burdened thereby technical as well as organisational integration of this information is useful. "Company modelling" - ie. data and function modelling - has proved to be a significant and suitable help in the setting up of an environmental information system.

EPE and LCA, as they are termed in Standards ISO 14031 and ISO 14040ff, must employ the same data structure. Data investigation requires identical procedures. Ecological evaluation procedures must, from the point of view of industrial usability, be workable; they must also be user-independent. The quantification of eco-performance provides for effective management and communication with the public.

7 REFERENCES

Ahbe, S., Braunschweig, A., Müller-Wenk, R. (1990) Methodik für Ökobilanzen auf der Basis ökologischer Optimierung: Schriftenreihe Umwelt Nr. 133. Bern: BUWAL.

Ahbe, S. (1995) Ökologische Bewertung als Instrument bei der Produkt- und Prozessplanung. Zürich: ETH-Diss. Nr.11'214.

Bleicher, K. (1992) Konzept Integriertes Management. 2.Auflage, Frankfurt/M., New York: Campus Verlag.

Büchel, A., Züst, R. (1995) Systems Engineering. Autographie. Zürich: BWI der ETH Zürich.

Frei, B. (1995) Integration des Umweltschutzes in die unternehmerische Wirtschaftlichkeitsbeurteilung. Zürich: ETH-Diss. Nr.11'234.

Meadows, D.; u.a. (1992) Die neuen Grenzen des Wachstums. Stuttgart: Deutsche Verlags-Anstalt.

Plötz, A. (1995) Die Ökologische Dimension der Unternehmensmodellierung. Zürich: ETH-Diss. Nr. 11'069.

Speerli, F. (1995) Unternehmens-Umwelt-Management-System. Zürich: ETH-Diss. Nr. 11'065.

Züst, R.; Wagner, R. (1992) Approach to the Identification and Quantification of Environmental Effects during Product Life: CIRP Annals 41/1/1992. Bern: Hallwag-Verlag.

Züst, R.; Plötz, A.; Caduff, G. (1995) Consideration of Environmental Aspects in Product Design. 10th International Conference on Engineering Design, Praha. Zürich: Heurista-Verlag.

Züst, R. (1995) Umwelt-Management. Autographie. Zürich: BWI der ETH Zürich.

8 BIOGRAPHY

Prof. Dr. R. Züst;
1986 Dipl. Masch. Ing. ETH; 1990 Doktorat in the aera of 'Computer Aided Process-Planning'; since 1990 leader of a research-group (topic: Environmental Compatible Product and Process Design); since 1993 Professor for 'Industrial Engineering and Management'.

Dipl. Ing. ETH G. Caduff;
1994 Dipl. Betriebs- und Produktionsingenieur ETH; since 1994 researcher and Ph.-D.-Student.

37

Development of Environmentally Friendly Products -Methods, Material and Instruments

H. Birkhofer, H. Schott
Departement of Machine Elements and Engineering Design
Technical University of Darmstadt
D-64289 Darmstadt
Magdalenenstr. 4
Telefon (06151) 16 2155
Telefax (06151) 16 3355
E-MAIL: birkhofer@muk.maschinenbau.th-darmstadt.de

Abstract

Most effective environmental measures demand an integral development of environmentally friendly products that has to deal with all corresponding processes in their life cycle. A new research program with an interdisciplinary co-operation between engineers, economists, scientists and social scientists will be set up to develop methods, working material and instruments to efficiently support the work of designers. Thus harmful influences on the environment in all phases of the life cycle can be comprehensively recognised and preventatively minimised under consideration of economic and technical requirements.

Keywords

Environmentally friendly products, integral development, life-cycle-design, environmental knowledge, life-cycle-processes, information model, design environment

1   Introduction

The earth's resources are limited and its capacity to cope with pollution and waste are restricted. The consumption of non-renewable resources and the flood of pollution and waste endanger the stability of ecological systems and thus the basis of life for humanity.

With an increasing degree of recognition, it is realised that present-day commerce must not be permitted to endanger or even destroy the basis of life for subsequent generations /11, 15, 16, 23/. The protection of the environment is thus given prime importance as it is a life-preserving target with regard to safeguarding the future. According to /13/, the environment is the totality of human beings, animals and plants, soil, water, air, climate and landscape, as well as cultural and other material objects.

Environmental protection is recognized and established nationally and internationally as an important social and political objective /12, 19/. It encompasses the protection of resources and the avoidance of detrimental alterations of natural systems by processes in the total life cycle of technical products /3, 8/.

2 Initial Situation

Environmental interference takes place during the overall life cycle of a product whereby the product itself is only indirectly involved. The „pleasure" obtained from the use of products conflicts with the „burden" caused by environmental damage from the corresponding processes (figure 1). They result from the diverse production, use, recycling and disposal processes through which a product, its components and parts, goes through during its life.

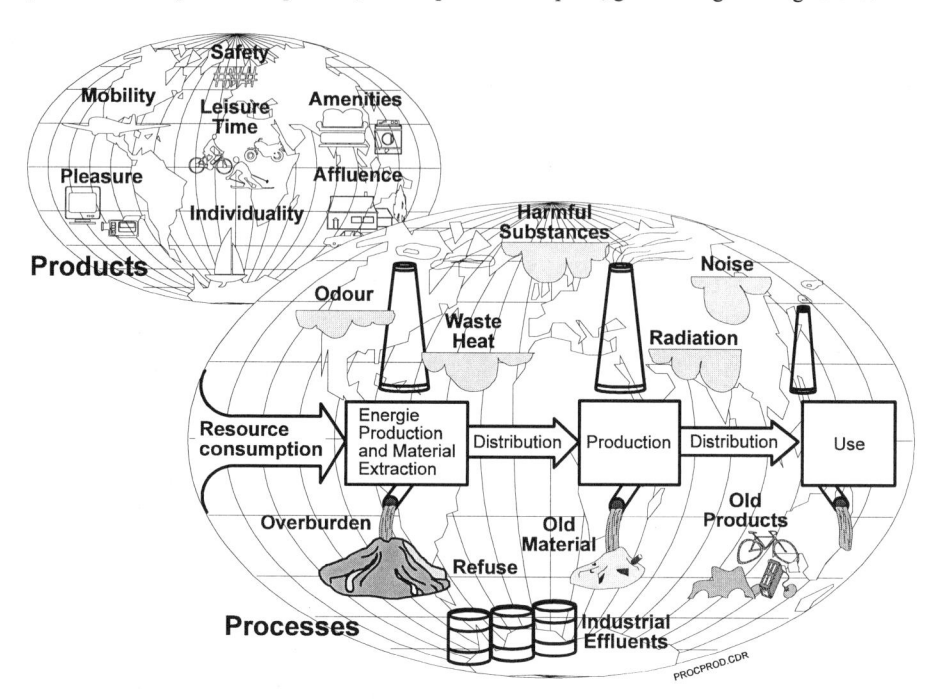

Figure 1: The „pleasure" of technical products, and the „burden" of technical processes

Environmental production in the sense of the principle that the party responsible is liable for the damage must mainly be applied in the area of industry. The measures which are common and propagated at the present time for environmental protection can be divided into four stages:

Measures to Eliminate Existing Environmental Damage
Examples here are the recycling of soils, the rehabilitation of contaminated building structures, or the disposal of dangerous waste from the past on to dumps.

Measures to Reduce Environmental Damage in the Case of Known Products
Examples here are the follow-up treatment of emissions from industrial processes to reduce the level of pollution content, the application of catalytic converters in motor vehicles to reduce the CO and NOx emissions. Also, the recycling of old products and materials /17, 18, 21/, by which means the consumption of primary raw materials can be reduced, as well as greatly reducing the increase in waste dump volume.

Measures to Reduce Specific Environmental Damage from New Products
Examples for measures to save resources are the use of regenerative energies such as water power, wind power or geothermic plants, the production of fuel and materials from regrowing raw products, as well as the realization of low emission, material and energy saving products /1, 3, 6, 9, 10, 18, 22/.

Measures for the Integral Development of Environmentally Friendly Products
This can be defined as an overall examination of all life cycle phases of a product with regard to potential damage to the environment, and a preventative reduction or avoidance thereof, already in the product development stage. A starting point here is made by the EC Regulations concerning the voluntary participation of industry in a common environmental management and company inspection control, which is known by the terminology „eco audit". According to this, the effect on the environment of each new product should be evaluated in advance, the client then being advised with regard to the environmental aspects during operation, use and disposal of the product /12/. Far-reaching definite directions for methodics in the assessment of environmental damage are missing in this regulation, as well as a conversion into company practice and inter-company realization which also includes suppliers in an environmental management.

3   Problem Situation

An overall preventative development of environmentally friendly products is, without doubt, the most effective measure to reduce damage to the environment. The claim of being overall cannot be met with by present-day design practice due to methodic and working technique problems and deficits. The problems can be mainly reduced to four causes:

Deficits in the Inventory of Environmental Knowledge and in Access Thereto
A wealth of environmental knowledge is required for the objective assessment of the harmful effect of products on the environment. This knowledge is extremely diverse, inter-linked, and is subject to great temporal changes. Environmental data banks do exist which document and

manage environmental knowledge but they only cover a limited spectrum of knowledge, are not always publicly available and contain often data which is not validated.

The variety of documented environmental knowledge is furthermore neither by no means complete nor directly applicable for the design.

In many areas a „design-relevant environmental knowledge" is missing which can be understood as a collective term for data, objectives, conditions, regulations, principles and guidelines, which not only describes ecological facts but also can be used for the development of environmentally friendly products.

Deficits in the Objective and Overall Assessment of Harmful Effects on the Environment
Ecological assessments provide methods to record, assess and compare harmful effects on the environment caused by products and processes, and are divided into the phases of life cycle inventory, environmental impact analysis and evaluation.

In the life cycle inventory, the main problem lies primarily in the limitation of the assessment area which is to be observed, apart from the work necessary for the extensive survey of the relevant environmental knowledge. In the setting up of an environmental impact analysis and the subsequent evaluation, basic questions also remain unanswered /3, 13, 20/. For instance, in many cases scientifically proven knowledge regarding the environmentally harmful effect of numerous materials is missing. Also unsolved is the comprehensive and objective aggregation of effects to valid environmental characteristic values (evaluation), which also must be formulated in a suitable manner in order to be capable of being used as a basis for decisions in product development.

Deficits in the Integral Evaluation of Products
The exclusive examination of products regarding their harmful influence on the environment is not sufficient to be able to develop marketable products which meet the technical and economical requirements made by the customer. Products geared to market requirements can rather result from the overall overview of ecological, technical and economical requirements /4, 5, 7/. There is also a lack of methods which permit an individual and defined time horizon to be determined for an optimum product depending on product type, customers and market conditions.

Deficits in the Implementation of Methods for a Preventative Development of Environmentally Friendly Products
A prerequirement for an efficient product development is the objective assessment of a product variant, which, in turn, assumes knowledge of its design and material characteristics as well as the thus required production, use and disposal processes with the related quantity frameworks. Simple usable methods are not available here, with which the designer has access directly to the varied basis of knowledge and which permits him to efficiently assess products and processes in all phases of the development processes. The normal working material available in design at the present time (e.g. industrial standards, data sheets) and instruments (e.g. CAD systems, data banks, calculation programs) can only be used with

limitations for this purpose. Differing hard and software equipment, operating surfaces and use philosophies form an insurmountable obstacle in everyday design in accordance with the model of an integral development of environmentally friendly products.

As well as the deficits in the methodical and computer-specific prerequirements, there are further deficits in the present-day organizational situation. The strong integration of companies within the net product chain and the increasing outsourcing strategies lead to ever-increasing demands on external goods and services.

With regard to their harmful influence on the environment, these external services are more difficult to assess for a company than their own services.

4 Objectives

A special research group sponsored by the DFG has been proposed at the TH Darmstadt, with the aim of working out methods, working material and instruments for the integral development of environmentally friendly products. A special research group is necessary as selective approaches on their own are not sufficient to cope with the complexity of the task in hand. They require new and extraordinary scientific effort which can only lead to success with an interdisciplinary co-operation between engineers, economists, scientists and social scientists.

Methods, working material and instruments should so support the integral development process that the potential harmful influences on the environment by products and the thus required processes in all phases of the life cycle can be comprehensively recognized and be preventatively minimized by specific design measures under consideration of economic and technical requirements (figure 2).

The designer, with all methods, working material and instruments -
• should be able to quickly determine and objectively assess the potential environmental
  influence which he causes by his design activities, in all phases of the design process
• should be able to determine, assess and use the potential for a development of
  environmentally friendly products in all phases of the design process
• should be able to compare the ecological influence of his design work with the
  technical and economic influence, and thus be capable of the overall optimizing of
  environmentally friendly products which are also geared to market requirements
• should be supported in the minimizing of influence on the environment by an efficient
  access to design-relevant environmental knowledge and the therein contained
  ecologically compatible product and process variants.

With the named objectives, the research program integrates the measures taken up to the present of the precautionary environmental protection in an overall way of looking at the situation and method of behaviour. The at present predominately subsequent environmental protection should thus be more and more replaced by an encompassing preventative environmental protection (prophylaxis instead of repair). It is to be expected that the necessity for „repairing" environmental technology will, in the long term, be reduced.

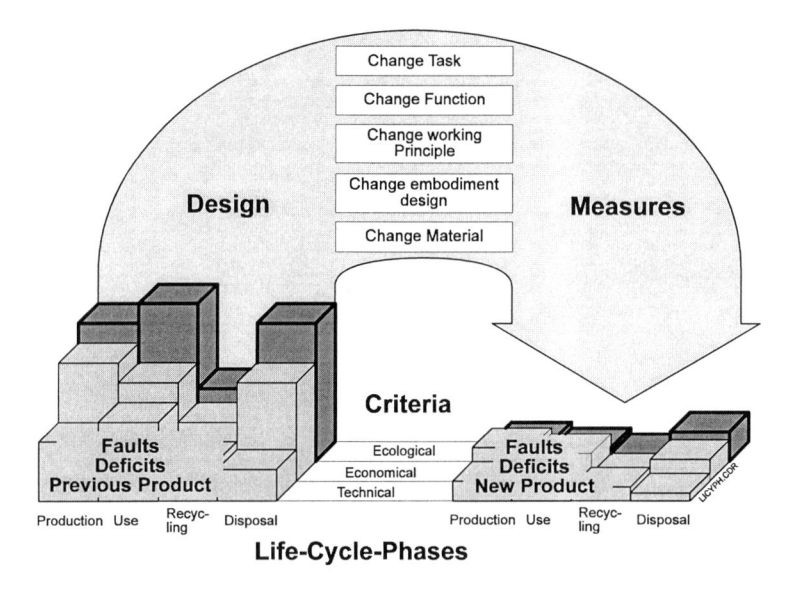

Figure 2: Integral Product Development

The results worked out in the special research program further present a basis for the formulation of standards to register, systemize and aggregate design-relevant environmental knowledge. These standards could then be used for a comprehensive environmental management for all companies within the next product chain. Thereby, each company determines only the characteristic environmental factors for those processes which occur within its own area of responsibility. With the initial products, the relevant characteristic environmental factors are passed on from company to company and aggregated.

5 Tasks

The various tasks of the research program are derived from the objectives.

5.1 Documentation and Representation of Environmental Knowledge

Not only the generally available environmental knowledge of processes and products are recorded (e.g. design rules) in the individual life phases but also specific environmental knowledge from the analysis of example products and example companies, e.g. the useful life of certain products. The relevant processes hereto are identified in each life cycle phase in accordance with standardized particulars, analysed and described on the basis of a common information model. Additionally, the environmental knowledge relevant for the design is documented, analysed, classified with regard to content, application and significance, and represented within the information system (figure 3).

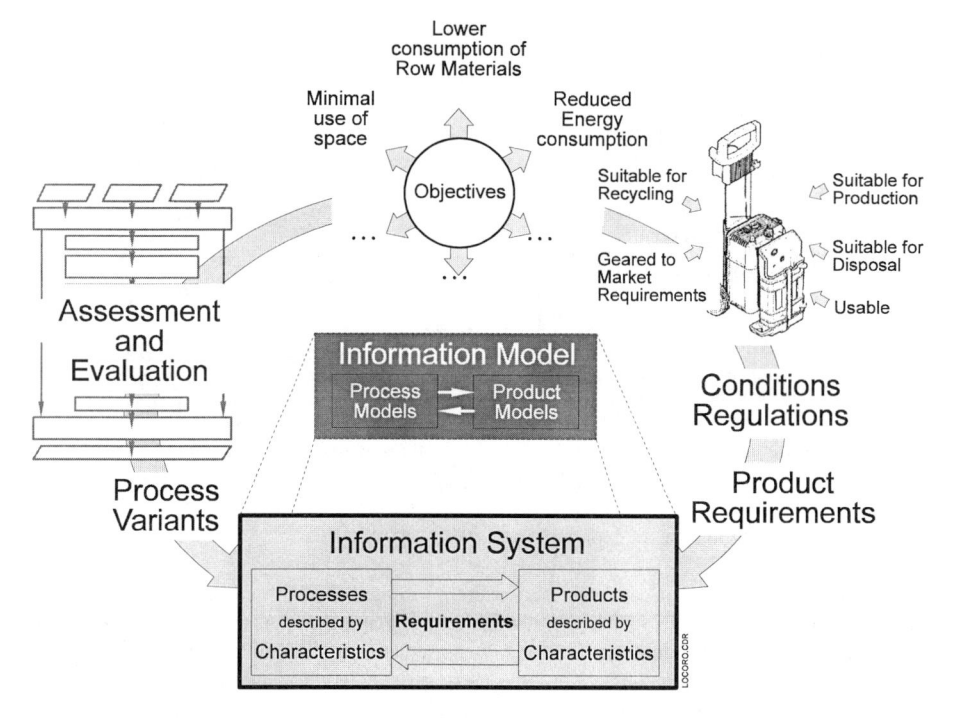

Figure 3: Documentation and Representation of Environmental Knowledge

5.2 The Development of Methods

New and modified methods are needed for the overall development of environmentally friendly products. Products and processes must be evaluated based on valid characteristic values as a measure for their influence on the environment; environmentally friendly products geared to the market requirements must be generated (figure 4).

Environmental knowledge must then be processed with regard to its representation, transformation and presentation for the relevant context and regarding the various levels of the designers. For this, existing methods are taken over, further developed and new methods worked out. The applied methods must be flexible and integrally formed, so that they can be adapted easily to a specific product, company or area of industry approach. They are integrated within the present design methodics and supplement these with regard to the overall development of environmentally friendly products.

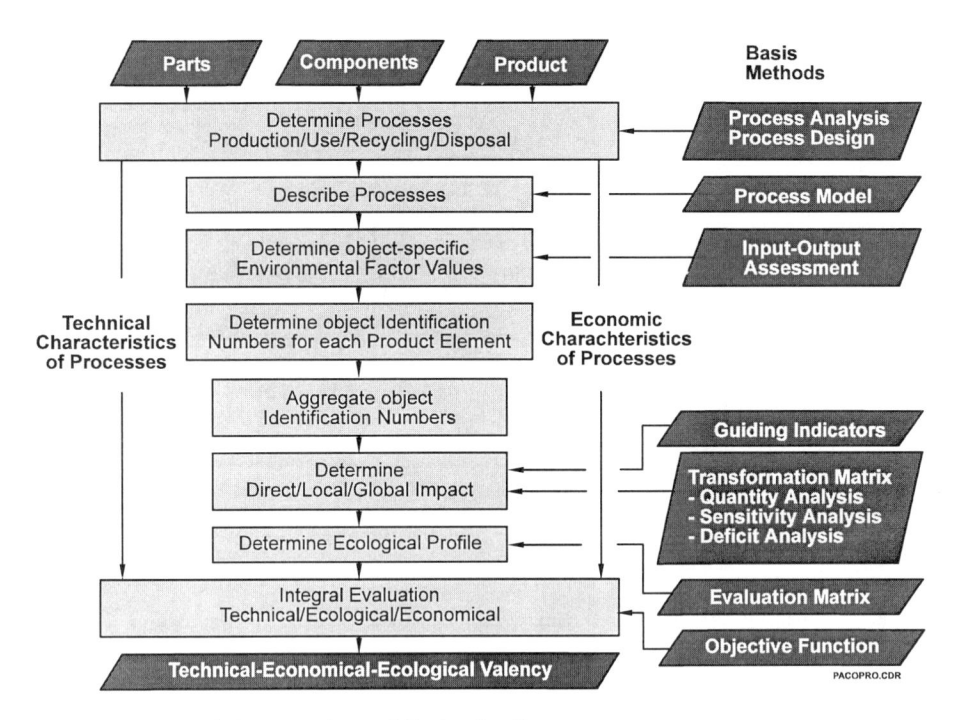

Figure 4: Approach for Assessing and Evaluating Processes

5.3 The Availability of Easy-to-Apply Working Material and Instruments

An efficient computer support is certainly necessary in order to properly support the design work, to fulfill the overall research formulation conditions, and to control the high complexity of the environmental knowledge made available.

It is realized in the form of a design environment which gives the designer an efficient access to the relevant required environmental knowledge, supports the assessment and generation of process and product variants, and simplifies and speeds up the design work by computer supported instruments with a common operating surface.

The design environment (figure 5) is based on an open integration platform, which permits a flexible co-operation of the instruments of the design environment newly developed in this research program and enables the integration of instruments.

Instruments to be newly developed for the design environment are an information system for environmental knowledge relevant to the design, an assessment system for products and processes, a uniform user interface, and an updating interface.

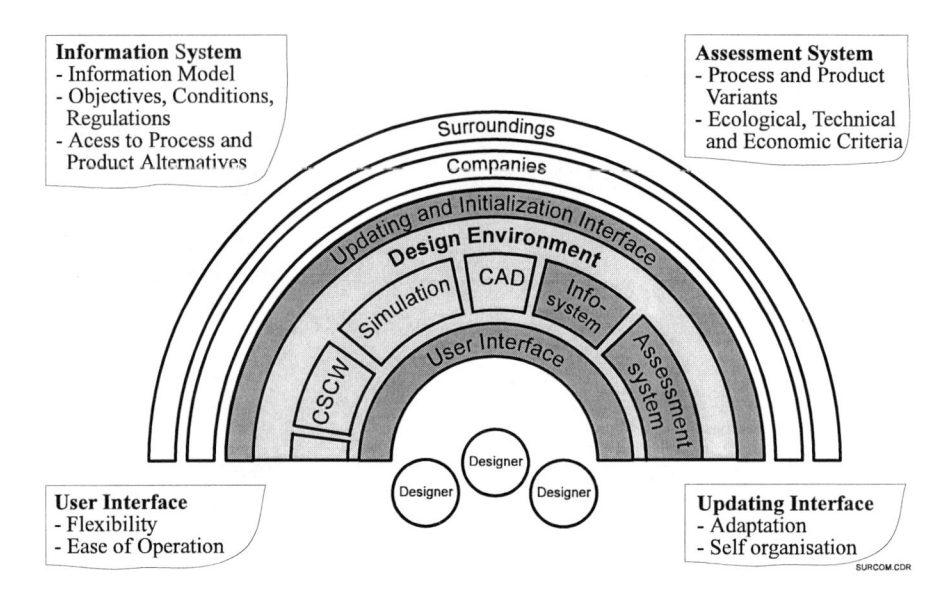

Figure 5: Design Environment and Instruments

Further design supporting systems are taken over into the design environment, which already exist within the company or are to be obtained from the software market, e.g. a CAD system to generate product models and a CSCW system for the rapid exchange of information between the company's personnel.

6 Structure of the Research Program

The tasks derived from the objective of the research program can be set in a coordinate system, with axes that concentrate similar types of tasks. In the following they are called expert tasks and cross-section tasks and define both the project areas A and B of the specialist research program (figure 6).

Project Area A: Expert Tasks
In this project area the environmental knowledge relevant for design regarding processes in all life cycle phases are collected and described in a standardized language, with the assistance of an information model developed in Project Area B. The environmental influences resulting from the processes are aggregated to environmental characteristic factors. The expert groups furthermore define the interfaces of their working areas to the design and formulate requirements to the computer supported design environment and the information management. The research carried out by the expert groups is on chosen product examples or their parts and components.

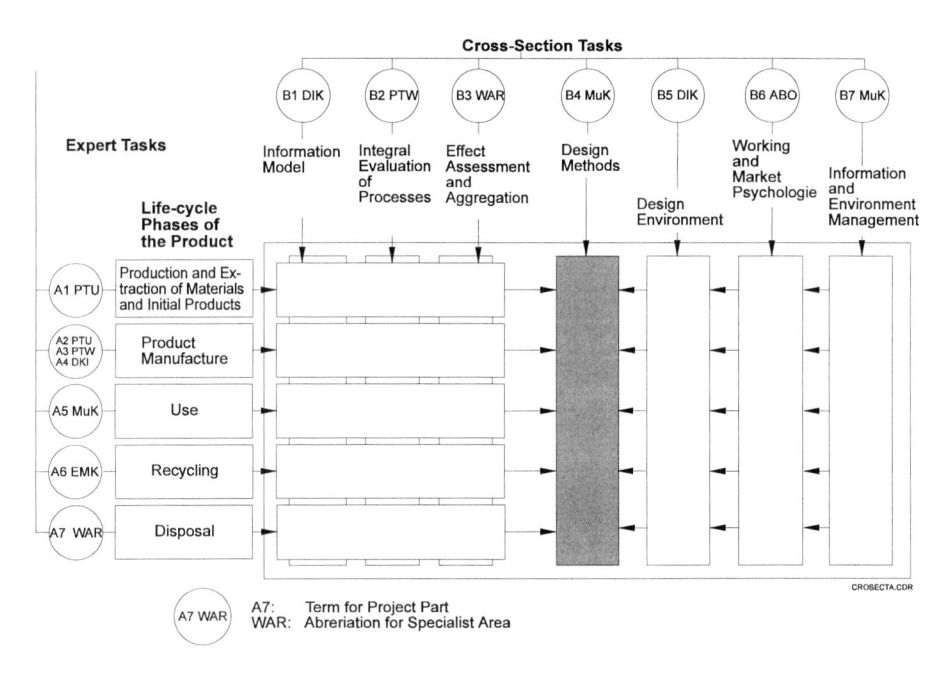

Figure 6: Structure of the Research Program

Project Area B: Cross-Section Tasks
Cross-section tasks include all tasks which process the results worked out in the individual expert areas in the product development in an overlapping manner and integrate them into an overall arranged design methodic. The processing of the cross-section tasks produces methods with which the expert groups analyse the processes in the life cycle phase and can describe the results in a standardized manner. Further to this, the cross-section tasks have as their objective to use these results for the development of environmentally friendly products and to work out a practical, computer supported design environment for the overall development of environmentally friendly products. Parallel to this, an inner company and external environment and information management is set up, which ensures a comprehensive updating of environmental knowledge.

7 Product Examples and Company Examples
The research work is carried out and verified on actual products or their components and parts in the sector of gardening equipment, household and cleaning appliances.

The reasons for chosing products from this group were -
- the considerable environmental influence, mainly in the use phase, due to ecologically problematic functions and impact principles

442   *Part Eight   Design for Environment*

- the large numbers and batches produced which contribute to damage to the environment and from which a relevant optimization potential results
- the large psychologically dominating influence of the expectations of the end-user made on the characteristics of the consumer goods
- the great variety of materials and production processes and
- the combination of mechanical, electrical, electronical and hydraulic components, which further the general validity of the results and knowledge of this research program.

It is planned to analyse high-pressure cleaners as a product example in the first research phase. The research work will, from the outset, be carried out in co-ordination with the example companies, in order to ensure a practical structuring of the methods, working material and instruments.

8   Conclusion

The increasing environmental awareness in society leads to ever-increasing demands from consumers for environmentally friendly products. For this reason, many companies proactively pursue more and more the environmental compatibility of their products, and propagate them as sole distinctive feature, in order to gain a competitive advantage therefrom.

The environmental compatibility of products has become an important competitive factor /2, 8/. Apart from the product itself, the know-how for the development of more environmentally friendly products becomes a more and more important marketing advantage in international competition. With approx. 30 % share of the market, German industry has a leading position in the international market for environmental technology. This position is to be consolidated for the preventative environmental protection and converted into a permanent advantageous position.

1.3.10 Literature

/1/	Beitz, W.:	Designing for Ease of Recycling. General Approach and Industrial Application. Schriftenreihe WDK 22, Heurista, Zürich 1993, S. 731-738
/2/	Byrne, G.; Scholta, E.:	Environmentally Clean Machining Process - A Strategic Approach. CIRP Annals 42/2 1993, S. 471-474
/3/	Environmental Protection Agency (Hrsg.):	Life-Cycle Assessment: Inventory Guidelines and Principles. Final Report, Cincinnati/Ohio 1993
/4/	Eversheim, W.; Böhlke, U.; Adams, M.:	Die Auswahl des "richtigen" Produktionswerkstoffes. VDI-Z 136 (1994) 4, S. 118-121
/5/	Eyerer, P.; Pfleiderer, I.; Saur, K.; Schuckert, M.; Parr, O.; Hesselbach, J.:	Ganzheitliche Bilanzierung von Automobilteilen aus Stahl, Aluminium und Kunststoffen am Beispiel Ölfilter für PKW-Motoren. In: VDI-K (Hrsg.): Kunststoffe im Automobilbau: Rohstoffe, Bauteile, Systeme. VDI-Verlag, Düsseldorf 1994, S. 5-42

| /6/ | Geißler, S.;
Harant, Ch.;
Hrauda, G.;
Jasch, Ch.;
Millonig, S.: | ECODESIGN - Ökologische Produktgestaltung. Anwenderfibel erstellt im Auftrag des österreichischen Bundesministeriums für Umwelt, Jugend und Familie, Eigenverlag des IÖW, Wien 1993 |
|---|---|---|
| /7/ | Hartel, M.;
Spath, D.: | Öko-Portfolio: Methoden zur Beurteilung der Recyclingeignung technischer Serienprodukte. Vortrag, Serienfertigung feinwerktechnischer Produkte, von der Produktplanung bis zum Recycling. VDI-Berichte 1171, VDI-Verlag, Düsseldorf 1994, S. 371-392 |
| /8/ | Hopfenbeck, W.: | Umweltorientiertes Management und Marketing. 2. Aufl. Verlag moderne industrie (mi), Landsberg/Lech 1991 |
| /9/ | Hübner, H.;
Hübner, D. S.: | Ökologische Qualität von Produkten - Ein Leitfaden für Unternehmen. Druckschrift der Hessischen Landesregierung, Wiesbaden 1991 |
| /10/ | Hunt, R.;
Sellers, J.D.;
Franklin, W.E.: | Resource and Environmental Profile Analysis: A Life Cycle Environmental Assessment for Products and Procedures. In: Susskind, L.; Hill, T. (Hrsg.): Environmental Impact Assessment Review. Elsevier Publ. Comp., New York 1992 |
/11/	Lovins, A. B.:	Negawatt statt Megawatt - Energiesparen als Energiequelle. Dokumentation der Manuskripte des Referenten A. B. Lovins (Rocky Mountain Institute). Dezernat für Umwelt, Energie und Brandschutz (Hrsg.), Frankfurt 1993
/12/	N.N.:	VERORDNUNG (EWG) Nr. 1836/93 des RATES vom 29. Juni 1993 über die freiwillige Beteiligung gewerblicher Unternehmen an einem Gemeinschaftssystem für das Umweltmanagement und die Umweltbetriebsprüfung. Amtsblatt der Europäischen Gemeinschaften, Nr. L 168/1
/13/	N.N.:	Grundsätze produktbezogener Ökobilanzen - German "Memorandum of Understanding" / "Conceptual Framework". DIN-Mitteilungen 73 (1994) 3, S. 208-212
/14/	N.N.:	Umweltrecht. Beck, München 1992
/15/	Nitsch, J.:	GEO - Studie - Energie im Jahr 2005. Deutsche Forschungsanstalt für Luft- und Raumfahrt. Stuttgart 1993
/16/	Schmidt-Bleek, F.:	Wieviel Umwelt braucht der Mensch? - mips - Das Maß für ökologisches Wirtschaften. Birkhäuser, Berlin 1994
/17/	Seliger, G.;
Kriwet, A.:	Demontage im Rahmen des Recyclings - Der Konstrukteur bestimmt Aufwand und Nutzen. ZwF 88 (1993) 11, S. 529-532
/18/	Steinhilper, R.;
Hudelmaier, U.:	Erfolgreiches Produktrecycling zur erneuten Verwendung oder Verwertung - Ein Leitfaden für Unternehmen. Rationalisierungs-Kuratorium der Deutschen Wirtschaft (RKW) e.V., Eschborn 1993
/19/	Umweltbundesamt	Daten zur Umwelt - 1992/93. Erich Schmidt, Berlin 1994
/20/	Umweltbundesamt	Ökobilanzen für Produkte, Texte 38/92, Berlin 1992
/21/	VDI 2243:	VDI-Richtlinie 2243: Konstruieren recyclinggerechter Produkte. Beuth, Berlin 1993
/22/	Vester, F.:	Ausfahrt Zukunft - Strategien für den Verkehr von morgen. Eine Systemuntersuchung. Heyne, München 1990
/23/	Wicke, L.;
Haasis, H.-D.;
Schafhausen, F.;
Schulz, W.:	Betriebliche Umweltökonomie. Vahlen, München 1992

Life Cycle Assessment (LCA) - A Supporting Tool for Vehicle Design?

C. Kaniut, H. Kohler
Dept. Technology, Environment and Traffic EP/VU, Mercedes-Benz AG
70322 Stuttgart, Germany
Tel.: ++49(0)711-17-26624; Fax: ++49(0)711-17-56475

Keywords
LCA Methodolgy; Life cycle costs; Application in vehicle design: utilisation, problems, consequences; LCA at Mercedes-Benz AG; Examples for the phases of use, concept and configuration, recycling; Conclusion

1 INTRODUCTION

Life Cycle Assessment (LCA) is a tool designed for the analysis of resource consumption and environmental impact of a product's complete life cycle, starting from raw material extraction via product manufacture and use up to recycling and disposal. The basic LCA steps which will be discussed later are as follows:

- Life Cycle Inventory (LCI)
- Life Cycle Impact Assessment
- (Life Cycle) Evaluation.

Using the LCA tool, Mercedes-Benz want to take ecological, technological and economic aspects into account and thus apply LCA as a supporting tool for vehicle design. The following chapters provide an overview of the LCA approach.

2 LCA METHODOLOGY

Figure 1 provides a general definition of the term and the basic LCA procedure. Figure 2 then shows the main LCA steps as an application example for vehicle parts.

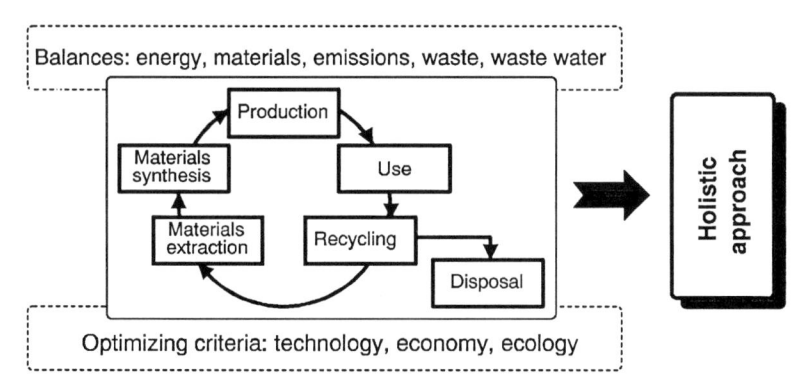

Life cycle assessment (LCA)

* Method to determine resource consumption and environmental impact of products during raw material extraction, production, product utilisation and recycling/disposal

Balances: energy, materials, emissions, waste, waste water

Materials synthesis → Production → Use → Recycling → Disposal → Materials extraction → (cycle)

Optimizing criteria: technology, economy, ecology

⇒ Holistic approach

* Steps: Inventory - Impact Assessment - Evaluation

Figure 1 Life Cycle Assessment (LCA) - Definition and basic approach.

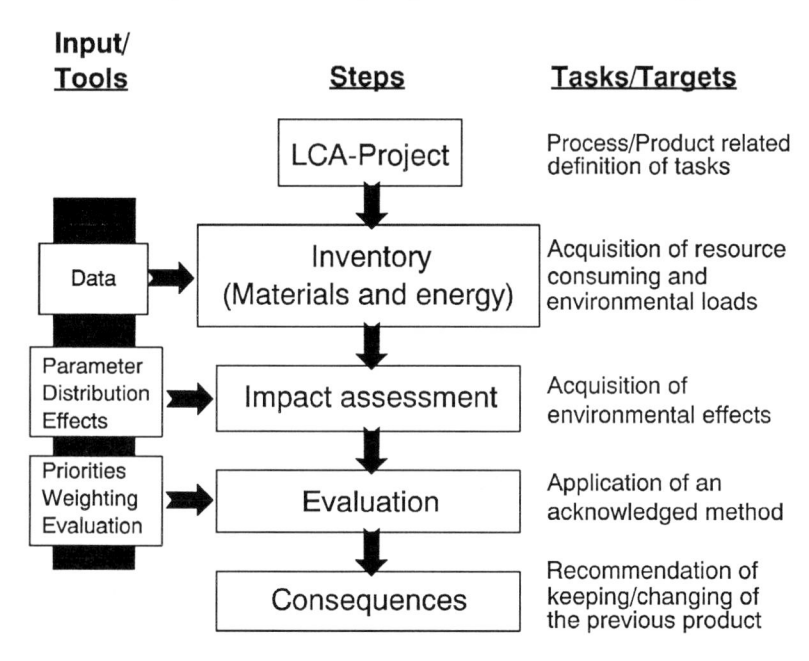

Input/Tools — **Steps** — **Tasks/Targets**

- LCA-Project — Process/Product related definition of tasks
- Data → Inventory (Materials and energy) — Acquisition of resource consuming and environmental loads
- Parameter Distribution Effects → Impact assessment — Acquisition of environmental effects
- Priorities Weighting Evaluation → Evaluation — Application of an acknowledged method
- Consequences — Recommendation of keeping/changing of the previous product

Figure 2 Main steps of LCA.

The actual "LCA project" clearly defines the task and in a detailed way determines the object to be analysed (e.g. the product or product alternatives), the relevant scope of the analysis (e.g. overall component life cycle), the limits of the analysis, the boundary conditions and the criteria to be taken into account (e.g. global warming potential). These parameters must be determined for all LCA steps (Inventory, Impact Assessment and Evaluation); in terms of type, scope and degree of detail they depend on

- the period of time specified for the LCA
- know-how and size of the LCA team
- maturity of the LCA tools used
- access to quickly available and high-quality data
- state of discussion of methods with recognised and binding LCA standards.

The "Life Cycle Inventory" acquires figures on resource consumption and environmental burdens and basically is an analysis of all materials and energies spent over certain life cycle phases or the entire life cycle of the analysis object. The result is available in the form of a table and/or a graphical illustration indicating all input (e.g. materials) and output variables (e.g. emissions).

As regards an LCA of automotive components, the "Inventory" mainly incorporates as realistic a presentation as possible of all production, utilisation and recycling processes including the extraction of materials, generation of energy, provision of energy, transport etc.. Product-specific data, i.e. no average data or data of other components, is required for product statements or recommendations so that the effort and expenditure of data acquisition in-house and at suppliers, subsuppliers and scientific institutes becomes in general an essential factor within the entire analysis process.

In the LCA step "Impact Assessment", material flows available as output result of material analyses are reduced to a relatively small number of reference substances. Thus, the impact of environmental burdens from different inventory analyses become comparable, which is important for the evaluation of alternatives. By way of example, Figure 3 shows how various emissions can be combined into a GWP criterion ("global warming potential") with carbon dioxide (CO_2) as reference substance. However, the large scatter of equivalent factors also makes it quite clear that as yet no unequivocal statements are possible.

"Evaluation" represents the most subjective LCA step since different criteria (corporate, sociopolitical, ideological ones etc.) of varying number and different weighting may come into play. Reasons may be (still) insufficient scientific confirmation of evaluation criteria or the fact that these criteria simply cannot be

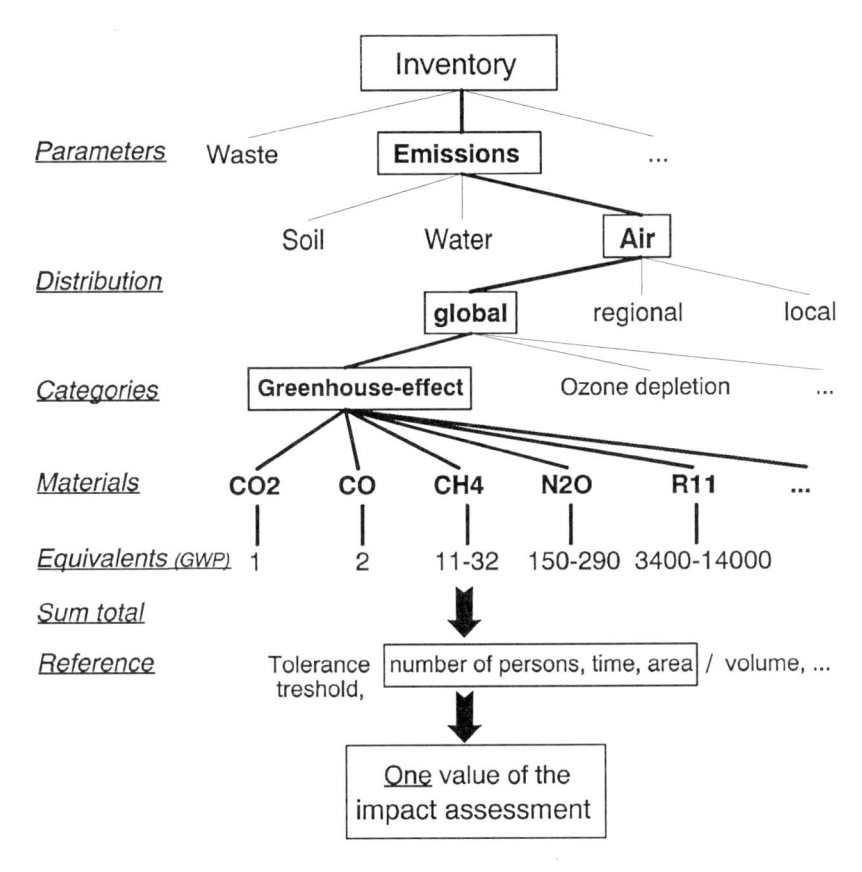

```
Parameters        Inventory
                    │
         Waste── Emissions ──...
                 ╱    │    ╲
               Soil  Water  Air
Distribution              ╱ │ ╲
                      global regional local
Categories         ╱       ╲
             Greenhouse-effect   Ozone depletion   ...
            ╱    │    │    │    ╲
Materials  CO2  CO  CH4  N2O   R11      ...
            │    │    │    │    │
Equivalents(GWP) 1   2  11-32 150-290 3400-14000
Sum total              ▼
Reference    Tolerance │number of persons, time, area│ / volume, ...
             treshold,
                       ▼
                  One value of the
                  impact assessment
```

Figure 3 Example of Impact Assessment concerning the determination of the Global Warmin Potential (GWP).

confirmed or put into objective terms. Moreover, industrial analyses require a technological and economic evaluation in addition to an ecological evaluation. This may for example be effected via a proposal from the University of Stuttgart/Germany, by means of which the individual (technological, economic, ecological) evaluations can be integrated into an overall evaluation.

As regards the economic component, an approach for life cycle costs in line with Figure 4 would be conceivable. Especially summand K_l will surely give rise to different conclusions, e.g. for an ecological tax reform. Basically, such a reform could be welcomed provided it does not increase the overall tax burden and is not misappriated for other purposes.

$$K_{LC} = K_M + K_D + K_P + K_U + K_R + K_I$$

Indices:

LC = Life cycle ("from cradele to grave")

M = Materials extraction and synthesis

D = Research and development

P = Production and sales

U = Utilisation including maintenance and repair

R = Reuse, recycling and disposal

I = Integration of "external costs", e.g. ecological tax

Figure 4 Product life cycle costs.

It remains to be hoped and expected that national bodies (e.g. DIN-NAGUS; NAGUS = Normenausschuß Grundlagen des Umweltschutzes: Standardization Committee on Basics of Environmental Protection) as well as international bodies (e.g. ISO TC 207 SC5; ISO = International Standardization Organisation, TC = Task Committee, SC = Sub-Committee) will work out solutions which will find acceptance and can be a basis for a comprehensive application of Life Cycle Impact Assessment and Evaluation.

"Conclusions" as the last of the major LCA steps in accordance with Figure 2 indicates implementation of the LCA results on the product, i.e. a recommendation to choose the "right variant". Especially as regards the integration of LCA results into the decision-making process for industrial development is concerned, comprehensive LCA know-how and tried and tested LCA tools are required.

3 UTILISATION, PROBLEMS AND CONSEQUENCES

Figure 5 shows potential major benefits of LCA application in automotive engineering. For example, it will be of interest to use resources in a pinpointed way for ecological measures based on LCA results instead of giving everyone a slice of the cake.

- **Further optimisation of future vehicles**
    * Overall integrated concepts (development, production)
    * "Right" choice of materials

- **Weak point analysis of production process**
    * Identification of main causes of environmental burdens
    * Module of ecological audit

- **More efficient use of resources**
    * Pinpointed optimisation of development steps and production processes
    * Minimising subsequent life cycle costs

- **Argumentation for public reasons**
    * Proactive line of reasoning for corporate measures
    * Basis of environmental political positions

- **Image gain** (product, company)

Figure 5   Expected major benefits of LCA.

Figure 6 shows the current major problems in connection with LCA for automotive engineering. It should be emphasised in particular that currently no specific methodo-logy exists for the automotive sector which takes the specific requirements of this branch into consideration (complexity, high time and cost pressure).

- **Inventory**
    * Amount and quality of data insufficient
    * Methods hardly standardised yet

- **Tools**
  Current state of development still unsatisfactory

- **Impact assessment**
  No consensus on procedures

- **Evaluation**
  Subjective assumtions and determinations,
  no consensus

- **Methodical approach**
  Still too complex for automotive applications
  (e.g. skeleton specifications/technical specifications)

Figure 6   Current major problems with LCA.

The conclusions and consequences arising from the potential and current problems associated with LCA in automotive engineering are summarised in Figure 7. In this context, it is particularly important that the automotive industry use practical experience to confirm its ideas of a "fast" LCA method, and that the LCA is further developed into a practically oriented tool to the benefit of the environment and the automotive industry instead of a purely theoretical tool because of the complexity of modern automobiles. All parties involved directly or indirectly in LCA method development can and should make a contribution.

- **Objective approach**
  Danger of subjective or ideologically results

- **Consensus**
  Analysis results are currently not broadly accepted

- **Danger**
  Ecologically unbalanced and regional environmental regulations which distort the competitive environment

Consequences: **The automotive industry must**

➡ built up its own know-how of life cycle assessment

➡ develop a "rough" methodology able to achieve consensus

➡ submit its approach to international bodies, agree on it and implement it

➡ communicate its procedure in a coherent way to the media, politics and customers

Figure 7 LCA - Conclusions and consequences.

4 LCA AT MERCEDES-BENZ

During past years, Mercedes-Benz AG - in co-operation with various institutions or on their own - have performed a number of analyses, at first pure LCI's, but increasingly taking into consideration the Life Cycle Impact Assessment and Evaluation. These analyses included car and commercial vehicle components such as

- bumpers
- fenders
- fuel tanks

- engine components
- etc.

but also production processes such as

- painting with solvent-based and water-based paints
- assembly processes
- etc.

These analyses basically constitute a "first LCA strategy" on the way to a "full vehicle LCA". They are based on existing components and production processes and aim at gathering experience with LCA and building up LCA know-how. These efforts are being increasingly continued with additional projects to gather "piece by piece", as in a mosaic, information on materials, production processes, components and recycling processes. This will eventually allow a detailed analysis of a full vehicle representing the complete mosaic.

The "second LCA strategy" for the "full vehicle LCA" aims at selecting "representative" materials, processes and components, performing an analysis and transferring the results to "similar" materials, processes and components. Thus, it would be possible to have an analysis for a full vehicle within a much shorter period of time than with the above mentioned "mosaic strategy", accepting a certain fault tolerance which currently can be only estimated with difficulty. Without doubt, however, the "right" choice of these representative materials etc. is decisive for the quality of the LCA worked out in accordance with this "representative strategy".

Ever larger practical experience and a combination of the "mosaic" and "representative" strategies can contribute to making a reliable full vehicle LCA available. And this not only for vehicles already in production but also for vehicles in the concept stage, the latter being an important prerequisite for the use of LCA as a supporting tool for the development of a vehicle optimised under technological, economic and ecological aspects.

Figure 8 shows the project management to build up and use LCA know-how at Mercedes-Benz AG. Through joint projects with internal and external partners and participation in DIN-NAGUS, in the EUCAR-LCA project (EUCAR-LCA = European Council for Automotive Research) and other bodies, the relevant LCA knowledge is currently being accumulated in the Environment, Technology and Transport department, which will then be transferred step by step to car and commercial vehicle development for independent analyses on site. The Environment, Technology and Transport department ensures that a uniform procedure is applied in the company and at the suppliers, provides methods, tools and data, initiates and implements LCA projects and increasingly supports implementation of such projects.

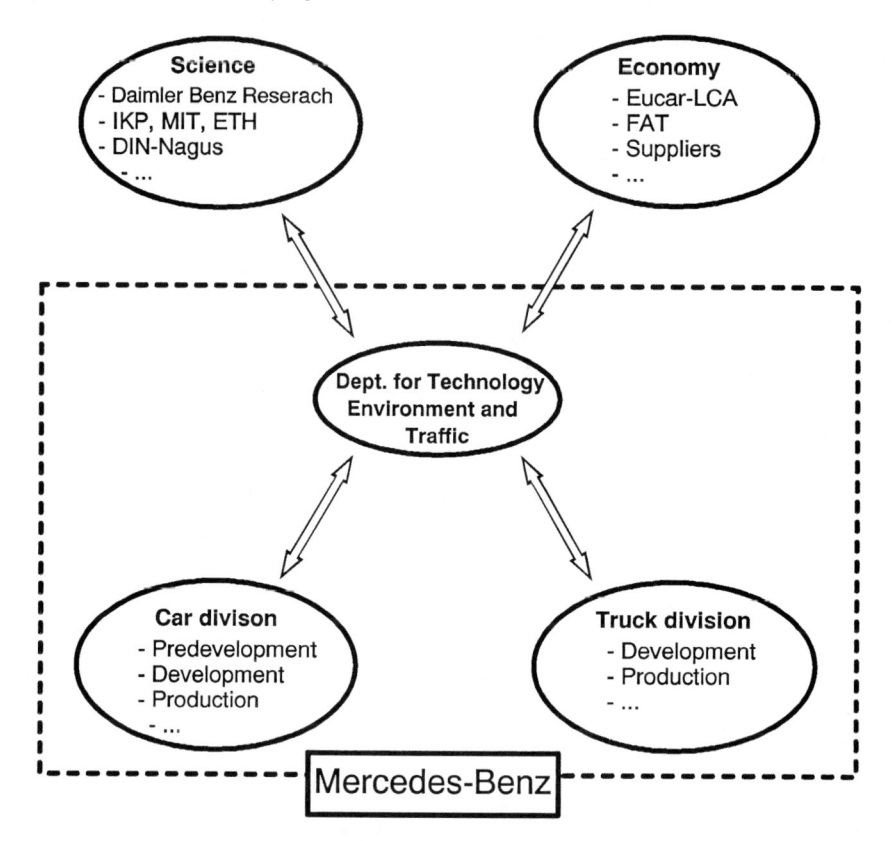

```
        Science                              Economy
 - Daimler Benz Reserach              - Eucar-LCA
 - IKP, MIT, ETH                      - FAT
 - DIN-Nagus                          - Suppliers
 - ...                                - ...

                  Dept. for Technology
                   Environment and
                       Traffic

      Car divison                        Truck division
   - Predevelopment                    - Development
   - Development                       - Production
   - Production                        - ...
   - ...
                       Mercedes-Benz
```

Figure 8   Management of LCA-projects at Mercedes-Benz AG.

5 LCA AS A SUPPORTING TOOL

Three examples will outline the possible application of LCA in automotive engineering as well as the problems and statements arising from it.

Phase of use:
The period of use (fuel consumption, maintenance, repair) constitutes an important phase for LCA's of automotive parts. Especially fuel consumption plays a major role in the overall analysis of automotive components with regard

to decisions on materials and procedures (e.g. the choice between a steel or an aluminium part), which means that determination of the "actual fuel consumption" and the corresponding "weight influence coefficient" is highly significant. In this context, the "weight influence coefficient" indicates the potential for enhanced fuel economy as a function of a reduction of the vehicle's weight. Fuel consumption figures in accordance with generally used standards (e.g. Euromix or NEFZ test: New European Driving Cycle) and a weight influence coefficient determined for example in accordance with the ADAC method (straight regression line in "Fuel consumption against vehicle weight" diagram) can only serve as reference values for the LCA of automotive components. Within the framework of the above mentioned EUCAR-LCA project, nine European automobile manufacturers are therefore working out a proposal which will take vehicle, user and traffic parameters into consideration, as shown in Figure 9.

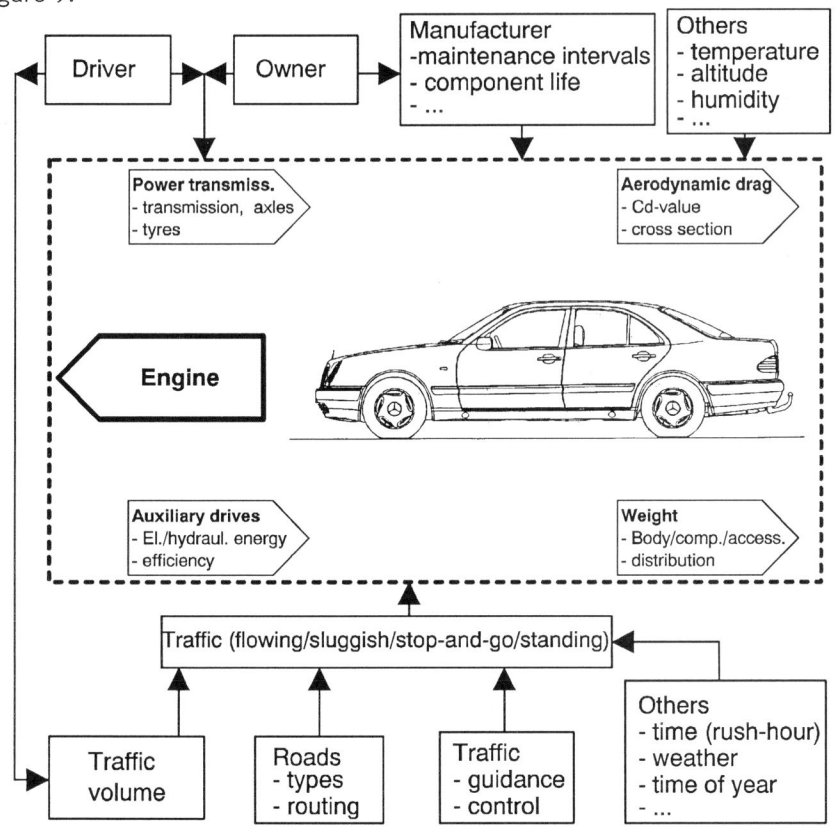

Figure 9 Parameters influencing fuel economy.

454   Part Eight Design for Environment

Concept/configuration phase:
The need for a fast LCA method was already emphasised in Section 3. Figure 10 shows that the automotive development process features various stages in which the LCA can be used. Whereas a relatively large amount of time is available for LCA during the deve-lopment process, for example to have a complete analysis of a component, the time available during the configuration phase and even more so in the concept phase is much shorter. Since, however, important decisions on the selection of materials and proces-ses are already made in these phases which can hardly be corrected later on, fast answers to the questions of vehicle developers are urgently required. If this is not feasible LCA can hardly play a role in the automotive development process.

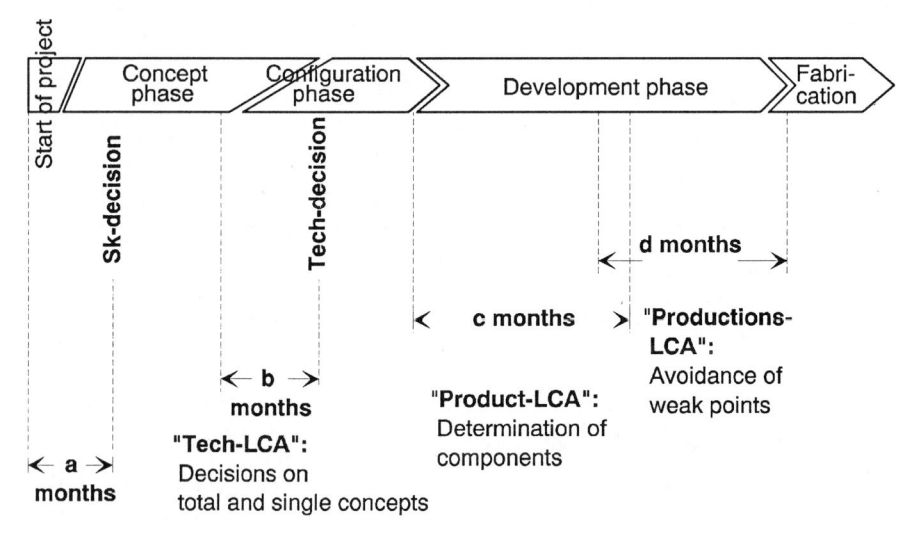

```
Start of project → Concept phase → Configuration phase → Development phase → Fabrication
                   Sk-decision      Tech-decision
```

← a → months
"Sk-LCA":
Pre-selection of alternative total concepts

← b → months
"Tech-LCA":
Decisions on total and single concepts

← c months →
"Product-LCA":
Determination of components

← d months →
"Productions-LCA":
Avoidance of weak points

➡ Necessity because of narrow time schedule:
Acknowledged "quick/lean" LCA-method

Figure 10 LCA implementation in milestone schedule of vehicle development (Sk = Skeleton specifications, Tech = Technical specifications; a < b < c).

Figures 11 and 12 show the potential and answers provided by the application of LCA in the concept and configuration phases.

Life cycle assessment (LCA)

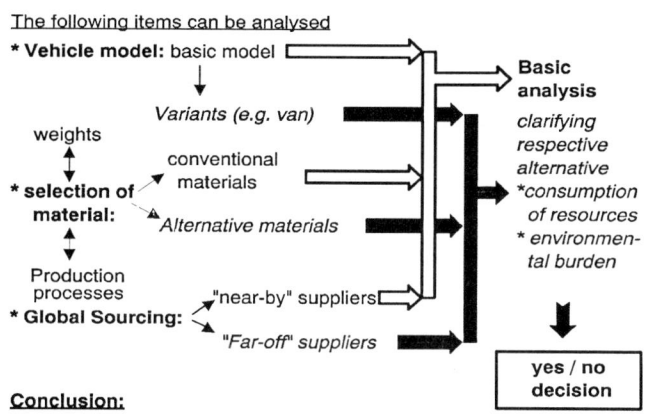

The following items can be analysed

* **Vehicle model:** basic model
  * Variants (e.g. van)
* **selection of material:**
  * weights
  * conventional materials
  * Alternative materials
  * Production processes
* **Global Sourcing:**
  * "near-by" suppliers
  * "Far-off" suppliers

Basic analysis
clarifying respective alternative
*consumption of resources
* environmental burden

→ yes / no decision

Conclusion:
- "Rough" inventory of alternative overall concepts
- Determining resource consumption (materials and energy) and important environmental burdens
- Recommendations as to the "do's" and "don't's" when using up resources and burdening the environment
- Experience with life cycle analyses will in the medium term lead to generally applicable results

Figure 11 LCA as a tool for the skeleton specifications.

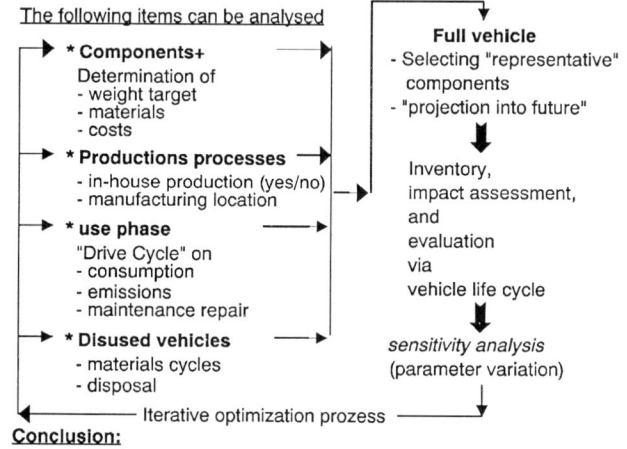

The following items can be analysed

* **Components+**
  Determination of
  - weight target
  - materials
  - costs
* **Productions processes**
  - in-house production (yes/no)
  - manufacturing location
* **use phase**
  "Drive Cycle" on
  - consumption
  - emissions
  - maintenance repair
* **Disused vehicles**
  - materials cycles
  - disposal

← Iterative optimization prozess →

Full vehicle
- Selecting "representative" components
- "projection into future"

↓

Inventory, impact assessment, and evaluation via vehicle life cycle

↓

sensitivity analysis (parameter variation)

Conclusion:
- As compared to the skeleton specifications analysis, this is an extended analysis of defined individual concepts and one overall concept
- Including all steps of a Life Cycle Analysis
- Precise definitions for components, materials, production processes, and manufacturing location
- Sensitivity analysis to reduce life cycle costs

Figure 12 LCA as a tool for the technical specifications.

Recycling phase:
Recycling has become an increasingly important element of the life cycle process of automotive parts. Figure 13 shows the major material cycles. On this basis, alternative analysis limits can be drawn for an LCA which are also indicated in Figure 13 as possible examples.

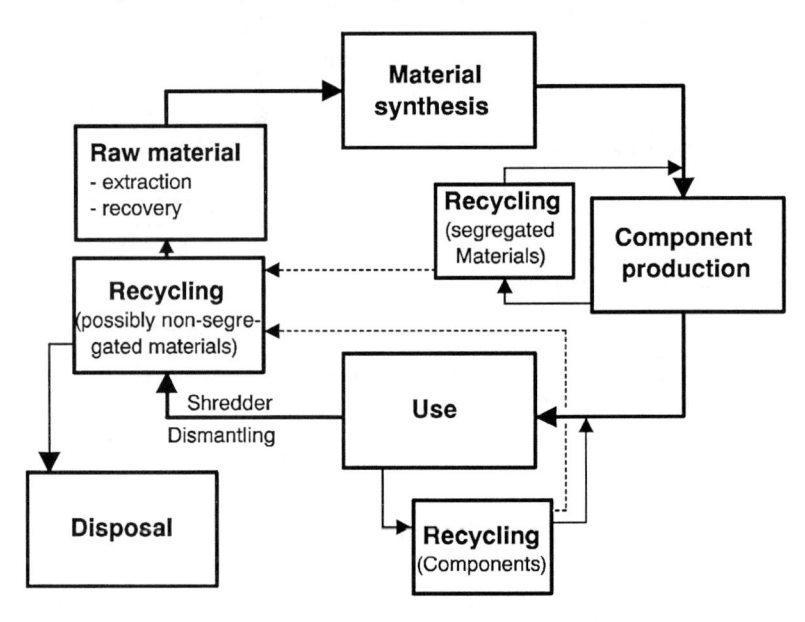

Figure 13 LCI (Life Cycle Inventory): Recycling within the product life cycle with alternative analysis limits.

Figure 14 uses the example of energy analysis of car fenders made of three materials (steel, aluminium and plastics) to investigate to what extent recycling - for which practically oriented recycling portions have been assumed - makes sense. Quite surprisingly, this specific example of component recycling shows that the energy saving is much smaller than would be expected from an energy comparison of the materials used. Main reasons are in particular the large difference between the component weight and the operational weight of the metal variants, as well as the low possible recycling portion of the high-quality plastic component. It should be clearly emphasised, however, that this example constitutes an exceptional case which is not generally applicable.

Material		Steel		Aluminium	Plastic	
Component weight Opertional weight		59 100		29 53	25 30	
Specific energy Consumption in MJ/kg for semi-finished products		16		100	79	
Energy consumption	Operational comp.	32		100	44	
Energy consumption	Production process	92		80	100	
Energy consumption	Sum (component)	45		100	57	Cycle 1
Energy consumption	Use *)	100		54	42	Cycle 1
Energy consumption	Recycling	85		100	93	Cycle 1
Energy consumption	Sum C + U + R	**100**		**95**	**63**	Cycle 1
Recycling rate (%) **)		20	100	100	20	
Energy consumption	Operational comp.	94	57	50	86	
Energy consumption	Production process	as for cycle 1				
Energy consumption	Sum (component)	96	72	58	91	Cycle 2
Energy consumption	Use *)	as for cycle 1				Cycle 2
Energy consumption	Recycling	as for cycle 1				Cycle 2
Energy consumption	Sum C+ U + R	**99**	**94**	**76**	**95**	Cycle 2
Energy saving cycle 2 versus cycle 1		1	6	24	5	

Data in relation to cycle 1

Figure 14 LCA of car fenders: Energy consumption of production, use, recycling, repeated use and repeated recycling (all data given in %)

6 SUMMARY

Even in the past, the automotive industry has spent vigorous efforts to design its products such that manufacturing, use and disposal are as environmentally compatible as possible.

The application of "Life Cycle Assessment" as an additional supporting tool for vehicle development has a high potential to make future vehicles even better in terms of ecology and technology but also with regard to economy based on a life cycle cost consideration. This requires on the one hand

- LCA know-how
- applicable tools
- a reliable data base on materials and procedures;

on the other hand, an LCA methodology is needed derived from general methods meeting the specific demands of the automotive industry. Such a "Car LCA" would needs be less detailed than previous approaches, but on the other hand it would be much more suitable for practically oriented analyses of a highly complex product such as automobiles.

In conclusion, some examples from practical analysis work indicate that LCA can be used as a supporting element in the automotive development process.

PART NINE

Specific Methods

39

Next Generation Product Development

Dieter Haban, Thomas Haase, Andreas Strobel
Computer Aided Engineering
Daimler-Benz Research and Information Technology, F3P
Wilhelm-Runge-Str. 11, D-89081 Ulm, Germany
Phone: +49/731/505-2834
Fax: +49/731/505-4210

Worldwide competition among manufacturing enterprises results in an increased pressure for high productivity, quality and flexibility. In addition, continuous and fast changing markets require rapid response to changes in demands, in product and manufacturing technology and in social environments. Manufacturing enterprises must operate highly flexible and efficient while avoiding wastage of resources. „Time-to-market", „lean management", „total quality", „just-in-time" are just a few examples of innovative concepts.

Facing these challenges, manufacturing enterprises improve their competitiveness by employing advanced information technology. Information technology supports enterprise operations from administration, construction, planning, engineering to manufacturing resulting in highly customized islands of computer aided systems. The physical integration of these islands is mainly concerned with communication networks and software archtectures. In this context, open systems play an important role since manufacturing and engineering systems are inherently distributed and heterogeneous. Despite all these efforts, information technology follows the philosophy of a functional-oriented structure, thus making it impossible to adopt more efficient flow-oriented structures. Each computer system uses different semantic models of the data which prohibits the direct exchange of information. A variety of standards, including IGES, EDIF are widely used to transfer data across applications. Investigations demonstrate a loss of information and dramatically increasing non-value adding activities along process chains, such as data conversion and data redundancy. A process chain comprises a comprehensive sequence of activities within an enterprise, such as a product life cycle from construction to manufacturing. Moreover, as the number of computer systems in an enterprise is increasing, the amount of data is growing to a critical point.

Only those enterprises capable of managing their distributed information systems can face the challenge of world-wide competition. The overall goal for the year 2000 is the entire integration of the enterprise: getting the right information, parts, processes, people to the right places at the right time. Advanced information technology and new data models tailored to flow-oriented structures support the continuity of information along process chains and lead to efficient joint application of information and manufacturing technology.

Within Daimler-Benz Research the division Computer Aided Engineering is focusing on two major topics:
- Continuous process chains in product development besed on integrated product data;
- Management of product data, especially Concurrent Engineering, for local and distributed product development.

Computer Aided Engineering is involved in projects for Mercedes-Benz, AEG, German Aerospace and debis. The projects center around product data bases for bill of material, geometry data, data exchange, process planning, integration of electrical and mechanical data, construction spaces. The demand for digital product development is increasing. Digital design, construction, simulation, evaluation enhanced with advanced visualisation techniques, such as Virtual Reality, promise time and cost savings.

In addition, product development is conducted in dispersed development teams and with external vendors. To support this distributed world-wide cooperation the use of federative databases and computer-supported teamwork via cyper-conferencing and multi-media will increase as the underlying techniques evolve.

The project Next Generation Product Development covers these objectives:
- build a visionary lab for entire digital product development using virtual reality;
- provide a CAD - CAE - CAM integration platform as a digital master throughout the product development process;
- improve the development process by applying Concurrent Engineering using Workflow, Multi-Media and Telecooperation.

CADVIEW is developed as a component within the project. CADVIEW is a system for the visualisation and the interactive manipulation of product data. The visualisation is done using shaded mode. Rotations and transformations are performed in real-time (with at least 6 frames per second). CADVIEW is capable of processing 10 to 100 times the amount of data compared to today's most poupar CAD-systems (until the end of '95, 1000 times the amount of data will be manageable by CADVIEW). This enables the user to keep the overview of even large models like necessary for the animation of a complete front section of a car.

Virtual Reality allows for a new vision of products. The designer is enabled to experience and judge the design using a spatial/three-dimensional view. This leads to a reduction in development time and to a better and more efficient information of the staff member. Data can be visualised in different ways:
- on the screen using Motif
- as a stereoscopic model on the screen using 3D-shutter-glasses
- as a stereoscopic model using a HMD (Head Mounted Display)

Mouse, spaceball or flightstick may be used as input devices. Socalled walk-throughs enable the designer to walk along the harness to identify possible collisions of the harness with mechanical parts (such as car body parts or ribs). The designer place himself into the engine compartment in order to verify his design „on the spot".

The CADVIEW-system supports two different mathematical models for the representation of CAD data: on the one hand as Advanced B-Rep in a CAD-kernel (ACIS), on the other hand using our specific data structure for fast visualisation of facets. Both data structures are linked for exact mathematical computations on the accurate B-Rep model. This is especially important when studying the packaging or Mock-Ups as the faceted representation lacks accuracy for the calculation of slices or collisions of the relevant CAD models. Using the B-Rep model it is furthermore possible to calculate lenghts and distances of different CAD models. The internal data is hierarchically structured in order to be able to represent e.g. bill of material information. The hierarchy or the positions within the hierarchical structure could be interactively altered.

STEP and ACIS-format serve as the data interface. Since the system is based on STEP AP214 and AP212, not only CAD data but also product data such as simulation or bill of material information and electric/electronic information can be visualised. The user interface enables engineers to access the product data via queries such as list of all wires of one harness or list of all devices connected to certain wires. Furthermore CADVIEW handles multiple data and therefore is capable of visualising data of different applications simultaneously. Another feature is the ability to take over 3D-scanner data of real objects and mix them together with the digital CAD data. As a consequence objects that lack CAD representation can be processed together with parts represented in the CAD system within a very short time-span.

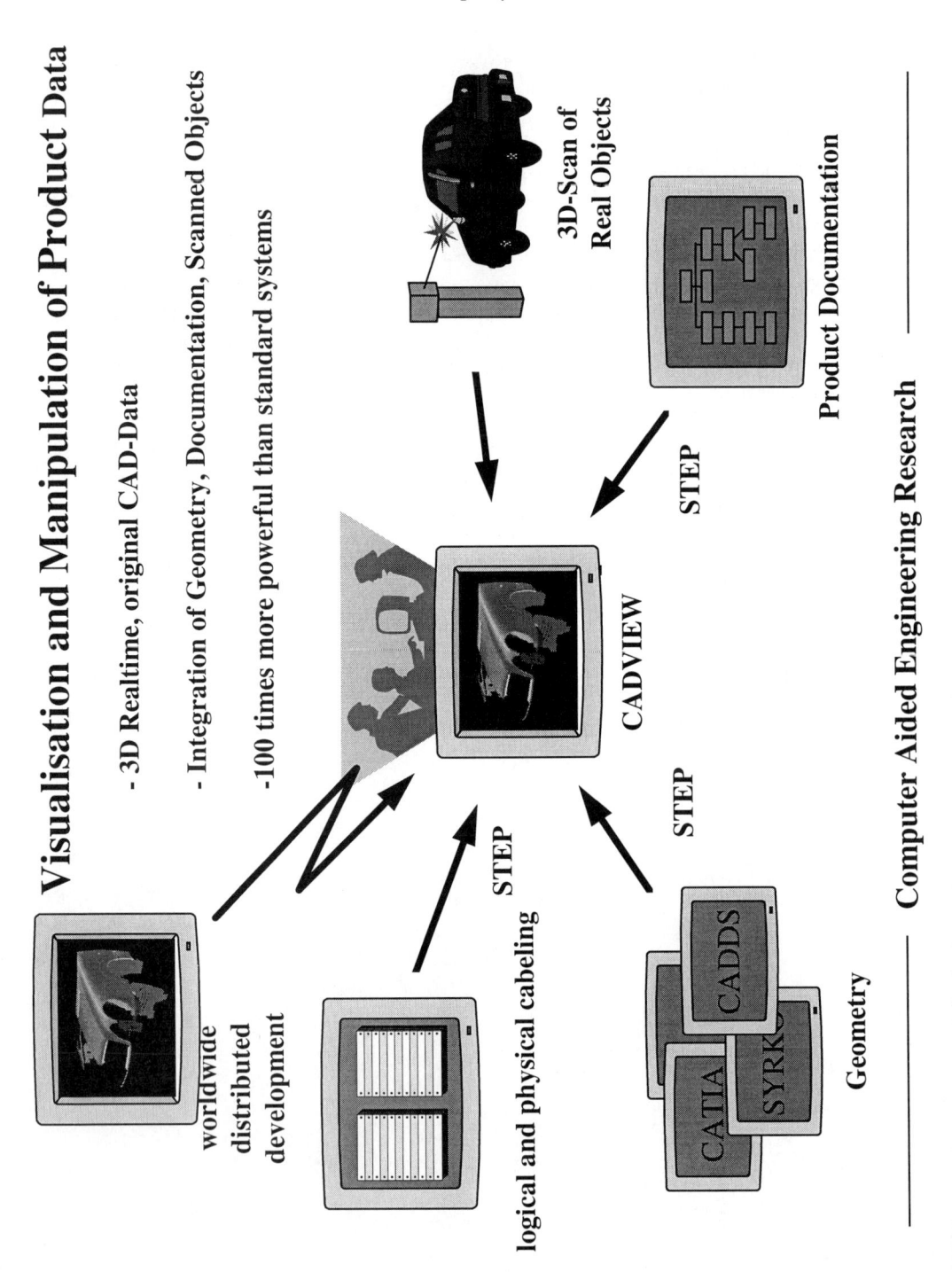

Visualisation and Manipulation of Product Data

- 3D Realtime, original CAD-Data
- Integration of Geometry, Documentation, Scanned Objects
- 100 times more powerful than standard systems

3D-Scan of Real Objects

Product Documentation

STEP

CADVIEW

STEP

worldwide distributed development

logical and physical cabeling

STEP

CATIA **CADDS** **SYRKO**

Geometry

Computer Aided Engineering Research

Visualisation and Manipulation of Digital Product Data with CADVIEW

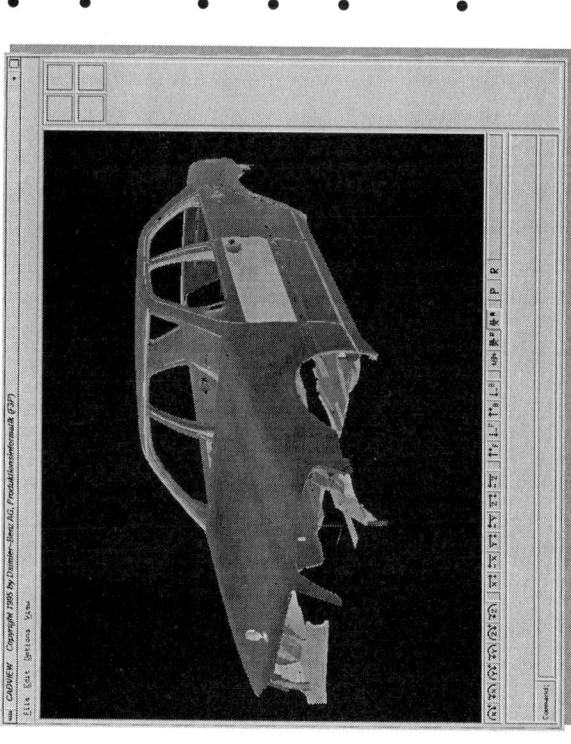

- 3D-Realtime, shaded image
- Mixing of CAD models, Product Structure, Documentation
- Input via STEP
- 2 Mio Polygons/sec (Workstation)
- Optional User Interface via Virtual Reality I/O Devices
- Support of Worldwide Distributed Engineering and Manufacturing

Computer Aided Engineering Research

Part Nine Specific Methods

40
Dynamically Modified Method of Data Model in the Product Development Process

Ruxin Ning, Bing Li
Beijing Institute of Technology.
No. 7 Baishiqiao Road, Beijing 100081, P.R.China
Telephone 86 - 10 - 8428281
Fax 86 - 10 - 8412889

Abstract
Concurrent Engineering requires a fast and user-friendly exchange of product information and of modified results between all employees involved in the process. So, it is very important to develop an integration environment which includes Feature-based product modeling, unified user interface and Data Management System. This paper presents such an integration environment to support the early stages of product development and possibilities for dynamic modifying of data model and maintenance method of data consistence.

Keywords
concurrent engineering, product model, dynamic modifying, meta-data

1 INTRODUCTION

Concurrent Engineering is recognized as a vital development strategy for modern manufacturing companies. The central theme of concurrent Engineering is the integration of the functions required for the development of a product so that the different design tasks can be processed simultaneously instead of sequentially.

In addition to creating a concurrent work team which consists of different field experts, it is important to provide an integration environment so that the simultaneous work of design tasks can be realized. This integration environment includes the feature-based integrated product modeling, concurrent engineering oriented unified GUI and Data Management System. Figure 1 illustrates the framework of this support system. It shows that the integrated product model is core to functionality of the system, the integrated data management system is the foundation of the system and the unified GUI provides powerful interactive tools for simultaneous work.

The research about the feature-based integrated product modeling has been discussed in numerous papers, and yet many problems in unified user interface and data management should be further studied and discussed. For example, the control of concurrent process, the dynamic modifying of data model, data consistence and traceability etc., are special considerations.

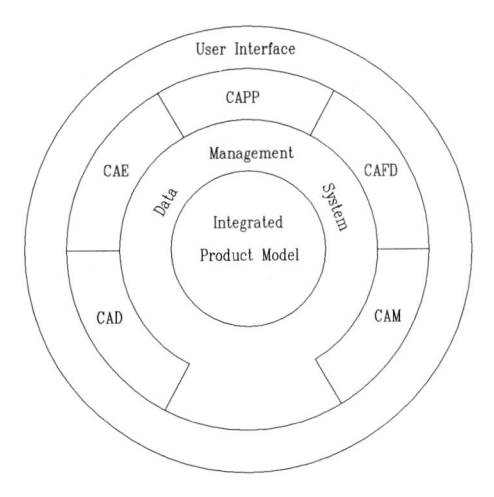

Figure 1 Integrated framework of information support system.

Feature modeling, object-oriented and visualization technologies provide the possibility for solution to the above problems. This paper will discuss how to use these new technologies on the basis of the currently commercial software system.

2 OVERALL STRUCTURE OF SYSTEM

This paper presents a Concurrent Engineering Oriented Support System which supports product development process, e.g. Computer Aided Design (CAD), Computer Aided Process Planing (CAPP) and Computer Aided Manufacturing (CAM). The overall structure of this system is shown in Figure 2.

This system is divided into five structure layers:
- computer, network and operation system.
- database, graphic base and knowledge base.
- integrated product model and product data management system.
- engineering application system.

- unified interactive interface and interface management system.

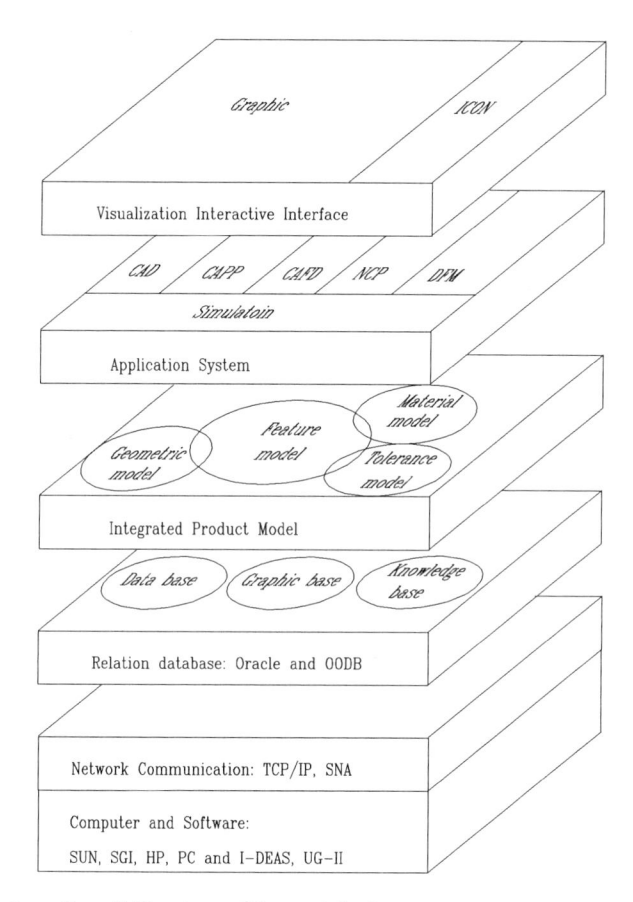

Figure 2 Overall Structure of Support System.

2.1 Unified User Interface

The highest layer in this support system is the Unified User Interface (UUI). The UUI was developed on OSF/MOTIF. All the subsystems are under the charge of the UUI. Here, it can be looked upon as a tool for controlling other activities required by an engineer. Using it, users can enter every subsystem, input initial designing data and modify data interactively in design process.

The UUI has partial intelligence. It not only provides all kinds of functions, but also can guide the users to operate correctly. By means of this interface, the user can select, modify or continue his own designing work, start a new work and look up other's result of design.

2.2 Unified DataBase Management System

Information sharing is the key to support multi-field team's concurrent design product and its related processes. In concurrent engineering, it's needed to process different types of information on a large scale. The explosion of concurrent engineering activity has created another bottleneck — managing the huge volumes of information. In addition, the need for rapid and accurate information handling drives the development of unified DataBase management System. In the view of contents, shared information includes product information and description information of developing process. In the view of types, it includes graphic information, structure information and text information, etc.

The traditional Relation DataBase cannot meet the requirements of complex object, dynamic modifying data model and consistence of data. So, we developed an OODB using C++ programming language on the base of Oracle 6. According to the characteristics of different data, the product structural data will be stored in OODB, the product geometric data in RDB and the documents in OS' file system. By the Unified DataBase Management System, all of the information in every stage can be shared by other related stages, and this supports the data consistence and traceability of the design process. Version control and trcebility are important for data consistence. It is a trend of concurrent engineering information system to integrate information on the basis of database.

2.3 Feature-based Integrated Product Model

The core of our research is to establish feature-based integrated product model in the view of design and manufacturing. Feature modeling is a kind of new product modeling technologies for the need of CAD/CAM integration. Compared with traditional geometric modeling, it can describe not only the shape of product, but also the manufacturing need of a product. So it meets the need of concurrent engineering [1]. It means that manufacturing can be considered while designing.

A form feature is a geometric shape defined by a set of parameters that have special meanings to design or manufacturing engineering. Manufacturing feature data provide constraints of how the feature should be produced with manufacturing knowledge in product model, so a feature carries implied engineering significance due to its shape. Thus feature based design and feature extraction play important roles to associate design and manufacturing information. With feature modeling technology, we can realize concurrent engineering oriented Design For Manufacturing (DFM).

Figure 3 shows an example with interacted feature model using the I-DEAS GEMOD. It uses product features described by both positive and negative form features. A positive feature is a geometric shape which encloses a material volume, such as a prism or a cylinder. A negative form feature is a geometric shape where material has been removed from the part, for instance a hole, a slot or a recess. In Figure 3, there is a convex cylinder in the middle of the rectangle concave part, and this shape is not defined in feature base. It needs two steps to create such a shape: first is to make the concave part and the second step is to attach the convex cylinder to the bottom surface of the chamber. In this example, the convex cylinder is a positive feature. The concave part and the sinks are negative features.

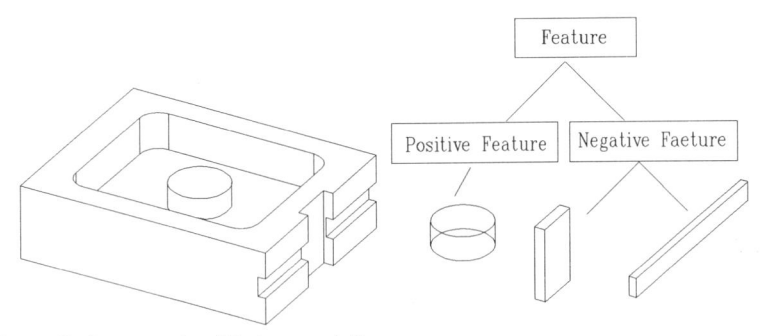

Figure 3 An example of feature modeling.

2.4 Concurrent Design Process

Integrated and structured large scale of product information provides a powerful tool for concurrent design and it allows the designers to start other related tasks such as CAPP, CAFD and simulation etc., as early as the beginning stage of design. Feedback information from CAPP, CAM to the designer is also essential because the designer must consider not only functions, but also manufacturability, serviceability, disassembly and recycling point view in every design stage.

At the design stage, major product cost is committed to choosing product features, material, tolerance and surface quality which are determined in CAPP and CAFD systems. In deciding on the design parameters, the designer should have option between the features required such as costly fitting or special tools and possibility to try other alternative features for manufacture.

The system we are currently working on will automatically produce machining plan information at the design representation phase. The part files that pass from process plan to NC programs automatically will be transferred to the database management system.

3   DYNAMIC MODIFIED METHOD OF DATA MODEL IN DESIGN PROCESS

In the product design procedure, the designer often modifies the design repeatedly to get a satisfactory result because of the product's complexity. This requires the system to support "trial-and-error" procedure. Object-oriented technology and meta-data are used to meet this need. By object-oriented technology, the complex entity can be described and the relation of components can be matched automatically. By meta-data, users can modify product model dynamically.

3.1 Object-Oriented Technology

The product model has complex structure and varied data types. It is difficult to describe it by traditional relation model. The object-oriented model provides the ability to describe complex entities and has varied data types. So it can meet the requirement of product model.

Besides the general advantages, the object-oriented model in this system uses the concept of relation match to satisfy the need of data consistence, especially automatic match in components assembling. The system guides the user to input the match relation of components while defining product model. This relation recorded in database could trigger the "match processor" and ask the user to modify the related data of related component if he (or she) wants to update one or more data in this relation.

3.2 Meta-data and Dynamic Modified Method of Data Model

Meta-data is a special kind of data used to describe other data, such as data of table structure. In this system, we can store the features' definition by means of meat-database. Because the product model is established on the basis of feature, how to define, select, add and modify features in feature database is important to the flexibility and adaptability of the whole system.

The structure of database management system is illustrated in Figure 4. All of the operations on product model are done by the meta-data processor.

When one operation of applications to store the product model takes place, the DBMS will invoke meta-data processor to fetch relational meta-data and then decide the actions to complete the operation. When the product model is modified, the meta-data processor will change the product model data stored in meta-database (it is very easy to do this because every DBMS has the ability) , but not update the structure of database (the traditional RDBMS has no ability to do this). Then, the meta-data processor will reload the data of product into database automatically at a right time. After the product model was changed, all accesses to it will remain the same. It is because the information of whole product model is stored in meta-database (not in applications).

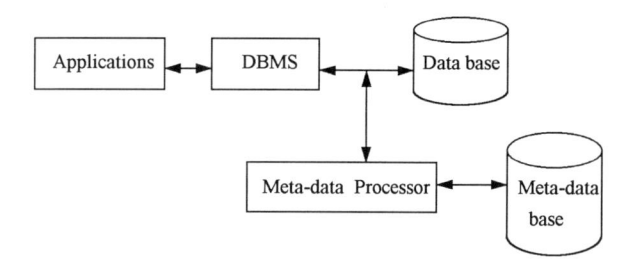

Figure 4 The structure of database management system.

The following is an example given to illustrate how to modify data base dynamically.

For instance, the feature counter hole is defined by the following three parameters: diameter 1 (D1), diameter 2 (D2) and height (H). It is obvious that a parameter is lack. These three parameters cannot describe the feature counter hole completely. There should be two parameters of height (H1 and H2) and two of diameter (D1 and D2). In a traditional system, it is difficult to deal with this matter because most parameter information is written in program and the only way is to modify the source code. In this system it is easy to do it. When the user found that he (or she) made a mistake in defining the feature, he (or she) could choose the "modify -> add" item from the system menu, and the system could guide the user to input the feature name: counter hole and the new parameter will be added: height 2 (H2). If the user want, he (or she) could also rename the defined parameter H as H1. After the user confirms the modification, the system record the new parameter information to replace of the older.

4 CONCLUSION

The integrated prototype system presented in this paper can complete preliminary product concurrent design at present. Feature model technology makes DFM possible, object-oriented technology provides an environment and tools for design process modeling. Visualization technology provides a user-friendly interactive interface. We accomplished integration and concurrent process of CAD, CAPP, CAFD, NC programming and simulation on the base of unified user interface and unified DBMS.

A large scale complex information will be processed in concurrent engineering, and compared with serial work, the information exchange is two-way, so unified DBMS is still a problem in the system, especially for control and trace of modifying data in design process. We will use artificial intelligence (AI) technology to establish knowledge-based DBMS. Additionally, we will make a cost accounting tool to establish a better environment for concurrent engineering.

REFERENCES

Ning Ruxin, Tang Chengtong, He Yongxi (1994) CAD/CAM Integrated Model Based on Feature. *Journal of Beijing Institute of Technology*, Vol. 14 No. 2, 181-5.

Duan Xiaofeng, Ning Ruxin (1995) Feature model and CAD/ACM information integration. *CAD/CAM*, 9, 28-30.

Sohlenius, G. (1992) Concurrent engineering. *CIRP Annals*, Vol. 41, 645-56.

About the Authors:

Ruxin Ning: Prof. Dr.-Eng. , vice president of Beijing Institute of Technology, graduated from Harbin University of Technology in 1965, obtained Dr.-Eng. degree from TU Berlin in 1987. Reasearch interests: CAD/CAM, CIMS and FMS.

Bing Li: doctoral student of Beijing Institute of Technology. Reaserch interests: CAD/CAM and Engineering DataBase.

41

Industrial methods for product and process development - a case study

J. Vallhagen
Chalmers University of Technology
Dept. of Production Engineering, 412 96 Gothenburg, Sweden
Fax: +46 31 772 38 19, e-mail: jova@pe.chalmers.se

Abstract
A case study was made in the Swedish industry, to record the actual methods used for development of products and assembly systems. It shows that the procedures usually have an organizational base with cross-functional project groups, prescribing which activities should be performed in different development phases. This provides a good base for parallel, integrated, and co-operative work in the spirit of simultaneous engineering. Opinions differ between designers and production engineers on how well these development procedures meet the needs. Production engineers often feel slighted and have few possibilities to influence the product design even when they feel it is warranted. No specific methods or systematic approaches were found that pertain to design theory or methodology. However, a need for such methods has been identified.

Keywords
simultaneous engineering, product design, assembly process planning, methods

1 INTRODUCTION

During the last few years the interest in product development processes has been very high. *Simultaneous engineering* is the framework for much research with subsequent changes in industrial activities. This paper reports a study of current conditions in Swedish industry. To define the issues studied, a review of previous work by the author is given as a background. A framework for the study is furnished at the end of this chapter (1.2).

1.1 Previous work

Reported here are parts of a long-term research project with the objectives of generating methods and models for integrated development of products and manufacturing systems. The aim is to identify and improve the decision processes of the early development stages in order to reduce lead-time. This can be achieved by integrating a well-structured methodology with engineering tools in a simultaneous engineering environment (Vallhagen, 1994a). The simultaneous engineering approach mainly wants to enable work in parallel. This, however, can be accomplished only after careful deliberation. The difficulty is that the nature of product development and process planning necessitates that some tasks are handled in a predetermined sequence. Therefore, one has to investigate what can be carried out *in parallel* and what must be done *in sequence*.

The literature already reports some work on integrating design and planning by combining *computer-aided design* (CAD) and *computer-aided process planning* (CAPP) (Allen et al, 1991) (Hird et al, 1988) (Wang, 1992). These methods are to be used in the *detailed design stage*, since the possibilities for parallel work and reduced lead-time in the *conceptual stage* are limited. Some work is reported on assembly systems and sequencing (Arpino et al, 1988), (Baldwin et al, 1991). Assembly planning, however, suffers from a scarcity of well-structured methods and procedures.

The product characteristics set by the designer is the input for the planning of parts manufacturing and assembly. In Figure 1, the principal stages in *product design* and planning of *parts manufacturing* and *assembly* are illustrated in conjunction with those product characteristics that affect planning. This model has been proposed (Vallhagen, 1994a) for the sequence of activities and information flow. One aim of the study is to confirm this model.

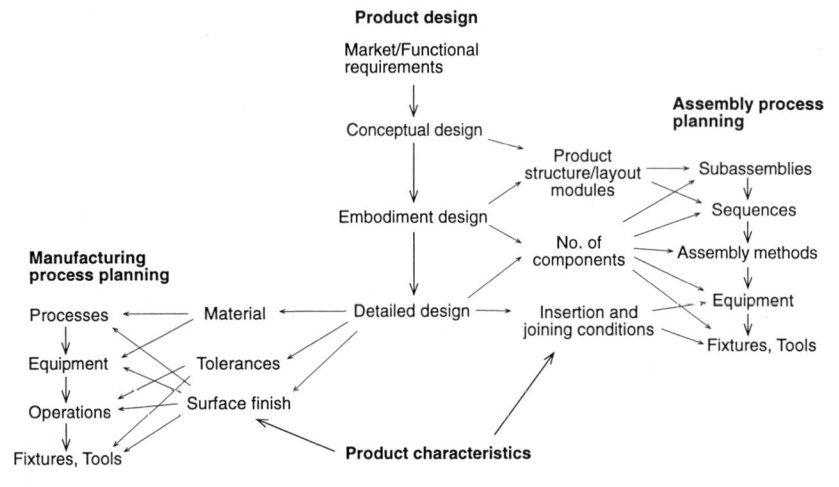

Figure 1 Design and planning sequence model.

Note that activities in this model can, or must, be carried out in parallel, since each step in the process must be taken with lateral inputs and is iterative in order to home in on the solution. The model illustrates how the sequences in the design stage affect parts and assembly planning. Also, parts manufacturing and assembly are dependent on each other, but their relation is not illustrated in this model. The arrows in Figure 1 represent, beside the information or decision flow, also the procedure for going from one node to the other. These steps involve the creative work to design the product and the manufacturing system. It should be supported by well-structured methods and guide the synthesis and the analysis of solutions.

For this work the *axiomatic design theory* (AD), developed by Nam P. Suh, stands as a theoretical base, being a systematic method for guiding the design process and analyzing the results. A model for the axiomatic design theory (Suh, 1990) is illustrated in Figure 2. In axiomatic design, mapping between *customer domain, functional domain*, and *physical domain* is concerned with the design of the product. Mapping between *physical domain* and *process domain* is concerned with the design of a manufacturing process or system. The axiomatic design theory has been analyzed for the case of designing multipart products made in automated manufacturing systems (Vallhagen 1994b). In that case, not just process variables for parts manufacturing are of interest, but parameters also are needed for functions related to assembly, such as materials handling, planning, and control, to fulfil the task. The conclusion is that the axiomatics in the original form cannot handle such a case properly. Therefore, a modified axiomatic design model has been suggested. For single parts manufacturing, however, the original method can be used, which is to find the appropriate process variables for realization of a component.

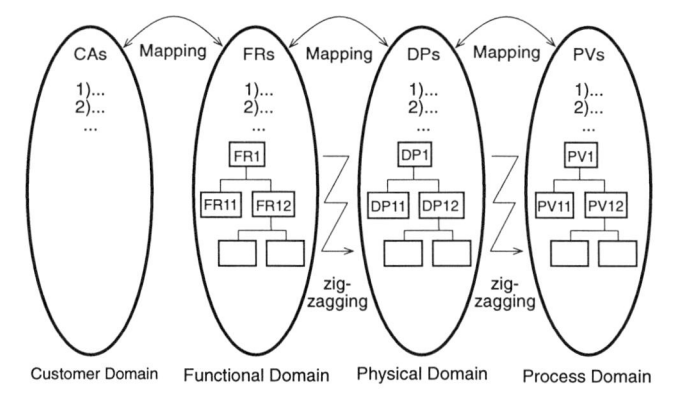

Figure 2 Model of the axiomatic design theory with its four domains of the design world.

1.2 Case study approach

When developing new methods, industrial needs should also be taken into account. Otherwise there is a risk for developing *ad hoc* methods which is not desirable. However, there is not much information in the literature about recent methods used for assembly planning. Therefore, a study was initiated to ensure that no important information for further development would be excluded. A framework was formulated in four question areas to cover the pertinent issues:

- What methods are used for product development and assembly process planning today?
- How and when do the designer and the production engineers communicate and cooperate?
- What factors affect the choice of solutions in their work?
- How does the proposed model in Figure 1 apply to industrial procedures?

The method used and the performed case studies are described in chapter two. In chapter three, the results are discussed and analyzed. In chapter four, conclusions are drawn.

2 METHOD

In this area it is difficult to collect quantitative data about work methods and procedures. Instead, practical insight into the work procedures and methods is needed in qualitative terms with which comparisons can be made. Therefore, simple questionnaires are inadequate. Instead more extensive studies have to be made through personal interviews. Guidelines for case studies are found in (Hellevik, 1987) and (Yin, 1988). It can be assumed that methods for product development vary between companies. Therefore, six studies were planned in different companies.

Three of the case studies were made in large-size companies with more than 5000 employees and three in small or middle-size companies with about 500 employees. The three largest ones were in the Swedish automotive industry and are referred to as companies A, B, and C. Among the smaller ones, D is a supplier of components to the automotive industry, E is manufacturing components for the electronics and computer industry, and F *desktop accessories*.

The case studies were designed to first list a set of questions arranged in a flow-chart structure. At this stage, two prestudies were performed. The questions and the structure were tested on selected persons with long experience in product and process system development. Structures of the proposed questions then were modified and two inquiries were developed. One pertained to development of products and one to development of assembly systems, covering the first three of the four issues previously mentioned. The question structures are illustrated in Figure 3 for product development, and in Figure 4 for assembly systems.

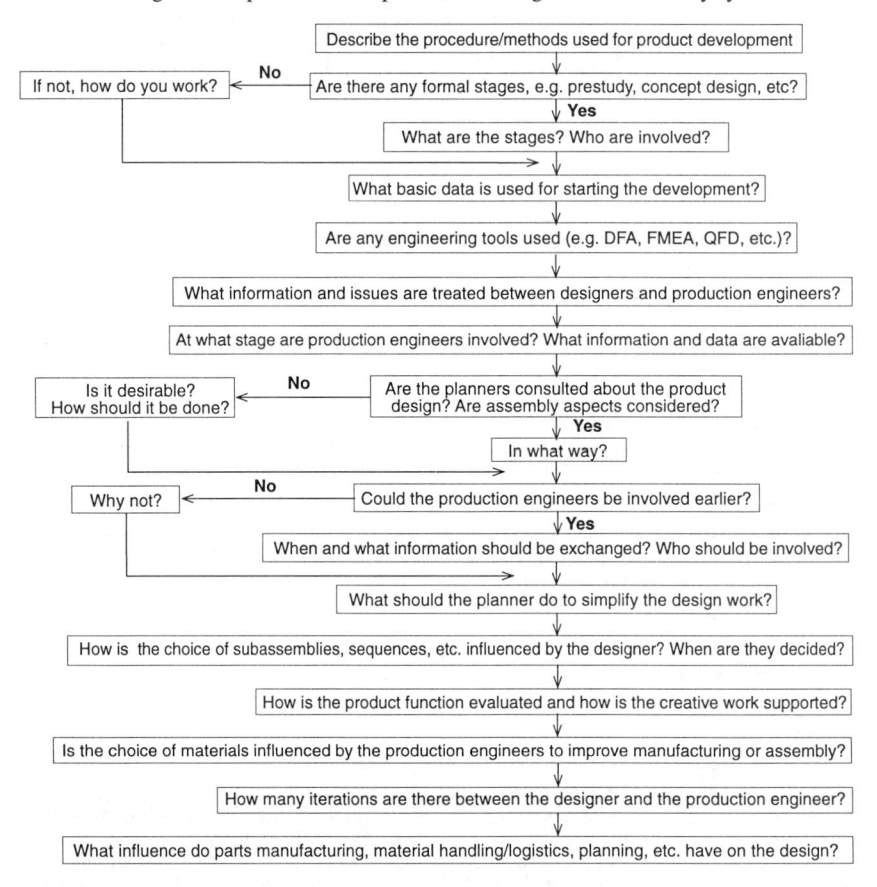

Figure 3 Question structure for study of product development.

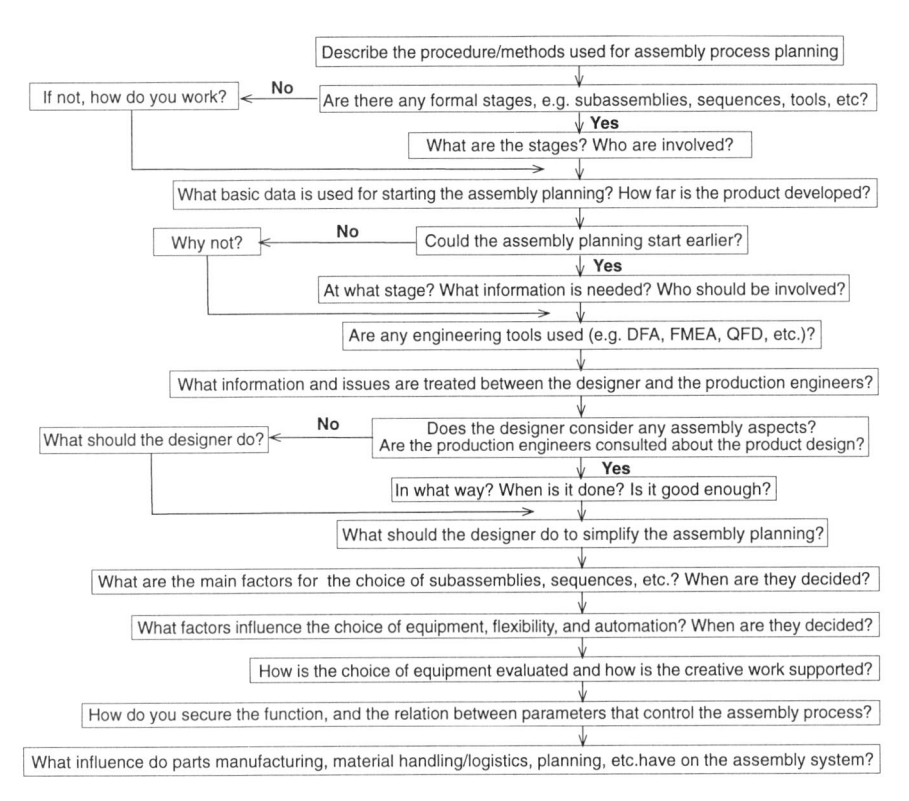

Figure 4 Question structure for study of assembly systems development.

To perform each case study, a contact person in the company first was found, who gave an introduction to the organization, products, and manufacturing methods. A development project was selected to be used as a reference. One or two designers and production engineers were selected for personal interviews. The inquiry was distributed in advance and was adhered to during the interview. At the end of the interview the sequence model in Figure 1 was presented and its relation to the industrial experience was identified.

3 DISCUSSION AND ANALYSIS

Due to the limited space available no direct reproduction of all results can be presented. Instead, a comprehensive, mainly qualitative discussion and analysis around an abstract of the most important findings is given. Even if the character of the results mainly is qualitative it is still useful to have some quantitative data to clarify what is good and what is missing in the development procedures. This also makes it possible to compare the different cases. With the results on hand, four key topics were selected as a measure of the product development capabilities and are discussed in sections 3.1 to 3.4.

- *Organizational conditions.* This is a measure of how well the organization meets the needs for a simultaneous engineering approach.
- *Integration and parallel work.* This is a measure of the success in performing the work in a parallel and integrated manner.
- *Methodology.* This is a measure of how much of the work is supported by tools and systematic methods.
- *Assembly considerations.* This is a measure of the effort spent on making the product easy to assemble.

For several reasons, statistical methods for analysis and presentation had to be disqualified. Instead, each topic has been classified in three categories: low, medium, and high, and displayed in bar diagrams. As opinions differ between designers and production engineers, the results from both are presented. This is more or less a relative measure to compare the case studies and not an absolute value of product development capabilities. In each case study the author has evaluated the situation and assessed the potential for improvements. The classification is based on the information obtained through the interviews and the author´s knowledge and experience from earlier research.

In section 3.5 a discussion of *product development processes* used in the companies is given and related to the four topics above. In section 3.6 some factors affecting the choice of solutions are discussed, and in section 3.7 the comments on the design and planning model from Figure 1 are summarized.

3.1 Organizational conditions

The product development work is, in most cases, based on cross-functional project groups having representatives from different functions in the company. In large projects, a hierarchical structure of project groups is used. Regular meetings is the main forum for information exchange, discussions, and decisions. The organization is more or less adhering to a plan (i.e. a *product development process*) for activities and documents in the different development phases.

The major difference between companies is that the larger ones have separate departments or functions dedicated to process development and planning (in cooperation with production engineers in the work shops). In the smaller ones, this work is sometimes done by production engineers from the work shops, which usually restricts the possibilities for a full time commitment in a development project. Even then, the organizational conditions are good in most of the studied cases (see Figure 5) especially in the larger companies, and a good base for a simultaneous engineering approach is on hand.

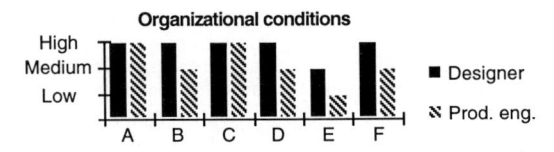

Figure 5 Comparison of organizational conditions in the studied companies.

3.2 Integration and parallel work

Since the development is organized and performed in project groups, almost all issues can be jointly discussed or handled. This should facilitate integrated and parallel work, since all functions take part in the different activities and phases. Co-operation and contributions are also usually formalized in the different phases of the development process. The company culture is very important for solidarity and team spirit. The different functions must respect each other as internal customers. Experience and personality are pointed out as especially important factors for close co-operation and good joint solutions. A comprehensive view of the problems and their solutions is vital especially for the designer, as he often has a key role in co-ordination and finding compromises.

Most designers have the belief that the production engineers are involved at an early stage with good insight and possibilities to influence the design. The majority is satisfied with the way it is presently handled, but some are also aware that more can be done. The production engineers, on the other hand, often feel that they are involved too late and too little in the early stages, when they would have had the best opportunity to be informed and to give input to the design work. Usually it is the lack of resources that limits the possibilities for parallel work in the early phases. The results have shown that the situation is better in large companies having a separate function for process planning, or otherwise being fully committed to the development project. In such cases the conditions for parallel and integrated work must be regarded as optimal (methodology aspects excluded).

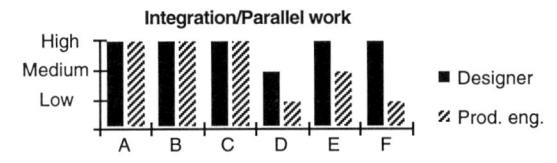

Figure 6 Comparison of integration and parallel work conditions in the companies studied.

3.3 Methodology

Early in the projects, the focus is on developing specifications based on studies and evaluation of technical, market, and economical prospects. Together with ideas and concepts of a proposed solution this is the basic body of information for starting a new project. The specifications cover desired functions and performance of both the product and the production system, which also includes suppliers and their processes.

Methods with *engineering tools* are used in most companies, mainly FMEA, DFA, and QFD but also a few others. SPC is sometimes used in the manufacturing process. These tools are proven and considered valuable, but also too inflexible and "bureaucratic", and this is the main reason for resistance to their introduction. When fully implemented they are usually accepted, though,, but the resources avaliable set the limit for how many tools can be used. Bench-marking or analyses of competitors' products are used to a large extent, both for finding new ideas and to confirm new solutions. The experiences from using *carry-over* are usually good, but sometimes this limits the possibilities for improvements in the production process. New technology and *off-the-shelf* solutions should be developed before a new product development project is started. It is very risky to develop new techniques *during* an ongoing project. Testing of prototypes and preproduction series is by far the most used method to verify and confirm functionality, both of the product and the assembly system, and has to be strictly documented.

The use of computers is widespread, especially for design of products or molding tools, and also smaller companies are well updated on these tools, making them less dependent on prototypes for early verifications and analyses. Computer models for simulation of function, and in some of the larger companies also of assembly, are being used. Product data model systems used are mainly administrative, with databases of drawings and parts lists describing what parts are used in a specific product or variant.

Perhaps the most interesting finding is that no methods seem to be used that pertain to the area of design theory or methodology. In academic research, much work is done on methods and models for conceptual design and functional modeling as a support in the earliest development phases. One opinion is that such methods and tools suffer from not being able to handle variants but only a single product. The ability to handle product assortments would be very interesting and valuable from an industrial viewpoint. Theories and methods developed must also be easy to understand, learn, and use, which is not always the case.

Thus, methods used in the industry are characterized by being applied in the later phases. There is a lack of systematic methods for the earliest stages, when specifications, system solutions, and concepts are being defined. They entail some difficulties in getting a good overview of all related requirements and functions. Therefore, the methodology conditions in the companies studied cannot yet be considered to have reached more than a mediocre level.

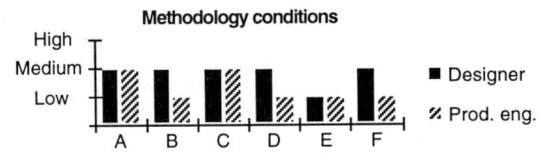

Figure 7 Comparison of methods used in the companies studied.

3.4 Assembly considerations

Most designers feel that they strive to make assembly considerations from the start, while still appreciating any feedback or suggestions from the production engineers. In the larger companies the latter usually agree; even if the final product is not optimal from an assembly point of view, it has been considered and evaluated from many aspects at an early stage. It is important, though, that they are creative in their suggestions and not just make complaints.

In the smaller companies, the production engineers often would like to be consulted or involved at an earlier stage, formulating assembly requirements as an input to the designer. They often experience that assembly is given less priority than parts manufacturing. In their opinion, by the time they are consulted the design is already set at a detailed level, where significant changes to simplify the assembly no longer are possible. Design changes in the later phases are usually characterized by short time limits, and assembly considerations may be neglected for that reason. Sometimes there is a general requirement that the product be designed for automatic assembly. However, if automatic assembly is not to be used from the start, the consequence usually is that this is not sufficiently provided for, which causes problems when automation is brought up after a few years.

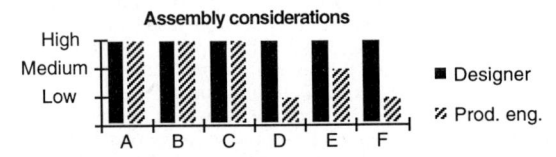

Figure 8 Comparison of assembly considerations made in the companies studied.

3.5 Development processes

For their product development most of the companies during the last few years have started using some model or process that they have experienced as valuable. This process usually contains 3 - 6 different phases: one prestudy phase, one or two concept or design phases, one or two phases for preproduction activities, and a production phase. For each phase, different activities are prescribed and time plans are established to describe when to do what, and what documents are needed for the decisions. This is more or less detailed and specified in the development processes used in different companies. Usually no special methods for problem solving are prescribed, except in some cases where certain engineering tools are used. The larger companies have used such processes for longer periods of time, compared with the smaller ones, but all are well aware of their importance for efficiency and economy. In the smaller companies the processes are less comprehensive, probably because they have fewer resources and smaller organizations.

In most cases, the designers feel that they are a natural part of the development process, but also sometimes that the procedure is not clear enough, making it difficult to know how to work in the different phases. This is probably due to the process not being comprehensive or detailed enough in prescribing work procedures for the different activities. The production engineers often have some difficulties identifying themselves and their work with the development process and do not always have the possibility, or are not allowed, to take part from the start. In their opinion the process focuses too much on the product. The reason for this is perhaps that a new development process is handled by people with a close relation to, or a background from, product development. Another reason is perhaps that the procedures are not yet fully implemented in all projects, as they are rather new to the smaller companies. There is also a difference in how designers and production engineers have responded to the questions. It seems the designers have given a more visionary picture of how they will or would like to work in the future. The production engineers are more likely only to express their experiences.

To get a better overview of conditions and results for product development, a summary is presented in the form of polar diagrams. In these diagrams, the four topics so far discussed in this chapter, are viewed in conjunction with how much support is given by the development process and at what time they are established or performed. The *time factor* indicates whether they are introduced or performed early, in the middle, or late in the project. The *process support factor* indicates whether the product development process gives little, average, or much support to fulfil the tasks. The estimated conditions from the view of both designer and production engineer are displayed.

In Figure 9, the pattern of good conditions in the larger companies can be recognized on almost all points. The agreement on these subjects, between designers and production engineers, is also fairly good. The situation in the smaller companies, which is illustrated in Figure 10, shows that they need further progress to reach an ideal environment, especially to bring the designers and production engineers to the same levels.

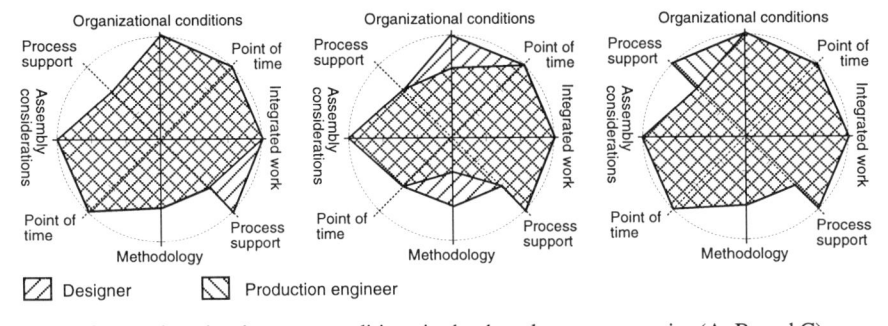

Figure 9 Product development conditions in the three largest companies (A, B, and C).

3.6 Factors affecting choice of solutions

The data for starting the assembly process planning vary from sketches or early drawings to more or less advanced prototypes or parts lists. The assembly process is usually decided or suggested by the production engineers and later discussed in the project group.They usually start by deciding the assembly sequence, estimating costs, calculations, and analyses for

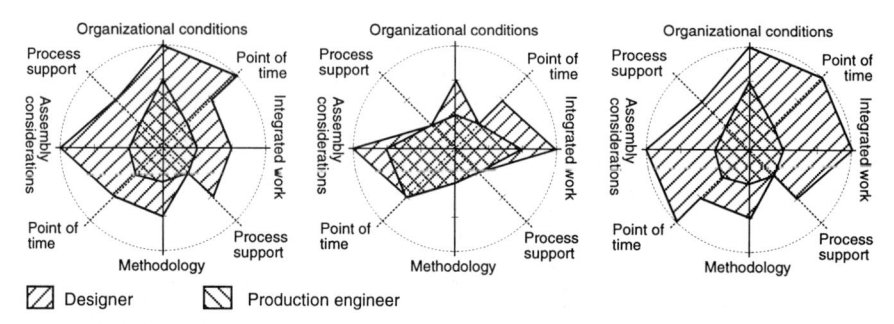

Figure 10 Product development conditions in the three smallest companies (D, E, and F).

different decisions and options. Usually it is the available space and the balancing of the assembly line that have the largest influence. Control functions to ascertain that parts are assembled as required are also included.

Regarding the choice of assembly sequences and subassemblies, a number of factors with influence can be identified. There are also different opinions on this matter. Several of those interviewed agreed that the design has a large influence and must be adapted to the desired process, which sets the limitations on how to design. Another opinion is that the choice of sequences and subassemblies is not much affected by the structure and concepts chosen by the designer, since he is not well versed in the assembly process.

The choice of manual or automatic assembly is very much dependent on the expected production volume but also at times on some strategic reason. Automation can also aim at increased volume if the design permits. The choice of equipment and assembly methods relies much on experiences from older equipment, or from *benchmarks*. The product design and its features set the conditions or requirements for what must, or can be done, manually or automatically, for economical reasons as well as to attain a safe and reliable process.

Material handling and logistics is an important part of the assembly process and must be considered carefully. It should be minimized to be able to concentrate on adding value to the product. Easily damaged parts with high surface finish must be handled gently which sets requirements on methods, tools, and equipment used. An overall view is important to anticipate how to handle all variants, philosophies of *just-in-time* deliveries, and production planning methods.

3.7 Comments on the model in Figure 1

The most frequent comment is that planning of parts manufacturing should start at the conceptual stage and be parallel to design and assembly process planning. Especially in those cases where the product is made of parts (e.g. sheet metal or injection-molded parts) manufactured with expensive tools having long lead-times for development, manufacturing, and delivery. Today, suppliers or subcontractors are used not only as purveyors of equipment and components, but are involved also in the design work as specialists, and their feedback must be included in the conceptual design.

The major reason for starting the planning early is to cut lead-times. This, however, usually results in many design changes, since the information avaliable when the planning starts, or when the molding tools are ordered, only is preliminary, and the detailed design is not finished and verified. Furthermore, in many cases the assembly process is tested and proven with parts made in the finished tools and not just prototype parts. Therefore, higher precision in the early phases is necessary to be able to decrease development costs and reduce the number of changes.

Positive comments are that this model would be the optimal way of working, since, in some cases, perhaps assembly should be considered first and set the requirements or conditions for parts manufacturing. Most important, it is perceived that assembly should have larger influence on the product design than it has today. The assembly method used is important for the priority and attention given to the planning. Automatic assembly is given more attention and priority than manual, due to the larger investments needed and the lead-time for development and implementation. If the development starts in the detailed design stage, with well-known concepts, planning for both parts manufacturing and assembly can start at the same time. In this case the minimum lead-time depends on the time for process planning, installation, and starting up the production.

Regarding details in the model, comments have been made that insertion and joining issues should be handled earlier, at the concept stage, as this affects the choice of subassemblies. The relations between these steps are quite complicated and a choice cannot be made without affecting others. It seems that a changed order of *assembly sequences* and *subassemblies* in the model is appropriate. Sometimes, however, it can be decided at an early stage that a part of the product should be a module or subassembly. In this case, the interfaces and joining methods must be considered and designed early. It is important, however, not to decide the detailed design too early, but to put a lot of effort into finding the best assembly structure and flow possible. Another frequent comment is that *process requirements* should be introduced at the same level as *market/functional requirements,* as an input to the design work.

To sum it up, the conclusion is that the type of product, production method, and magnitude of the project affect how the planning is done and when it starts for different processes. Automatic assembly requires more efforts at an early stage and a product designed for such processes. The question of early attention and priority is also a matter of where the largest value is added: in parts manufacturing or in the assembly process. The model in Figure 1 needs some further development, and a modified model is illustrated in Figure 11. In this model, alternative sequences, or procedures, have been suggested and are illustrated with indications for when they should be used. Thus, the model is an attempt to generalize the sequence of design and planning activities in a development project. For a specific company or product, however, the details in such a model will need customization

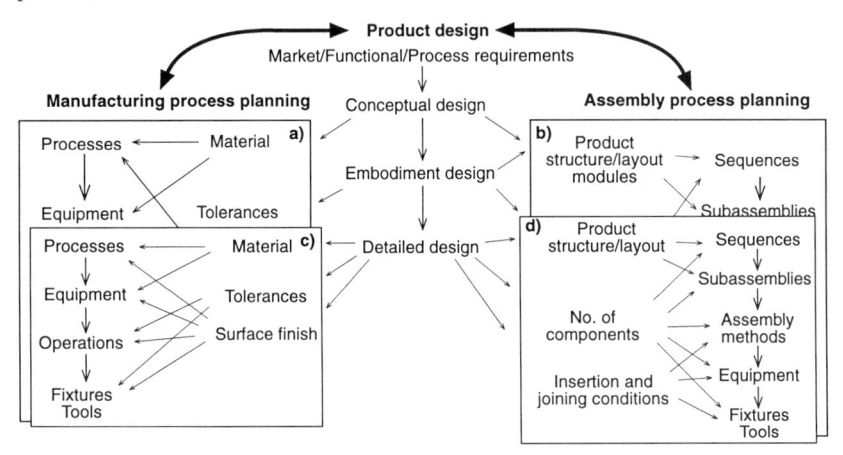

Figure 11 Modified sequence model with alternative procedures. a) and b) are used in case of long lead-time for development, delivery, and heavy investments. c) and d) are used in case of detailed design changes or for processes that need less exertion.

4 CONCLUSIONS

The results show that in most companies some model or procedure is in use for product development. It usually springs from an organizational base and consists of a document describing how all activities are to be performed in different phases. Thus, the conditions for a simultaneous engineering approach are good, but work methods and procedures need improvements.

Large companies are ahead in this area and show good control of their product development process. The smaller companies are aware of its importance but are using less comprehensive or strict development models or have not yet approached their full implementation. The study shows a significant difference of opinion between designers and production engineers on how well their respective development process meets the desired conditions for close co-operation. The production engineers feel that they are still slighted in some cases and cannot influence the design as desired.

Except for the use of some engineering tools and computer support, no specific assembly process planning has been found that engages methods and models from the area of design methodology or theory. However, since there is a clear and strong relation between the product, its function, its physical realization, and the functions needed in the manufacturing system, there is a need for such a methodology. Therefore, a model or method is desired that describes all relations and supports synthesis and analysis of the evolving design. Experience from this study will be valuable for further development since it has presented a comprehensive overview of industrial activities and needs in this area.

5 REFERENCES

Allen A. J., Bielby M. S. and Swift K. G. (1991) Development of a Product Manufacturing Analysis and Costing System. The Int. J. of Advanced Manufacturing Technology, 6 (2).

Arpino F. and Groppetti R. (1988) ASSYST: a consultation system for the integration of product and assembly system design. Developments in Assembly Automation, IFS Ltd.

Baldwin D. F. et al (1991) An Integrated Computer Aid for Generating and Evaluating Assembly Sequences for Mechanical Products. IEEE Trans. on Robotics and Aut., 7 (1).

Hellevik O. (1987) Forskningsmetoder i sociologi och statsvetenskap. Bokförlaget Natur och Kultur, Lund, Sweden, (in Swedish).

Hird G., Swift K. G., Bässler R., Seidel U. A., Richter M. (1988) Possibilities for Integrated Design and Assembly Planning. Developments in Assembly Automation, 155-166.

Sohlenius G. (1992) Concurrent Engineering. Annals of the CIRP, 41 (2), 645-655.

Suh N. P. (1990) *The Principles of Design*. Oxford University Press, New York, 1990.

Vallhagen J. (1994a) *Assembly and Process Planning in Axiomatic Design*. Licentiate Thesis, Department of Production Engineering, Chalmers University of Technology, Sweden.

Vallhagen J. (1994b) Aspects on Process Planning Issues in Axiomatic Design. 1994 ASME Design Technical Conferences – the 20th Design Automation Conference, Minneapolis, Minnesota, USA.

Wang Y., Hsieh L.-H., Seliger G. (1992) Knowledge-Based Integration of Design and Assembly Process Planning. CIRP Int. Sem. on Manuf. Systems, 24,135-145, Copenhagen.

Yin R. K. (1988) *Case study research - design and methods*. SAGE Publications, USA.

ACKNOWLEDGEMENTS

Financial support by grants from *the Volvo Research Foundation and the Volvo Educational Foundation*, and from the *inpro*-program of *the Swedish National Board of Technical Development* (NUTEK) is gratefully acknowledged. I would also like to thank all those who contributed to this study.

PART TEN

Recycling

42

Design for Ease of Recycling

Wolfgang Beitz

Technical University Berlin
Phone: +49/30/314-23341
Fax: +49/30/314-26481
E-mail: beitz@kt10.tu-berlin.de

Abstract
The paper will give a survey about steps and rules to realize recycling friendly products. Further it will explains that requirements and product characteristics are necessary for a life cycle product modelling.

Keywords
Recycling, methodical design approach, life cycle product modelling.

1   INTRODUCTION

In the future the designer has to consider the utilization or the reusing of full products, subassemblies and parts in connection with a recycling process. He has to design for ease of recycling, but he must also consider the other requirements for marketable technical products, f.e. requirements according function, safety, ergonomic, operation, manufacturing and assembly.

We differentiate between so called Pre-Consumer Recycling (immediate recovery of production waste and material during product manufacturing), Post-Consumer Recycling (recovery of materials after product use) and Consumer Recycling (reuse of reconditioned products after their first using period) (VDI 2243, 1991).

The paper will give a survey about steps and rules to realize recycling friendly products as well as some typical applications. Further it will explains that requirements and product characteristics are necessary for a life cycle product modelling.

2 GENERAL STAGES OF DESIGN

The guideline 2221 (VDI 2221, 1986) and (Pahl/Beitz, 1993) give proposals for a systematic design process. Fig. 1 shows the main steps of such systematic approach combined with typical recycling orientated tasks for the designer. Very important for a recycling friendly product development is a complete specification with requirements for all fields of a marketable product quality. To consider recycling requirements during the design process it is necessary to have a definition of an effective recycling procedure before starting the product development. This definition is not easy because at most of products we have a longer period of time between the product development and the recycling process. During this period the recycling technologies can be changed and then the product structure are not more optimal for these technologies. Nevertheless it is possible to give general rules for designing recycling friendly products (VDI 2243, 1991).

3 GENERAL RULES

Consumer recycling (product recycling)
The goal of consumer recycling is to use products again after a first operation phase by reconditioning them. To manufacture such products, we need a product layout which enables or eases the reconditioning operation.

Designing for ease of disassembly. A suitable structuring of assemblies allows an economically dismounting of the product without damaging the parts. This can also be aided by selecting joints which can simply be accessed and separated.

Designing for ease of purifying. The layout of the parts to be reused should permit those purifying process which are unproblematic to the environment.

Designing for ease of testing an classifying. The embodiment should provide an easy and clear recognition of the state of wear and corrosion of parts to be reused. To ease the classifying of disassembled parts with identical functions, they should be standardized, and should be marked according to their structure, connecting dimensions and materials.

Designing for ease of reconditioning. The reprocessing of disassembled parts can be supported by providing additional material as well as gripping and adjusting facilities.

Designing for ease of reassembly. The layout should favor easy and customary assembly for both reconditioned and new parts.

Post-consumer recycling (material recycling)
The essential objective is to keep the quality of the old material as high as possible. There fore, a compatible mixture of old material, meeting the requirements of the utilization technology, is necessary. All parts must fulfill such a mixture or must be separated by means of preparation.

The following rules are general: Designing for ease of

- minimizing the waste;
- reducing the spectrum of used materials;
- using reutilizable material in compatible mixtures for dismountable assemblies;
- disassembling incompatible material combinations.

To fulfill these aims the designer needs at first informations about wastage groups, their allowed mixtures together with a suitable usable source. Further he needs informations about the joining and detaching behavior of connections.

For product and materials recycling it is important to realize a easy disassembly (Schmidt-Kretschmer/Beitz, 1991). Fig. 2 and Fig. 3 shows guidelines for this aim.

4 EVALUATION CRITERIA FOR RECYCLING

At all steps of the design process it is necessary to evaluate and select solution variants. Fig. 4 gives for a hand drill relevant criteria for material and product recycling as an example. The recycling properties of a product must have a good integration with the other technical and economic properties. It is necessary to define a overall value (Fig. 5).

5 PRODUCT MODELING

A product model must integrate all phases of the product generation process (Fig. 6) and all product data. Such model could be combined of a basic model and attached partial models. Fig. 7 shows the structure of a overall product model and Fig. 8 partial product models for the part *recycling*).

6 EXAMPLES

Fig. 9 shows a modern dish washing machine with good disassembling quality. All high tech components like pumps, drives or electric installations are founded in the basic box without special connections. Through move the housing into the normal position the components shall be fixed. At preparation or regeneration the components are easily to disassemble.

Fig. 10 shows a gear box of a hand drill. After losing banded brad, the shaft of the electric drive can be disassembled without problems.

7 REFERENCES

VDI 2243 (1993) Konstruieren recyclinggerechter technischer Produkte. Düsseldorf: VDI-Verlag.
VDI 2221 (1993) Methodik zum Entwickeln und Konstruieren technischer Systeme und Produkte. Düsseldorf: VDI-Verlag. English-Edition: Systematic Apparoach to the Design of Technical System and Products. Düsselforf: VDI-Verlag 1987.
Pahl, G. and Beitz, W. (1993) Konstruktionslehre-Hanbuch für Studium und Praxis, 3. Aufl., Berlin: Springer-Verlag. English-Edition: Engineering Design - A Systematic Apprach. London, Berlin: Design Council an Springer-Verlag 1988.
Wende, A. (1994) Integration der recyclingorientierten Produktgestaltung in den methodischen Konstruktionsprozeß. VDI-Fortschrittberichte, Reihe 1, Nr. 239, Düsseldorf: VDI-Verlag.
Beitz, W. and Grieger, S. (1993) Günstige Recyclingeigenschaften erhöhen die Produktqualität. Konstruktion 45, 415-422.
Schmidt-Kretschner, M. and Beitz, W. (1991) Demontagefreundliche Verbindungstechnik - ein Beitrag zur Produktrecycling. VDI-Berichte Nr. 906, Düsseldorf : VDI-Verlag.

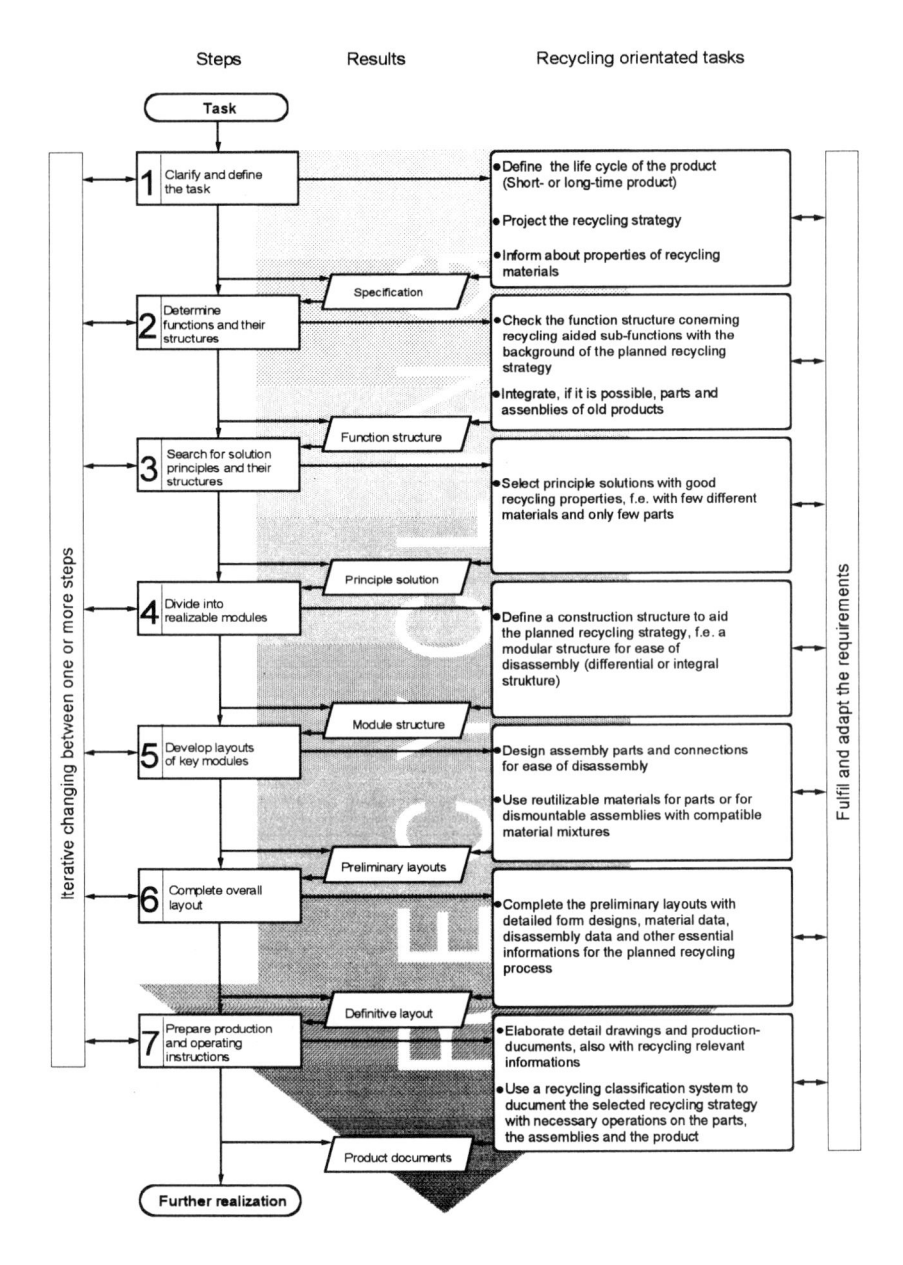

Figure 1: Steps of systematic design and recycling orientated design tasks (VDI 2243, 1991; Wende 1994).

Guidelines	adverse to disassembly	easier to disassembly
manufacturing structure		
- Structure to disassemblies, whose parts and materials are reutilizable	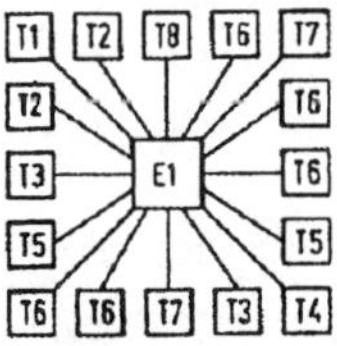	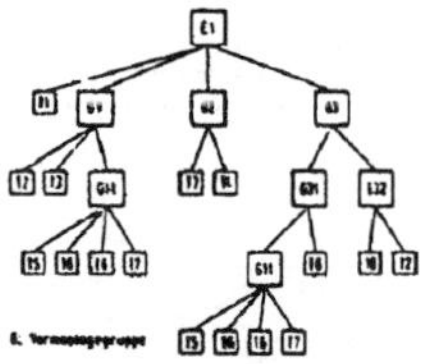
- Attach a base part of a disassembly group to an easy reutilizable wastage group	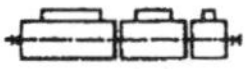	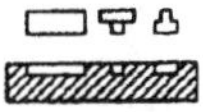
- Avoid dismountable composite constructions with uncompatible materials	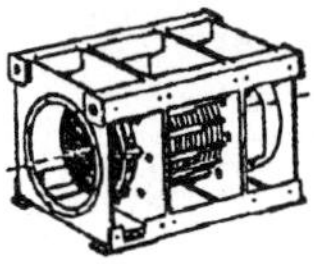	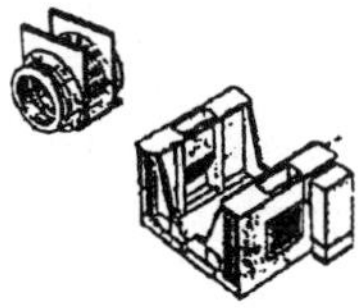
- Reduce joining points	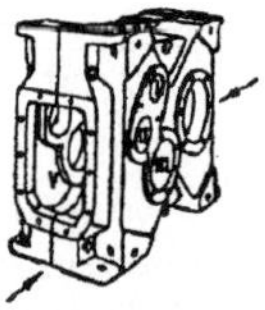	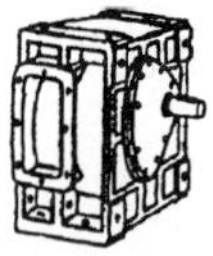
- Aiming for constrained disassembly-operations (Rule of clarity)	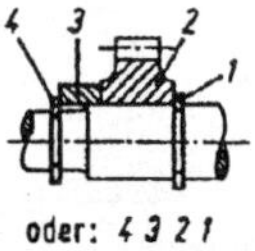	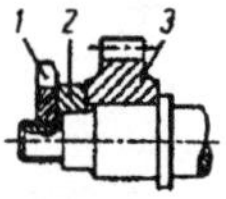

Figure 2: Guidelines for disassembly of manufacturing structures (Schmidt-Kretschmer/Beitz, 1991).

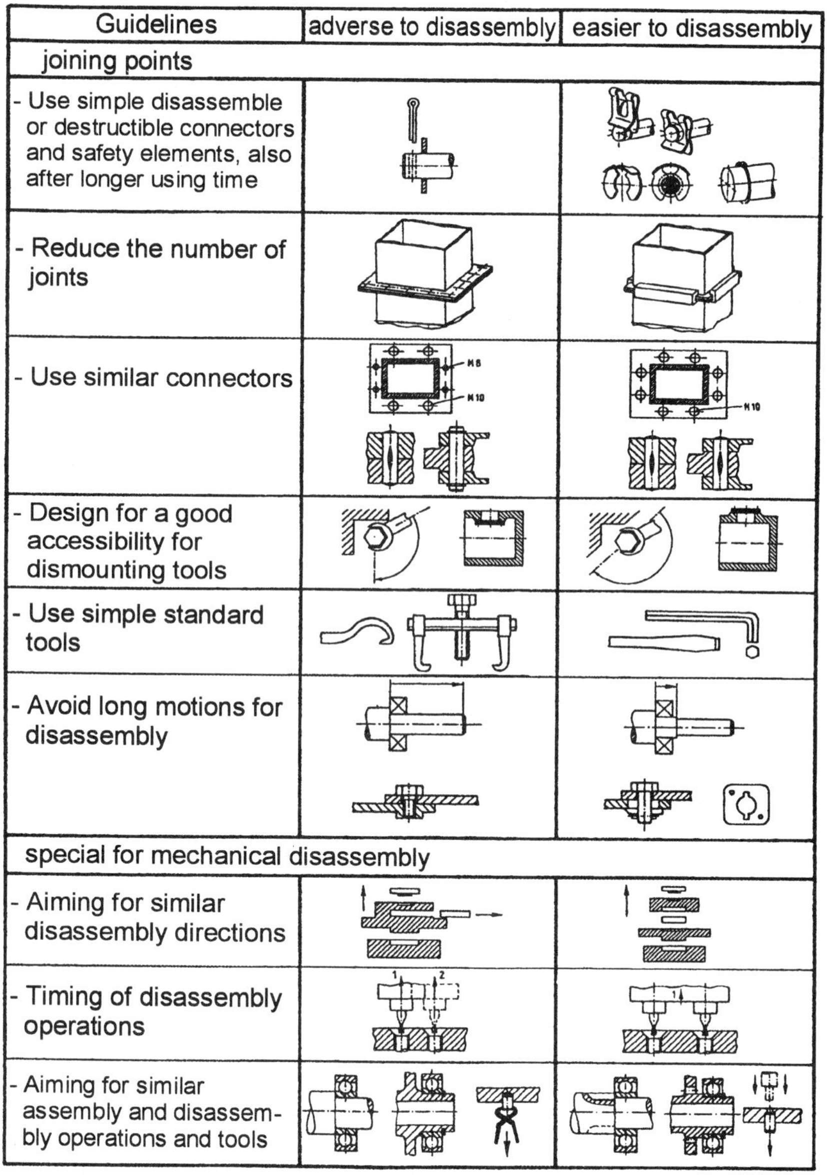

Guidelines	adverse to disassembly	easier to disassembly
joining points		
- Use simple disassemble or destructible connectors and safety elements, also after longer using time		
- Reduce the number of joints		
- Use similar connectors		
- Design for a good accessibility for dismounting tools		
- Use simple standard tools		
- Avoid long motions for disassembly		
special for mechanical disassembly		
- Aiming for similar disassembly directions		
- Timing of disassembly operations		
- Aiming for similar assembly and disassembly operations and tools		

Figure 3: Guidelines for disassembly of joining points (Schmidt-Kretschmer/Beitz, 1991).

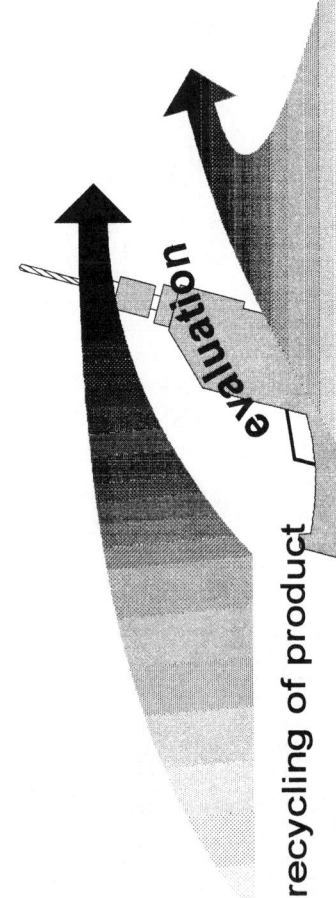

recycling of product

structure of product concerning function
pre-disassembling
disassembling
nondestructive disassembling
easy testing
easy identification
easy classification
reconditioning
easy reassembling
easy exchangeability of components
for upgrading
modular structure
easy identification of wear
using standard components
few parts
automatisation of the steps of recycling

recycling of material

disassembling:

number of steps of disassembling
number of directions of disassembling
number of different steps of desassembling
number of joining elements
number of different joining elements
easy access for disassembling
automatisation of disassembling
work to detach the connection
no special devices for disassembling
small number of necessary tools

separation process:

number and expenditure of necessary
steps of separation processes
identification of material
number of materials which are to
separate or which are not reuseable

possibility to reuse material:

possibility to reuse metal
possibility to reuse plastics
recycling process
upgrading of material
grade of reclamation
quality loss of material
grade of impurity or dirt

Figure 4: Recycling criteria for a hand drill (Beitz/Grieger, 1993).

Design for ease of recycling

technical rating
- simplicity, clarity
- small size
- low weight
- sufficient insulation
- clear positioning of parts and components
- clear joinig positions
- reutilization of joining elements
- sufficient stiffness of actuation
- sufficient stiffness and strength of the housing
- sufficient safety of overload
- compatible and safe handling
- low vibrations

economic rating
- few parts
- simple components
- few an usual manufacturing processes
- easy automatisation of manufacturing
- significant division in subassemblies
- easy parallel assembly
- short distances for joining
- simple junction areas
- easy testing
- easy automatisation of testing

recycling rating
- overall layout design
- reusable parts and components
- disassembling
- separation processes
- possibility to reuse metal
- possibility to reuse plastics

→ overall value

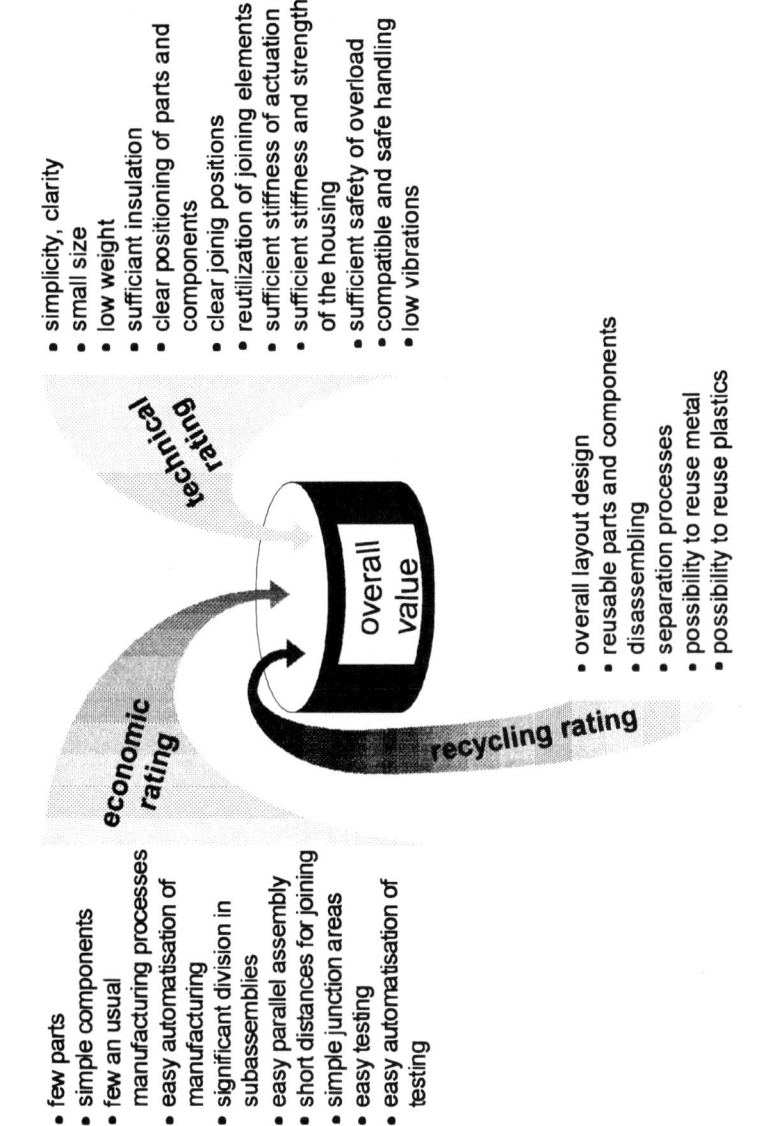

Figure 5: Technical, economic and recycling criteria for the overall value of technichal products (Beitz/Grieger, 1993).

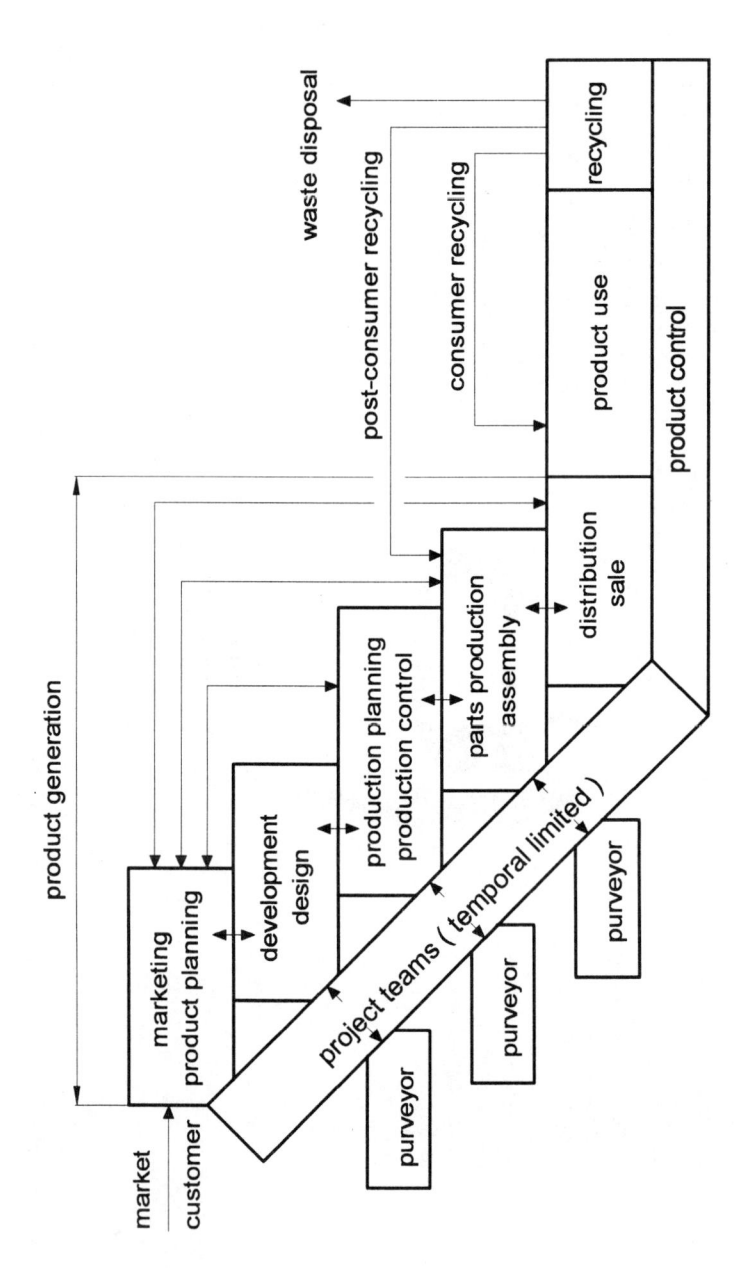

Figure 6: Process of product generation with Simultaneous Engineering.

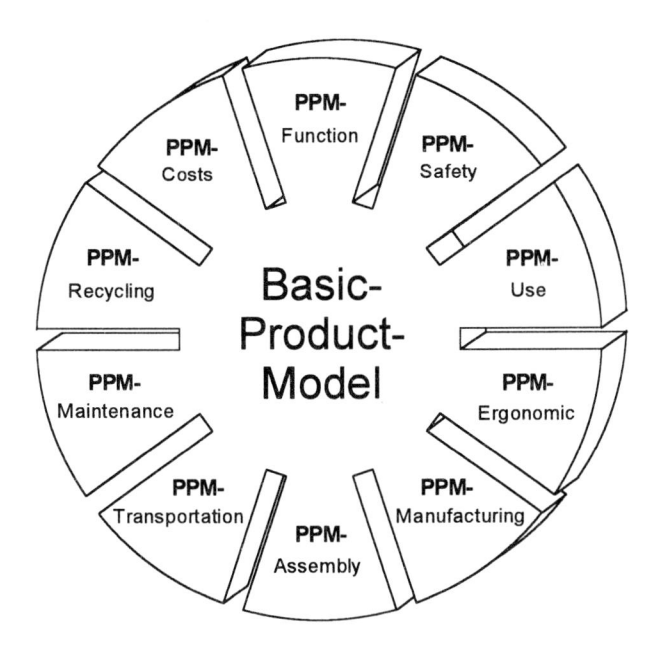

Figure 7: Basic product model with Partial Product models for all fields of requirements (Wende, 1994).

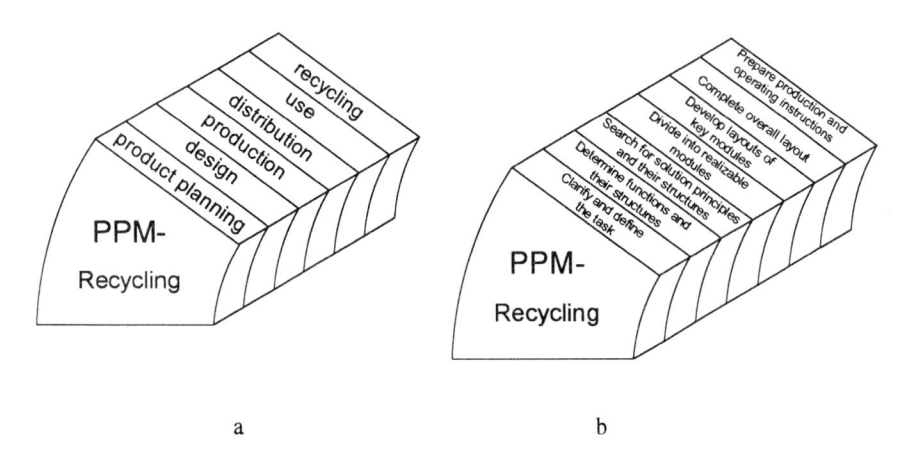

Figure 8: Partial Product models; a) for all product life phases (Wende, 1994), b) for the steps of product development (Wende, 1994).

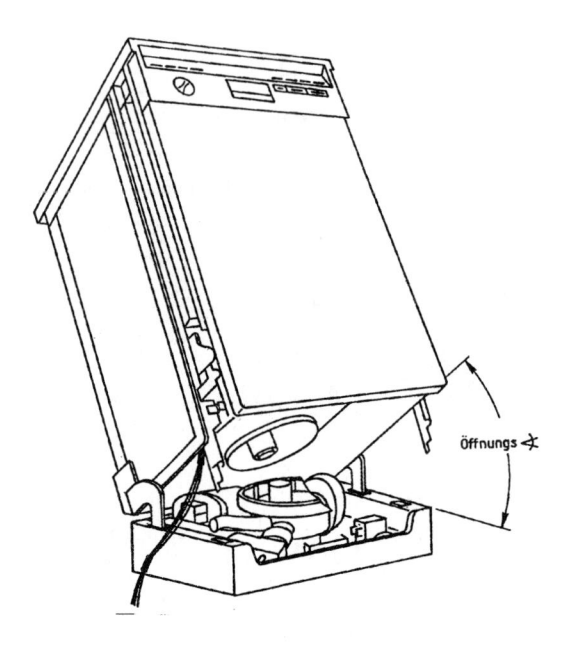

Figure 9: Dish washing machine (Bosch-Siemens)

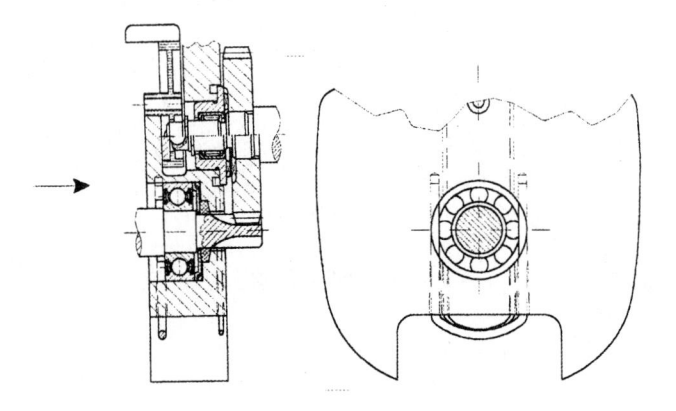

Figure 10: Gear box of a hand drill (Bosch)

43

Design for Recycliability- an Analysis-Tool in the "Engineering Workbench"

Harald Meerkamm, Johannes Weber
Institute for Engineering Design, University of Erlangen-Nuremberg
Martensstraße 9, Germany, D-91058 Erlangen
Tel.: ++49 (0) 9131 / 857986; Fax.: ++49 (0) 9131 / 857988
Internet: weber@mfk.uni-erlangen.de

Abstract
In this paper the Design System mfk is introduced as an assistance system for the engineer, which has been developed at the institute for Engineering Design (KTmfk) for 8 years. This system, which is based on an integrated, semantic and relation based product model includes a wide range of analysis tools, which support the designer in all fields of demands, e.g. design for production, cost, stress, etc.. Therefore this system can be seen as an approach to an "Engineering Workbench".
An important objective of this paper is a detailed discussion of design for recycliability on the one hand and the presentation of the computer integrated analysis tool RecyKon for supporting the designer in finding out the best recycling-strategy on the other hand. The general concept of this tool will be presented as well as the knowledge preparing part, the process modules and the different interfaces. This tool can be integrated in the a.m. Engineering Workbench.

Keywords
Engineering Workbench, Design System, Design for Recycliability

1 INTRODUCTION

The designer's work has not become easier within the last years, in spite of improved hardware- and software-tools and basic design methods. The reason for it is a very tough time restriction in its work and completely new and additional demands in products (Figure 1). The number of restrictions in product development is increasing. In former times it was sufficient to design for function, for production and for cost /1/. Actually "Design for Recycliability" has become an additional restriction.

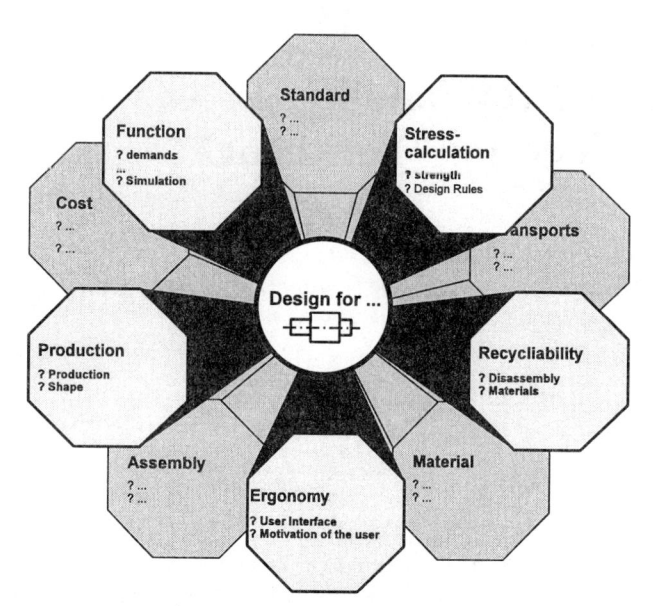

Figure 1 Restrictions and Demands in Product-Development.

As the engineer has to fulfill simultaneously the great quantity of demands, which partly are in contradiction to each other, we should provide him with effective analysis tools.

2 NECESSITY AND TARGETS OF DESIGN FOR RECYCLIABILITY

Design for recyclability is an almost new subject in the field of design for X (DFX). The necessity to consider more intensively this aspect as one part of the design for environment, has its roots not only in the actual problems of waste management, but also in the shortrunning of ressources. The social interest i.e. the environment responsibility, furthermore economical reasons in the field of waste management are the two reasons for a company to aim the the following targets (Figure 2):

- According to "Neuwertwirtschaft /2/" and "Upcycling" of products, reusable components and assemblies have to be to identified and removed easily.
- The integration of harmful materials - if really necessary - has to be carried out in a way that provides separation by simple means. Thereby the remaining materials can be reprocessed much easier.
- Those materials and compounds have to be prefered in product development, which can be reprocessed to raw materials economically.
- By using combinable materials, which can be regenerated to new fractions, their reuse can be ensured.

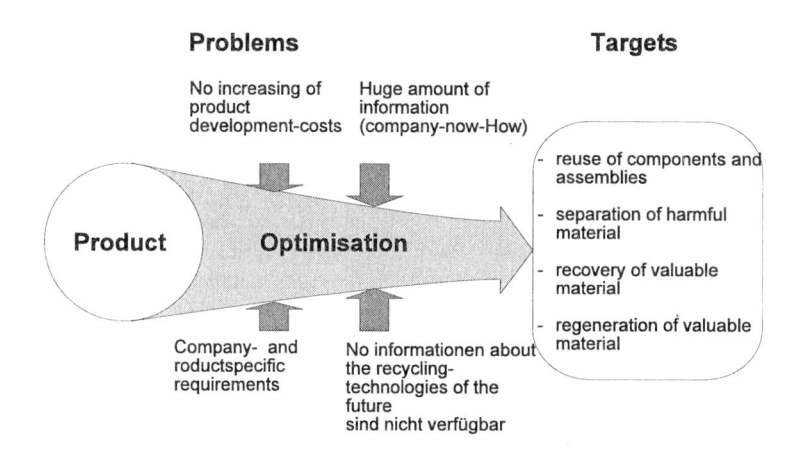

Figure 2 Problems and Targets of the Design for Recycliability.

For reaching the a.m. aims, the following fields of problems have to be considered:

- Using tools and methods of design for recyclability may not increase the cost of product development.
- The huge amount of information, which must be considered, has to be structured as well as possible and adapted to the requirements of the designer.
- Interfaces have to be created in order to provide the integration of company-specific requirements and product-characteristics in an universal tool.
- The designer's limited knowledge on recycling topics has to be considered. There is not all data of waste management technology available, because the time gap between product developement and the end of use is too big.

3 CONCEPTS OF SOLUTION

In order to assist the designer in solving these difficult problems, an effective analysis tool should be developed. Figure 3 shows the architecture of the computer aided tool **RecyKon,** that supports the designer in the Design for Recycliability. It is developed at the institute for Engineering Design (KTmfk) at the University of Erlangen-Nürnberg under the leadership of Prof. Dr.-Ing. H. Meerkamm.

This "Design for Recycliability-Tool" consists of the modules:

- **Recyclinggraph** as the representation of the product and therefore as the Product Model
- Recyclinggraph-Editor **REGRED** as a user interface
- Analysis modules
- Technology and process knowledge base
- Internal and external interfaces

The individual parts of this analysis tool are developed and improved in a project together with the Institute of Manufacturing Automation and Production Systems (leadership Prof. Dr.-Ing. K. Feldmann). This cooperation offers good possibilities to prove the requirements of disassembly to the software tool, that has to be developed.

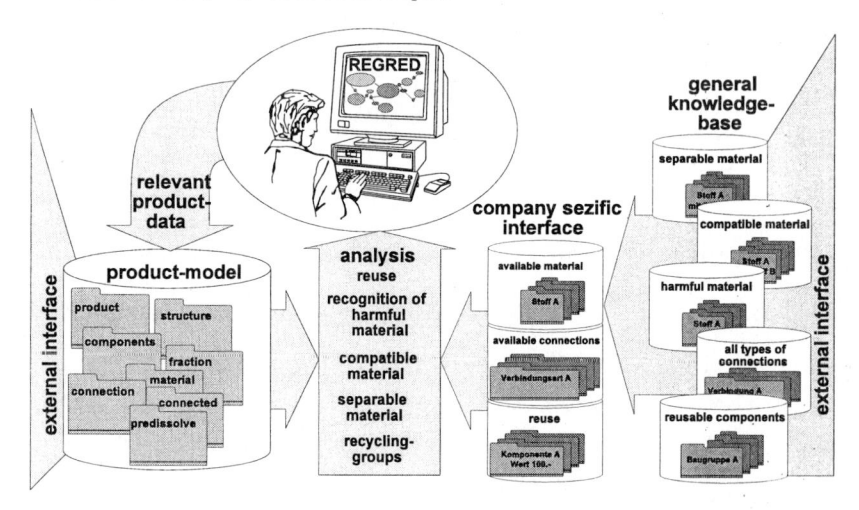

Figure 3   Overall Structure of the RecyKon.

In chapter 4 the integration of RecyKon in the Design System mfk will be explained. This system is realized as an Engineering Workbench, including all analysis tools necessary for the demands on product development.

3.1 PRODUCT MODEL

Core of the system is a semantic, relational, object-orientated product model (Figure 4), containing all essential product-information regarding to questions of recycling and disassembly. According to /5/ the information consists of the subjects geometry, function, technology and organisation for the product but also for each component.

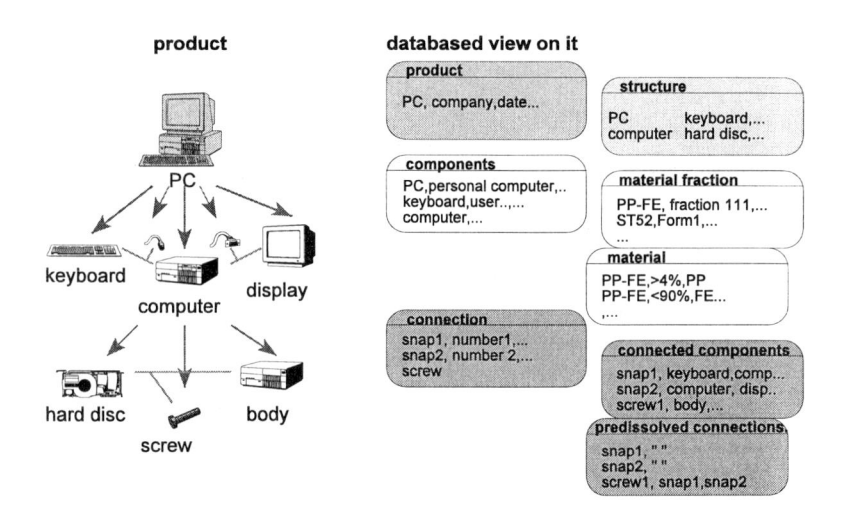

Figure 4   The Product Model / Recyclinggraph.

Besides these terms for describing all parameters of the single components, the description of the building structure of the product ist the most important aspect for demands on disassembly. This is effected in the recyclinggraph, which containes all parts, connections and recyclinggroups. This graph has got all informations, which later are necessary for defining the recycling strategy and the way of disassembly.

Therefore two structuring criterias are included: on the one hand, the real connections between parts are pointed out explicitly (e.g. the harddisc is attached to the body by six screws), on the other hand, the topolocigal hierarchy of the complete product is built up, e.g. the PC consists of a display, a keyboard and the computer.

The detailed description of the material fraction as part of the technolocigal information will be the second very important aspect as a base for the analysis modules, which will be explained later.

3.2 THE USER INTERFACE REGRED

As already mentioned, the large amount of knowledge, which influences the strategy of how to recycle a product, is one of the main problems for the designers work. So when conceiving a user interface for the planned tool, one demand was the abstract representation of the analysis specific knowledge. Figure 5 shows the recyclinggraph-editor REGRED as our solution. The components are displayed as ovals and the connections are shown as circles.

These connections only describe the logical function, the connection-elements are viualized as "normal" components.

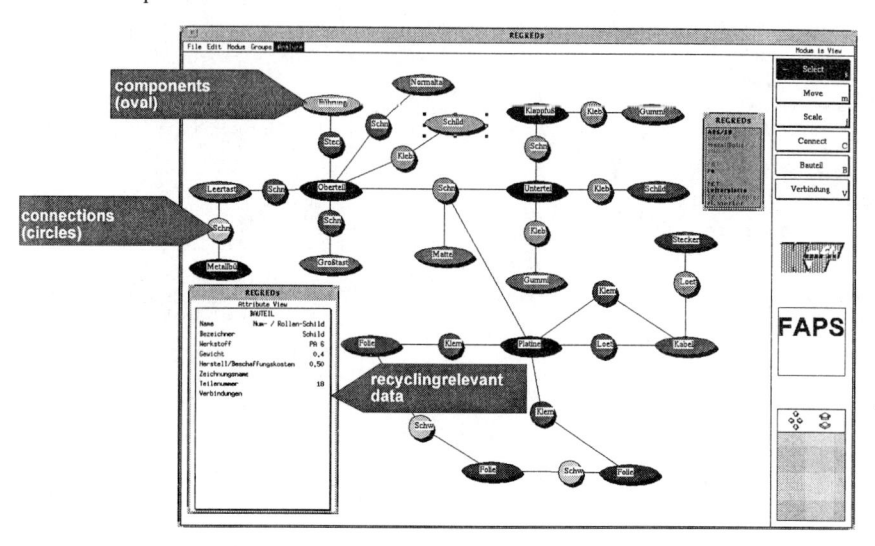

Figure 5   The Recycling Graph Editor as the User-Interface of RecyKon.

Figure 5 also shows the way, in which all recycling specific component information is displayed for one single part, or on the other side, a specific information, e.g. the material fraction, is represented in a specific colour. In addition to the analysis of weak points, the graph offers the possibility to carry out changes in the product model, which is directly linked to the model of the CAD-system, in which RecyKon is based on.

3.3 STEPS OF THE ANALYSIS

When the designer starts the analyis for the recyclability of his product, he has the choice of six analysis steps, which usually have to be passed through as shown in the arrow in Figure 6.
The sequence of these single steps is given by the different strategies of disposal. The most economical way of the recovery is the reuse of the whole product or some components of it. Therefore the system at first has to detect those components, which migth be reused. This detection will be done by a comparision of the datastructure of the product with the models of reusable components in a knowledge base.
As a result of this first analysis step, the detected parts are marked and displayed as recylinggroups "reuse", Thereby the designer and also the system henceforth knows, that for these parts the recycling-strategie "reuse" is the best.The modules for the following analysis steps work in the same or in a similar manner.

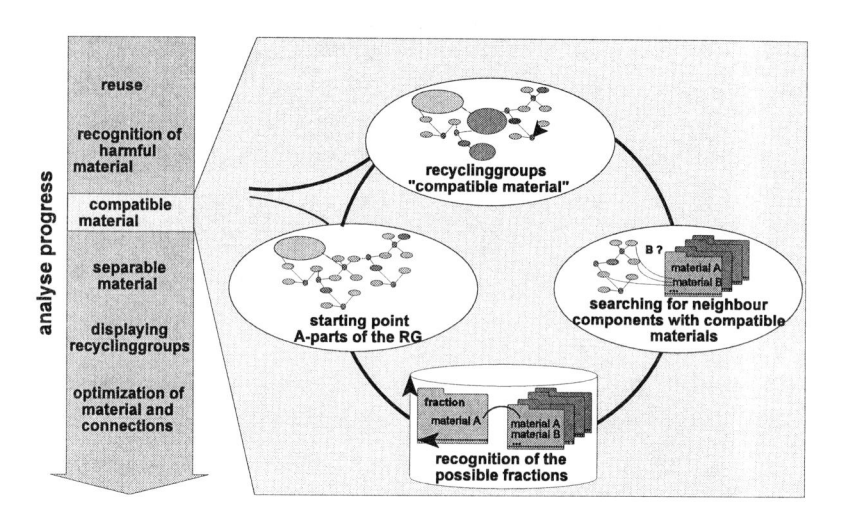

Figure 6   Modul "Compatible Material" as one part of the Analysis Tool.

In the second step, the components including harmful materials are searched for and fixed in a recyclinggroup "harmful material". This is an important information for the designer, for these harmul materials will probably cause higher costs for the recovery of the product.

For all components, which do not belong to the recyclinggroups "reuse" and "harmful material" one has to find out now, which ways of material recycling are available and are significant for these parts. Hereby it is important for the disposal costs, that the number of disassembly steps for removing parts is as low as possible. Therefore the analysis tool has to search for connected groups of parts, which might be removed very easily. There are two criterias for establishing groups, i.e. the possibilities for separation or combination of materials.

The parts in the recyclinggroups "compatible materials" (Figure 6) allow the reproccessing of the included parts to a new material fraction, which will be used in the future for new purposes. The group "separable materials" contains parts with a high mass flow ability, by which the pure fraction of the raw materials can be recovered.

The analysis modules searching for these two kinds of recyclinggroups start at the A-parts of the product, which are this parts with a sum weight, that makes 80 % of the whole product. Then the material of this A-part is compared with fractions of the knowledge base, that fulfill the criterias "possibility for separation or combination". Then neighbouring parts of the starting component are looked for, which are built of any material of this fraction.

After all groups have been found out by the a.m. steps of the analysis, the designer has to optimize his product by three aims:

- Increasing the size of the single recylinggroups
- Redoucing the number of parts with harmful material,
- Optimizing the connections, which have to be removed in case of separating the groups during disassembly for the purpose of recovery.

For this he gets offered lists of alternative materials/connections, which fulfill similar functions.

3.4 THE KNOWLEDGE BASE

The building up of the knowledge base was done by following the determinations of our industrial partners and the efforts of the disassembly planning experts at the Institute for Manufacturing Automation. Figure 7 shows the different information clusters of the knowledge base in combination with the information of the product model.

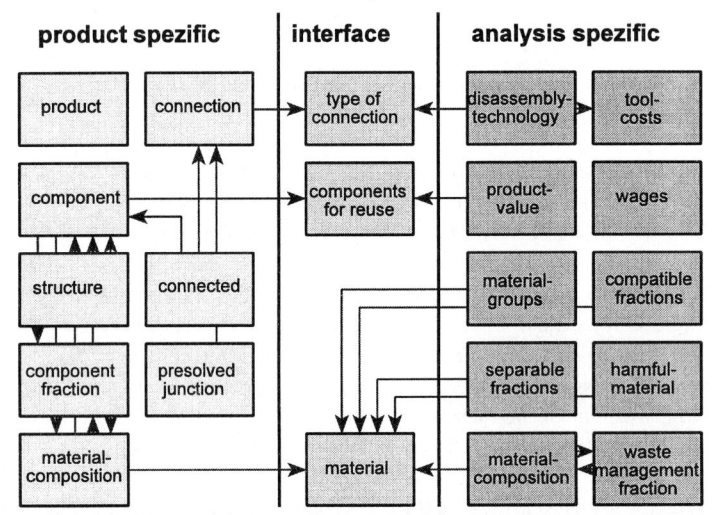

Figure 7   The Architecture of the ORACLE-Database.

For the adaptation of the system in industry, that means producing industry and also disposal industry, an interface is integrated, that is connected to the *product modell* on one side, and to the general *analysis-specific knowledge* base on the other side.

The Tables of the *interface*, displayed in the middle of Figure 7, must be built up by the company, which wants to use RecyKon as an analysis tool. The general information also can be filled by the company, but it is possible, to use the general basic information, e.g. by connecting the interface to world wide web.

Usualy, the disposer has to give informations into the *analysis specific* tables, which content costs of recovery, available tools, techniques of removing connections etc. These informations are the basis for the analysis tool RecyKon to prove the Recycliability and the Disassembly ability of a product in the product model.

4 ADAPTING RECYKON INTO THE DESIGNERS WORLD

The first concept for the Design System mfk was made eight years ago. This concept was realized and complemented in various cooperations and prototypes together with partners from industry, so that we consider the actual prototype-stage as a well proved tool for assisting the designer on all demands on product development. As it includes analysis modules in the same way, as it does a workmans tool box, the name "Engineering Workbench" has been defined.

4.1 THE CONCEPT OF THE DESIGN SYSTEM MFK

Figure 8 shows the general structure of the Design System /8/. It is built up out of the modules:
- Synthesis Part
- Product Model
- Analysis Part.

The Synthesis Part is combined with an usual CAD-System by a programming interface.

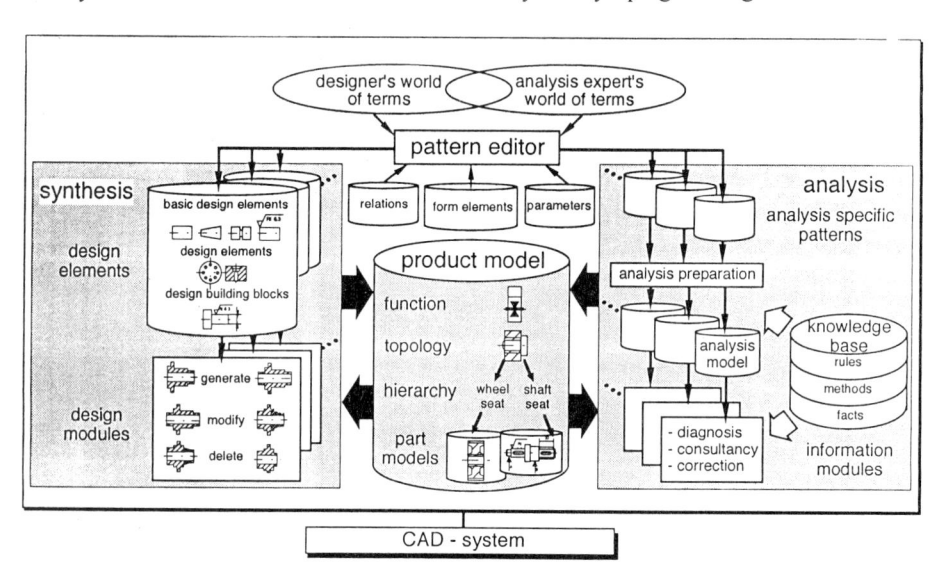

Figure 8 The General Structure of the Design System mfk.

The **CAD-System** is only used for visualization and user-dialog. All operations on the product will be documented in the product model, so that the CAD-Data-Structure is not used by the Analysis Part.

In the **Synthesis Part** the Design Modules, which allow the creating, changing and deleting of Design Elements, are provided to the designer. These Design Elements correlate to production-techniques, e.g. turned parts, sheet metal or casted parts. It is possible to build up an hierarchy with these Design Elements and to synthesize parts of higher level by combining basic Elements (e.g. cylinder, champfer of turned parts) to higher valuated components that regards to processes of production or functional solutions. In /9/ the "Beschreibungsmerkmal" is defined as the most complex feature. Basic of all functionality of the Design System mfk is the **Product Model**, including not only geometrical information, but also terms of technology, function and organisation. All the analysis modules work based on these information. For complex products the model is divided in component models.

The modules of the **Analysis Part** at least allow the designer the valuation of his design solution. At the beginning of each analysing process the necessary product information is extracted of the product model and prepared for the information module, in which results are generated by downworking rules, facts and methods of the analysis knowledge base.

Its very important to point out, that the designer decides the start of an analysis, as the whole Design System has to be understood as an assitance system, not as a "automatical product development machine".

4.2 INTEGRATING RECYKON IN THE ENGINGEERING WORKBENCH

The extreme modular conception of the Engineering Workbench allows a very comfortable integration of new, separably developed analysis tools. Figure 9 shows, how the Design for Recycliability-Tool RecyKon will be integrated.

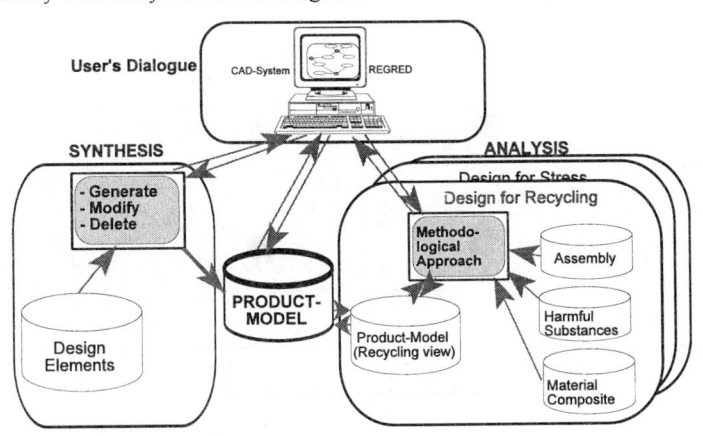

Figure 9   Integration of RecyKon in the Engineering Workbench.

The general product model includes all information, which is necessary fulfilling the efforts of each analysis tool. The general interface also allows the users dialog for all kinds of analyses, independent from the chosen CAD-platform.

The special knowledge can be added into the overall knowledge base. For that and all questions of knowledge maintenance this knowledge base was structured in a modular way.

5 CONCLUSION AND OUTLOOK

In this paper an analysis tool for "Design for Recyclability" is presented, embedded in a complex engineering workbench. The aim and the influences of Design for Recyclability are discussed, before the concept of RecyKon as one assistant tool is described. The structure of this tool is completed, the knowledge base has to be filled up with detailed information. By providing and testing the single analysis components as prototypes together with partners from industry and universities, the depth of knowledge will be extended.

After testing and improvement of a seperate tool, its modular structure allows a very easy integration in the Engineering Workbench. Figure 10 for that describes a vision of the future:

For each Design for X-Demand there is a suitable corresponding assistance tool available in the Engineering Workbench, which supports designers work during all stages of product development.

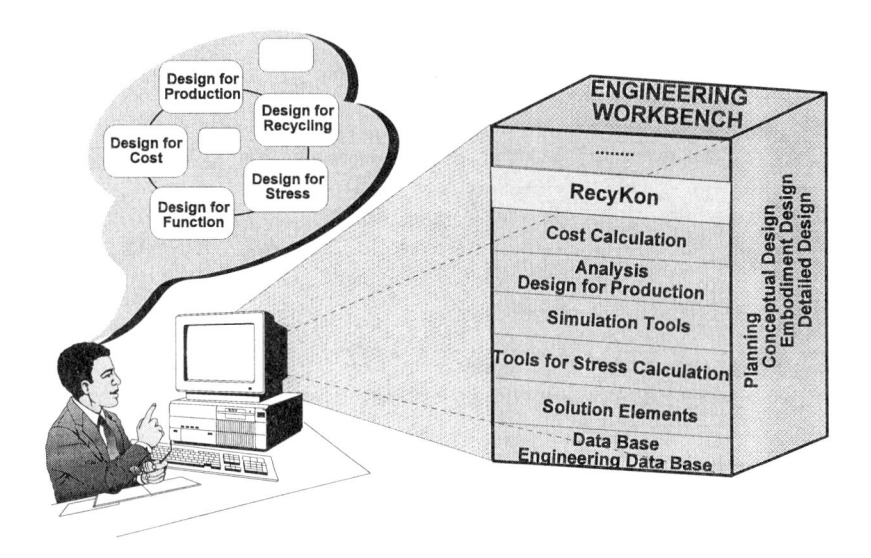

Figure 10   DFX versus Engineering Workbench.

6 REFERENCES

/1/ Meerkamm,H.; Krause,D.; Rösch,S.; Storath, E.: Anforderungen an integrierte Konstruktionssysteme - Auswirkungen auf die Architektur. VDI-Berichte Nr. 1079, S.299-319, VDI-Verlag Düsseldorf, 1993.

/2/ Hubka, V.: Theorie Technischer Systeme - Grundlage einer wissenschaftlichen Konstruktionslehre. Springer-Verlag, Berlin 1973.

/3/ Konschak,K.: Forschungsstand und Forschungsbedarf im Rahmen der Neuwertwirtschaft, Kolloquium Neuwertwirtschaft, 02.06.1994, Lauchhammer.

/4/ Krause,D.: Rechnergestützte Bewertungsmethoden zur recyclinggerechten Produktgestaltung - Stand der Technik, neue Wege. Konferenz "Elektronikschrott-recycling und -verwertung", 17/18.02.1994, Regensburg.

/5/ Meerkamm,H. ; Bachschuster, S. : Semantisches, relationsbasiertes Produktmodell für Gußteile im Konstruktionssystem mfk. Entwicklungsmanagement: Simultaneous Engineering, Oranien-Verlag, 1994, Herborn.

/6/ Finkenwirth, K.-W.: Fertigungsgerecht Konstruieren mit CAD - Konzept eines Konstruktionssystems zur Informationsverarbeitung mit CAD-Systemen. Dissertation an der Universität Erlangen-Nürnberg, 1990.

/7/ Meerkamm,H.: Design System mfk - an important step towards an engineering workbench. Journal of Engineering Manufacture, Vol. 207, p.105-116,IMechE 1993.

8/ Meerkamm, H. ; Weber, A.: Konstruktionssystem mfk - Integration von Bauteilsynthese und -analyse. VDI-Bericht Nr.. 903, VDI-Verlag, Düsseldorf 1991, S.231-248.

/9/ Weber, A.: Ein relationsbasiertes Datenmodell als Grundlage für die Bauteiltolerierung. Dissertation an der Universität Erlangen-Nürnberg, 1992.

7 BIOGRAPHY

Prof. Dr.-Ing Harald Meerkamm
Born 1943. Studies in Mechanical Engineering at the University of Stuttgart. Promoted to Dr.-Ing. at the Institute of Machine Tools there. Eleven years industrial experience in a company producing machine tools and plants.Head of design and development department.
Since 1984 Professor at the University of Erlangen-Nürnberg, Technical Faculty, Institute of Production Engineering. Owner of the chair of Engineering Design. Main fields in research: problems in rolling/sliding contact of bearings; Computer Aided Design; Design assistance system/Engineering Workbench; Design for X.

Dipl.-Ing. Johannes Weber
Johannes Weber, born in 1967, reached his Master of Production Engineering at the University of Erlangen-Nuremberg in 1992. Since 1993 he is sientific collaborator at the Institute for Engineering Design (head: Prof. Dr.-Ing. H. Meerkamm) in Erlangen. After working two years on the field of tolerancing and NC-Programming his research area is now the Design for Recycliability. Currently he is working on a project for developing computer supported tools and methods for Design for Recycliability, sponsored by the Bavarian goverment and from industry.

PART ELEVEN

Feature Technology

44

SESAME - Simultaneous Engineering System for Applications in Mechanical Engineering

H. K. Tönshoff, T. Baum, M. Ehrmann
Institute of Production Engineering and Machine Tools
University of Hannover
Schlosswender Str. 5
30159 Hannover
Germany
Phone: ++49-511-762-5220
Fax: ++49-511-762-5115
e-mail: baum@ifwsn3.ifw.uni-hannover.de

Abstract

The simultaneous engineering system SESAME employs feature technology for product development. The system comprises three modules: Feature Based Design, Automatic Process Planning and Interactive Process Planning with NC-Code-Generation. The integration of the SESAME prototype is based on a common feature based workpiece model. Feature based workpiece descriptions are transferred between the modules. The implementation on a single platform and the involvement of a common geometric modeller, the ACIS kernel, are the prerequisites for overlapping engineering activities.

Arbitrary user defined features can be easily added to the design system. Dimensions and a selection of geometric tolerances are implemented in the design system. Workpiece models are transferred to process planning where manufacturing operations and tools are related to features. Finally the tool paths are generated largely automatically in the NC-module.

Keywords

Feature Technology, Simultaneous Engineering, Dimensions, Tolerances, CAD, CAPP, NC

1   INTRODUCTION

Currently industrial companies have to solve two key problems in product development:
- To retain competitiveness they need to reduce development time.
- The component complexity is ever increasing.

To meet these demands new techniques have been developed, e.g. Simultaneous Engineering, Concurrent Design or Teamwork. Simultaneous Engineering means to perform the various engineering activities more in parallel rather than in sequence (Sohlenius, 1992). Looking at the classical waterfall principle (Figure 1), an activity has to be finished before the next one can be started. The simultaneous approach however leads to time savings in the compound of design, process planning and NC-programming tasks. Thus the goal of a new generation of mechanical engineering systems is to support these working techniques (Krause, 1991). The SESAME system is a prototype demonstrating the power of a Simultaneous Engineering approach (Tönshoff, 1994a).

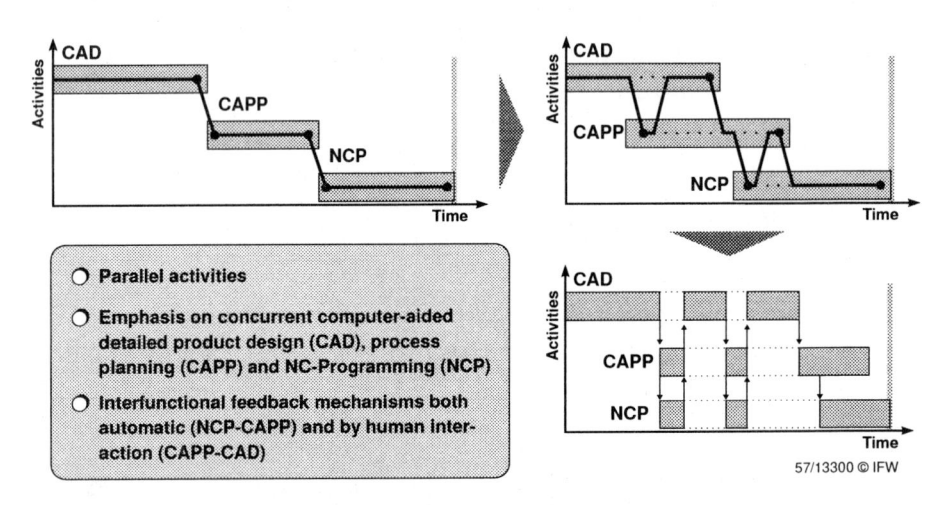

Figure 1 Classical vs. Simultaneous Engineering Approach

2 THE SESAME SYSTEM ARCHITECTURE

The SESAME system consists of three main modules (Figure 2):
- Feature Based Design System (FBDS)
- Automatic process planner with Plan Space Generator and Optimisation module (PSG/OPT)
- Interactive Feature Planner and NC-Programming System (SFP/NC).

All the modules are based on the geometric modeller ACIS and are implemented on a Unix workstation. The user can access the modules via a common man-machine-interface. Design models are transferred between the FBDS and the process planning modules using a formal description language. The Automatic and the Interactive Planning module produce process plans which can be exchanged amongst them. Manufacturing related data such as operation sequences and tools are stored in a Technical Database, which can be accessed by the planning modules. In the last step process plans are processed to NC-programs by the NC-module.

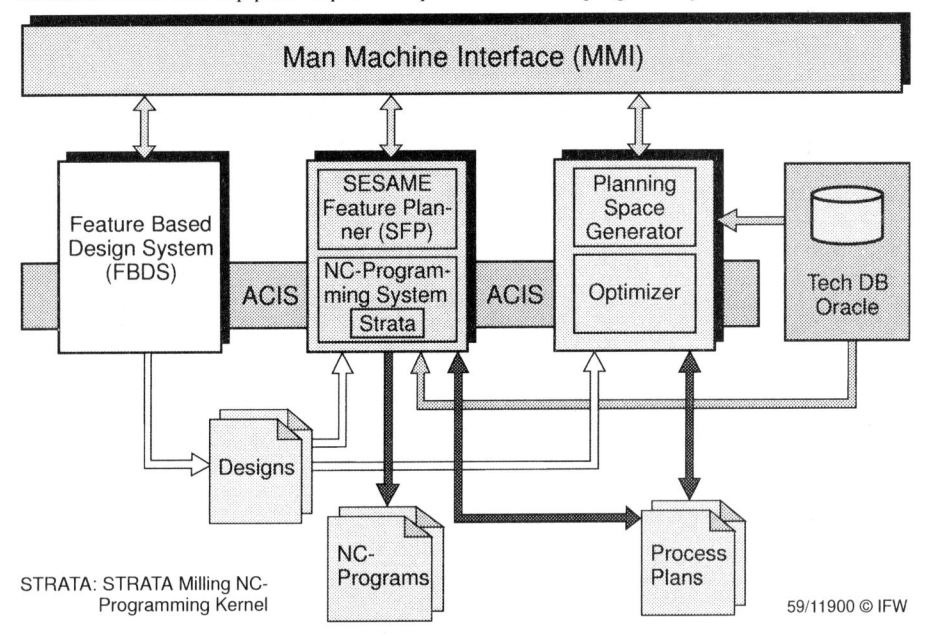

Figure 2 The SESAME architecture

3 THE FEATURE BASED DESIGN SYSTEM (FBDS) WITHIN SESAME

The FBDS allows the detailed design of complex mechanical parts. The use of feature technology facilitates the design process by supporting the designer's functional way of thinking. Several classes of features are defined in the FBDS:
- Design Features containing information meaningful in a design context.
- Auxiliary Features including construction geometry like centre lines or mirror planes.
- Dimension Features representing ISO dimensions.
- Dimension Tolerance Features representing ISO plus/minus tolerances applied to dimension features.
- Geometric Tolerance Features representing ISO geometric tolerances.
- Surface Roughness Features representing ISO surface roughness tolerance applied to a face.

3.1 Design Features

Besides pure geometric information Design Features may include other data: Additional geometric data, technological data, functional data. In SESAME eighteen design feature types were specified such as holes with different end types, pockets, slots etc.. Each of these features is associated with a geometric body. Features like chamfers or blends form a subclass within the design feature class (according to the above mentioned feature classes). They merely modify existing geometry by pointing to geometric entities of other features. All Design Features are described in feature templates. The templates include the feature parameters which are instantiated during the design session. Using a property sheet, the designer can enter and change the parameter values of each feature. If the current feature set does not meet a companies' requirements, new feature templates may be added by the user. The templates are described in a text file which is read when the system is started. No recompilation is necessary when new feature types are defined.

The geometric engine of the FBDS is ACIS, a commercial threedimensional boundary representation modeler. The FBDS, written in C++, provides a CSG-like structure which is necessary for feature modelling. Moreover it offers a programming interface using the macro command language MCL+, which is an interpreter language based on C++. Nearly all the ACIS functionality can be addressed using MCL+.

In order to achieve a better user acceptance, the concept of technical views was introduced. Different fixed views like topview or sideview of the monitor layout similar to a conventional technical drawing are used. Hence the designer is able to work as he is used to do. Technical views are very useful to apply and visualise dimensions and tolerances.

Design Features are a means to support a functional design. In addition, accuracy data is required to design a part completely. This accuracy information is the basis for the selection of appropriate manufacturing operations.

3.2 Dimensions and Tolerances

Traditionally dimensions and tolerances are used in twodimensional drawings and express the size and allowable deviations in workpiece accuracy. This is necessary because a model, especially a threedimensional one, has a perfect shape, but manufacturing methods are incapable of producing parts with perfect geometry. Hence tolerances are applied to denote an allowable variation of the manufactured component from the nominal geometry (Tönshoff, 1995).

Besides Dimensional Tolerances a second tolerance class is defined: Geometric Tolerances. They specify variations of form, attitude, location and run-out of geometric elements of a workpiece. The set of parts, which are in the stated tolerance range, form a variational class. The mathematical model describing the variational class is known as variational model (Pedley, 1995).

The tolerancing related research work can be divided in four fields (Salomons, 1995):
- Tolerance Specification
- Tolerance Analysis
- Tolerance Synthesis
- Tolerance Representation.

Tolerance Specification is the first step in tolerancing a part. According to international, national and company specific standards tolerance types and values have to be specified. Existing standards cover all tolerance cases, yet the way how to apply tolerance types is not covered by those standards.

Tolerance Analysis aims to check if stated tolerances have the intended effect. Known applications deal with the evaluation of assembly feasibility and clearances of assembled components. The analysis can be performed either by statistical tolerance analysis or by worst case analysis.

Tolerance Synthesis is an area not well defined up to now. It intends to improve tolerance specifications considering manufacturing and quality assurance aspects. Both Tolerance Analysis and Synthesis base on a variational model.

Tolerance Representation focuses on tolerance descriptions adapted to the demands of computer systems. A few approaches can be found in literature (Requicha, 1986), but most of them pose problems by not confirming to existing tolerance standards. The SESAME approach for tolerance representation is to adapt existing standards to the use in a threedimensional model.

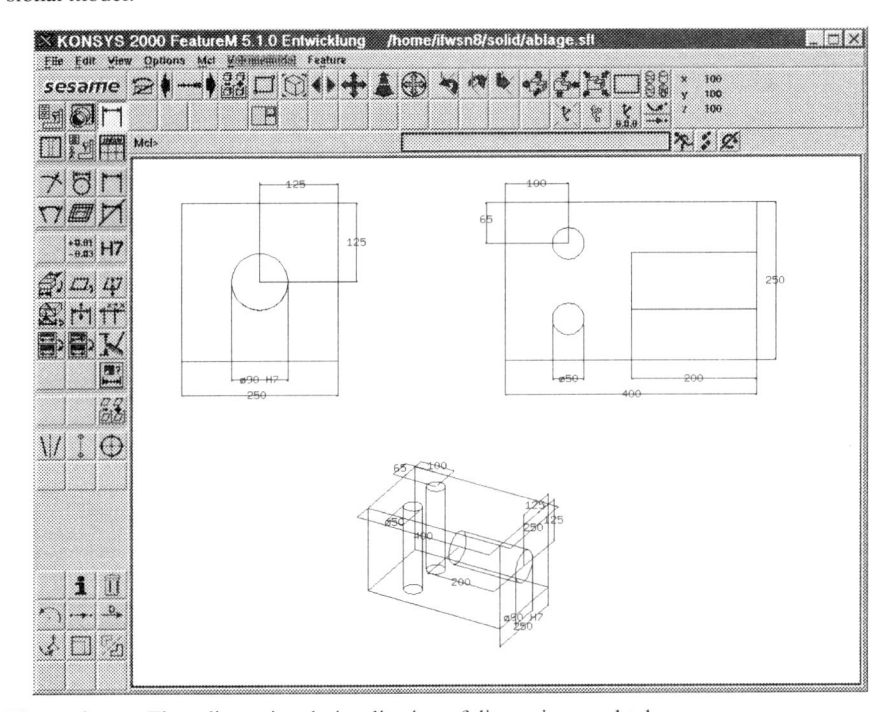

Figure 3 Threedimensional visualisation of dimensions and tolerances

In SESAME, dimensions and tolerances are modelled as features using the same mechanism that enables design features to be modelled and controlled, however they belong to a separate workpiece of which geometry is created. This is useful for protection, classification, monitoring and saving. A separate template is required for each different dimension or tolerance type. However it is not necessary to specify the exact type of entity that a dimension or tolerance is to be related to. It is sufficient to say that it will be an entity (face, edge or vertex) in the model. Dimensional tolerances are modelled as attributes of the respective dimension in the same feature template. In general dimensions may be classified into three types:

- Non-associative dimensions, which are used for measuring a model during the design process.
- Associative dimensions, which follow the model if changes of design features occur.
- Bidirectionally associative dimensions, that can be used to change the model.

Geometric Tolerances currently behave as associative dimensions.

Much of the information stored in the tolerance templates is to enable graphical representation. As opposed to twodimensional drawings where tolerance zones usually refer to the drawing plane, i.e. the plane of the paper sheet, an expanded tolerance description has to be developed for threedimensional models. The tolerance has to include parameters for:
- The direction in which the tolerance should be calculated, if the tolerance zone is ambiguous.
- The root and where appropriate the normal of the dimension plane.
- The position of the dimension line in the dimension plane.

Consequently there are dimension lines, InPlane leader lines and ToPlane leader lines in a 3D tolerance representation. Figure 3 shows an example of SESAME dimensions and tolerances in a threedimensional model.

3.3 Coordinate systems

To provide a flexible graphical representation of dimensions and tolerances different frames of reference are used:
- World Coordinates as the absolute reference frame.
- Workpiece Coordinates as the reference frame of the base feature.
- Feature Coordinates as the reference frame of each individual feature.
- Local Coordinates of the current working coordinate system.

If the workpiece is moved, the position and direction of dimension and tolerance symbols have to be recalculated. If this data is stored relative to world coordinates, the symbols will not follow the repositioned model. Hence the dimension and tolerance data must be stored relative to workpiece coordinates. Alternatively transformation matrices could be provided to transform coordinates.

4 PROCESS PLANNING

The development of the process planning subsystem within SESAME has taken into account two key aspects. Firstly a tight integration with the areas of design and NC-programming. Secondly an extensive support for the human process planner. The strong integration has been achieved using a common feature-based workpiece description in CAD, CAPP and CAM (Tönshoff, 1994b).

The process planning subsystem comprises two different components. The automatic and the interactive process planning system. Process plans are generated automatically as far as possible. The rest of the planning has to be performed with the interactive process planning system. Both systems are independent but process plans can be transferred from the automatic system to the interactive in order to complete them (Figure 2).

4.1 Automatic process planning system and technological database

The automatic process planning system comprises two modules, the planning space generator and the optimiser (Figure 2). Feature based workpiece descriptions generated in the Feature Based Design System are transferred to the planning space generator which relates possible manufacturing operations to features. The optimiser searches for an optimised plan in the generated planning space.

Before the assignment of manufacturing operations to features, implicit feature interactions as intersections and thin walls are detected employing a geometric reasoning process. Information on tolerances and surface finish is transferred together with the feature based workpiece description to process planning. Besides explicit feature interactions as tolerances and surface finish, implicit feature interactions play an important role in process planning. Feature interactions determine the selection of possible manufacturing operations. After the geometric reasoning process, a network of manufacturing possibilities for the component is generated by relating technologically possible microcycles to each feature. The microcycles define alternative operation sequences for the manufacturing of a feature. Thus, several alternative manufacturing possibilities are assigned to a manufacturing feature.

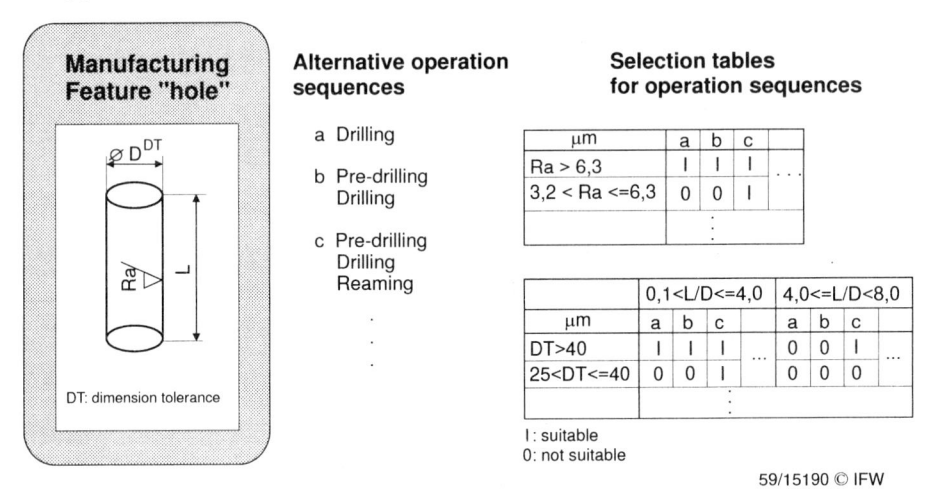

Figure 4 Selection of operation sequences

The assignment of microcycles is performed considering feature parameters as well as explicit and implicit feature interactions. A technical database has been produced containing tables where technologically possible microcycles can be selected depending on feature dimensions, tolerances and implicit feature interactions. The end-user within the project, an Italian machine tool manufacturer, provided the necessary data. The end-user made up the tables with suitable manufacturing strategies and tools considering dimensional parameters, tolerances and implicit feature interactions. Figure 4 shows an example for the structure of these tables. The information summarised in these tables has been stored in an Oracle-database which can be accessed from the planning modules via a C++-interface.

After the selection of microcycles, tools and cutting parameters are selected for each microcycle. This data is also selected from the Oracle-database.

The network of manufacturing possibilities generated by the planning space generator serves as input for the optimiser. An optimised process plan considering tool costs, the number of set-ups and the machining time is generated by the optimiser using genetic algorithms (Husbands, 1992).

The automatic planning module does not support user-defined features. This is mainly due to the fact that the system developer has to define specific geometric reasoning algorithms for each feature. Therefore only a limited number of feature types can be planned automatically within the SESAME system. The rest of the features has to be planned using the interactive feature planning module.

4.2 Interactive planning module and NC-programming system

The interactive process planning module has been integrated into the NC-programming system NcS. Originally the NcS-System works with pure, non feature-based geometry. The integration enables the user to work with features in NC-programming.

NcS is a 2 1/2D NC-programming system supporting the manufacturing technologies turning, milling and drilling (Figure 5). The kernel of the NcS datastructure is the process plan where the list of manufacturing operations is kept. The process plan is visualised and manipulated with an editor. Each operation is described with 9 forms where parameters and strategies as well as the related geometry are defined. Examples are tool diameter and length, speeds, feeds, cutting depth and tool approach and retract strategies. When pure geometry is loaded into the NcS-System, the user has to fill the forms manually with the necessary data. Using a feature based workpiece description as input, the integrated interactive feature planner is used to assign manufacturing operations and tools to each feature. The necessary operations are generated and the forms are filled automatically with the data connected to the features. If the user wants to change the predefined data, he has to employ the process plan editor. When all forms are filled with the necessary information, the NC-paths can be generated automatically. The NC-part program is generated with a post-processor. Alternatively, CL-Data can be generated.

The process planning activity in the interactive planner is based on a manufacturing feature workpiece description. A manufacturing feature is a piece of geometry or topology, which has a set of possible manufacturing operations (microcycles) associated with it. The manufacturing feature workpiece description is produced employing a feature mapping process. This process transforms the design feature workpiece description into a manufacturing feature workpiece description. The relation between design and manufacturing features can be 1:1 or 1:many. A design feature can have a corresponding manufacturing feature or is subdivided into several manufacturing features (Figure 6). The mapping of several design features into one manufacturing feature is not considered since this requires the whole range of geometric reasoning in order to find related features. Geometric reasoning is not employed in the interactive planner because this is not necessary when the user interacts with the system. Only feature intersections, which may influence the selection of operations and tools are detected automatically.

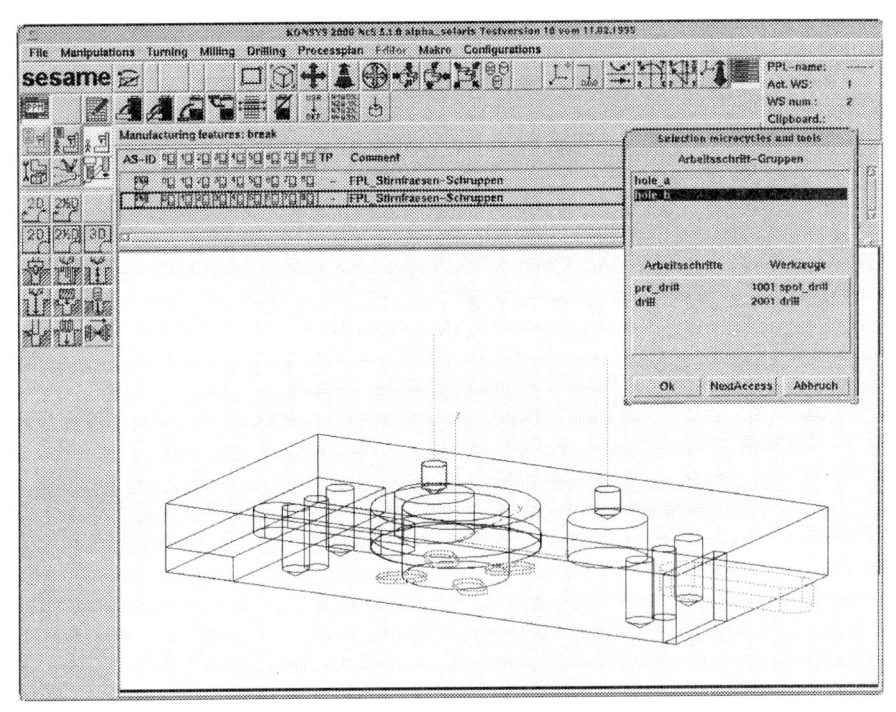

Figure 5 Process plan and NC-program generation in NcS

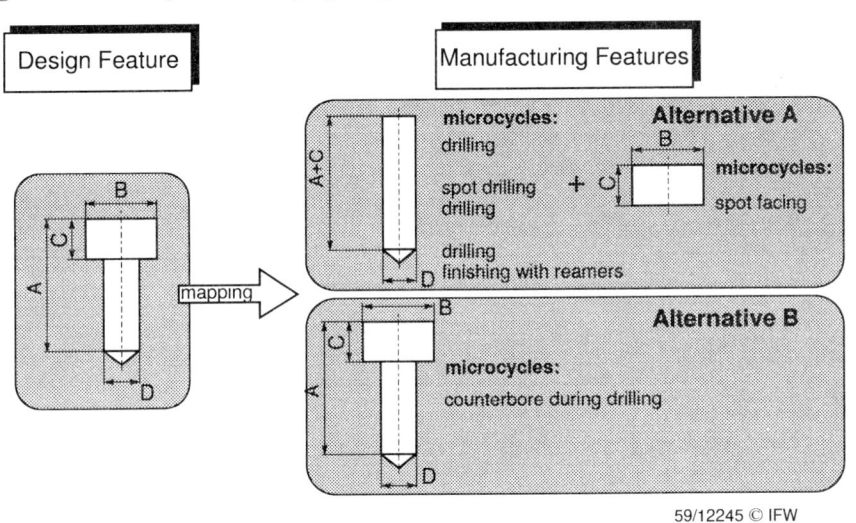

Figure 6 Mapping and alternative microcycles

```
-- *******************************************************************
-- DESIGN FEATURE : count_hole
-- *******************************************************************
SCHEMA count_hole;
    ENTITY count_hole;
        SUBTYPE OF (design_features);
        REF_POINT      : ARRAY [1.3] OF REAL:- [0.0, 0.0, 0.0];
        DIRECTION      : ARRAY [1:3] OF REAL:= [0.0, 0.0, 0.0];
        FEATURE_NAME   : STRING:= "counthole_eins";
        diameter       : REAL:= 10;
        depth_-z       : REAL:= 10;
        ...
        angular_tol    : REAL := 0;
        surface_finish : REAL := 12;
        ...
        VOLUME_OPERATOR : STRING(1) := "-";
        DERIVE
        INSTANCE_ID  : INTEGER:= create_count_hole (diameter, ... );
    END_ENTITY;

    FUNCTION create_count_hole (diameter: REAL, ... ): INTEGER;
        create_sesame_count_hole (ID, "+", 0.0, 0.0, 0.0, 0.0, 0.0, 0.0, diameter, depth_-z,
                                  countdiameter, countdepth, handle, level);
        return (ID);
    END_FUNCTION;

    FUNCTION mapping(diameter: REAL; ...):LOGICAL;
        ENTITY MFcounterbore;
            radius : REAL:=countdiameter/2;
            height : REAL:=countdepth;
            dx     : REAL:=0.0;
            ...
        END_ENTITY;
        ...
        mapping_alt1("count_hole->MFhole+MFcounterbore", MFcounterbore, MFhole);
        mapping_alt2("count_hole->MFcount_hole", MFcount_hole);
        success := map2MF(); -- function that performs the mapping
        return (success);
    END_FUNCTION;
END_SCHEMA;
```

Figure 7 TEBES example for a counterbored hole

The interactive planning module can be initialised with user-defined features. The feature templates are described in a feature library using a formal language (TEBES) (Tönshoff, 1993). TEBES is based on the EXPRESS language which is part of STEP. Parameters and the geometry evaluation function of design and manufacturing features are defined in the feature library. An example for a counterbored hole is given in Figure 7.

Furthermore, TEBES is used for the assignment of manufacturing features to design features, i.e. the definition of the mapping rules. This information is stored together with the design feature templates in the feature library used to configure the planning module.

The relation of microcycles is based on the manufacturing feature workpiece description. If alternative microcycles are selected from the technical database, these alternatives are displayed and user interaction is requested (Figure 5). When microcycles and tools are chosen, the workpiece model is enhanced with this information and an operation is generated and visualised in the editor of the NcS-System as described above.

5 CONCLUSIONS

The presented prototype system has validated the approach of feature technology for the integration of CAD, CAPP and CAM. Using a feature based workpiece model and a common geometry modeller, tasks can be performed partly in parallel. Incomplete workpiece models can be transferred from design to process planning. Process plans for incomplete workpieces can be transferred from process planning to NC-programming.

The user acceptance largely depends on the possibility to create user-defined features. The described feature based design system as well as the interactive process planning system support user-defined features.

Tolerances are important for the selection of manufacturing operations and tools. Tolerance specification and representation are part of the feature based design system. The tolerance information is transferred together with the feature based workpiece description to the process planning systems.

6 ACKNOWLEDGEMENTS

The presented research work is being funded by the Commission of the European Community (CEC) under the Brite/EuRam Programme (Project BE #4539).

7 REFERENCES

Husbands, P. (1992) Genetic algorithms in optimisation and adaptation. In: Advances in Parallel Algorithms, *Blackwell Scientific Publications*, Oxford, p 227-276.

Krause, F.L. et al. (1991) IMPPACT - Integrated Modelling of Products and Processes using Advanced Computer Technologies. *Proceedings of the IMPPACT workshop,* Esprit-Project 2165, 26-27 Februar 1991, Berlin.

Pedley, G. and Ehrmann, M. (1995) Explicit Feature Interaction Modelling in CAD and CAM Systems. *Proceedings of the 11th International Conference on 'Computer Aided Production Engineering',* 20-21 September 1995, IMechE Headquarters, London, UK.

Requicha, A.A.G. and Chan S.C. (1986) Representation of geometric features, tolerances and attributes in solid modelers based on constructive geometry. *IEEE Journal of Robotics and Automation,* Vol. RA-2, No. 3, 1986, pp 156-166.

Salomons, O.W.; Jonge Poerink, H.J.; van Slooten, F.; van Houten, F.J.A.M. and Kals, H.J.J. (1995) A Computer Aided Tolerancing Tool based on Kinematic Analogies. *Proceedings of the 4th CIRP Seminar on 'Computer Aided Tolerancing',* 5.-6. April 1995, Tokyo, Japan, pp. 53-72.

Sohlenius, G. (1992) Concurrent Engineering. *Annals of the CIRP*, Vol. 41, Nr. 2, 1992, pp 645-655.

Tönshoff, H.K. and Ehrmann, M. (1995) Quality Features in CAD - and CAPP - Systems. *Proceedings of the 4th CIRP Seminar on 'Computer Aided Tolerancing'*, 5.-6. April 1995, Tokyo, Japan, pp. 131-141.

Tönshoff, H.K., Aurich, J.C., Hamelmann (1993) Formale Elementbeschreibung für Konstruktion und Arbeitsplanung, *VDI-Z* 11

Tönshoff, H.K., Baum, Th. and Ehrmann, M. (1994a) SESAME: A System for Simultaneous Engineering. *Proc. 4th Int. FAIM Conf. 'Flexible Automation and Integrated Manufacturing'*, 8.-11. Mai 1994, Virginia, USA, pp. 380-389.

Tönshoff, H.K.; Aurich, J.C.; Baum, Th.; Hamelmann, S. (1994b) A Workpiece Model for CAD/CAPP Applications. To be published in: Production Engineering - *Annals of the German Academic Society for Production Engineering*, Vol.1, Issue 2, Hanser, München.

8 BIOGRAPHY

H.K. Tönshoff is full professor of production engineering at the University of Hannover, Germany. After receiving his Ph.D. degree in mechanical engineering in 1965 he worked as a designer, as technical manager and vice president in a machine tool company. In 1970 he became director of the Institute of Production Engineering and Machine Tools at the University of Hannover. In 1992 his work was honored by bestowing a Dr.-Ing. E.h. degree on him. He is vice president of the German Research Council (DFG) and member of the International Institution for Production

Thomas Baum studied production engineering at the University of Hannover, Germany. Since 1992, he works as research engineer at the Institute of Production Engineering and Machine Tools in the CAD/CAPP group.

Marc Ehrmann studied production engineering at the Technical University of Munich, Germany. Since 1993, he works as research engineer at the Institute of Production Engineering and Machine Tools in the CAD/CAPP group.

45

Feature-based modelling of conceptual requirements for styling

F.-L. Krause, J. Lüddemann, E. Rieger [*]
Fraunhofer-Institut für Produktionsanlagen und Konstruktionstechnik (IPK),
Institut für Werkzeugmaschinen und Fertigungstechnik (IWF-TUB)
Pascalstr. 8-9, 10587 Berlin, Germany
Tel.: +49(30)39006 244   Fax: +49(30)3930246
E-mail: frank-l.krause@ipk.fhg.de

Abstract
Industrial design plays a key role during the product development cycle in the integration of conceptual product design and engineering design. Innovative product design depends on the appropriateness of computer-aided tools for the support of the creative and intuitive design process using a highly interactive modelling tool. Efficient use of computer-aided industial design systems necessitates meeting the requirements of the designer. A prototype system is presented which aims at both integration of industrial design with adjacent product development processes employing feature-based modelling techniques of conceptual data models as well as intuitive modelling for a rapid shape definition using the approach of virtual clay modelling.

Keywords
Computer-aided industrial design, conceptual requirements, rapid shape definition, feature modelling, virtual clay modelling

1   INTRODUCTION

The use and integration of computer-aided processes in the early phase of conceptual specification and design of a product is a prerequisite for an efficient product development process. In current industrial design, the documentation and exchange of progressive product concretization is still achieved using conventional media such as 2D sketches, package drawings or physical models and, increasingly, due to the introduction of 2D paint and industrial design systems (CAID), computer-aided representations, Figure 1.

Medium gaps result from the multiple transition of product data between 2D- and 3D-representations, computer-aided formats and physical models. It would be desirable to employ a continuously computer-supported design process starting with the formulation of the inital

---

[*] Modelling of the conceptual requirements for telephones has been performed by cand.-ing. J. Neumann (IWF)

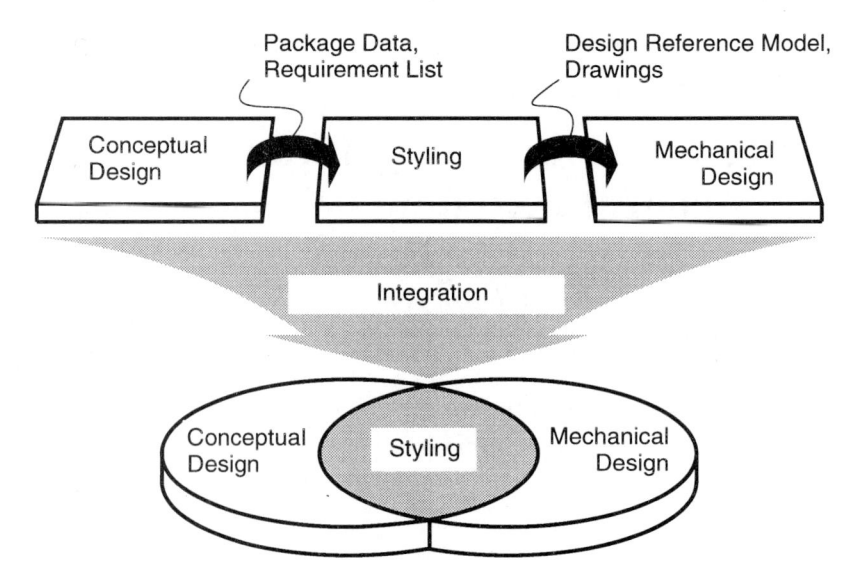

Figure 1 Integration of Industrial Design

product idea, (Mischok, 1992) and (diGuisto, 1993). The objections of designers to the use of computer-aided design systems for the creative and intuitive phase of early shape definition (Tovey, 1994) and (diGuisto, 1993) constitute an obstacle to this goal. The primary reason for these objections is the restriction on the designer's expressive freedom imposed by the modelling tools and abstract geometric modelling techniques offered by industrial design systems. CAID systems only partially support an intuitive and sketch-like working style since representation schemata have been adopted from traditional surface-oriented CAD Systems.

A computer-aided approach to the early phase of shape definition in industrial design should offer the following (Krause, 1994a):

- Closure of the gap in the process chain between the initial layout of a product and the subsequent design by establishment of an information model suited to the styling process for sharing and communicating the conceptual requirements between conceptual design and mechanical design within the styling process.
- Meeting the designer's requirements by providing sketch-like 3D modelling capabilities which take conceptual reqirements into account but limit the solution space of design as little as possible.

The research approach presented in this paper focusses on ab initio design, a phase where numerous ideas are developed in a sketch-like manner and varying conceptual requirements have to be considered. A generic conceptual data model for consumer goods has been defined which is extended by the introduction of feature modelling techniques to allow for the representation of conceptual data in a manner both structured and meaningful for industrial design. The development is based on the virtual clay modelling (VCM) approach resting on strong analogies to conventional generation of physical models, hereby utilizing conceptual data to define the core of the shape to be designed.

Feature-based modelling of conceptual requirements for styling 529

2 CONCEPTUAL DATA MODEL FOR THE REPRESENTATION OF STYLING REQUIREMENTS

The concept model gathers and structures styling-relevant requirements defined in the conceptual design phase (Krause, 1994a). In conformance with STEP, the concept model is derived from a redefined activity model of the styling process. To provide for general applicability, the concept model is defined on the basis of a modelled automotive styling and general consumer good process. The following aggregated data items, Figure 2, represented in the EXPRESS language are distinguished independently of product type:

- Technical specification data comprise styling-relevant technical aggregates, hardpoints and reference systems in geometrically explicit form. They define the minimum space reqirements.
- Ergonomic data represent templates for ergonomic measurements, anthropometric data, the intended user profile by heritage, percentile and gender as well as the time of planned product usage.
- Measurement data comprise all product-relevant main distances and angles. They explicitly define the to-be sizing of the product.
- Legal constraints refer to external guidelines and standards which restrict the freedom of definition of the shape of the product.

A concept data item may be limited by minimum, maximum or interval sizing constraints for any dimensional entry of a concept data item. Organizational attributes comprise the release status, ownership and a modification status to indicate the request for a redefined interpretation of the concept model.

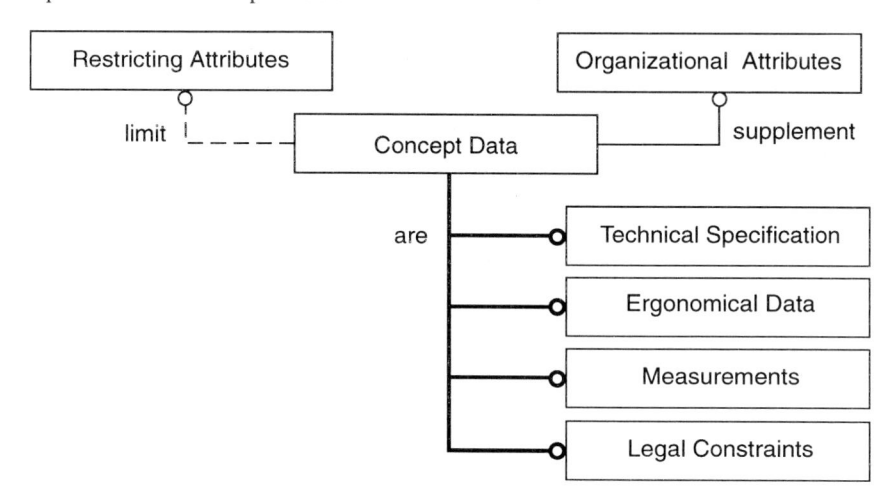

Figure 2 Structure of the Conceptual Data Model

Conceptual data may act as commonly agreed-upon place holders for various product components concurrently under development in other departments, such as aggregates, but also for hull volumes of moving product components and reserved functional or empty spaces.

An instance of a concept model is always specific to a defined product development project and the product type. Within an enterprise, the structure of the concept model can serve as a reference for representing product development requirements of product families. In inter-enterprise collaboration, modifications in concept can be communicated for an unambiguous interpretation of constraints defined by the concept model.

3 SYSTEM COMPONENTS

3.1 Feature Modeller FEAMOS

In the early design phases, product development is mainly determined by non-geometric information. A continuous support of the full product development cycle therefore requires the additional capability of using modelling elements which represent the semantics of described information independently of geometry. The introduction of features as semantically endowed objects, which can also possess geometric shape, makes the capture of this information possible (Rieger, 1994). The feature modelling system FEAMOS (FEAture Modelling System) was implemented based on this feature approach. In FEAMOS, semantic, non-geometric information therefore constitutes the central computer-internal representation. The explicitly geometrical model is a subsequent enhancement and the geometric modeller employed can easily be replaced by others. This differentiates the present approach from conventional CAD systems using the geometric representation as the central model.

All features are defined in a graphic- interactive fashion by the user. Thus, greatest flexibility in the configuration of the modelling system with respect to design-task-specific needs is achieved. For the storage of newly defined features, the textual feature description language PDGL (Part Design Graph Language) has been developed which allows for the management of features independently of the system (Krause, 1991). Using the syntactic base EXPRESS of PDGL, it is also possible to map information models as specified in IDEF or EXPRESS-G directly onto features. These features can then be used for modelling an instance of the information model as initial basis for product development.

Since at this stage the modelling of this information is free of any geometrical shape, a user interface for structure-oriented modelling has been implemented. In this mode, modelling elements are formed by features, combination and assembly nodes. The view represents the structure of the product to be modelled. Available user functions are structure operations, such as combination, copying or deletion of subtrees of the product structure.

The advantages of the feature approach and the corresponding implementation of FEAMOS have also been demonstrated for the realization of a process chain supporting the full product development cycle, from specification of customer requirements to generation of the NC code for machining (Krause, 1994b). Integration of the application modules within this process chain is performed via the exchange of semantic product information, thus avoiding the deficits of standard interface formats.

3.2 Virtual Clay Modeller (VCM)

High interactivity and intuitiveness of use are the main goals of the virtual clay modelling approach presently under development. The basis for meeting the goals of the modelling approach is provided by a change in the modeling paradigm as against the conventional surface modeling scheme for aesthetic shapes, Figure 3. The design model is represented using a discrete representation scheme of elementary volumetric items, voxel data. As a result, modeling tools can be defined as complexly shaped tools. The virtual clay modelling approach offers exact representation of tools by spline profiles, Krause (1995) , comp. Gaylean (1991) and the approach by Wang (1995).

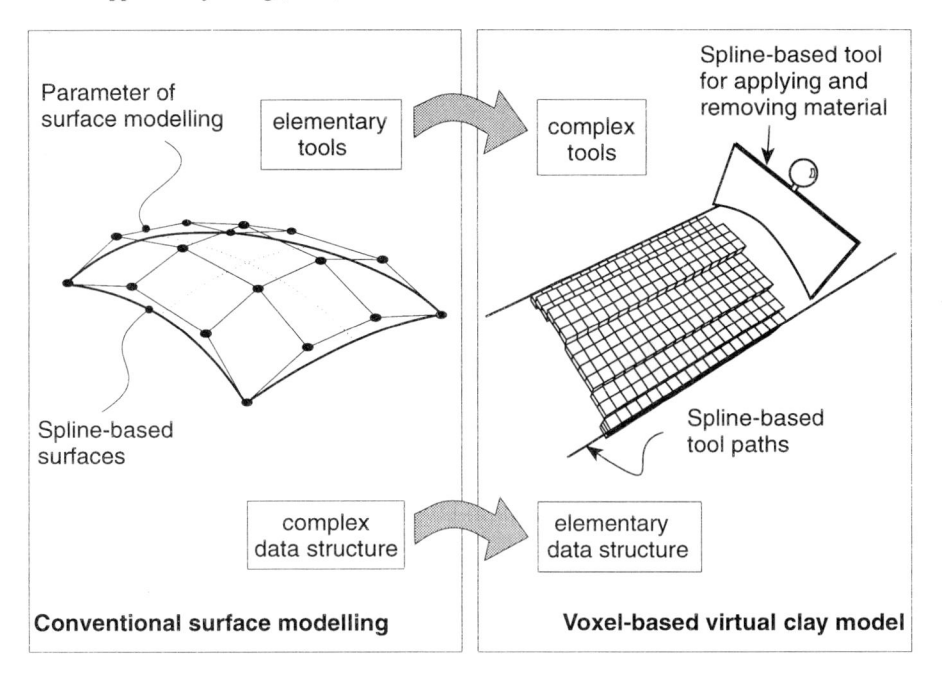

Figure 3   Principle of Cirtual Clay Modelling

The motivation underlying the notion of „virtual clay modelling", VCM, is the provision of modelling tools defined by rational B-splines such as to simulate their real world counterparts, conventional clay modelling tools such as true-sweeps, templates and rakes, Yamada (1993), Figure 4. The selection of a tool type and its functional parameters as well as the setting of the allowable translation and rotation of the tool are the preparatory steps for a modelling operation. Analog to the conventional process „cut", modelling operations correspond to a subtraction of the swept spline volume of the moved tool. The paste operation in a contrary fashion adds material to the clay model. The operation „compensate" simulates the pile-up techniques of conventional clay modelling.

Real-time modelability and high-resolution representation of freely formed shapes in a design model occupied the foreground during the development of the virtual clay modelling approach. These in general contradictory goals are tackled by the provision of a compressed data model using run length encoding to record material changes of air and various clay materials along the encode axes in a clay model. The flat data structure allows for rapid modelability for highly discretized data sets, since modelling operations can be performed locally and on the basis of encoded material changes by interval arithmetics, as used by Menon (1994) for analytical geometries. Aliasing in object space is avoided by imprinting the residuum of converted spline volumes of the moved tool into the virtual clay data model

(Krause, 1995). Direct manipulative interaction techniques are thus made possible for rapid shape definition, yielding complexly shaped design models depending on user interaction and the selected tools.

The ability to record different materials in the data model is utilized to create a core as a minimum hull of the desired shape defined by the concepual data model. Core material data is treated during modeling as fixed volumes and reference volumes surrounding the geometric representation of aggregates.

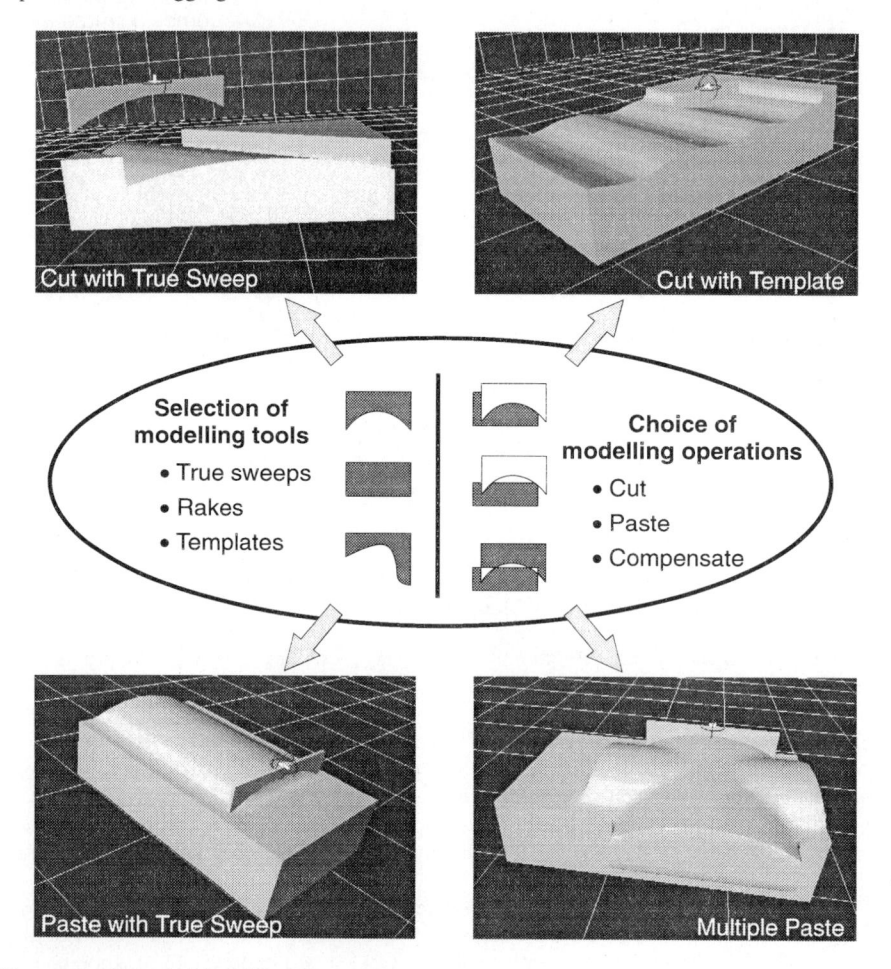

Figure 4   Sample Modelling Operations

4 CONCEPT FOR CAID INTEGRATION

Communication of concept data with pre-, post- and concurrent product development steps in product conception and mechanical design can be achieved purely on the basis of exchange of concept data via physical files. More flexibility in concept and shape development, integrity of data models and modification management of concept versus shape is achieved by integration of the relevant CA-Systems. The following functional enhancements may result from the provision and use of a database of concept models:

- Cross references of styling-oriented and engineering-oriented concept data allow for integrated modelling of comprehensive concept models as a consistent and progress-dependent frame of reference for a product development project.
- Utilization of external computer-aided engineering methods is made possible by relating their results to concept data items.
- Analysis on demand is feasible for the current state of concept specifications and the intermediate design model under development.

For the latter, the integrated use of FEAMOS and the virtual clay modeller has been defined, Figure 5. The analysis within the feature modeller provides functionality to verify the integrity of the concept model. Feature modelling algorithms can be utilized to resolve parametric dependencies, detect inconsistencies in the concept data structure due to current values of concept data and check the validity of concept data references. Modelling operations for the concept or the incorporation of modifications requested by virtual clay modelling are possible sources for a redefinition of concept data.

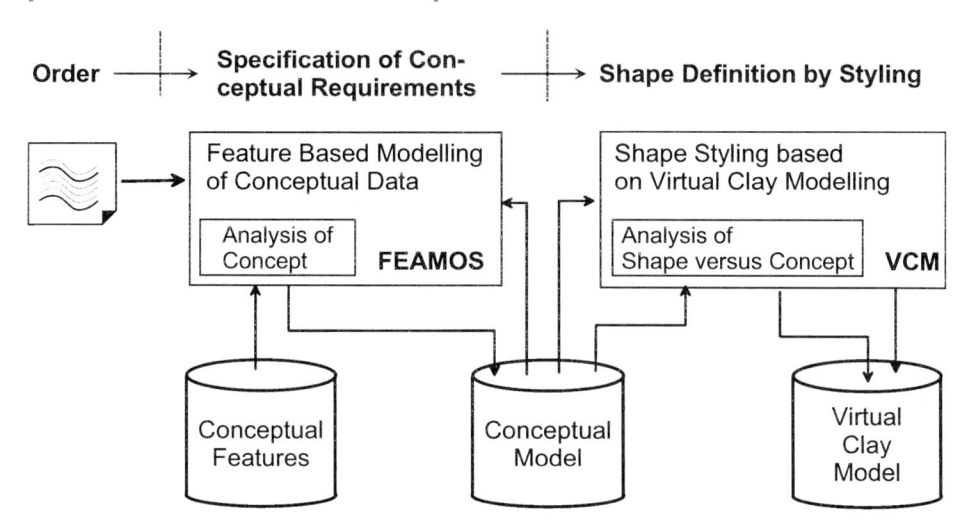

Figure 5 Conceptual Data Model as Integration Factor

Within the virtual clay modeller, analysis comprises the verification of consistent concept data versus shape. For this purpose, analysis functions are introduced as a first step which allows for verification of the core defined by technical aggregates and hardpoints against the shape of the design model. Modelling operations compare the movement of a tool with the material of voxel positions and thus inhibit shape alterations in the proximity of the core. The use of ergonomic templates for a visual inspection is straightforward.

5 FEATURE-BASED CONCEPTUAL MODELLING

Within conceptual modelling, the requirements from a conceptual point of view, expressed as entities of an information model, are transformed into a corresponding geometric shape describing the resulting geometric boundary conditions for styling. This geometric shape then represents the degrees of freedom to be observed during the creative styling phase, which is used as input for virtual clay modelling.

The modelling phase for generation of the conceptual model comprises the following steps, Figure 6:

- Specification of Conceptual Requirements as performed in conventional product development using information modelling techniques such as EXPRESS-G. The result of the specification is a structured representation of boundary conditions for the new products.
- Definition of conceptual features in the form of mapping of the previously described information structure onto features for modelling. Due to the use of EXPRESS as the syntactical basis for feature description, this step is straightforward. The result is a feature library containing the modelling elements for the conceptual model.
- Modelling of requirements, where the product is structured into its basic components with respect to valid requirements. For performing the modelling task, the structure-oriented modelling functions as described for FEAMOS are provided. The result is a product structure taking consideration of requirements. This activitiy must be performed for each product or product family.
- Instantiation of geometric view of the conceptual model, thus generating the geometric shape of requirements for the new product. The shape is generated by reinterpretation of features using the parameters resulting from conceptual information.
- Discretization of the conceptual model for styling, which serves as input to the virtual clay modeller for the definition of the core.

In order to provide functionality for handling the entities of conceptual information, the previously described feature approach is used. Entities as specified in the information model are described in terms of features. On the other hand, the conceptual components of the intended product are also described as features. During modelling, these components are structured and brought into relation with the specific conceptual requirement features.

The features describing the conceptual requirements can also be structured according to their validity as follows:

- Branch-specific features containing information whose validity is independent of different companies. Examples are requirements resulting from standards, such as the plug connector required for any telephone device or requirements for a product-specific function, such as the space required for the speaker in the case of a telephone.
- Company-specific features, which contain requirements resulting from the goal of market differentiation or compatibility with other company products. An example might be the space provided for a rechargeable battery as produced throughtrout the entire company.

- Product-type-specific features containing information about characteristics of a product type within a product family. An example would be the limitation of the maximum size for a mobile phone in order to keep it handy.
- Project-specific features containing information about a specific product in accordance with the development order. This also includes competing requirements for product alternatives.

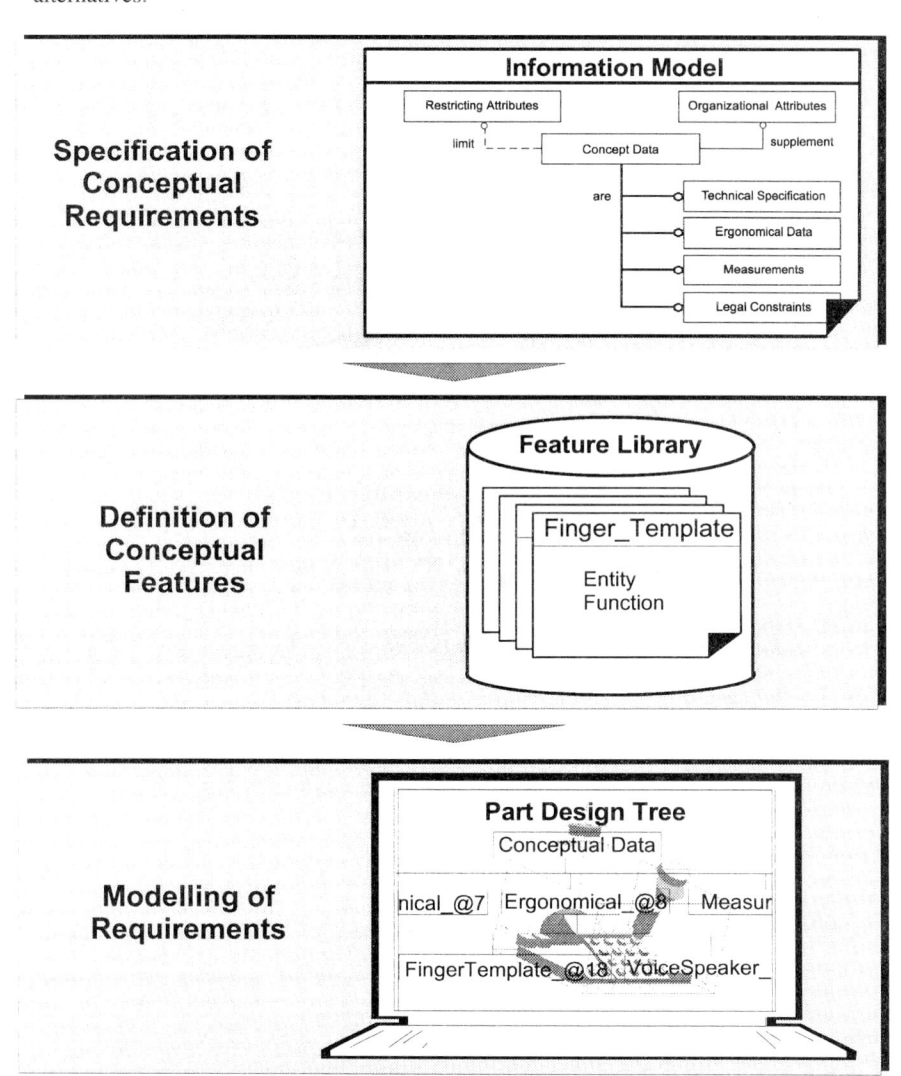

Figure 6   Process of Conceptual Modelling

The sample depicted in Figure 7 represents an instance of a concept model which is product-, branch- and project-specific, but company-generic in its application. For comparison, the discretized concept model is assigned as output for the definition of the core within the virtual clay modeller.

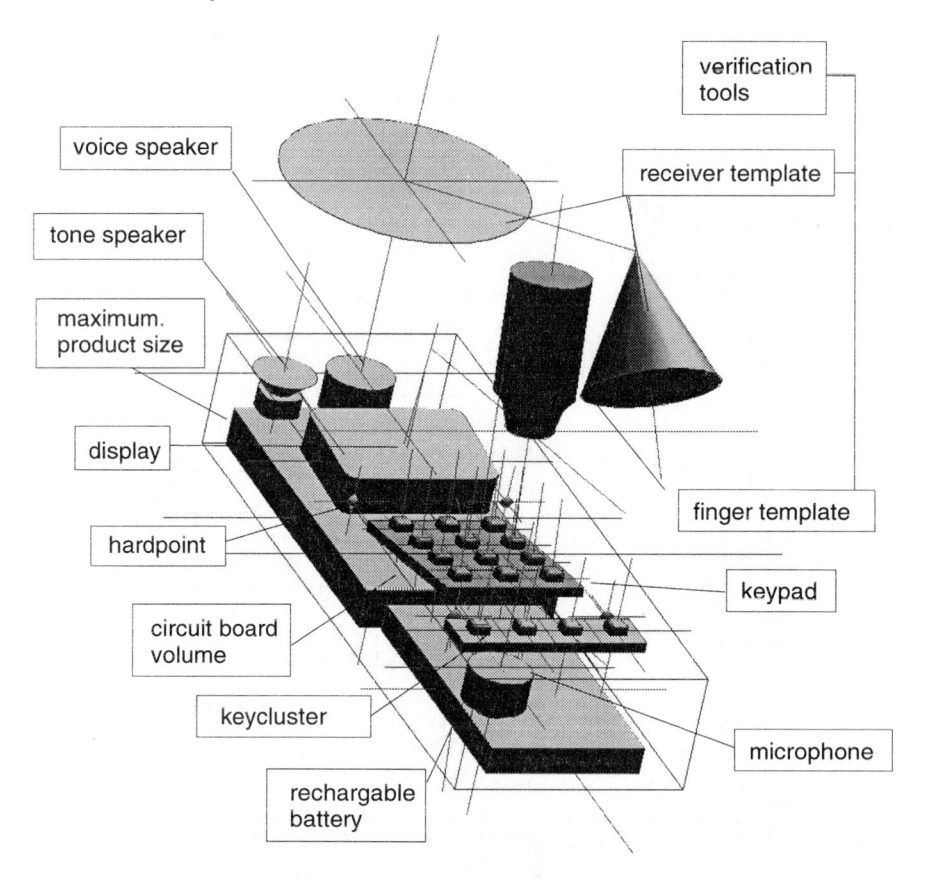

Figure 7 Conceptual Data Model of a Mobile Telephone

6   APPLICATION OF VIRTUAL CLAY MODELLING

For the realization of a design model, repetitive modelling operations are to be performed. In the virtual clay modelling approach, they can be characterized as modification operations rather than operations which construct the model. The intial virtual clay model may be defined in depency on the core, proceeding from an intial shape which limits the the allowable size or from scratch by adding material. At any time in the course of detailling the design model, the user has — aside from the selection of a tool type, its sizing and the choice of the working mode — the freedom to limit tool movement by defining allowable translation and rotation.

An important supporting feature for impressing the intended shape alteration is the distinction between active and inactive modelling tools. The inactive status of a tool offers the user the opportunity of evaluating the relative position and orientation of tool and shape while the tool is being moved by the user, from which the expected modification to the model is deduced. Thus, all parameters can be set to prepare for the modelling operation in an iterative manner. In the active mode, the intended modification of shape can be achieved from the users viewpoint in a single operation. Movements are split and evaluated to spline-based volumes of a swept profile by the modelling kernel while the tool is dragged.

This iterative setting procedure for all parameters reduces the number of modelling oprations considerably, so that a few are sufficient to yield a first expressive model, Figure 8.

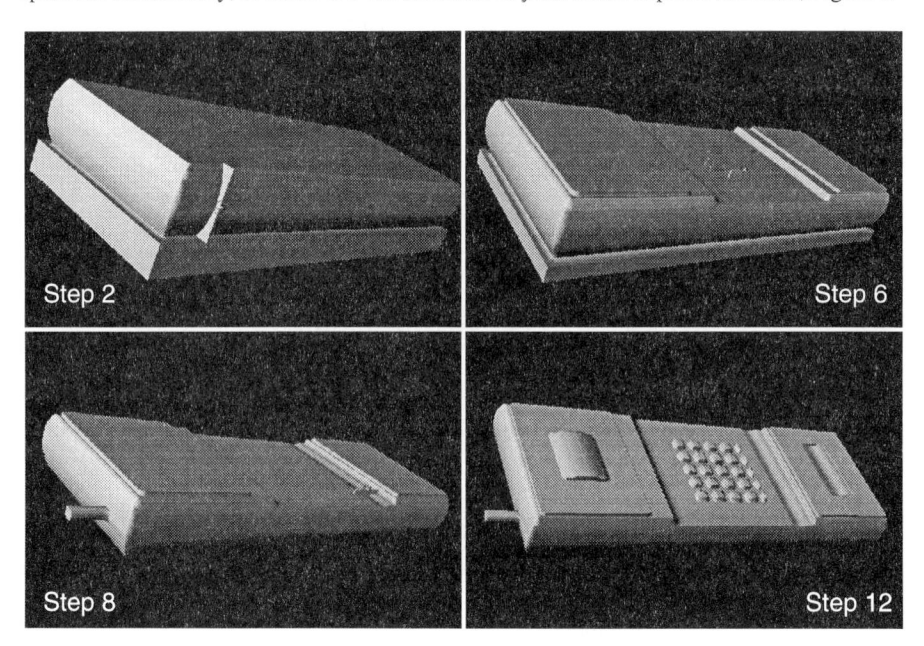

Figure 8 Mobile Telephone Design by VCM

Real-time modelling performed in the sample design of a mobile telephone is based on a virtual clay model with a resolution of 160x80x80 voxel. Resolution will be raised for ongoing development to provide for the ability to model very small details as well.

7 FUTURE RESEARCH

The present contribution introduces a new approach for the consideration of conceptual requirements within the styling phase. The approach is based on semantic features, independent with respect to geometry, which make possible the capture of conceptual modelling information and provide for the virtual clay modelling of the geometric representation of enriched conceptual package information.

Future investigations will concentrate on a functional integration of the described system components. Based on the presented concept for integration, it will be possible to relate the legal conceptual requirements in styling directly to corresponding semantic concept information. Furthermore, styling and mechanical design processes can be parallelized. Required modifications to mechanical design can thus be directly transmitted to styling and vice versa.

A promising approach to integration is the exploitation of the independence of the feature modeller from the geometric modelling kernel. Thus the geometric representation of the conceptual requirements can be instantiated either as an analytical or a voxel model and modelling functions can be used for styling and mechanical design.

Further benefits from the approach here presented can be expected from its application to spatial analysis. Here the voxel representation of product components or their expected space will be used to verify the layout and configuration of the assemblies. This results in requirements to be taken into account within ongoing product development. The input for this analysis could also be captured in terms of semantic features as part of conceptual information. The instantiation of geometric templates for non-volumetric conceptual requirements leads to a considerably extendend analysis functionality.

8 REFERENCES

di Giusto, N. and Robinson, M. (1993) Computer Aided Styling at Stile/Design Fiat. International Conference on Time to Market in the Automotive Industry, Torino, Italy, 15.6.-17.6.1993, 227-33

Galyean, T.A. and Hughes, J.F. (1991) Sculpting: An Interactive Volumetric Modelling Technique. Proceedings SIGGRAPH '91, Computer Graphics, 25(4):267-74

Krause, F.-L.; Lüddemann, J. and Kehler, T. (1994) Rechnerintegriertes Industriedesign. VDI-Berichte Nr. 1148. VDI Verlag: Düsseldorf, 1994, 501-16

Krause, F.-L.; Lüddeman, J. and Striepe, A. (1995) Conceptual Modelling for Industrial Design. Annals of the CIRP 44/1:137-40

Krause, F.-L.; Kramer, S. and Rieger, E. (1991) PDGL - A Language for Efficient Feature Based Product Gestaltung CIRP Annals, Vol. 40/1, 135-38

Krause, F.-L.; Rieger, E. and Ulbrich, A. (1994) Feature Processing as Kernel for Integrated CAE Systems, Proceedings of the IFIP International Conference, Feature Modelling and Recognation in Advanced CAD/CAM Systems, Vol. II, Valenciennes, 693-716

Menon, J.; Marisa, R.J. and Zagajac, J. (1994) More Powerful Solid Modelling through Ray Representation. IEEE Computer Graphics and Applications, May 1994, 22-35

Mischok, P., Alber, S.; Robb, D.: Anwendung neuer CA-Techniken im Automobildesign der BMW AG. VDI-Berichte Nr. 993.2: VDI Verlag: Düsseldorf, 1992, S. 141-157.

Rieger, E. (1994) Semantikorientierte Features zur kontinuierlichen Unterstützung der Produktgestaltung, Reihe Produktionstechnik Berlin, Band 158, München Wien, Carl Hanser Verlag

Tovey, M. (1994) Form Creation Techniques for Automotive CAD. Design Studies, 15(1):85-114

Yamada, Y., 1993, Clay Modelling: Techniques for Giving Three-Dimensional Form to Idea, Car Styling Extra Issue, Vol. 93 ½

Wang, S.W. and Kaufman, A.E. (1995) Volume Sculpting, in: Symposium on Interactive 3D Graphics, Montery CA: ACM SIGGRAPH, 1995, 151-56

9 BIOGRAPHY

Prof. Dr.-Ing. Frank-Lothar Krause, born 1942, studied Production Technology at the Technical University Berlin. In 1976, he became Senior Engineer for the CAD Group at the Institute for Machine Tools and Production Technology (IWF) of the TU Berlin and earned his doctorate under Prof. Spur. Since 1977, he has been Director of the Design Technology Department at the Fraunhofer Institute for Production Systems and Design Technology (IPK Berlin). He earned the qualification as a university lecturer in 1979 and has been University Professor for Industrial Information Technology at the IWF of the TU Berlin since 1990.

Dipl.-Ing. Jörg Lüddemann, born 1959, studied meachanical engineering at the Technical University of Aachen with a concentration on Automotive Engineering. Work on his diploma thesis was conducted at Michigan State University with scholarship funding. Since 1987, he has been Research Engineer at the IWF in the Department for Industrial Information Technology and since 1993 head of the Geometric Modelling Group at the IPK.

Dr.-Ing. Erik Rieger, born 1963, studied Aeronautics and Astronautics at the Technical University Berlin. After finishing his diploma, he joined the R&D Department of Norsk Data GmbH. Since 1989 he has been research engineer at the IWF/IPK in the Department for Industrial Information Technology. In 1994 he obtained his PhD in the area of feature modelling.

46

Flexible Definition of Form Features

R. Geelink, O.W. Salomons, F. van Slooten, F.J.A.M. van Houten, H.J.J.Kals
University of Twente, Department of Mechanical Engineering, Laboratory of Production and Design Engineering,
7500 AE Enschede, The Netherlands, tel. X–53–4892532, fax. X–53–4335612, e–mail: r.geelink@wb.utwente.nl

Abstract
Present feature based Computer Aided (CA) systems do not offer a facility to easily define one's own form features, if they offer such a facility at all. The reason why present software tools are inadequate for feature definition is that defining features requires a lot of programming skills, making feature definition an error prone and difficult task. 'The people in charge of defining new features usually can be regarded as domain experts in the fields of geometry modelling or applications that use geometry information or both and not as programming experts. This paper therefore elaborates on a philosophy of an improved, flexible feature definition.

Keywords
Feature definition, form features, feature validation, feature recognition, feature based design

1. INTRODUCTION

In most Computer Aided (CA) systems that process geometry, features are commonly used in combination with solid modeling techniques. Features are regarded as notions of form, combined with some engineering meaning. The form aspect of features is usually handled by means of either B–rep or CSG solid modeling representations.

When modeling geometry (e.g. in CAD) while using features, these can often be regarded as CSG primitives, or combinations thereof. However, these features are insufficiently related to the application for which they are used: their instantiation behavior is insufficiently pre–defined, parameter change propagation and geometric constraints are often not considered. Moreover, the way in which features are defined is laborious and error prone: it involves programming.

When processing geometry in some application domain, like CAPP where feature recognition may be employed, features need not necessarily be CSG primitives or combinations thereof. For instance in feature recognition a solid model is interrogated for specific application features that once more have been pre–defined by means of programming. In this case the feature recognition algorithm often implicitly includes the feature definition. In this domain programming is also laborious and error prone.

Feature definition is therefore a candidate for improvement, both in the field of geometry modeling and in the field of geometry processing.

1.1 Feature definition

Geometry modeling (e.g. CAD) systems which have embedded the feature paradigm should support the definition of design form feature classes in order to allow end–users to model with these pre–defined design form features. Geometry processing systems, like CAPP systems, need feature definition functionality in order to support the creation of templates for recognition of application form feature instances, if they cannot handle feature based product models as an input.

Feature based systems generally do not offer a facility to easily define new or additional generic form features. Most geometry processing systems hardly allow the definition of application features. CA systems which do offer a facility for feature definition, usually provide some kind of programming interface to the geometric modeler. Often, the programming interface has limited access to the geometry processing functions of the modeler. The person in charge of defining new features usually will be a domain expert in the field of the particular application. Present feature definition functionality requires this person to be a programming expert as well. In practice, this is seldom the case. Therefore, language based approaches such as advocated in for instance (Krause et al., 1991a,b), (Laakko and Mäntylä, 1992) and (Sreevalsan and Shah, 1991) should not be preferred in the current context. This paper elaborates on the problems of current feature definition techniques and provides a philosophy for interactive feature definition. The latter should facilitate the definition of features with as little programming as possible.

There are some differences between the definition of form features for geometry modeling and for application purposes. In geometry modeling, the feature's behavior is important, e.g. in case of feature instantiation or modification. In geometry processing, like feature recognition, it is important to be able to relax the definitions of features in order to cope with different kinds of possible feature intersections.

As a part of the process of feature definition, feature class hierarchies are usually created. These class hierarchies are company, application and/or product dependent. Therefore, a user friendly feature definition functionality is needed to adapt features for specific application domains, products, companies, etc..

In order to avoid an explosion in the number of features, it is required that the feature definition functionality is only available to a limited number of users. This type of users will be referred to as system users in the following, in contrast to normal users, which we will refer to as end–users.

Figure 1 reflects the major role of feature definition in the link of feature based modeling and subsequent down–stream applications. Components can be modeled using incomplete (abstract) features, geometry modeling related form features or application related form features (see figure 1). An incomplete feature model needs completion before any down–stream action on the component can be performed. Feature completion should preferably take into account application specific knowledge (design by least commitment). If components are modeled using application specific features, feature recognition can be skipped, provided that the features have been validated. Feature validation, although not shown, is an essential function for all feature model information flows in figure 1. Feature validation, feature recognition and feature mapping should make use of similar geometric reasoning functionality. As will be shown in the following, all these techniques are closely related to feature definition.

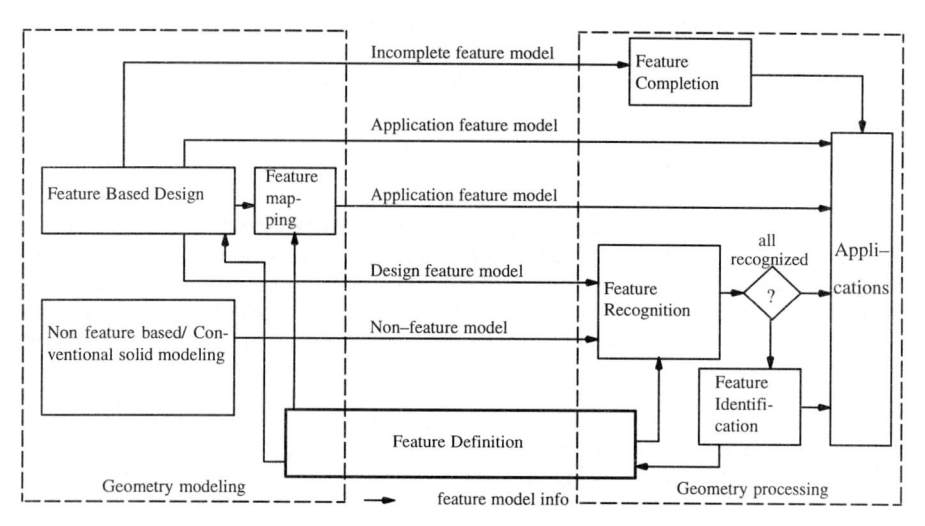

Figure 1 : The central role of feature definition, driving feature based modeling and subsequent down-stream applications.

In case components are modeled using design form features, either feature mapping or feature recognition can be applied to derive an application specific feature model. When a conventional solid modeling system (non-feature based) is used, feature recognition has to be performed.

1.2 Organization

The organization of the remainder of this paper is as follows. Section 2 briefly summarizes prior work. Section 3 elaborates on feature and solid representation issues. In section 4 functions are identified for which feature definition should provide. In section 5 a discussion is provided. Finally, in section 6 conclusions and recommendations are presented.

2. PRIOR WORK

A brief overview of prior work is provided. More elaborate overviews can be found in (Salomons et al. 1994a, 1995a) and (Geelink et al. 1995). A distinction is made between prior work in feature representation (section 2.1) and feature definition (section 2.2).

2.1 Prior work in feature representation

Feature representation is regarded as how all feature information is stored in computer internal memory: it includes representing both the individual (form) features and the relations between the individual features that constitute a component. Pratt (1988 and 1990) proposed a non-manifold feature representation based on the belief that B-rep is the preferable solids representation scheme and that features should be volumetric. Shah and Rogers (1988) used a hybrid CSG/B-rep feature representation scheme for their feature based modeling system. Wang and Ozsoy

(1991) also propose a hybrid representation scheme. Joshi and Chang (1990) proposed the use of an attributed adjacency graph for feature representation. Fu et al. (1993) propose a graph grammar for feature representation. None of the cited references has addressed feature representation from a flexible form feature definition perspective, however.

2.2 Prior work in feature definition

Billo et al. (1989a, b) tried to bring some more rigor into feature definition by using conceptual graphs. It is not clear, however, in what respect the concepts have been elaborated in order to realize interactive feature definition.

Sheu and Lin (1993) present a definition scheme that is suitable for representing and operating on form features. However, although the concepts are promising, it seems that an interactive feature definition functionality has not been realized.

Duan et al. (1993) employ generalized sweeping techniques for feature definition. The solid model definition that Duan et al. use is a hybrid B–rep/CSG representation scheme.

A unification of form feature definition methods for integrating feature based design, automatic feature recognition and interactive feature identification has been discussed by Sreevalsan and Shah (1991). Some problems were noted with regard to the unification of the three modes of feature definition. These problems are related to the lack of dynamic interfaces to geometric modelers and the restricted access to the functions and data of geometric modelers. The problems can be alleviated by the use of open architecture geometric modeling kernels like ACIS^{TM}.

In recent papers Shah et al. (1994a, b) provide a method of feature definition which is very similar to the one presented in this paper. In their approach, called declarative feature modeling, they apply constraint based feature definition using an adapted form of degrees of freedom analysis by Kramer (1992).

In (Salomons et al. 1994a, 1995a) and (Geelink et al. 1995) conceptual graphs according to Billo et al. (1989a, b) have been used to realize interactive feature definition. Similar to Shah et al. degrees fo freedom analysis is used. The main differences of this approach with respect to the work by Shah et al. (1994a, b) are:
- A greater focus on the interactive user interface functionality.
- A somewhat different use of degrees of freedom analysis.
- The automatic extraction of feature recognition rules.

However, this approach has not been proven for geometrically complex features.

In (Geelink et al. 1995) feature intersection is dealt with during feature definition; features are defined in a relaxed way so that they are as generic as possible. An example of a feature that needs a relaxed feature definition is provided in figure 2. Faces can be defined to be allowed to be missing, to be essential for the feature etc.. Subsequently, an automatic transformation of the graphs that represent their pre–defined features into feature recognition algorithms is performed.

3   FEATURE AND SOLID REPRESENTATION ISSUES

Defining features using CSG primitives seems advantageous at first sight because of its simplicity. An example is shown in figure 3 where a pocket feature has been defined by means of CSG primitives. However, each CSG primitive has its own dimensional parameters, which do not necessarily have a relation with the feature parameters and the parameters of the other primitives defining the feature. Therefore modification of feature parameters is not evident.

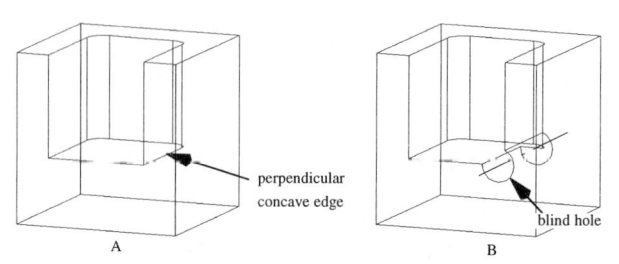

Figure 2 : Feature examples needing relaxed feature definition
    A : Generic shape of slot partial feature (4 planar and 2 cylindrical faces)
    B : Intersection of a slot with a blind hole which caused
        the deletion of the perpendicular concave edge

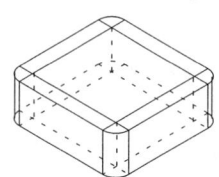

Figure 3 Defining a pocket feature using CSG primitives.

By using a half space CSG approach, the final feature faces and their dimension parameters could be addressed directly. However, in such an approach it will be difficult to directly address edges and vertices. This may in some cases be necessary for feature definition, e.g. in case of defining weld or bend lines for sheet metal products.

Feature definition thus involves the topological elements: faces, edges and vertices. So a B–rep solid representation seems to be favorable from a feature definition point of view. However, the B–rep topological elements alone are not sufficient. Apart from topological constraints, geometric constraints need to be part of a feature definition as well. Figure 4a shows 2 cylindrical

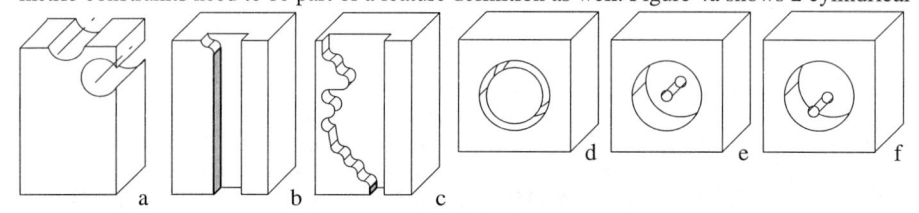

Figure 4 Influence of dimensions on the interpretation of form features.

holes which all have the same topology but different geometry. However, depending on the application, the 'hole' at the top could be considered differently compared to the other hole because it has different geometry. In Figure 4b & c a free shaped slot is presented with two parallel side faces. In 4b the two side faces have almost the same size, in 4c one of the parallel faces is much smaller compared to the opposing side. Figure 4d represents a circular groove. However, in Figure 4e, with other dimensions, the groove can be seen also as a blind round pocket with a concentric island. In Figure 4f it is shown that even the position of the island can become important. In all these cases the dimensions have to be considered in order to enable a proper feature type determination.

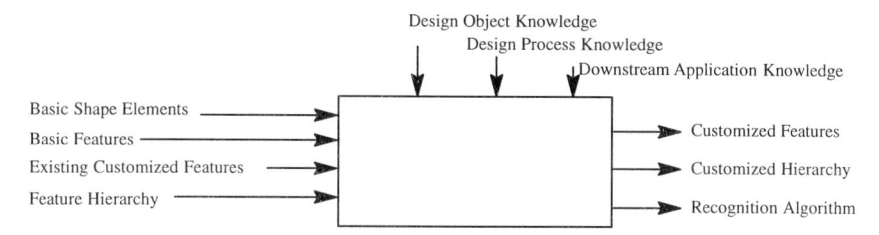

Figure 5 Black box description of the interactive feature definition functionality.

Other topological elements than just faces edges and vertices may be useful as well. For instance compound features may easily be defined by using other previously defined atomic features.

4 INTERACTIVE FEATURE DEFINITION FUNCTIONALITY

The functionality for feature definition, depicted as a black box, is shown in figure 5. A decomposition which has been derived in (Salomons, 1995a) is shown in figure 6. Each distinguished

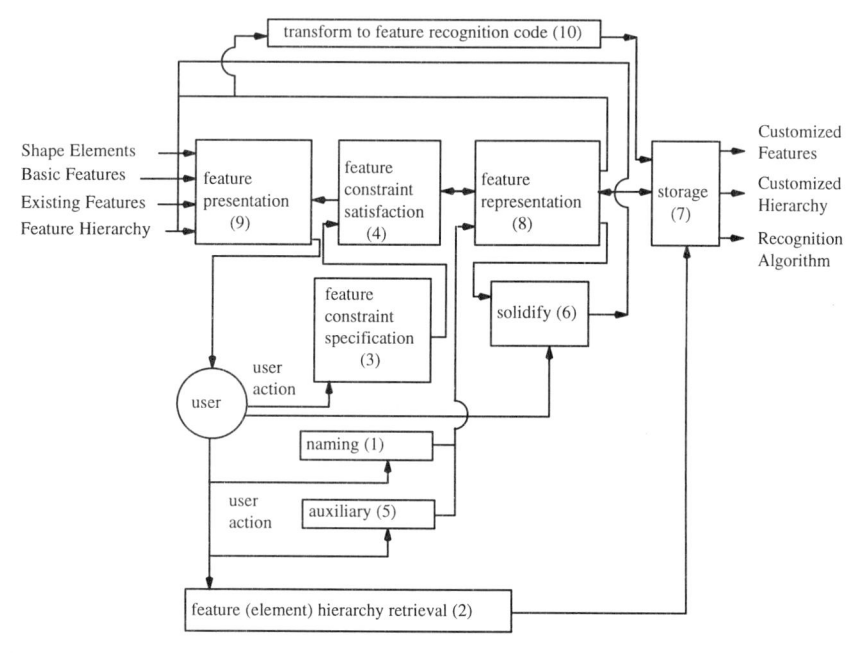

Figure 6 A functional decomposition of the feature definition function (redrawn after Geelink et al. (1995)).

sub function is briefly described in the following.

The feature to be defined has to be given a proper name (1) to be able to identify it later. This holds especially for feature based design purposes. Naming includes placing the feature somewhere in the feature hierarchy and thus inheriting properties of the parent feature. Features, feature elements and feature hierarchies have to be retrieved(2) from storage(7).

Topology and geometry definition can both be seen as constraint specification(3). Topology describes the number and types of feature elements (faces, edges and vertices) and how they relate to each other. Geometry constraints determine the actual shape, the dimensions. Constraints are regarded as desired relationships between two or more objects. Sketching can be used for the specification of geometric constraints (Geelink et al. 1995). In the case of sketching, geometry and topology constraints can be specified easily and quickly.

Features can also have algebraic constraints and symbolic constraints, not necessarily relating to geometry. These constraints should also be included. Thus, a constraint editor for relating feature parameters is necessary.

Salomons (1995a) has used a geometric constraint solver(4) based on a combination of the theories by Kramer (1992), Liu and Nnaji (1991) and Clément and Rivière (1993, 1994).

The auxiliary feature definition functionality(5), includes the definition of the intent of a feature and defining affinity relations that may or may not exist between features.

Solidification (6) reflects making a solid model of the feature elements. This is necessary for feature based design and in feature validation.

Feature representation(8) has been addressed in section 3. Feature presentation(9) reflects relevant feature representation information, therefore it addresses the user interface needed for feature definition.

Transformation(10) of the feature representation resulting from the feature definition into feature recognition code for down–stream and feature validation purposes is important. By doing so, programming is avoided once more, feature validation can be automated to a great extent.

5. DISCUSSION

One of the main difficulties of defining geometrically complex features is that they often depend on some piece of geometry which should be created by the end–user prior to instantiating the feature. This means that feature geometry is not known at the time of feature definition. An example of such a feature is the flange feature in Figure 7. This flange feature can be created only after

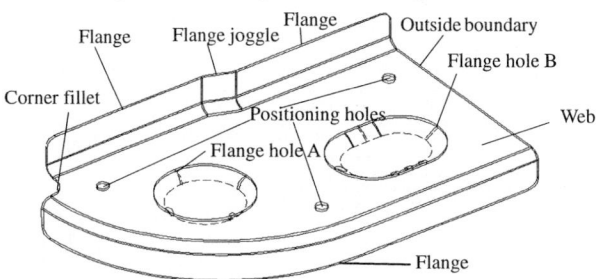

Figure 7 Examples of geometrically complex features on a component typical for application in aircraft and made by means of rubber pad forming (redrawn from (Kappert et al., 1993)).

the web surface is known (as well as the flange supporting surface). Then the user has to provide a number of feature parameters like flange height and bending radius. Both are not necessarily constants, but can be functions of some other parameter. The flange joggle is an even more complicated feature: one needs to know the outside boundaries of the web and the two flanges which it connects in order to be able to make an instantiation. In (Kappert et al., 1993) a prototype system has been described for feature based design and manufacturing of the type of components as shown in Figure 7. However, the feature definitions were hard–coded on top of an existing CAD system and not defined interactively. The example of Figure 7 shows that interactive feature definition as presented in (Salomons, 1995a) and (Geelink et al., 1995) is not yet generally applicable. To obtain a larger applicability it needs some extensions.

6. CONCLUSIONS AND RECOMMENDATIONS

This paper has focussed on a philosophy for interactive feature definition. In this process topology and geometry play an important role. The paper has indicated some interesting developments in feature definition. The required functionality for feature definition has been identified. In the future a lot of work still has to be done in the field of feature definition, such as the definition of geometrically more complex features.

Acknowledgements
This research is partly supported by the Technology Foundation (STW: a Dutch foundation).

7 REFERENCES

Billo, R.E., Henderson, M.R., and Rucker, R., (1989a), "Applying conceptual graph inferencing to feature–based engineering analysis", *Computers in Industry*, Vol. 13, pp. 195–214.

Billo, R.E., (1989b), "An object–oriented modeling methodology utilizing conceptual graphs for form feature definition", Ph.D. Thesis, Arizona State University.

Clément, A. and Rivière, A. (1993) "Tolerancing versus nominal modelling in next generation CAD/CAM system", *Proc. CIRP Seminar on Computer Aided Tolerancing*, Paris, pp. 97–113.

Clément A., Rivière A., Temmerman M., (1994), "Cotation tridimensionelle des systèmes mécaniques, théorie & pratique", PYC Edition, Yvry–Sur–Seine Cedex (ISBN 2–85330–132–X), in French (English version is in progress).

Duan, W., Zhou, J., and Lai, K., (1993), "FSMT: a solid modelling tool for feature–based design and manufacture", *Computer–Aided Design*, Vol. 25, No. 1, pp. 29 – 38.

Finger, S. and Safier, A., (1990), "Representing and recognizing features in mechanical designs", *Proceedings, 2nd. Int. conf. on design theory and methodology DTM '90*, Chicago, pp. 1–14.

Fu, Z., De Pennington, A., and Saia, A., (1993), "A graph grammar approach to feature representation and transformation", *International Journal of Computer Integrated Manufacturing*, Vol.6, No.1, pp. 137–151.

Geelink R., Salomons O.W., Slooten F. van, Houten F.J.A.M. van, Kals H.J.J. (1995), "Unified feature definition for feature based design and feature based manufacturing", *Proceedings ASME Computers in Engineering Conference*, Boston.

Han J., Requicha A.G., (1994), "Incremental recognition of machining features", *Proceedings ASME Computers in Engineering Conference*, Vol.1, pp. 143–151.

Han J., Requicha A.G., (1995) Integration of feature based design and feature recognition, proceedings ASME Computers in Engineering Conference, Boston.

Henderson, M., Srinath, G., Stage, R., Walker, K., and Regli, W., (1994), "Boundary representation–based feature identification", *Advances in Feature Based Manufacturing*, J.J. Shah, M. Mäntylä, D.Nau, eds., Elsevier, pp. 15–39.

Joshi, S., and Chang, T.C., (1988), "Graph–based heuristics for recognition of machined features from a 3–d solid model", *Computer–Aided Design*, Vol.20, No. 2, pp. 58–64.

Joshi, S., and Chang, T.C., (1990), "Feature extraction and feature based design approaches in the development of design interface for process planning", *Journal of Intelligent Manufacturing*, Vol.1, pp. 1–15.

Kappert J.H., Houten F.J.A.M. Kals H.J.J., (1993), "The application of features in airframe component design and manufacturing", *Annals of the CIRP*, Vol 42, No.1, pp. 523 – 526.

Kim, Y., (1994), "Volumetric feature recognition using convex decomposition", *Advances in Feature Based Manufacturing*, J.J. Shah, M. Mäntylä, D.Nau, eds., Elsevier, pp. 39–65.

Kramer, G., (1992), "*Solving Geometric Constraint Systems, a case study in kinematics*", The MIT Press, Cambridge, Massachussetts, London, England.

Krause, F.–L., Kramer, S., and Rieger, E., (1991a), "PDGL: A language for efficient feature–based product gestaltung", *Annals of the CIRP*, Vol. 40, No. 1, pp. 135–138.

Krause, F–L, Ulbrich, A., and Vosgerau, F., (1991b), "Feature based approach for the integration of design and process planning systems", *Product Modelling for Computer–Aided Design and Manufacturing*, J.Turner, J.Pegna, M.Wozny, eds., Elsevier Science Publishers B.V.(North Holland), pp. 285 – 297.

Laakko, T., and Mäntylä, M., (1992), "Feature–based modelling of families of machined parts", *Human aspects in computer integrated manufacturing*, G.J. Olling, F. Kimura, eds., PROLAMAT '92, pp. 351 – 360.

Marefat, M., Kashyap, R.L., (1990), "Geometric reasoning for recognition of three–dimensional object features", *IEEE transactions on pattern analysis and machine intelligence*, Vol. 12, No. 10, pp. 949–65.

Menon, S. and Kim, Y.S., (1994), "Cylindrical Features in Form Feature Recognition Using Convex Decomposition", *Proceedings, IFIP WG 5.3 Conference on Feature Modeling and recognition in advanced CAD/CAM systems*, Valenciennes (F), Vol.1, pp. 295–314.

Pratt, M.J., (1988), Synthesis of an optimal approach to form feature modelling, *Proceedings, Computers in Engineering Conference*, pp. 263–273.

Pratt, M.J., (1990), "A hybrid feature based modelling system", *Advanced geometric modelling for engineering applications*, F.L.Krause, H.Jansen, eds., Elsevier Science Publishers, pp. 189–210.

Sakurai, H., and Chin, C.–W., (1994), "Definition and recognition of volume features for process planning", *Advances in Feature Based Manufacturing*, J.J. Shah, M. Mäntylä, D.Nau, eds., Elsevier, pp. 65–80.

Sakurai H., (1994), "Decomposing a delta volume into maximal convex volumes and sequencing them for machining", *Proceedings ASME Computers in Engineering Conference*, pp. 135–43.

Salomons, O.W., Jonker, H.G., Slooten, F. van, Houten, F.J.A.M. van, and Kals, H.J.J., (1994a), "Interactive feature definition", *Proceedings, Conference on Feature modeling and recognition in advanced CAD/CAM systems, IFIP WG 5.3*, Valenciennes (F), Vol. 1, pp. 181–204, available on http://utwpue.wb.utwente.nl/stw–doc/papers/paper–ifd.ps.

Salomons, O.W., (1995a), "Computer support in the design of mechanical products, constraint specification and satisfaction in feature based design for manufacturing", Ph.D. Thesis, University of Twente, Enschede (NL).

Shah, J.J., and Rogers, M.T., (1988), "Expert form feature modelling shell", *Computer–Aided Design*, Vol. 20, No.9, pp. 515–524.

Shah, J.J., Balakrishnan, G., Rogers, M.T., and Urban, S.D., (1994a), "Comparative study of procedural and declarative feature based geometric modeling", *Proceedings, Conference on Feature modeling and recognition in advanced CAD/CAM systems, IFIP WG 5.3*, Valenciennes (F), Vol. 2, pp. 647–671.

Shah, J.J., Ali, A., and Rogers, M.T., (1994b), "Investigation of declarative feature modelling", *Proceedings, ASME Computers in Engineering Conference (CIE'94)*, Vol.1, pp. 1–12.

Shah, J.J., Shen, Y. and Shirur, A., (1994c), "Determination of machining volumes from extensible sets of design features", *Advances in Feature Based Manufacturing*, J.J. Shah, M. Mäntylä, D.Nau, eds., Elsevier, pp. 129–157.

Sheu, L–C., and Lin, J.T., (1993), "Representation scheme for defining and operating form features", *Computer–Aided Design*, Vol. 25, No. 6, pp. 333–347.

Sreevalsan, P.C., and Shah, J.J, (1991), "Unification of form feature definition methods", *IFIP WG 5.2 workshop on Intelligent CAD*, Columbus, Ohio.

Tseng, Y–J. and Joshi, S.B., (1994), "Recognizing multiple interpretations of interacting machining features", *Computer–Aided Design*, Vol. 26, No. 9, pp. 667–688.

Vandenbrande, J.H., Requicha, A.A.G., (1993), "Spatial reasoning for the automatic recognition of machinable features in solid models", *IEEE transactions on pattern analysis and machine inteligence*, Vol. 15, No 12, pp. 1269–1285.

Wang, N., and Ozsoy, T.M., (1991), "A scheme to represent features, dimensions and tolerances in geometric modelling", *Journal of Manufacturing Systems*, Vol. 10, No.3, pp. 233–240.

8   BIOGRAPHY

R. Geelink is a PhD. candidate working at the Laboratory of Production and Design Engineering at the University of Twente. He obtained his MS. degree in mechanical engineering in 1990 at the same university. Presently he is working in the field of feature based process planning and its link with CAD. His research interests include feature technology and CAD–CAPP integration.

O.W. Salomons obtained a PhD. degree on his work on the FROOM system at the University of Twente. He obtained his MS. degree in Mechanical Engineering in 1990 at the same university. Presently, as an assistant professor, he is performing research at the laboratory in the field of design support systems as well as their link with process planning systems.

F. van Slooten holds a BS. in Computer Science, which he obtained in 1990. Since 1991 he works as a system analyst/ technical software developer at the laboratory of Production and Design Engineering of the University of Twente.

F.J.A.M. van Houten is associate professor of the laboratory of Production and Design Engineering at the University of Twente. Professor Van Houten obtained his MS. degree in Mechanical Engineering at the Technical University of Eindhoven in 1977. He has been working at the laboratory of Production Engineering at University of Twente since 1978. Van Houten has worked in the field of CAD and CAPP; he has been closely involved with the development of several CAPP systems. In 1991 he obtained a PhD. degree on his work on the PART system. He is memeber of CIRP and IFIP WG 5.3.

H.J.J. Kals is professor and head of the laboratory of Production Engineering at the University of Twente. He is also part–time professor at the Technical University of Delft. Professor Kals obtained his MS. degree in Mechanical Engineering in 1969 at the Technical University of Eindhoven. In 1972 he obtained his PhD. degree. In 1977 he became professor of the Laboratory of Production & Design Engineering at the University of Twente. Professor Kals is a member of CIRP. He is active in the fields of CAD, CAPP, CAM, workshop– and work station control. Currently one of his main scientific interests is concurrent engineering.

PART TWELVE

Distributed Product Development and Manufacturing

47

A Conceptual System Support Framework for Distributed Product Development and Manufacturing

Hirsch, B. E., Kuhlmann, T., Maßow, C., Oehlmann, R. and Thoben, K.-D.
Bremen Institute of Industrial Technology and Applied Work Science at the University of Bremen (BIBA)
Hochschulring 20, 28359 Bremen, Germany, Tel.: +49 421 218 55 31, Telefax: +49 421 218 5530, e-mail: km@biba.uni-bremen.de

Abstract
This paper is to device a framework for distributed and interorganisational product development and production (in short: distributed manufacturing). Purpose of the framework is to identify and relate key problem and solution domains of distributed manufacturing. It is based on experiences from a number of research and industrial projects focusing on various organisational and technological aspects of distributed manufacturing.

Keywords
Distributed manufacturing, standardisation, control, autonomy, integration, communication, cooperation, coordination, information technology, communication technology

1 INTRODUCTION – RATIONALE AND REQUIREMENTS FOR DISTRIBUTION

Pressures
Today enterprises operate in a tremendously competitive environment characterised by a number of changed business conditions. These are the trend to global and transparent markets, the rise of mass customisation, the reduced product live cycles, and the demand for environmentally benign products and processes.

The market conditions require a consequent strategic streamlining of the enterprise. Basis for this strategic planning has to be the own principal abilities. The abilities, which are mainly a matter of the company's staff must be recognised and transferred systematically into core competences. With the individual core competences in mind, it is possible to perform an efficient diversification. Objective of this diversification around the core competences is to minimise risks and costs. As a result there is an individual and competitive scope of economies for the company. A logical consequence of the limitation towards company specific core competences is that a considerable less amount of services can be offered by one single enterprise. The collaboration with customers, suppliers, and fellowship companies becomes in this respect a central aspect of competitiveness.

Opportunities

The globalisation of the world's markets steps forward in a rapid speed. Actual agreements and rules between countries (GATT etc.) are leading to growing economical spheres of influence. Globalisation not only induces increased pressure but also offers new opportunities. Companies are able to get access to new markets concerning both cheap procurement of recourses and new market potentialities for their products.

Modern information and communication technology developing faster than any other technology ever before enables totally reshaped organisational approaches. Fast and flexible communication between application systems and people is the key in distributed settings.

Transaction Cost Theory (Williamson, 1995) identifies the costs involved in the interactions within processes as a key criteria in order to determine size and scope of an individual enterprise. Accepting this interpretation, information and communication technology is one principal driving force for the phenomenon of distributed manufacturing.

Approaches

Lean production refers to the idea of getting rid of all excess in the production process including the organisation and the level of inventories in the process. Goods are manufactured to customer orders in production cells as small lots if not as one-of-a-kind. Suppliers are co-operating partners.

Virtual production is an approach claiming to be a step beyond lean production. A virtual company is a strategic alliance of companies. The virtual products are either concrete products or services and the clients of a virtual company are seen as members of the product development process. A virtual product is manufactured to customer orders. The relationship between the virtual company and its suppliers is a codestiny – a very strong partnership.

Mass customisation refers to the manufacturing of varied products customised to specific requirements at the cost of standardised, mass produced goods. For mass customisation a company has to form a dynamic network of relatively autonomous operation units. Each unit is assigned a specific process or a task. The modules interact with each others without precise plan (Pine, 1993).

Value-added partnership networks can be another logical approach to above changes: A temporary, collective enterprise involving several independent companies, integrated and aligned, combining capabilities to create new business opportunities, which no company could possibly gain alone, would be the outcome. The orientation would be external, customer-oriented and based on the relationship of people and processes, constantly being redefined and reconfigured. The assumption under which the scheme would operate is that technology and customer preferences are changing constantly. The goal is to quickly capture the business opportunities and collaborate with those best suited for this scheme (Hirsch, 1994).

2 THE CONCEPT OF DISTRIBUTED MANUFACTURING

The concept of distributed manufacturing refers to a structure of two or more organisational units related for the purpose of collaborative product development and/or collaborative order processing. The appropriate combination of specific capabilities and capacities of individual actors is to optimally fulfil specific market and/or customer requirements.

Distribution refers to the distribution (division) of work together with the division of risks and responsibilities.

Distribution in terms of the division of work between agents (contracted actors acting on behalf of someone else) usually implies, that an agents covers to some degree the risks and responsibilities bounded to his task. This in turn implies, that the agent himself requires freedom to execute the task according to his own conditions and objectives except for those specifications predetermined in his contract (dates for milestones and delivery etc.). In other words, the agent requires to be independent (i.e. autonomous). On the level of different organisations, autonomy is specifically motivated through

- legal issues: transfer of responsibility through a contract defines the rights and obligations of the agent in a national or international legal system.
- competitive issues: in order to maintain his competitiveness, an agent wants to avoid the disclosure of his know how.
- control issues: an agent cannot be controlled beyond the contract provisions. Even if the contracting organisation has the power to control the agent beyond the contract specification (e.g., by threatening him to not consider the agent for future orders), this effectively undermines the principle of the transfer of responsibility. If something goes wrong, the agent may claim it was because of unjustified outside influence.

Autonomy as opposed to integration is considered here the fundamental design factor for distributed manufacturing. The design goal is to identify the right balance between the „looseness" of the relationship between actors, and the strength and quality of these relationship with respect to again a single, integrated unit. In addition, the aspect of the distribution of locations indicates a separate major factor for the identification of specific problem and solution domains with respect to distributed manufacturing (see Figure 1).

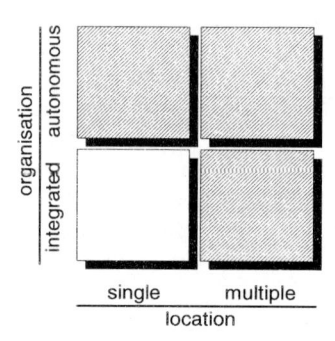

Figure 1 Principal factors of distributed manufacturing.

3 FRAMEWORK

3.1 Distributed Manufacturing as Design Challenge

3.1.1 Problem Domains
The following subsections discuss major problem domains, which are the result of a shift from a traditional centralised business paradigm towards the paradigm of distributed manufacturing as characterised in section 2.

3.1.1.1 Autonomy
Although organisational autonomy provides principal advantages like simplified control as indicated in the concepts of self-organisation, it also implies some severe disadvantages, which can be summarised as lack of integration. More specifically, this lack of integration is apparent as the following problem domains:

Divergent Goals
Distribution among multiple agents may lead to the fact, that each agent has an own, „private" agenda of goals, which may or may not coincide with the goals of the other agents and the contracting organisation. Nevertheless, shared, common goals are a prerequisite for effective collaboration. Lack of common goals require, that the behaviour of an agent can be completely controlled though the agent's contract. This requires, that all possible requirements and circumstances are known and incorporated in the contract – a condition which can only be fulfilled to some degree but never completely.

One effect of divergent goals together with autonomy and thus lack of control is the unpredictability of the behaviour of the agents involved. Consequently, this leads to lack of trust whether the other actors are really committed to explicit or implicit agreements.

Heterogeneity

An organisation is (usually) designed for maximum efficiency of its major manufacturing processes. Considering a process to be the transformation of data[1] along individual tasks performed by actors (people, systems), this requires *compatibility* between these tasks respectively the actors if considered information processing systems. Major elements which require compatibility are:
- (data) artefacts, i.e. the things explicitly exchanged between tasks. Since systems are predominantly involved to (pre-) process such data, this is mainly an issue of system-compatibility.
- concepts, i.e. the models people make about reality. Effective communication between people requires a shared understanding of the meaning of things. Different to system compatibility, such understanding is not subject to any automatism (of error detection or transformation), but relies on overlapping cultural and professional domains, and awareness about potential problems.
- technical communication means, i.e. the hard- and software (protocol implementations etc.) enabling the communication of data.
- human communication means – the human language.
- partial processes. An overall process split between agents may involve partial processes already established with individual agents, which are incompatible. Such process incompatibility beyond the points already mentioned above may be due to incomplete outputs, missing feedback, overlapping or missing subtasks etc.

A major consequence of the involvement of more or less autonomous organisations is that heterogeneous environments are brought together which have not been designed to efficiently perform a combined process. Resulting incompatibility may prohibit certain communication altogether, lead to poor quality communication, or requires investment to meet desired communication standards.

Limited Vertical Integration
Autonomy (see above) implies limited control over agents, i.e. limited vertical integration. The design possibilities provided by a central and top down form of control are limited to the degree agents are granted autonomy. It should be noted, that autonomy is a major constituent of distributed manufacturing as envisaged here (see section 2). Reducing autonomy reduces at the same time the benefits which are tight to this principle: the ability to transfer risks and responsibility.

Limited Horizontal Integration
The same holds for the aspect of horizontal integration, i.e. the level of communication, coordination and cooperation between individual tasks respectively actors having „equal rights". In principle, an autonomous agent has no interest for such integration unless the fulfilment of his own task is dependent on some other task, or unless he shares common goals with other agents.

---

[1] Although material logistics does also imply a compatibility issue, it is not considered here due to the focus on information technology.

3.1.1.2 Process Complexity

Distributed Locations
Involving two or more locations in the manufacturing process lead to the following problems:
- new or higher costs for transportation of people, data and material
- poor informal communication, if people are distributed and not co-located
- in case of truly global manufacturing the crossing of time zones such that direct communication (via phone) is restricted to certain time slots.

Multiple Identity
To be able to operate independently and locally, data artefacts are in many cases simply „copied" or even reinvented and thereafter maintained independently. This is especially the case with processes which do not execute strictly sequential such that an artefact can be treated as a token (e.g., the paper folder) which is passed on form on task to the next. If not controlled and supported properly, any kind of „concurrency" leads to a multiple identity of things, which should exist only once. This is either not controlled at all, leading e.g. to a large variety of versions e.g. of parts, or requires effort to subsequently merge artefacts. Although this effect may sound trivial, it may amount to a severe problem simply through its nature as an every-day phenomenon.

An obvious solution for that problem is to use a shared database. Nevertheless, the autonomy requirement prohibits the usage of a centrally controlled database across different organisations. In addition, a shared database solves the „read only" case, but does not provide an ultimate solution for truly simultaneous work.

3.1.2 Solution Domains

The following points discuss solution domains identified not as immediate remedy for the above mentioned problems, but on basis of the more general question, what means may support distributed manufacturing. Consequently, there is no one-to-one relationship between problem and solutions domains, but at least in some cases, a multiple relationship (see Figure 2). In this context it should be noted that the domains introduced are conceptually not totally disjunct. E.g., a standard is in principle a coordination mechanism. The mapping between problem and solution domains below tries to reflect only major relationships.

Standardisation
A standard is a model (to be) used by two or more actors. A standard is effectively „the" solution for heterogeneity (Rada, 1993). A standard is either used to eliminate heterogeneity altogether, or to encapsulate heterogeneity through communication filtered via a „neutral" interface from respectively to a proprietary solution (as with neutral product data models like STEP). With respect to the autonomy principle, the latter approach enables an agent to maintain his specific internal solution, but requires – as a cost to be paid – the effort to map to the external standard. Considering the example of product data exchange, such cost is evident through the acquisition of translators, processing time, loss of semantics through the translation etc.

Although a standard is in itself not an information technology, it is a facilitator to enhance and exploit the benefits of such technology in that it makes systems interoperable. Depending on the semantics of a standard, such interoperability addresses pure communication (network protocols etc.) or – by means of capturing application specific models – facilitates coordination.

Coordination

Coordination is to manage dependencies between tasks (see Malone, 1994). Such dependencies may exist with respect to the product (e.g., potentially interfering design spaces) and the process (e.g., limited resources). A dependency may be mutual, i.e. the involved tasks require a mutually accepted handling of such a dependency, or a dependency expresses a hierarchical relationship.

System support for coordination can be categorised as follows:
- support for the „visibility" of dependencies in order to make the involved actors aware about the need to resolve dependencies.
- support for dependency resolution. Typical examples are time and capacity planning systems.
- support for dependency reduction. Here an example are information systems which avoid the creation of multiple instances of the same artefact (see point *multiple identity* in section 3.1.1)

Coordination as a solution to distributed manufacturing supports the following problem domains:
- The quest for vertical integration and autonomy can be addressed by multilevel planning and control concepts which provides agents with the freedom for own detailed planning and control. Only coarse grain planning and control are performed by superiors.
- Horizontal integration under the constraint of autonomy is facilitated through concepts which reduce the impact of dependencies on the internal design of the processes within a task. As an example, data sharing instead of data exchange supports the pull instead of just the push principle. As a consequence, an actor can determine at least to some degree on his own, when he requires certain information. The initiation of some activity is thus not solely dependent on some other task (although the issue of information completeness has to be taken into account).
- The creation of multiple instances can either be avoided through data sharing as an implicit coordination mechanism, or controlled through versioning concepts or systematic replication.

	Divergent Goals	Heterogeneity	Limited Vertical Integration	Limited Horizontal Integration	Distributed Locations	Multiple Identity
Standardisation	X					
Coordination			X	X		X
Cooperation	X			X		
Communication					X	
Encapsulation		X	X			

Figure 2 Relationships between problem and solution domains of distributed manufacturing.

Cooperation

Cooperation is acting towards a common goal, i.e. the motivation of an actor is not (only) induced by his individual goals but also through goals shared with other actors.

Thus cooperation defines a certain quality of communication respectively coordination, where the behaviour of each actor is aligned according to shared goals.

Obviously cooperation characterises a state beyond divergent goals. Enabling means for cooperation are limited to basically those supporting the identification of goals shared or of areas with necessary goal alignment. Information systems can play in that respect some, yet only a limited role.

Communication

Communication is the most generic means for any synchronisation between actors. Communication is the prerequisite for coordination and cooperation. Distinct to these concepts, the concept of communication does not impose any kind of quality: people may communicate, but may misunderstand each other.

Beyond those problem domains already addressed by coordination and cooperation, (technical) communication means enable communicate across distributed locations.

Encapsulation

Instead of just managing dependencies, another approach is to confine dependencies through small and well designed interfaces, encapsulating the details of some process or product related artefact. In other words, encapsulation reduces the need for coordination to a limited number of control points (interfaces). Thus encapsulation facilitates vertical or horizontal integration under the requirement of autonomy.

3.2 Support Through Information Technology - Examples

3.2.1 MARITIME

The ESPRIT III project MARITIME addresses the coordination and communication requirements of multiple organisations participating in the development of ships. The development process pre design – design – class approval – production is characterised to date by islands of product data. This fragmentation is a reflection of different organisations using different application systems which a) use different internal schemas, i.e. are incompatible and b) are not integrated on instance level, i.e. do not form an integrated product data model.

The solutions developed by the MARITIME project address the following solution domains (Mehta, 1994):

Standardisation: MARITIME has developed neutral product models towards the ISO product model standard STEP. Specifically, an Application Protocol draft is proposed for ship steel structures. In a larger context, a suite of 5 shipbuilding related Application Protocols are under development. Emphasise is to ensure interoperability across the domain of each AP in terms of a truly integrated product model.

Communication: The transfer of product data across different locations is supported by a CORBA (Common Object Request Broker architecture) based communication mechanism linking client and server applications. Such communication is totally transparent to the end user.

Coordination: Beyond just communication a coordination mechanism has been developed as so called Repository Services. These services implement a virtually central information system providing location independent access to coarse grain product data artefacts. In addition the solution provides relationship services implementing version control and other relationship concepts which address the coordination needs between actors (here different organisations). Despite the appearance of the information system as a central, shared repository, the implementation based on CORBA comprises actually a set of interacting servers. This concept facilitates true autonomy with respect to the system architecture. Each participating organisation may setup and control its own server, yet their interaction provides the end user with an integrated view on the product data artefacts available within a context spanning two or more organisations.

3.2.2 CMSO

Objective of the ESPRIT project CIM for Multi Supplier Operations (CMSO) was to optimise interorganisational operations and logistics chains by means of advanced communication systems and logistics coordination procedures. The main results with respect to solution domains are as follows:

Standardisation and Communication: An EDI solution (CMSO/1, 1992) applies a mechanism to conceptionally bundle all interorganisational communication functions in one EDI communication server. In this server all functions related to the analysis, preparation and transmission of EDI messages are located. In order to structure the required EDI functions a multilayer reference model is used.

Coordination: With regard to the interorganisational product development between manufacturer and supplier or sub-supplier a Technical Information Management System (TIMS) (CMSO/2, 1992) optimises and coordinates the information flow and its control between the associates and their applications used in the development process. The system thus consists of a development coordinator, neutral interface selection, order database manager, temporary database manager, and a customer database manager.

3.2.3 MUSYK

Main objective of the ESPRIT project Integrated Multi-level Planning and Control System for One-of-a-Kind Production (MUSYK) was the development of an integrated concept for project/production planning and control. This covers the various control levels within a CIME environment and within multi-site production facilities. Apart from developing different coordination modules to support planning, control and harmonisation of engineering and production activities in distributed manufacturing environments, the approach integrates an information archive as a generic communication means. In particular, the following solution domains have been addressed (MUSYK, 1993):

- Standardisation and communication: The information archive enables data sharing in distributed environments. It provides a means for coupling proprietary application systems, thus supporting the communication flow within an overall business process.
- Standardisation and coordination: The archive embraces a common product and process model for exchanging planning and control data based on the STEP/EXPRESS technology. The exchange itself is supported by using an ODETTE-Engdat-Messagetype as container.

3.2.4 DECOR

The ESPRIT project Decentralised and Collaborative Production Management via Enterprise Modelling and Method Reuse (DECOR) is to create an integrated toolbox for the development of multi-level, distributed, hierarchical and autonomous decision-making systems in the domain of production management. The approach is to integrate existing solutions and apply them within a problem-solving framework thus allowing their distributed coordination across multiple software and hardware platforms. The coordination framework itself will provide a solution along both the computational (i.e. the processing level) and the behavioural (i.e. the problem solving level) level of distributed management (Hirsch, 1995). The package of software tools for the construction of DECOR-like distributed decision making subsystems is a layered and modular collection of object-oriented tools. Categorically, the tools should support the implementation of three classes of management knowledge:

Standardisation: Domain knowledge covering the representation of all the physical and conceptual entities in the distributed manufacturing environment (Domain Modelling Module). This module models the underlying production data and knowledge relevant to the particular decision making subsystem in question.

Coordination: Problem-solving knowledge defining the actual production management procedures and problem-solving rules (Problem Solving Module). The framework supports the integration of heterogeneous production management software of a specific decision making subsystem.

Communication, standardisation and encapsulation: Communication knowledge establishing coordination links within a distributed set of loosely coupled decision-making subsystems (Message Management Module). It enables message communication and the interpretation of standard messages. The specification of the message protocol is open to allow customisation of end-users needs, future growth, and heterogeneity in the management network. The DECOR problem-solving message mechanism is able to respond to the following problem situations/questions: When to send messages to other nodes; what are the contents of the messages and how to react to incoming messages.

3.2.5 ITiS-3

The national research project ITiS-3 (Information Technology in Shipbuilding) deals with organisational and technical aspects of Electronic Data Interchange (EDI) in the shipbuilding industry. ITiS-3 aims to shortening the "non swimming time" in a ship's life cycle by supporting distributed information management (ITiS, 1994).

Standardisation: Until now the necessity to convert data between different applications has prevented the shipbuilding industry from efficiently using EDI. Therefore ITiS-3 is developing a "ship-building-industry-oriented message standard" based on EDIFACT structures as well as necessary interfaces to handle messages based on this standard.

Communication and Coordination: The corresponding EDI-System includes functions to coordinate the information flow by using dynamic models of the cyclic, iterative and parallel development and production processes, and to ensure the security of product and process data.

4 REFERENCES

CMSO consortium (ed.) (/1, 1992) *Integrated EDI Architecture*; Deliverable 10; ESPRIT Project 2277.

CMSO consortium (ed.) (/2, 1992) *CMSO-Box and CIM-IAS for Product Development*; Deliverable 13; ESPRIT Project 2277.

Hirsch, B. E. and Crom, S. (1994) *Management Challenges in Globally Distributed Production Environments: Necessities and Strategies for Reorientation of Production Management Methods*; IFIP WG 5.7 Working Conference on Evaluation of Production Management Methods; Gramado, Brazil.

Hirsch, B.E., Kuhlmann, T. and Maßow, C. (1995) *Decentralized and collaborative production management - A prerequisite for globally distributed manufacturing;* Computer Applications in Production and Engineering (CAPE), Beijing.

ITIS-3 consortium (ed.) (1994) 2. report, BMFT Project ITIS-3, Bremen, Germany.

Malone T.W. and Crowston, K. (1994) The Interdisciplinary Study of Coordination; *ACM Computing Surveys*, Vol. 26, No. 1.

Mehta S. and Lehne M. (1994) *Product Data Technology Benefits – A perspective of yard and classification society*; ICCAS, Bremen, Germany.

MUSYK consortium (ed.) *Prototype of Information Archives;* deliverable D22; ESPRIT Project 6391, 1993.

Pine II, B. J., Victor, B. and Boynton, A. C. (1993) Making Mass customisation Work; *Harward Business Review*; Sep.-Oct.

Rada, R. (1993) Standards: The Language for Success; *Communications of the ACM*, Vol. 36, No. 12.
Williamson, O. E. (1995) Transaction Cost Economics and Organization Theory; in: *Organization Theory* (ed. O.E. Williamson), Oxford University Press, Oxford, New York, ...

48

Virtual Product Development for Total Customer Solutions

Dr. K. Preiss
Director, Agile Enterprise Projects, and
Editor-in-Chief, Publications Division
Agility Forum, Bethlehem, PA 18015-3715
and
Sir Leon Bagrit Professor of Computer-Aided Design
Ben Gurion University of the Negev, Israel
Phone + 1 610 758 5510
Fax   + 1 610 694 0542
e-mail preiss@menix.bgu.ac.il

Abstract

The industrial competitive environment has changed from being that of isolated processes separated by warehouses of material, known as the mass production system, to that of coupled intercompany processes dealing with dynamic, time-dependent change, known as the agile production system. This new system provides a new context for design work.

Companies with work processes so connected are known as virtual organizations, since the work is done in distributed fashion among many companies and teams in many locations, often spread around the entire globe.

The paper gives a description of the new environment and describes models used to understand the new influences on companies, and the new structure of companies, and a model which enables rational planning of business and technical connections between companies.

Keywords

Design, customers, virtual organization, agility, competitiveness.

1 FROM ISOLATED COMPANIES TO THE VALUE-ADDING CHAIN

During the 1980's and 1990's, companies around the world, influenced by technology and the pressures of international competition, have undergone a fundamental restructuring. Efficiency measures have squeezed from companies almost all activity which does not give a sure short-term economic financial reward. At the same time, companies have taken advantage of two significant opportunities, made possible by technology. These are empowerment of individuals and burgeoning electronic communication.

1.1 Technology Empowers Individuals

Technology empowers individuals. An engineer today with a modern computer-aided design system, even a young engineer, has at his fingertips a capability exceeding that of a whole design office 20 short years ago. This capability is constantly becoming cheaper and hence more accessible to more companies and individuals around the world. The CATIA-type capability, which was once in main-frame computers for a cost of $50,000 and more, became available in personal computers in Autocad at a cost of $5,000; and the latest generation of PC-based computer design tools such as XCAD and XCAM make even more powerful software available for $500. The individual designer can therefore do more powerful work than could a whole design office a decade ago. This factor opens up the competitive environment to many more people than before, from around the world.

With the spread of education around the world, anyone in any country, using standard software, can today do routine design. The routine work of standard design is not any more a competitive capability.

The same effect has occurred in production work. The cost of computer-controlled production machinery constantly drops, even as the capability becomes more powerful. For example, the capability of a milling machine, which a decade ago cost $500,000, is now available for $50,000 or less. Similar changes are seen in many fields. Twenty years ago printing production required a large printing job shop or factory with many typesetting and printing machines and specialized workers.

Today a computer network and color printer exhibit a capability exceeding that of the old printing works, and at much lower cost.

Twenty years ago only those organizations which could enlist significant amounts of capital could produce product; today the limitation to producing product is not access to machines, but knowledge. The reducing cost of capability and the world-wide spread of education have opened up to many people and companies around the world the capability to produce goods, including good quality goods.

The capability to produce products, by itself, does not anymore give a competitive advantage. If one makes a product with significantly better function, then only in rare cases can such a product be protected by keeping knowledge secret or by robust patents. Usually, as soon as a profitable product is seen, someone in the world makes a competitive product. In the recent past, the price, quality and function of a product used to be enough to maintain competitive capability. Today, those factors by themselves are not enough. Function, price and quality are necessary, but by themselves are not enough to ensure competitive success.

1.2 Communication Connects Individuals

The technology of communication has almost overnight created the universally recognized 'global village'. For product design this means that all of a sudden one can obtain instant access to any significant company or professional individual in the industrial world. Companies have 'downsized' many activities, including those related to product development and design. Simultaneously, they have found access to many companies, external consultants and collaborators of all kinds. It therefore becomes accepted that a company expand its capabilities not by hiring people, but by connecting its work processes with the work processes of other companies, using the capabilities of modern communication. This has changed the context within which we work, including the context of product design and development.

1.3 The Context has Changed

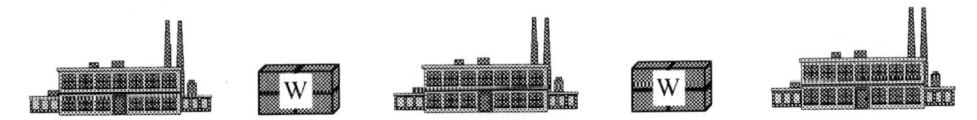

Factories Separated by Warehouses
Processes Insulated by Inventory
Processes Managed Independently
Products Separate Processes
Manage by Managing Inventory
Capable Only of Slow Change
Metrics are Product-Focused
    Cost per product
    Units per hour

Figure 1 Factories Separated by Warehouses.

Because a classical mass production system turns out a given constant product ("any color as long as it is black"), the demands are unchanged and the processes are unchanged for a long period of time. Both the products made and time rates of output of product, for instance items produced per day, are constant. Changes in time rates do not change the system. Information is not transferred between processes; this information could be used to modify the behavior of the system, but that is not at all required. Only product is transferred between processes. The mass production system is therefore static. The system can manage a certain amount of change, but the system is not designed for dynamic factors, and will not withstand highly dynamic loadings. We can say that classical mass production consists of static, decoupled processes.

Factories Dependent
Processes Dependent
Manage by Managing Processes
Manage Processes Together
Cannot Manage by Product Inventory
Capable of Quick Change
Metrics are Process-Focused

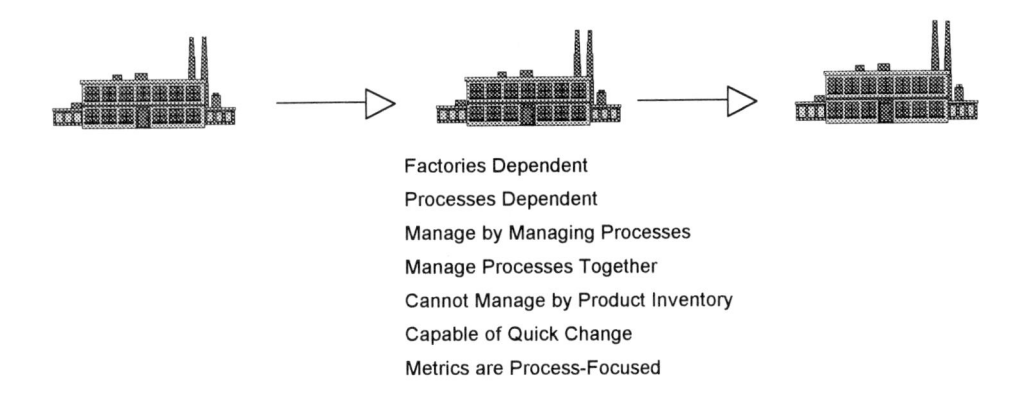

Figure 2   Factories with Direct Flow of Material.

The nominal elimination of warehousing has created a system of coupled processes. The mass production system was relatively easy to manage because each production line or factory was managed independently of all others. Since the total system is now coupled, one has to manage production taking account of that fact; factories, production lines and other work processes are no more independent. The information available for management decisions made in one factory or production line has to take into account the information, status and behavior of other factories or other production processes in the system. This is a new environment. The availability of inventory, the information and behavior of a process in any one production line, business unit or company, is affected by all the other processes to which it is connected.

This is a new context. In the old context which was with us for a century, the product was the aim and all the work processes revolved around that. In the new world of industrial competition, called the Agile Competitive Environment, the aim is to connect work processes.

Product, of course, flows from that, but the reason a company is chosen is because of its work processes, not its products.

2   FROM PRODUCT TO PROCESS

In the old industrial world, to which many of our practices and work processes are still wedded, the aim of the entire enterprise was to produce competitive product. The aim now is to connect one's work processes to those of customers and of suppliers. Faced with the choice of maintaining competitive capability either by being a chosen supplier of

high-quality, competitively priced product, or by becoming an integral part of a customer's work processes, the choice is obvious:
- it is safer to be a chosen team member of a customer's processes. For an industrial customer, this means being part of the work processes in the customer's company. For a consumer, this means being part of his or her life-style process. This is the lesson learned by Allied Signal under Larry Bossidy, by General Electric under Jack Welch, by Ross Operating Valves under Henry Duignan, and by many other companies. It is the lesson being learned by Mercedes, Ford and GM in their interactions with consumers, and by the US textile industry under the technical leadership of the Textile/Clothing Technology Corporation in North Carolina.

In the move from product to process, many concepts change. We may continue to use the old words, but they will have new meanings.

- The *aim* of business moves from supplying competitive product to finding new customers and keeping them forever.
- A *product* is not an aim in itself, but becomes a platform for a continuing customer relationship.
- A *customer* is not any more the recipient of product in an instantaneous event, but a subscriber to the problem-solving and development services of the company.
- A *supplier* becomes an associate and member of one's problem-solving team.
- A *sale* is not anymore a one-time event, but becomes a continuous life-long relationship.
- The *supplier* provides not just a product, but rather a total solution to the customer's needs. This solution will often be customized and will be a fusion of product, information, knowledge and service.
- *Pricing* of the solution is not by cost plus margin, but is set by the market and will reflect the relative value to the customer of the solution provided.

To deal with this new context, we have available to us models of the enterprise and of the connection between enterprises.

3 THE MODEL OF A SINGLE COMPANY

The starting point for the model of a modern industrial enterprise is the diagram shown in Figure 3.

External Influences

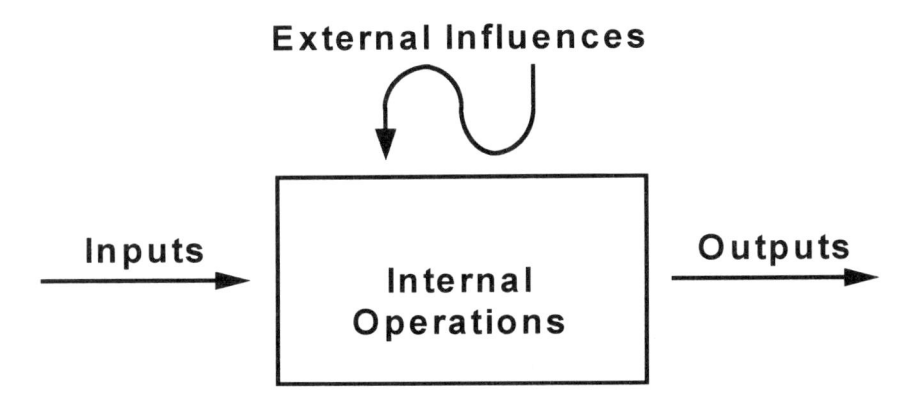

Inputs → **Internal Operations** → **Outputs**

Figure 3   An Engineering System in General.

The diagram shows the inputs, the outputs, and the internal details of a system, together with all the external influences on it. The system may be an organization such as a factory, a company, or a business unit; a man-made physical system; or a natural system. Figure 3 will now be used, not to list all the components of a business unit, but to list only those items that are new and different in each of the four dimensions in leading businesses. These well-known items shown in Figure 4 are:

- for outputs: beyond product to individualized, comprehensive customer solutions;
- for inputs: cooperative production and customer-supplier interaction methods;
- for the internal structure of a company: a constantly adapting, entrepreneurial, knowledge-driven organization;
- for the external influences: relentless change.

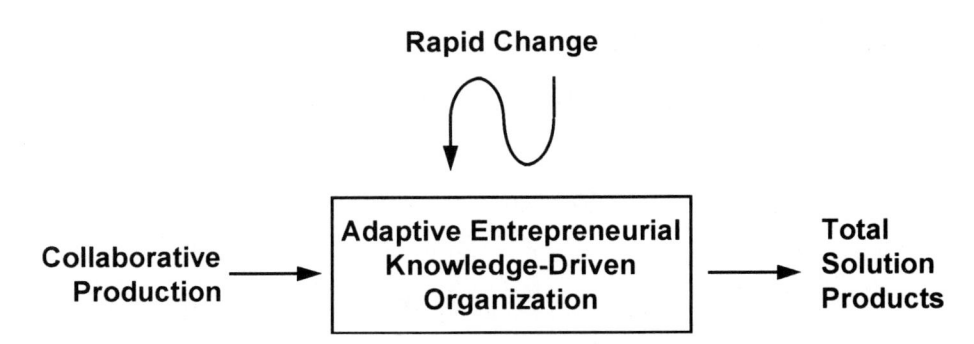

```
                    Rapid Change
                         |
                         v
Collaborative  --->  ┌─────────────────────────┐  --->  Total
 Production          │ Adaptive Entrepreneurial│        Solution
                     │    Knowledge-Driven     │        Products
                     │      Organization       │
                     └─────────────────────────┘
```

Figure 4   An Agile System.

The four dimensions of agility have been recorded in various publications, including *Agile Competitors and Virtual Organizations: Strategies for Enriching the Customer* by Goldman, Nagel and Preiss. These dimensions give what is the most predominant new item for each dimension of a business system and clarify the differences between agile and mass production systems.

This model shows that companies now are faced with change in all three important dimensions of activity
- new interactive methods of working with customers
- restructuring of the company internally
- new methods of cooperation and interaction with suppliers.

4 THE MODEL FOR INTERCOMPANY RELATIONS

A model which has been developed for understanding agile inter-company relationships, is shown in Figure 5.

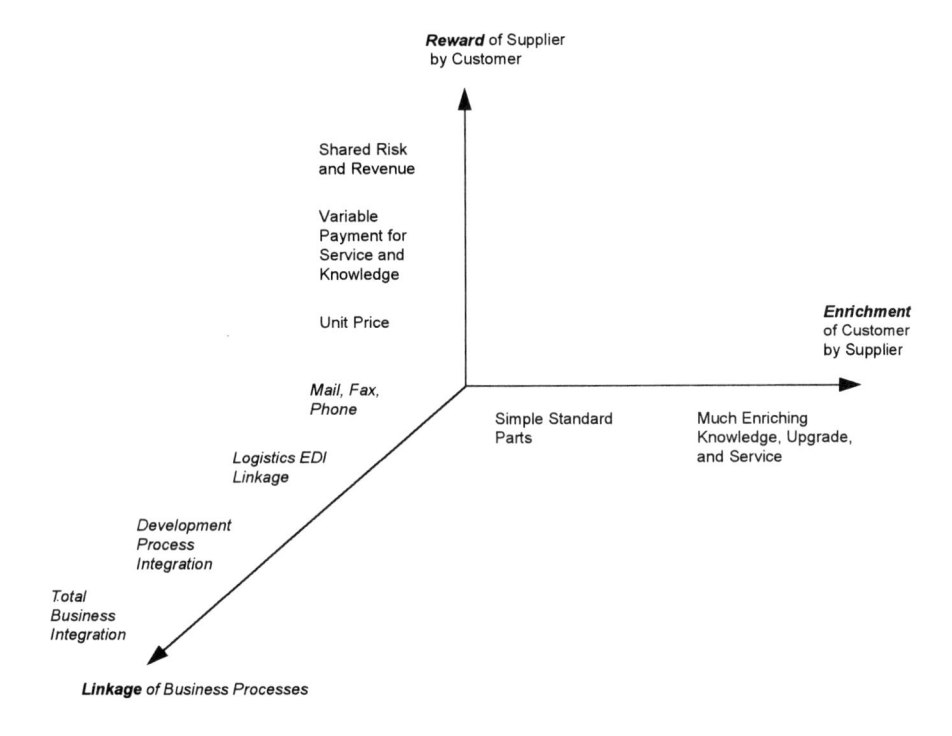

Figure 5 The Enrichment-Reward-Linkage Diagram.

This model was developed by the Agility Forum's Supplier Support Focus Group and reported in *Agile Customer-Supplier Relations* (Preiss and Wadsworth, eds.). This figure illustrates what the supplier organization gives the customer organization (X-axis), the payment flowing from the customer organization to the supplier organization (Y-axis), and how they work together (Z-axis). These three items cover all the flows that connect two systems or businesses.

It should be noted that in mass production a supplier takes input from a warehouse and places output in a finished goods warehouse. The inventory in the warehouses insulates systems one from another. Also, in mass production, many identical products are made so change is slow. Mass production was therefore a static, decoupled system. In agile competition, companies do not depend on inventory isolating the processes, and so the processes become coupled. In addition, they need to deal with accelerations in rate of demand. Agile systems are hence coupled, dynamic systems. A dynamic system with time-dependent influences does not behave as an amplified static system, but behaves quite differently. Given a system subject to dynamic influences, one cannot infer its behavior from how it behaved under static, time-independent influences. Managing a business unit or designing product in the agile environment is therefore not a simple extrapolation of management in the old, static environment, doing what one did before, but more intensely; it requires quite different approaches.

Figure 5 shows in a compact and quantifiable way all of the interactions between businesses or business units. The X-axis shows the value added by a supplier to a customer's outputs, and that is a definition of what a supplier provides to the customer. The Y-axis shows the method of payment to the supplier by the customer, from fixed price for a unit of product all the way up to shared risk and revenue. The Z-axis gives a measure of how closely the business systems of the two companies are tied together.

The old mass production system was a point at the origin of these axes. Competition is now moving the envelope further and further up the axes. Design work has thus moved from the aim of making product to the aim of tying profitably, and for a long period, with one's customers. To do so effectively means that one has also to tie effectively to one's suppliers, and the product development process has to work as one, integral, coupled process over all the value-adding industrial "food-chain".

Investment in design tools and methods in the past emphasized the quality of the product obtained. To this is now added the emphasis on the ability to team across and within company boundaries. Development of this interactive capability between companies should be advanced in a balanced way, coordinating the added value services and information given the customer, and the market-driven reward method, with the advancing investment. The optimal path for progress along the three axes of Figure 5 is along a three-dimensional, 45 degree diagonal relative to those three axes.

5 CONCLUSIONS

The context of product design has advanced from being part of decoupled, static processes which were called "mass production", in which the product was the focus and the aim of the work, to a coupled, dynamic agile environment, in which the design process and its seamless connections to both customers and suppliers are the focus.

The change from product to process focus has changed companies from hierarchic organizations built around work function to empowered customer-focused teams. In Germany these are often called "fractal organizations". This has changed the entire context of design work.

Product, is of course, designed, but the organization of the work, the investment in computer and information systems, and the training of designers and their support staff now include emphasis on the teaming and cross functional connections within and between companies.

6 REFERENCES

Goldman, S. L., Nagel, R. N., Preiss, K. (1995) Agile Competitors and Virtual Organizations: Strategies for Enriching the Customer. Van Nostrand Reinhold, New York.
Preiss, K. and Wadsworth, W. (Editors) (1994) Agile Customer-Supplier Relations. *Perspectives on Agility Series.* No. RS94-01. Agility Forum, Bethlehem, Pennsylvania.
Preiss, K. (1995) Models of the Agile Competitive Environment. *Perspectives on Agility Series.* No. PA95-03. Agility Forum, Bethlehem, Pennsylvania.
Preiss, K. (1995) Mass, Lean and Agile as Static and Dynamic Systems. Perspectives on Agility Series. No. PA 95-04. Agility Forum, Bethlehem, Pennsylvania.
Warnecke, H. J. (1993) The Fractal Company: A Revolution in Corporate Culture. Springer-Verlag, New York.

7 BIOGRAPHY

Professor Kenneth Preiss holds the Sir Leon Bagrit chair in Computer-Aided Design at Ben Gurion University in Beer Sheba, Israel, and is a senior fellow of the Agility Forum at Lehigh University, Bethlehem, Pennsylvania. He is an Honorary Member of the American Society for Mechanical Engineers. His extensive list of published work includes well over 100 original research papers and reports; he is co-author with Steven L. Goldman and Roger N. Nagel of the recently released book title -- Agile Competition and Virtual Organizations: Strategies for Enriching the Customer", Van Nostrand Reinhold, 1995. He is now Director of Agile Enterprise Projects at the Agility Forum, as well as Editor-in-Chief of Publications, including the new John Wiley journal, *Agile Enterprise Quarterly.*

49

Distributed and multicriteria management tools for integrated manufacturing

D. Trentesaux[1,2] - J.F.N. Tchako[1] - C. Tahon[1]
[1]Laboratoire d'Automatique et de Mécanique Industrielles et Humaines.
Université de Valenciennes et du Hainaut-Cambrésis
Le Mont Houy, BP 311
59304 Valenciennes cedex, France
Tel:(33).27.14.13.54.
fax:(33).27.14.12.88.
e-mail: chip@univ-valenciennes.fr

[2]Laboratoire d'Automatique de Grenoble.
E.N.S.I.E.G., BP 46
38402 Saint Martin d'hères cedex, France

Abstract
Existing regulation structures still not provide sufficient tools for designing effecient production structures, because of the control structure itself: that is, hierarchical control requires strong couplings and huge amount of data for algorithmic decision making, which is inconsistent with the need of self-adaptability of this control structure.
  In this paper, we present a control structure based on distributed problem solving techniques, embedding real time monitoring and management. Distributed problem solving is a sub-field of distributed artificial intelligence. This structure is based on the negotiation approach. An entity of the system is considered as an autonomous agent, able to collaborate with other agents to achieve a production program. Cooperation is performed through message passing between agents. Main interest of this concept is that the control can be setup individually and even after installation, changes and improvements can be easily made. Most of the existing methods for decision making are using monocriterion algorithms (OR approach) causing drastic simplifications of the data amount to propose a particular action.
  On the other hand, a multicriteria algorithm aims at establishing (with or without the human operator) a single or a multiple selection of choices (ordering them for example) among a set of choices using a list of criteria to evaluate relative preference. It may be coupled with a Decision Support System (DSS) responsible for the human-computer cooperation for decision making.
  An example of industrial application (bottling up line) is given: the introduced concepts have been successfully applied to a high-speed packaging line structure, which needs very intensive dynamic control. Estimations of forecast dates are based on calculus and mathematical relations of accumulation conveyers, which was previously unreleased.

Keywords
Production control, distributed problem solving, multicriteria algorithms, packaging line

1. INTRODUCTION

The most commonly used hierarchical structure is based on the principle of levels of control organised in a pyramidal structure: hierarchization and arborescence decomposition of the tasks. In this approach a controller at a higher level of control sends commands to and receives the status information from controllers at lower levels of control. Each decides in accordance with the orders received from the immediate upper level. This creates a master/slave relationship between different levels of control. Thus it is not possible to deal adequately with the issues of fault-tolerance, extendibility, resource sharing, robustness and dynamic adaptative control. The heterogeneity of the components of a system forbids instantaneous control and coordination, reducing the control system ability to manage the interactions in real time.

This structure has most certainly some advantages, for example: a very short turnaround to all requests. However its inflexibility makes any modification and even its conception extremely expensive. Moreover, the complexity of the problem to be solved is highly increased when fault-tolerance has to be managed. As a matter of fact, a given level in the hierarchy must have a substantial knowledge of the immediate upper and lower levels when this fault-tolerance must be incorporated.

On the other hand, an egalitarian based structure in which each control unit can hold a dialogue with others, offers an alternative to these problems. Localization of the raw data and those to be processed, and the control at the lowest levels, reduces the complexity of the system. This kind of structure offers modularity, maintainability and fault tolerance aspects. It allows a dynamic elaboration of dispatching orders according to the real state of the cell.

The paper shows that the use of distributed problem-solving techniques could be applied to the management of a production system. First, we present a specification of a distributed architecture applied to a production structure. The distribution occurs at the production activity control (PAC) level. Then we introduce the multicriteria concepts for dynamic and distributed decision making. The study focuses on the management of a packaging line. Actual centralized control reduces global flexibility and reactivity. Moreover, experience shows that such a production structure is often disrupted. We study then possible obvious solutions (same speed for each of the resources, etc.) for managing the line and show that these solutions are not acceptable due to production constraints. This leads us to propose a distributed control coupled with a multicriteria approach for decision making.

2. DISTRIBUTED CONTROL OF A PRODUCTION STRUCTURE

Basic concepts of a distributed PAC structure have been first developed in (Bakker, 1988). (Tchako et al., 1994) deepened and applied such concepts to the managing of a manufacturing cell. We sum up the main characteristics of the distributed PAC structure.

The distributed production activity control structure inherits from the agent concepts: an agent (Ferber et al., 1988) is defined as an abstract or a physical entity able to act on itself and on its environment and to communicate with other agents. It aims at performing a set of *tasks*, parts of a global problem. To execute these tasks, the agent may use a set of *objects*. *Control* defines the cooperation between agents, the group organization, and its evolution. The cooperation is defined by a cooperation degree, which ranges from fully cooperative to antagonistic agents. *Communication* between agents depends on the selected protocol, that is the set of rules that specifies the way to synthesize messages to make them significant and correct.

Each agent controls a set of production resources (objects as a set of production tools such as mill, lathe, robot, etc.). Thus, tasks are operations to be performed by these production tools. Two kinds of cooperation level can be defined: horizontal (between agents who cooperate) and vertical (between each agent and the human operator associated to this agent as a supervisor). The human operator has the responsibility for the task resolution and allocation (vertical cooperation through a DSS). The task allocation consists in allocating responsibility for process on products in a dynamic way.

3. THE IMS

The Integrated Management System (IMS) represents an agent responsible for operation processes (in that case, objects are production facilities such as mill or lathe) and operation allocations, including an operational-level DSS for task allocations (Trentesaux et al., 1994b). The DSS model for task allocation is based on Sprague's concept for a DSS defined as a set of three sub-systems (data, modeling and dialog systems) (Sprague, 1987).

An IMS is composed of:
- A decision system based on a representation of the different agents and supports the multicriteria tools for decision making. This sub-system supervises the local queue of jobs and participates through cooperation with other agents in order to attempt to process tasks.
- A control system responsible for the command orders for the handling automation (e.g., command orders for a conveyer, or a control signal to motor).
- A communication system responsible for the information exchanges through a local network.
- An information system supporting required information for other sub-systems (local database).
- An interface system insuring the dialogue with the human operator, and the interactions between the different sub-systems.

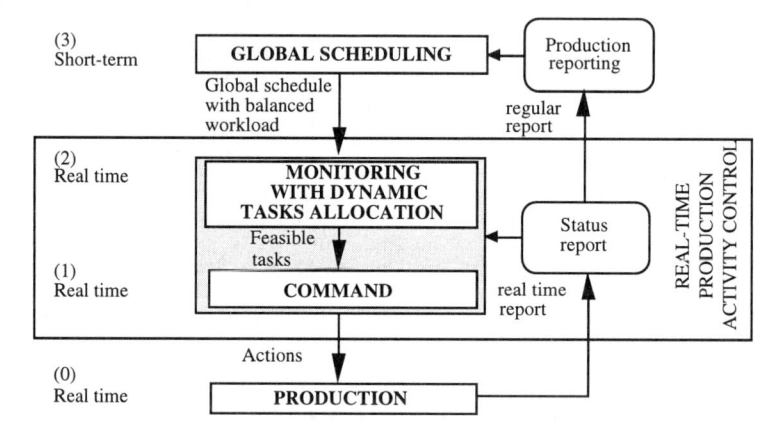

Figure 1 The production management structure.

Figure 1 presents the integration of the dynamic task allocation in the global plan of production. Tasks are dynamically realized. A global production planning is first created regarding the whole set of constraints of the manufacturing orders (workshop capacity smoothing, etc.). This planning does not describe the sequence of jobs on resources, but only provides to the production system a set of feasible manufacturing orders according to major constraints such as: lead times, average workload, etc.

The scheduling function is incorporated in the real time production control part. Hence, it is established dynamically through the task allocation process.

The distributed PAC structure is responsible for the work process without any planned scheduling. A set of manufacturing orders established by the production planning level is provided to the distributed PAC structure that manages this whole set as a sum of sub-tasks (operations, parts of a manufacturing order) to be allocated and performed.

The vertical control is managed by the operational-level DSS responsible for task allocation support and queue management. The horizontal control supports fully cooperative data exchanges and cooperation through negotiation. To be as satisfactory as possible, the operation allocation must be performed by a set of cooperative IMS that has to support global (static) and local (dynamic) constraints.

The basic principle of the communication protocol (negotiation paradigm) is the following: when an operation of a particular manufacturing order has been performed, the IMS sends a **request** about the next operation to process, according to the routing list of this manufacturing order. Each of the IMS able to perform this operation returns an **acceptance**. The requesting IMS selects one of the proposed IMS (through human-computer interaction supported by the DSS if required) and sends to this IMS a **reservation** and a **release** to the others. A **discharge** from the selected IMS concludes the protocol. The complete protocol has been modeled using colored and temporized Petri nets (Trentesaux et al., 1994a).

4. DYNAMIC DECISION MAKING: A MULTICRITERIA APPROACH

This part aims at showing that a multicriteria (MC) approach provides efficient tools for operational level decisions. Such decisions concern the real-time monitoring of the production system.

A MC algorithm aims at establishing a single or a multiple selection of choices (ordering them for example) among a set of choices using a set of criteria to evaluate relative preferences. Decision making with MC algorithms and theories has been widely developed and published in (Korhonen et al., 1992) (Pomerol et al., 1993), (Roy et al., 1993), (Vincke, 1989), (Zeleny, 1982). A various number of industrial applications of the MC algorithms can be found (Bui, 1987), (Gravel et al., 1992), (Karpak, 1989), (Kendall, 1985), (Tabucanon, 1988). Few other papers deal with the impact of MC algorithms in industry (Belton et al., 1992), (Pomerol et al., 1993). We sum up the main characteristics.

4.1. Theoretical issues

MC theory aims at helping decision making using very simple algorithms based upon more complex hypothesis. Analytical theory proposes complex algorithms based upon simple hypothesis (Roy et al. 1993). Such algorithms provide an optimum value from objective functions, which requires drastic simplifications of the model, else it would not be possible to provide solutions in acceptable times. They also support with difficulty the integration of the human operator into the decision making processing.

On the other hand, MC algorithms suit better reality (which requires complex hypothesis). Moreover, the decision making algorithms are interfaced with the human operator who better trusts in simple and interactive methods and in the possibility of interaction (Pomerol et al. 1993). Another reason for using MC algorithms is that at the level of production activity control, the operator will face the problem of conflicting objectives (such as minimizing the lead time vs. the production costs and vs. the global quality of production), which is supported by the MC approach. Thus, solutions are satisfactory, not optimum, since a compromise is realized. They can also support either quantifiable or unquantifiable criteria. Of course, the choice for the set of criteria has to be pertinent, otherwise the decision making will be distorted (Pomerol et al. 1993).

4.2. Practical issues

Most of the industrial applications concern the implementation of goal-programming algorithms and decision making for the strategic (plant location, management choice, marketing, etc.) and tactical (global production planning and scheduling, etc.) levels. There is almost no application of MC algorithms to the operational level.

The distributed and dynamic aspects of the decision imply three major constraints:
- The time allowed for decision making should be short enough considering the average operating time, else it would considerably reduce the global performance.
- The multicriteria algorithm coupled with a DSS should be clearly understandable by the operator, which is not obvious to fulfill. Most of industrial applications concern the tactic/strategic level where an expert is provided to get required data and to proceed the decision processing. We must note that on the other hand, decisions at the operational level

are event-based (thus periodic) and decision needs should occur very often. Thus, the multicriteria should be efficient enough to avoid operator's weariness. In fact, according to Pomerol, the decision maker, which is not expert should appreciate simple methods, which hardly constraints the choice and the development of the method.
• Computerization of the method should be possible if required, e.g. when no extra event occurs, the distributed system should be able to regulate the system itself or when decisions are to be very frequent.

It would be interesting for the DSS to support qualitative data, e.g., production quality, operator qualification, average operating time. For example, this could allow sensitivity tests to evaluate robustness of a solution.

5. CASE STUDY: PERRIER PACKAGING LINE

5.1. Structure

The packaging line (Tchako et al., 1993) has the following structure (Figure 2).

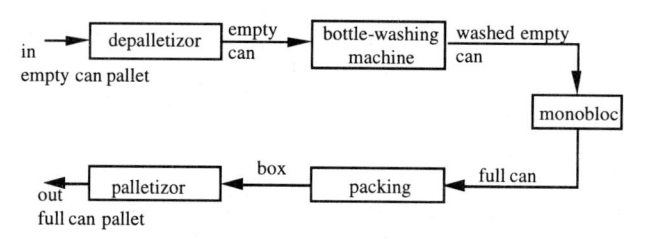

Figure 2 Lay-up of packaging line.

The conditioning shop is constituted of packaging lines connected to a container transport system (bottle, pack, metal cop), and downstreamed to a products storage handling system and planning order. The set is complex and is subject to many and various disturbances, such as temporary or long break of the packaging line, conveying system failures (conveyer or truck), and so on. The upstream part can be considered as flow-shop, processing one product, whereas the downstream part responds to job-shop laws, processing several products.

5.2. Constraints

The less flexible machine is the Monobloc where constraints are:
• No speed variation is allowed during working,
• Reduce as far as possible the number of Start/Stop to insure pressure and temperature stability of the racked product.

6. PROBLEMS

The processing of the packaging arises great problems. The classical packaging line is made of workstation manufacturing or controlling products (washing, filling in, packing) linked together by multi-band stretch accumulation conveyer. The workstation can be classed as conveyer without accumulation and then be modelled as dead time.

On the other hand, the accumulation conveyers are not well known components though they are often used. In a production line they insure simultaneously two functions:
• the products transportation in and between workstations;
• a storage buffer allowing to reduce the incidence of workstation speed variations on their up and downstream.

They are generally modelled by a queue and a dead time, which corresponds to a steady state. A fine observation shows that their behaviours are more complex (Demongodin et al., 1993), accumulation corresponding to a transient state. This means that delays of product propagation always change, without being masked by accumulation.

On the other hand, when two machines of a packaging or a conditioning line are linked together by a conveyer, the latter suffers from random events that damage products and equipment. Origins of such problems are principally clashes, lack of products (famine), excessive accumulation (jamming), both famine and jamming resulting in pumping phenomena. This implies the following effects :
- products falls;
- products deterioration;
- noises and disturbances;
- mechanical efforts on products and equipment.

The frequency of these events and the seriousness of their consequences are closely bound to products and equipment's characteristics, but more over to the production rate of the concerned packaging lines.

7. FIRST STUDIES FOR SOLVING THE PROBLEM

Before invoking regulation, it is necessary to quote other approaches that have been proposed.

7.1. Problem suppression

By adopting compact groups performing several functions and avoiding the conveyer links, problem could be suppressed. The effectiveness of this solution is limited in many cases by the actual technological knowledge, which forbids such an approch.

7.2. Machine synchronisation

For example, mechanical or electrical coupling shafts may be used. This solution, beyond their prohibitive costs, only solves one part of the problem. For example these solutions do not allow to compensate the possible and gradual elimination of rejected products. This degrades production with the appearance of famine. In fact, accumulation states are compulsory, and the control must deal with such accumulation.

7.3. Current control structure

The aim is to maintain a constant level of machines upstream accumulation. It is possible to measure for a single conveyer, accumulation of products by sensors so as to act in boolean logic command on the production process. Whole decisions (centralized control) focuses on the Monobloc, which is the less flexible machine (Tchako et al., 1993).

Several major objections are made to this simple solution especially, rigidity and inflexibility that forbids real time control of the line. The system becomes rapidly unstable when the production rate increases. Moreover this lack of flexibility and reactivity implies the necessity of disposing of large distance between machines because of pumping phenomena, skyrocketing the line costs. In addition, a constant accumulation should imply a uniform production speed, which is not the real situation due to machine characteristics. This method, which is actually used, also induces pumping phenomena that harm equipment productivity and availability.

Such solution list is not exhaustive, but shows the necessity to go towards more intelligent, distributed and adaptative solutions, and we already feel intuitively that its multi-criteria part must not be negligible (Trentesaux et al., 1995b).

8. DISTRIBUTED AND MULTICRITERIA CONTROL APPROACH

8.1. Management of a "normal processing zone"

The accumulation front is the exact position where products transported by a conveyer stand from a free state (i.e. spaced) to an accumulation state (Demongodin et al., 1993). Rather than try to regulate the level of accumulation between machines, the principle consists in deciding of an ideal accumulation front on conveyers in order to find strategies that allow the real front to oscillate as less as possible around the ideal front. This authorized interval is called "normal processing zone". This zone is designed so that the corrections have a very short turnaround and high flexibility. Therefore those corrections are sufficient to increase the stability level of the system.

8.2. Means

The monitoring of each line can be split up on four levels that principally deals with:
- the determination of the speed of the resources (machines and conveyers) according to the global state of the line and the production goal;
- a local adjustment of the resource speed according to the up and downstream stocks;
- a speed adjustment of each conveyer according to its up and downstream;
- the regulation of the line.

8.3. Proposed architecture

Distributed control
In the remaining of this paper, we propose a particularisation of the distributed and mulitcriteria approaches to integrate such concepts into the control structure of the packaging line. The distributed control is structured as shown on Figure 3. Including the notion of the normal processing zone into the line managing strategy, two kinds of IMS can be distinguished (cf. Figure 4):
- A Conveyer management IMS, which is responsible for the control of a conveyer, split into three parts. We can see that if the normal state is available, one extreme sub-conveyer is always in transient state, with no accumulation and the other extreme is full (jamming state).
- A production resource management IMS, which is responsible for the control of a machine.

As the lay-up of the line is always known, each IMS has only one up- and downstream and the communication protocol needs to be simplified. No bidding is done and no task allocation is made. The simplified protocol is dedicated to data exchange as following:
- **Request** : is a task announcement to request information from an agent;
- **Acceptance** : is a message corresponding to the answer of a requested agent;
- **Transfer** : it is not a message but the real transfer of data.

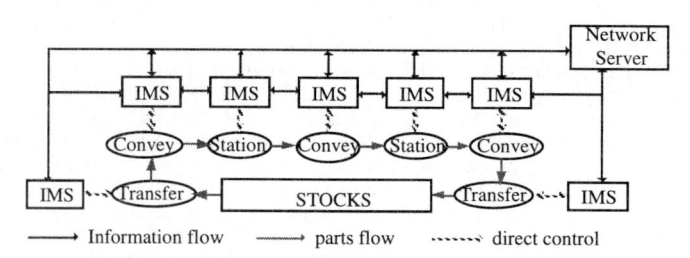

Figure 3 The entire distributed PAC architecture for the packaging line control.

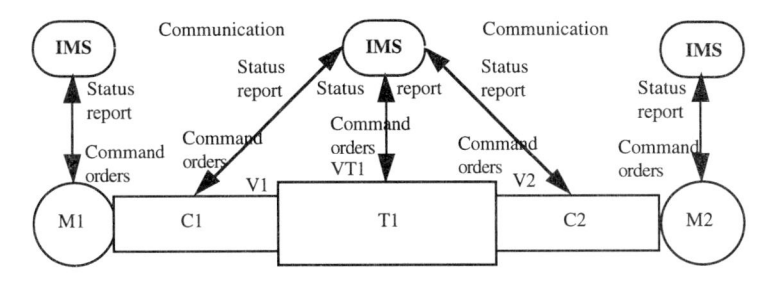

[Figure 4 diagram: IMS nodes connected via Communication/Status report/Command orders to M1—C1—V1—T1/VT1—V2—C2—M2]

Figure 4 Focusing on local management of two machines and a conveyer. The decomposition of the conveyer in three parts is shown.

The following example shows how a distributed structure is able to guarantee a good global stability level.

Assume that throughput of IMS_i decreases abnormally, its input buffer will increase. The decision system, whose order is also to minimise the stock level, informs the upstream IMS about the new situation. The decision sub-system of the upstream IMS (IMS_{i-1}) defines the strategy to adopt: if its throughput is too high it will probably saturate the input buffer of IMS_i, it can either decrease its speed (in safety margin) while respecting due dates, or anticipate IMS_i recovering. In case of problem IMS_{i-1} can inform IMS_{i-2} about the situation and so on. This multicriteria method for such decision making is detailed in the following part.

Multicriteria and dynamic decision making
Figure 5 shows the information flow inside and outside the decision sub-system of the IMS_i. X_i is the set of data returned by sensors. X_{i+1} and X_{i-1} are available according to the communication protocol. U_i is the command order according to global objectives. M_i is the command vector for the controlled resource (speed, etc.).

[Figure 5 diagram: Decision sub-system with Data acquisition (decision matrix) and Multicriteria algorithm; inputs U_i, X_i Sensor data, X_{i+1}, X_{i-1}; output M_i Command order for resource]

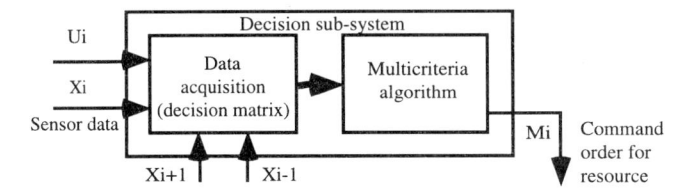

Figure 5 Decision sub-system and information flow.

Since the need of high speed decision, the DSS is reduced in that case to an automatic decision making through an automatic multicrtieria algorithm. The decision making is realized in three steps:
- Data is obtained through communication dialogs and data capture from local IMS sensors (up- and downstream resource speeds, conveyer states, and so on). The requested data is concerned with specific criteria, detailed below.
- A filtering operation (constraint satifaction method) cancels unrealisable solutions and provides a set of feasible actions.
- The decision matrix (criteria, alternatives and evaluations) is provided to the multicriteria method that selects a satisfying solution. A complete study has been realized (Trentesaux et al., 1995a) and an interesting method (according to real-time constraints, need of an automatic anf fast method, and so on) has been selected: Promethee method developed by (Brans et al., 1985).

584 Part Twelve Distributed Product Development and Manufacturing

A complete list of possible (but not necessary feasible) alternatives is:
- increase/decrease upstream or downstream conveyer speed (discrete values according to resource specifications);
- increase/decrease upstream or downstream machine speed (discrete values);
- stop specific conveyer or machine;
- modify characteristics of the normal processing zone (discrete values);

The actual method is based on a classical OR method using a single criterion (forecast jamming date). Unfortunately, other criteria have to be taken into account. The complete list is detailed (the used list does not necessary contain all the criteria listed below).
- time to recovering normal processing (to minimize);
- speed variation (machine or conveyer) (to minimize);
- forecast jamming/famine date (to maximize);
- noise evaluation (to minimize);
- absolute value of the gap between the speed (when it is to be modified) with the normal operating speed (to minimize).

Of course, one can notice that these criteria are conflictual.

Example of dynamic, local and multicriteria decision making

A model of a significant part of the line has been developed to study different regulation strategies. This model consists of three machines and six conveyers (Figure 6) and was developed in (Tchako et al., 1995). The product flow is inputted in Machine_A.

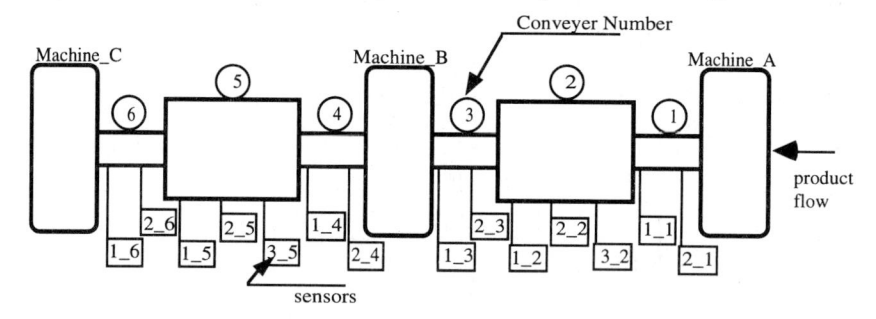

Figure 6 The study model.

The first version of control rules was formulated for two-machines case using only current information about sensor states. Figure 7 and 8 shows the results of corresponding simulation run. "Conv_2.cur_quant" corresponds to the total quantity of bottle on the conveyer n°2. "Machine_A.cur-rate" shows the real-time rate of the Machine_A.

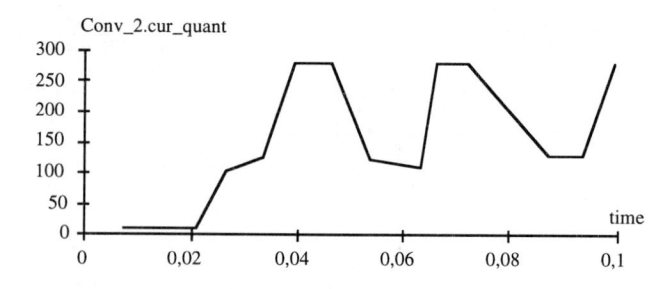

Figure 7 Real-time quantity of bottle on conveyer 2.

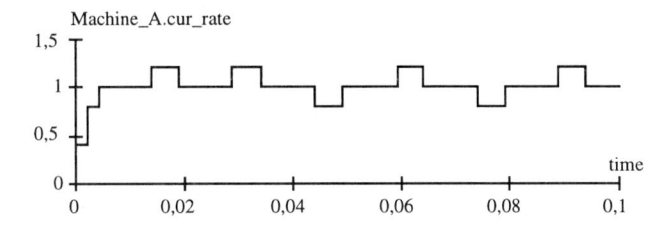

```
                Machine_A.cur_rate
          1,5

           1

          0,5
                                                              time
           0
            0    0,02    0,04    0,06    0,08    0,1
```

Figure 8 Real-time rate of Machine_A.

It is clear that these rules are rather bad (there are many switchings that occur very often) and it is necessary to improve them by using more realistic methods (integration of a multicriteria approach).

For example, at date 0.056, the IMS monitoring the conveyer 1-2-3 detects a jamming state for this conveyer and informs the IMS monitoring Machine_A of the situation. IMS_A establishes a decision matrix that is composed of several criteria (belonging to the previous list), weight for these criteria (weight of relative importance), possible alternatives (filtered) and relative evaluation of these alternatives regarding each criterion. The values are obtained from the model developed in (Demongodin et al., 1993) and (Tchako et al., 1995) and presented in table 1.

Weights of criteria depend on global objectives. In our case, the "time to recover normal processing state" is favored. However, the two other criteria are not negligible. The nature of criteria is established regarding the preference and the experience of managers. The quasi criterion with threshold q=0,01 means that the difference between two evaluations is significant when this difference is greater than q, e.g., alternatives "decrease -20%" and "stop" are indifferent regarding the first criterion. A true criterion is a particular quasi-criterion with q=0. That is, a preference exists, whatever the difference is. Indifference is only possible when the two evaluations are identical. See (Roy et al., 1993) for other details.

Promethee outputs a "net flow" that can be used to create a complete transitive relation between two alternatives (Promethee part II). The net flow values are summed up in Table 2. The relation is: higher a value, better the alternative.

Table 1 Decision matrix

Criteria	Time to recover normal processing state for conveyer_2	Speed variation of Machine_A	Forecast jamming/famine date for conveyer_2
Objective nature of criteria	minimize quasi-criterion threshold: 0.01	minimize true criterion	maximize true criterion
weigth of criteria	8	5	6
Alternative for rate of Machine_A		Evaluations	
nothing	0,17	0	0 (jamming)
Stop	0,0015	1	0,01 (famine)
Decrease -10%	0,012	0,1	0,048 (famine)
Decrease -20%	0,0072	0,2	0,032 (famine)

The most satisfying solution is to decrease the rate by 10% (maximum value for the net flow: 0,404). Thus, the IMS_A orders the machine to decrease the production rate by 10%. The results are shown on the two following figures (Figures 9 an d 10).

Table 2 Net flow calculations

Alternative for machining rate of Machine_A	Net flow
nothing	-0.474
Stop	-0.088
Decrease -20%	0.158
Decrease -10%	0.404

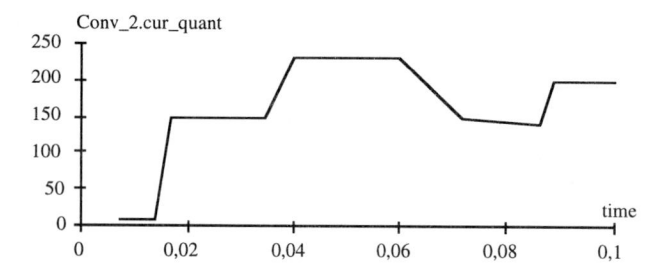

Figure 9 Real-time quantity of bottle on conveyer 2 (multicriteria method).

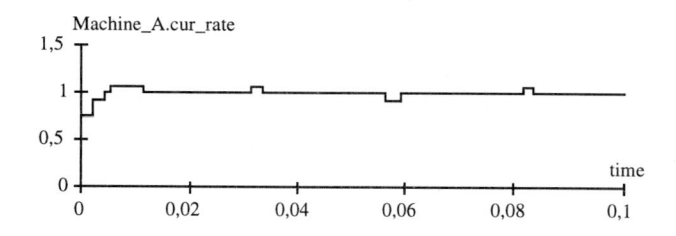

Figure 10 Real-time rate of Machine_A (multicriteria method).

9. CONCLUSION

This study has shown the interest of the distributed and multicriteria control for the managing of a packaging line. Distributed architecture improves reactivity and multicriteria methods improve the integration and the support of conflicting objectives.

Other studies concern the specification of the nature of criteria and their weight. An extension has been proposed to better suit reality by integrating qualitative or estimated data to the Promethee method. This will allow to support the notion of forecasted times that are uncertain values. Thus, robustness will be estimated.

The distributed control should be improved by coupling the multicriteria method to an expert system to help to anticipate events. Decision quality should be then increased.

10. REFERENCES

Bakker, H. (1988), DFMS : A new control structure for FMS, *Computers In Industries*, **10**, 1-9.
Belton, V. and Elder, M. D. (1992), Can Multiple Criteria Methods Help Production Scheduling?, *10th international conference on M.C.D.M TAIPEI'*, Taïwan, July, 171-178.
Brans, J. P. and Vincke, P. 1985, A preference ranking organization method: the Promethee method, *Management Science*, **31**, 647-656.
Bui, T. X. (1987), *Co-oP: A Group Decision Support System for cooperative Multiple Criteria Group Decision Making*, (Lecture notes in Computer Science n° 290 Springer-Verlag, Berlin).
Demongodin, I. and Prunet, F. (1993), Simulation modelling for accumulation conveyors in transcient behaviour, *Proc. of Compeuro conference*, May, 29-37.
Ferber, J. and Ghallab, M. (1988), Problématique des Univers Multi Agents Intelligents, *Journées Nationales du PRC/IA Cepadues-Edition*, Toulouse, France.
Gravel, M., Martel, J. M., Nadeau, R., Price, W. and Tremblay, R. (1992), A multicriterion view of optimial resource allocation in job-shop production, *European Journal of Operational Research*, **61**, 230-244.
Karpak, B. (1989), *Multiple criteria decision support systems in production planning: a micro experience*, (Karpak & Zionts-Springer-Verlag-NATO Series, Berlin), **F56**, 341-355.
Kendall, K. E. and Schniederjans, M. J. (1985), Multi-product production planning: A goal programming approach, *European Journal of Operational Research*, **20**, 83-91.
Korhonen, P., Moskowitz, H. and Wallenius, J. (1992), Multiple criteria decision support- A review, *European Journal of Operational Research*, **63**, 361-375.
Pomerol, J. C. and Barba-Romero, S. (1993), *Choix multi-critère dans l'entreprise* (collection Informatique-HERMES, Paris).
Roy, B. and Bouyssou, D. (1993), *Aide multi-critère à la décision: Méthodes et cas* (Economica, Paris).
Sprague, R. H. Jr.(1987), DSS in context, *Decision Support System*, **3**, 197-202.
Tabucanon, M. T. (1988), Multiple criteria decision making in industry, *Studies in production and engineering economics*, **8**.
Tchako, J. F. N. and Tahon, C. (1993), Distributed management system for a packaging line, *Proceedings of International Conference on Industrial Engineering and Production Management*, Mons, Belgium, 833-845.
Tchako, J. F. N., Beldjilali, B., Trentesaux, D. and Tahon, C. (1994), Modeling with coloured Petri nets and simulation of a dynamic and distributed management system for a manufacturing cell, *International Journal of Computer Integrated Manufacturing*, **7**(6), 323-339.
Tchako, J. F. N., Lenclud, T., Tahon, C. and Yasinovsky, S.(1995), Knowledge based control system for manufacturing lines, submitted to Production, Planning and Control.
Trentesaux, D. and Tahon, C. (1994a), Modèle de communication inter-agents pour une structure de pilotage temps réel distribuée, *Revue d'Automatique et de Productique Appliquées*, **7**(6), 703-727.
Trentesaux, D., Dindeleux, R. and Tahon, C. (1994b) A MultiCriteria Decision Support System for Dynamic task Allocation in a Distributed Production Activity Control Structure, *European Workshop on integrated Manufacturing Systems Engineering*, IMSE'94, INRIA, Grenoble, France, 383-393.
Trentesaux, D. and Tahon, C. (1995a), Dynamic and distributed production activity control: a multicriteria approach for task allocation problematic, *International conference on industrial engineering and production management*, **1**, Marrakech, Morroco, 137-154.
Trentesaux, D. and Tahon, C. (1995b), DPACS: a self-adaptative production activity control structure, *INRIA/IEEE conference on Emerging technologies and factory automation*, Paris.
Vincke., P. (1989), *L'aide multicritère à la décision* (Université de Bruxelles-Ellipses, Bruxelles).
Zeleny, M. (1982), *Multiple Criteria Decision Making* (MC Graw-Hill, New-York).

11. BIOGRAPHY

D. Trentesaux is an engineer graduated of an electrical and automation control engineering school at the Institut National Polytechnique de Grenoble (INPG), France. He is actually Ph.D. student in the field of distributed and mulitcriteria decision making for production control in collaboration with the Laboratoire d'Automatique de Grenoble (LAG), France and the Laboratoire d'Automatique et de Mécanique Industrielles et Humaines de Valenciennes (LAMIH), France.

DR. J.F.N. Tchako received his Ph.D. thesis in Automation control and human-computer interaction from the University of Valenciennes. His current research interest is in various fields of distributed decision making for production systems.

Prof. Dr. C. Tahon is graduated ESTP school engineering in Paris, France and is Agrégé in Applied Physics in 1979. He received his Ph.D. thesis in Automation control and mechanics. He is actually professor of industrial engineering and management at the University of Valenciennes. His research interest is in the area of manufacturing automation, production control, project management and logistics systems based on IT tools.

50

Integrated Enterprise Modelling for Business Process Reengineering

Dr.-Ing. Kai Mertins, Dipl.-Ing. Roland Jochem

Fraunhofer Institute of Production Systems and Design Technology

(IPK) Berlin, Division of Systems Planning

Pascalstrasse 8-9, 10587 Berlin, Germany,

Telephone: ++49 30/39006-195,

Telefax: ++49 30/3911037,

E-mail: roland.jochem@ipk.fhg.de

Abstract
The authors describe a methodology for business process modelling and modelling of related organisational structures based on an object-oriented approach which is in discussion at CEN TC 310/WG1 and ISO TC 184/SC5/WG1 for standardisation. Application areas with industrial examples are represented as well as a supporting modelling tool prototype.

Keywords
Business Process Reengineering, Object-Oriented Approach, Integrated Enterprise Modelling, Modelling Tool

1 INTRODUCTION

All methods such as Lean Management, Simultaneous Engineering, Total Quality Management and Continuos Improvement Processes aim at strengthening the competitiveness and productivity of the company by improving the product quality, reducing lead times and optimizing the marginal pricing [14].
To improve the competitiveness all efforts are traditionally concentrated on optimizing single functions. The traditional way of managing an enterprise is to subdivide it into a number of

separate functions which are easier to overview and control. This method results in numerous 'interface' problems regarding the organization and optimization of single functions at the expense of the manufacturing process and the organization as a whole [1].
When approaching the mentioned targets companies start to concentrate

- on their main business processes,
- on gearing all functions and resources to their processes and
- on improving the communication by widely sharing information within the processes.

The integration of separated functions, the optimization of the main business processes and the specification of a suitable information flow require a higher degree of transparency within the organization. In consideration of the complex relationships - looking at the manufacturing enterprise as a network of functions - models or modeling methods have to be applied in order to support, to ease and to systematize the planning and integration of functions into business processes and to describe the related organizational structure. Such a concept ensures a common understanding of business processes and an understanding of how the required information and the organizational structure needs to be organized [3,4,10,11].
In the following, the method of Integrated Enterprise Modeling (IEM) is presented. IEM uses the object-oriented modeling technique for modeling business processes, related organizational structures and required information systems. It provides a model for planning and optimizing the processes and organizational structures within the enterprise.
It will be introduced as a methodology and will be exemplified as a planning tool following an example which describes the determination of the potentials of future CAD/CAM-applications as a communication instrument within the process of developing gears in automotive industry.
Models developed according to the IEM method give a transparent representation of planning information and are therefore the basis for discussion between project participants. In order to evaluate the variety of planning information and description requirements it allows different views on one consistent model.
IEM models provide the means to precisely assign the value of planning goals, such as improvements in time, costs, or quality, to each business process and resource and, therefore, to optimize the process organization.

2 OBJECT-ORIENTED MODELLING

2.1 The Approach

Object-oriented techniques are extensively used for the development of applications in various fields [5,12]. The main advantage of this approach is the entirety of data and functions operating on these data. Provided with the powerful inheritance mechanism it yields models which are more stable and easier to maintain than those based on other modeling approaches. However, a method providing an entire, object-oriented approach to enterprise modeling has not been heard of so far.
In order to utilize its advantages and to provide acomprehensive and extendible enterprise model, the IEM method uses the object-oriented modeling approach, thus allowing the integration of different views on an enterprise in one consistent model as well as the easy adaptation of the model to changes within the enterprise [1,7,8].

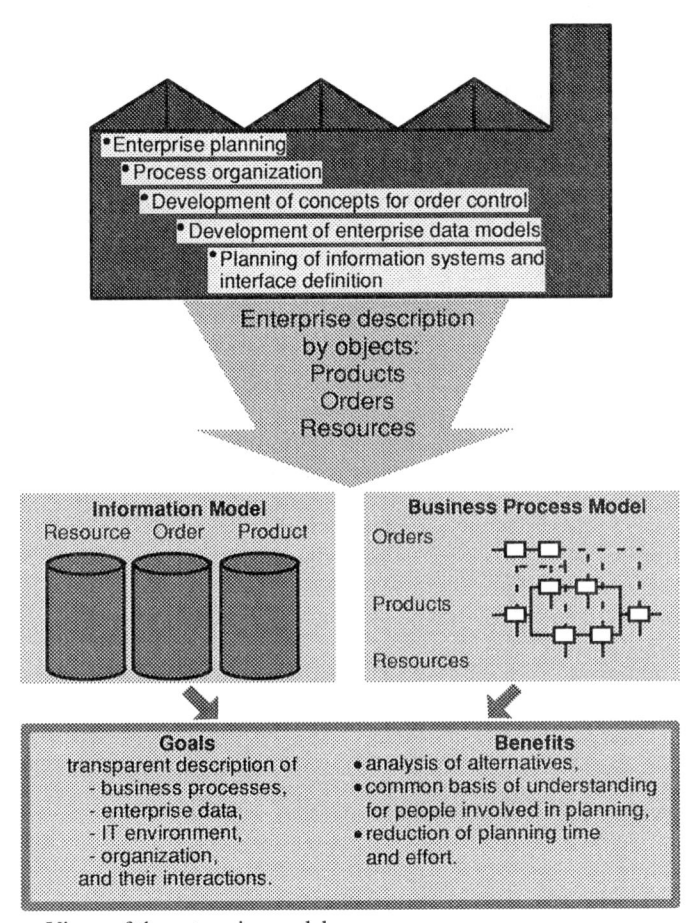

Figure 1 Views of the enterprise model

2.2 Generic classes of objects

The generic classes 'Product', 'Resource' and 'Order' are the basis of Integrated Enterprise Modeling for developing models from a user's point of view. They will be specified according to the specifications of an individual enterprise [2,3,7,8]. Each generic class prescribes a specific generic attribute structure, thus defining a frame for describing the structure and behavior of objects of its subclasses (cf. Fig. 4). Real enterprise objects will be modeled as objects of these subclasses.
Required enterprise data and the business processes, i.e. the tasks referring to objects, are structured in accordance with the object classes (see below). Furthermore, the relations between objects are determined. The result is a complete description of tasks, business

processes, enterprise data, production equipment and information systems of the enterprise at any level of detail [1,6].

The model core comprises two main views. The tasks, which are to be executed on objects, and the business processes are the focal point of the Process Model View, whereas the Information Model View primarily regards the objects describing data (Figure 1). Therefore, the core of the enterprise model consists of the data and process representation of classes of objects. The views are interlinked by referring to the same objects and activities, although they represent them in different ways, levels of detail and context. Any view on the model can be derived from this standardized model core. Additional features may be attaches to the core if necessary.

2.2.1 Business Processes as Interactions of Objects

Everything that happens in a manufacturing enterprise as part of the manufacturing process can be described by activities. In general, activities process and modify objects which were classified above as 'Products', 'Orders' and 'Resources'. The execution of any activity requires direct or indirect planning and scheduling. It is executed by resources which posses the required capacities. The IEM method suggests three levels when describing the essentials of an activity.

- The **Action** is an object-independent description of any task or business, a verbal description of some task, process step or procedure.
- The **Function** describes the processing of objects as a transformation from one determined (beginning) state to another determined (ending) state.
- The **Activity** specifies the order, which controls the execution of the function, and the resource(s), which is (are) in charge of executing the function.

Figure 2 graphically represents the Generic Activity Model. The beginning and ending states are connected with the action rectangle by arrows from left to right. The controlling of the activity is represented by an order state description and a dashed vertical arrow from the top; the required or actually assigned capability for executing the function is represented by a resource state description and a dashed vertical arrow from the bottom.

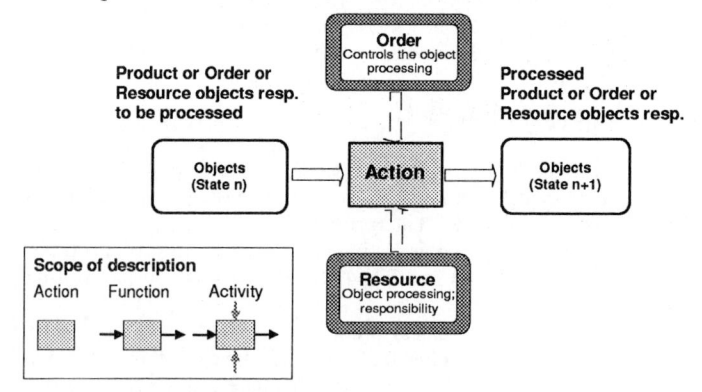

Figure 2 Interactions of Objects Described by the Generic Activity Model

Integrated enterprise modelling for business process reengineering

The Generic Activity Model represents the processing of objects of the product, order or resource class respectively indicating the interactions of objects while processing. The related organizational structure is described by specific resource classes along with their interrelations. Using special linking constructs (cf. Fig. 3), actions, functions and activities are combined to represent business processes. The decomposition and aggregation of processes is also supported. The IEM modeling constructs of the Process Model View are shown in Figure 3.

Product classes represent the principal results of the entire enterprise process - the products. Resource classes represent all means, including organizational units, which are necessary to carry out any activity in the enterprise. Order classes represent planning and control information. Figure 4 describes the attribute schemes of the generic classes [3,4,6].

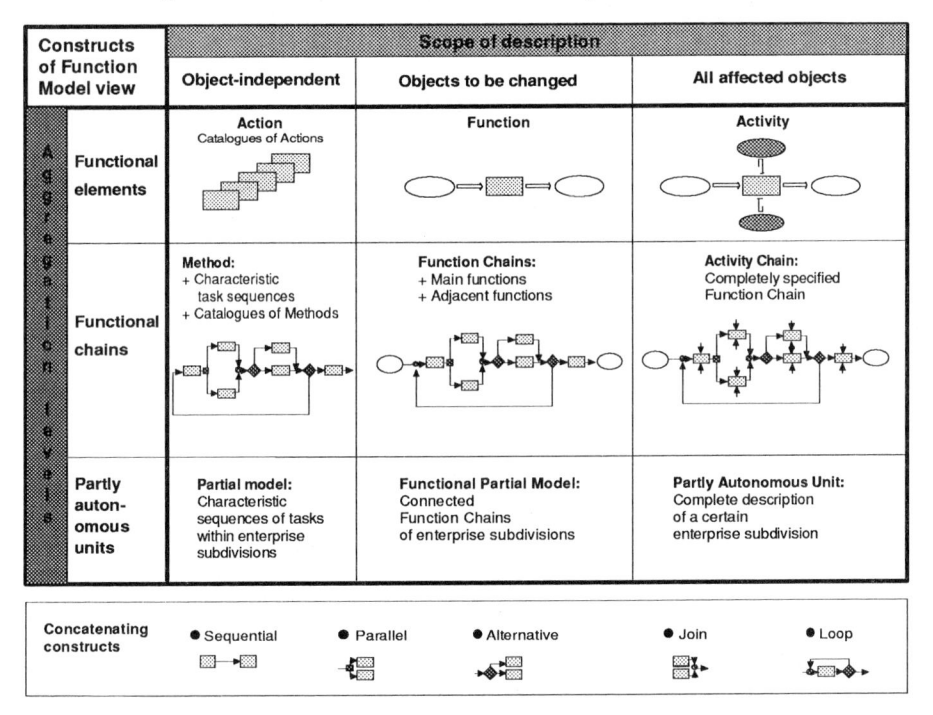

Constructs of Function Model view		Scope of description		
		Object-independent	Objects to be changed	All affected objects
	Functional elements	Action Catalogues of Actions	Function	Activity
	Functional chains	Method: + Characteristic task sequences + Catalogues of Methods	Function Chains: + Main functions + Adjacent functions	Activity Chain: Completely specified Function Chain
	Partly autonomous units	Partial model: Characteristic sequences of tasks within enterprise subdivisions	Functional Partial Model: Connected Function Chains of enterprise subdivisions	Partly Autonomous Unit: Complete description of a certain enterprise subdivision
Concatenating constructs		● Sequential ● Parallel ● Alternative ● Join ● Loop		

Figure 3 Constructs of the Process Model View

Further modeling aspects related to special modeling purposes may be integrated as additional views on the model. Examples of such special views include representations of control mechanisms, organizational units and costs. The two main views are the basis for the development of application-oriented views to describe all relevant aspects for meaningful decision making.

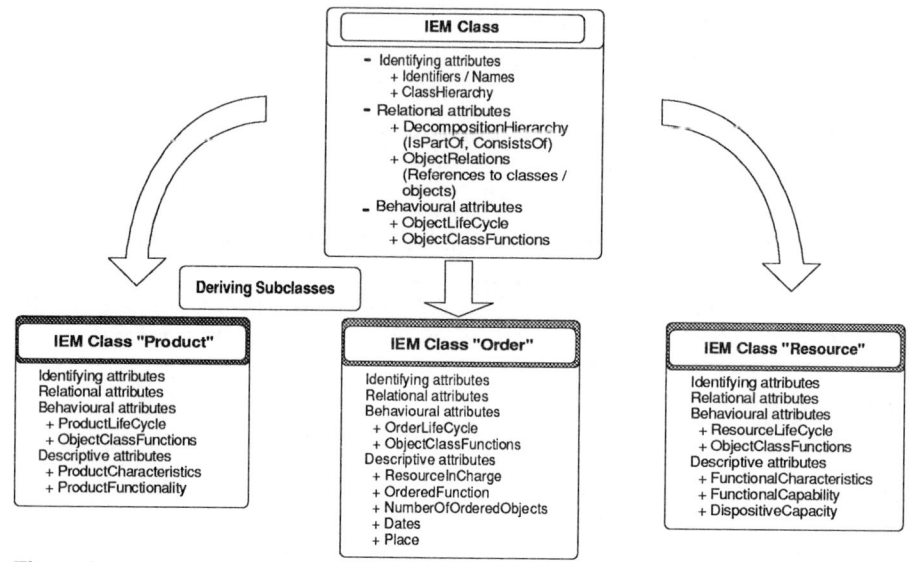

Figure 4 IEM Object Class Attributes

3 THE PROCESS OF DETERMINING THE POTENTIALS OF AN ENTERPRISE

The application of a modeling method should not only support single steps of factory planning. It has to guide and support the whole project. Therefore, it is necessary to reveal all aspects which are relevant to clarify weak points, to show potentials for optimization and their contribution to the objectives of the enterprise. The following modeling phases can be distinguished [13]:

- system delimitation,
- model design,
- model evaluation and utilization and
- model modification.

The purpose of the system delimitation is a well-aimed selection and limitation of the model which is to be prepared. It is indicated which enterprise-specific object classes and function areas have to be considered. The motto is 'fewer equals more'.

The model design is the phase of the composition of models. In operational planning projects enterprise-specific models are usually used to illustrate the actual state with so-called 'actual state models'.

The model design is carried out in three basic steps:
1. Identification of the relevant enterprise-specific class structures for products, resources and orders within the delimited system;
2. Development of the two main perspectives: information model (identification of the class-describing attributes as far as necessary) and function model (description of functions and processes performed on the objects);
3. Derivation of partial models to clarify relevant planning aspects.
In the phase of model evaluation the of the developed models are identified. The potentials for improvement are estimated and possible suggestions for optimization are evaluated. Improvement potentials, which can be realized by means of abolishing weak points, should be evaluated according to

- the degree of an improved accomplishment of enterprise goals and
- the expenditure necessary to abolish weak points.

The phase of model modification transfers the actual state into an ideal state. In general, the extended use of actual state models is far more effective and entails less expenditures. (Figure 5 shows the modeling sequences.)
For the participants, the results of a model-based business process reengineering project appear to include an effective support of teamwork, a fairly simple opportunity to participate and an easiness to follow and understand the entire project. However, it also has an effect on costs, time demand and quality of the project work and the concepts to be compiled.

project phases	system delimitation	model forming	model evaluation	model modification
project specification	determination of required model delimitation and detail			
goal determination		model of goals, i.e. in a goal tree		
analysis of the actual state + inquiry of the actual state		actual state description + main modelling views + special views		
+ identification of weak points			analysis of actual state description	
estimation of improvement potentials			comparison: actual state /goals	
development of concepts			comparison: actual state/ideal concepts	description of ideal state
realization				description of realized state

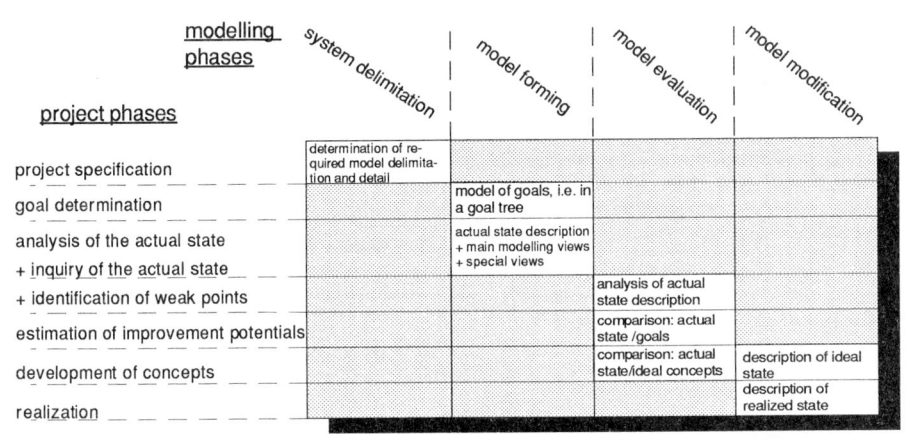

Figure 5 Relation between the Phases of Projects and Modeling Phases [16]

4 THE SUPPORTING SOFTWARE TOOL

Most modelling tools based on traditional approaches to enterprise modeling such as SSA (Structured Systems Analysis), SADT (Structured Analysis and Design Technique) and E/R (Entity/Relationship) complicate the design of business processes. Often, they are dependent on the existing organization structure. Data and Functions can never or only rarely be integrated. In the course of the process modeling functions are often associated directly with the existing organizational units. Process-organizational alternatives are difficult to describe. The models are not easily accepted in the different departments of the companies.

The object-oriented approach including prestructured model constructions facilitates the organization-overlapping analysis and optimization of business processes. Functions are not related to organizational units anymore, but to those objects that are to be processed. Data and Functions are integrated in one model. For example, the enterprise control, resources, the system support, the manufacturing process as well as their connections may all be portrayed integrally in one model. The easy to understand and transparent description of the business processes leads to a higher degree of acceptance in the departments concerned.

The tool MO²GO supports the objectoriented modeling with the IEM method. MO²GO stands for 'Method for the Object-Oriented Business Process Optimization'. The universal tool for the description and analysis of operational structures and business processes allows the comfortable description and the purposive analysis of products, orders, resources as well as the business processes belonging to them. The main advantages of the application of the tool include the systematization of the reengineering and optimization process and the reusability of the enterprise model for later projects with different objectives and optimization tasks.

Reengineering requires discussions between different project groups, within the respective project group, and between experts and managers in the enterprise and the project members. Graphic and text-based documents are provided which can serve as basis for communication between the participants of the project. The documents include directories of all modeled functions, objects, their documentation and their graphic representation [9].

To obtain immaculate printouts of the model different printing configurations are supported. To do justice to the multitude of relevant information and display requirements of individual areas concerned different views on an integrated model of the company may be selected. Business processes and the necessary information is described in a model core. The information and the model structure is stored in the core of the tool (Figure 6) as object classes and instances with their relations. User views related to the model core include information systems, the process organization, quality requirements and qualification profiles. These views are available in libraries of class structures and models. They are supported by the evaluation functions of the tool. Process-organizational alternatives and changes can described with regard to their changes of control, quality, system support, organizational structure and the qualification profiles of personnel. For example, the gradual transition from the actual state to the desired state was pointed out to a medium-sized company of the automobile industry. The transition from a central manufacturing control to a decentralized Kanban-controlled production with immediate customer-supplier-relations was described.

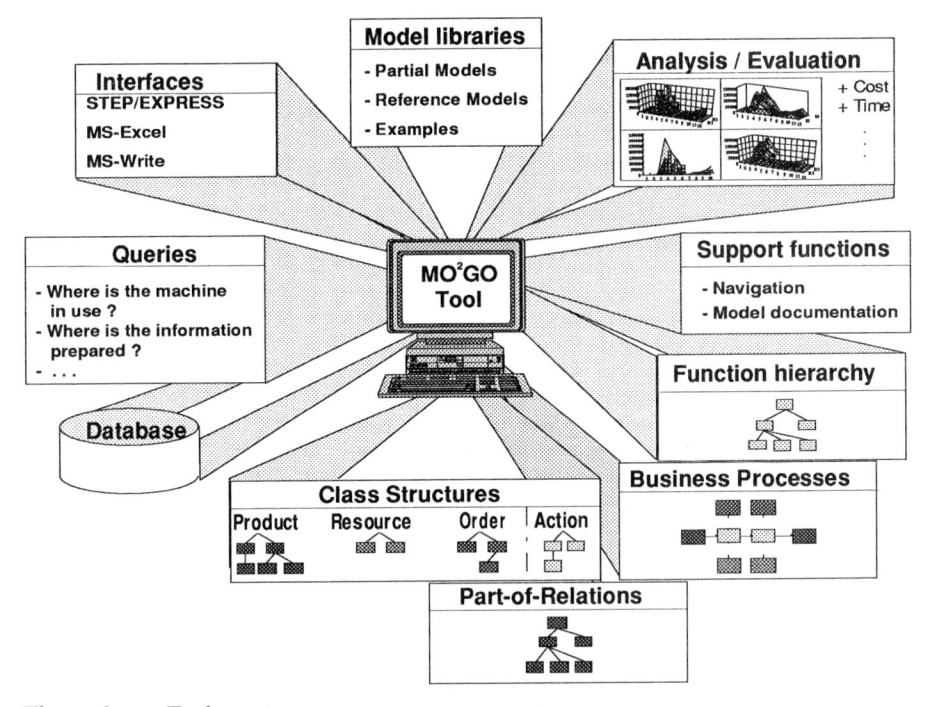

Figure 6 Tool structure

The user interface of the tool enables the simple, interactive design of enterprise models. Business processes and their connections are represented in appropriate windows where they can be refined (Figure 7). Mechanisms to design the models 'bottom-up' or 'top-down' in any combination are implemented. Class editors allow the description of company-specific characteristics of products, orders and resources. The user is enabled to define his own classes and descriptions of the characteristics. The description of the components of an object occurs at the appropriate classes as well, for example to generate bills of material.

The object-oriented approach supports the continuous reusability of partial models as modules of new models and the development of corresponding model libraries. Reference models and exemplary models for certain applications can be provided. An automatically generated model description enables the connection of different partial models and the development of interfaces with other tools. Consistency checks support the local consistency of the model.

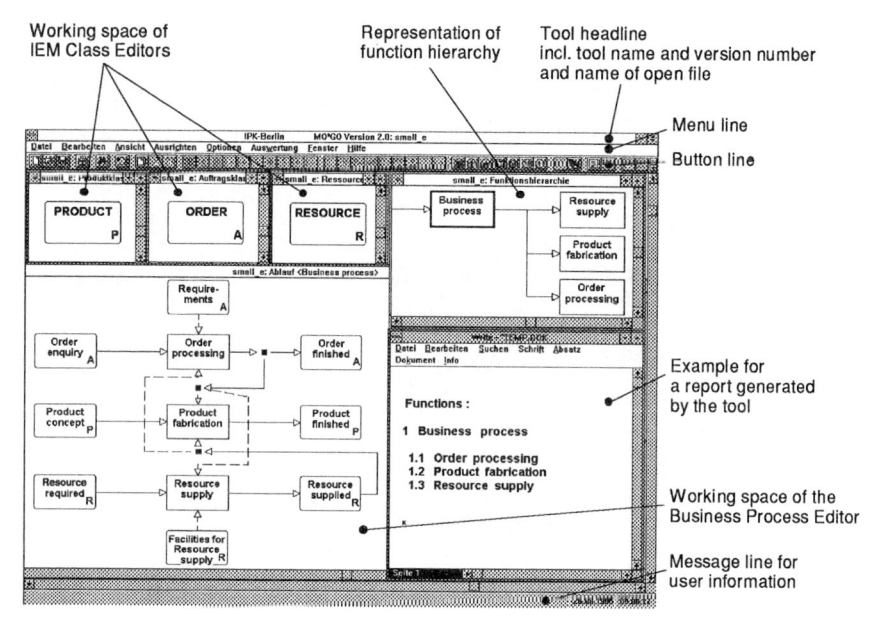

Figure 7   User interface of MO²GO

In the modeling process within the tool it is only necessary to model those things that are in the focus of interest. The user can employ the default structures. For example, it is only necessary to design classes if the user needs them in his approach. He can also use the generic classes 'product', 'order' and 'resource' directly. During the modeling and analyzing process changes occur every time for both the class and the process structure. The tool supports these changes by navigation and changing functionalities as well as by consistent checks.

IEM enables the modeling of product, order and resource processes within one model. Real models are typically large; the different process sequences and the relations between them could make the model complex. To handle this complexity the tool provides a functionality to fade-out model parts. Therefore, the user can focus on the process sequence of his interest and is also enabled to look at the entire model.

The analysis based on the model is supported by the evaluation functionality of the tool, e.g. the generation of specific tables or the measuring of an attribute such as 'process time' within a process sequence. Examples for specific tables include a resource, an order and a life phase table. The order table describes the modeled orders and the processes which produce an order. It also describes the processes which are to be controlled by an order. The same table can be generated for resources. The life phase table describes the values of the attributes of objects from a beginning state to their last state in a modeled process sequence. Tables are shown to the user by using a interface of the Microsoft-Windows program EXCEL.

In correspondence with the information represented in the IEM model within the tool, quality manuals , e.g. the structure of ISO9000ff documents, can be generated. The object-oriented approach enables the generation of these manuals by including additional class sets in an existing model and linking them with the process description. This is supported by the library functionality of the MO²GO tool.

The system environment for the employment of the tool MO²GO should include a personal computer with a 486 processor, at least 8 MB main storage, 10 MB free on the hard disc and MS-DOS/Windows 3.1® as the operating system. For the implementation of the user interface a special commercial class set is used which supports different platforms. The system core is implemented in pure C++, which enables an easy and fast movement to other platforms such as UNIX. The system architecture enables the availability of an external programming interface. The training costs for the tool functionality should be low because the user interface is oriented towards other MS-Windows software, e.g. WinWord.

The next versions will focus on additional evaluation mechanisms, an interface to a simulator and an interface to data base systems.

An interface to existing, actual enterprise data is being developed. It should make the process model available to other tools which are used in the entire enterprise. It could, for example, be used for operations scheduling. The use of actual data would reduce the modeling time for analysis and simulation. This would save time for the transfer of parameter values into the model. The interface specification EXPRESS/STEP (ISO 10303) is used to obtain a common interface to different enterprises and tools. STEP stands for 'Standard for the Exchange of Product Model Data'.

The presented tool MO²GO is available at the IPK-Berlin. The described method and tool is suitable for many planning and structuring tasks in companies. The application includes the design of material flows and information flows. In projects, the systematic and transparent description of business processes as communication base between the departments and between the different hierarchical levels proved to be successful. Among other things, time saving potentials were made clear. The distribution of costs was improved with regard to the respective 'initiators', the deployment of personnel was improved with regard to qualifications. Method and tool has been employed in various industrial projects of the IPK Berlin and it is also use by customers.

5 REFERENCES

[1] Spur, G.; Mertins, K.; Jochem, R.: Integrierte Unternehmensmodellierung. Beuth-Verlag, Berlin. 1993.

[2] Mertins, K.; Jochem, R.: An Object Oriented Method for Integrated Enterprise modelling as a Basis for Enterprise Coordination. International Conference on Enterprise Integration Modeling Technology (ICEIMT). Hilton Head (South Carolina), US Air Force-Integration Technology Division, June '92.

[3] Mertins, K.; Süssenguth, W.; Jochem, R.: Integration Information Modelling. In. Proceedings of Fourth IFIP Conference on Computer Applications in Production and Engineering (CAPE '91). Bordeaux, France. Elsevier Science Publisher B.V. (North Holland).

[4] Mertins, K.; Jochem, R.: Planning of Enterprise-Related CIM-Structures. In: Proceedings of 8th International Conference CARS and FOF. Metz; France, 17.-19. August 1992.

[5]   P. Coad, E. Yourdon: Object Oriented Analysis. Yourdon Press/Prentice Hall, Englewood Cliffs, NJ, 1990.

[6]   Mertins, K.; Süssenguth, W.; Jochem, R.: Modellierungsmethoden für rechnerintegrierte Produktionsprozesse (Hrsg.: G. Spur).
Carl Hanser Verlag. München, Wien. 1994.

[7]   Mertins, K.; Jochem, R.: Integrierte Unternehmensmodellierung - Basis für die Unternehmensplanung. DIN-Tagung. April 1993.

[8]   Süssenguth, W.: Methoden zur Planung rechnerintegrierter Produktionsprozesse. Dissertation. Berlin 1991.

[9]   Mertins, K.; Jochem, R.; Jäkel, F.-W.: Reengineering und Optimierung von Geschäftsprozessen. In. ZwF 89 (1994) 10, S. 479-481. Carl Hanser Verlag, München 1994.

[10] Scheer, August Willhelm:Wirtschaftsinformatik, Referenzmodelle für industrielle Geschäftsprozesse. Springer Verlag, Berlin 1995

[11] Scheer, A.; W.: Architektur Integrierter Informations Systeme. Springer Verlag, Berlin 1991

[12] Jacobson, I.; Ericsson, M.; Jacobson A.: The Object Advantage. Addison-Wesley, Bonn 1994

[13] Mertins, K.; Schwermer, M.: Modellierungsregeln zur Anwendung der Integrierten Unternehmensmodellierung, Version 1.1, QCIM-Projektbericht, IPK-Berlin 1994.

[14] Mertins, K.; Edeler, H.; Schwermer, M.: Model Based Analysis and Reengineering of Buisness Processes. In. Proceedings of IEPM'95, Marrakech, April 1995.

6   BIOGRAPHY

Dr.-Ing. K. Mertins
Born in 1947, Education in Electro-Mechanic, Study of Electrical Engineering at the Engineering School of Hamburg. Several Industrial Experiences as Electrical Engineer. Study of Economical Engineering at TU Berlin. 1984 Doctoral Thesis at TU Berlin. Since 1982 Head of Department and since 1988 Director of the Department Systems Planning at the Fraunhofer-Institute for Production Systems and Design Technology (IPK Berlin).

Dipl.-Ing. R. Jochem
Born in 1962, Study of Mechanical Engineering at TU Berlin. Industrial Experiences as Mechanical Engineer. Since 1988 resracher and since 1991 Group Leader for Manufacturing Integration at the Fraunhofer-Institute for Production Systems and Design Technology (IPK Berlin), Department Systems Planning.

INDEX OF CONTRIBUTORS

Alting, L. 31
Anderl, R. 395
Asama, H. 408
Astinov, Il. 183

Baba, Y. 106, 369
Bagrit, L. 565
Baum, T. 515
Beitz, W. 489
Birkhofer, H. 93, 432
Böhlke, U. H. 71

Caduff, G. 420
Caudill, R. J. 383
Cheng, J. 337
Chryssolouris, G. 131
Cook, H. E. 146
Cugini, U. 313

Deng, Z. 349

Ehrenkrantz, E. 383
Ehrmann, M. 515
Erb, J. 281
Eversheim, W. 71

Feldmann, K. 233
Fink, A. 158
Foy, M. 170

Gausemeier, J. 158
Geelink, R. 540
Geiger, K. 281
Grabowski, H. 281

Haase, T. 461
Haban, D. 461
Hadjijski, P. 183
Handel, D. 356
Hentschel, C. 56
Hiraoka, H. 408
Hirsch, B. E. 553
Houten, van F.J.A.M. 540

Itterheim, C. 356

Jacquet, L. 116
Jacucci, G. 170
Jochem, R. 589

Kals, H. J. J. 540
Kaniut, C. 444
Katzenmaier, J. 395
Kitiyama, T. 106
Kimura, F. 80
Kind, Chr. 14
Kiolen, I. 301
Kis, T. 195
Klocke, F. 325
Kochan, D. 293
Kohler, H. 444
Kölscheid, W. 71
Krause, F.-L. 14, 527
Kress, H. 301
Kuhlmann, T. 553

Leu, M. C. 383
Li, B. 467
Lüddemann, J. 527

Magee, R. 383
Mandorli, F. 313
Marchese, M. 170
Márkus, A. 195
Maßow, C. 553
Meedt, O. 233
Meerkamm, H. 501
Mertins, K. 589
Mourtzis, D. 131

Ning, R. 467
Nöken, S. 325

Oehlmann, R. 553
Ohkura, K. 256

Papakostas, N. 131
Preiss, K. 565

Rieger, E. 527
Rix, J. 301
Rommel, B. 356

Saito, D. 408
Sakao, T. 369
Sallez, Y. 116
Salomons, O. W. 540
Scharke, H. 221
Schlake, O. 158
Scholz-Reiter, B. 221
Schott, H. 93, 432
Sebastian, D. H. 383
Seliger, G. 56
Sirkar, K. K. 383
Slooten, van F. 540
Soenen, R. 116
Spath, D. 246
Spur, G. 3
Storm, T. 209
Storr, A. 356
Strobel, A. 461
Ströhle, H. 356
Suzuki, H. 80

Tahon, C. 576
Takata, S. 408
Tarassov, V. 90

Tchako, J. F. N. 576
Thoben, K.-D. 553
Tipnis, V. A. 43
Tomiyama, T. 106, 369
Tönshoff, H. K. 515
Trentesaux, D. 576
Tritsch, C. 246

Ueda, K. 256
Umeda, Y. 106, 369

Vallhagen, J. 475
Váncza, J. 195

Wagner, M. 56
Wang, J. 256
Wang, N. 337
Warman, E. A. 268
Weber, J. 501
Willemse, M. A. 209
Wirtz, H. 325

Xanthos, M. 383

Zussman, E. 221
Züst, R. 420

KEYWORD INDEX

Agility 566
Application in vehicle design 444
Architecture 349
Assembly 183, 209
  process 256
  process planning 475
  sequences 256
  task 256
Automated disassembly 246
Automation 116
Autonomy 553

Bayesian network 221
Business process reengineering 589

CAD 515
CAD design 268
CAD/CAM technologies 325
CALA 71
CAM 356
CAPP 515
Cellular machines 370
Clean production 293
Coating manufacturing 170
Communication 553
Communication technology 553
Competitive strategy 43
Competitiveness 566
Computer
  aided industrial design 527
  aided lifecycle analysis 71
  models 408
Concept and configuration 444
Conceptual requirements 527
Concurrent engineering 116, 467
Consequences 444
Control 553
Cooperation 553
Cooperative manufacturing 195
Coordination 553
CPR-Graph 256
Customers 146, 566

Decision making 183
Decision support systems 170

Decision support tool 146
Design 116, 566
  coordination 209
  environment 93, 432
  for assembly 209
  for environment 432
  for recyclability 501
  support 209
  system 501
Deterioration 80, 408
Dimensions 515
Direct slicing 313
Disassembly 56, 233
  planning 233
  planning and optimization 246
  for recovery 221
  tools 233
Distributed
  manufacturing 553
  problem solving 576
  scheduling 195
Durables 383
Dynamic modifying 467

Eco design 93
Eco-performance 420
Engineering data management 395
Engineering workbench 501
Environment orientated production 71
Environmental
  assessment 31
  information system 420
  knowledge 432
  management system 420
  performance evaluation (EPE) 420
  performance indicators (EPI) 420
  stewardship 43
Environmentally friendly products 432
Environmentally sound CAD 395
Evaluation method 71
Examples for the phases of use 444
Expert systems 170

Feature
  based design 540

definition 540
modelling 527
recognition 540
technology 515
validation 540
Form features 540
Future-robust visions 158

Genetic algorithms 170
Global engineering network 93
Goal state 256
Green
 browser 107
 information sharing 107
 life cycle design 107
 life cycle model 107

Hypermedia guidelines 93

Information
 conversion 93
 management 246
 model 432
 technology 14, 553
Initial state 256
Integral development 432
Integrated enterprise modelling 586
Integration 553
Interactive graphical user interface 356
Interdisciplinary research 56
ISO 14000ff 420

Kinematic motion 408
Knowledge intensive engineering 370
Knowledge systematization 370

Layer manufacturing 293
LCA at Mercedes-Benz AG 444
LCA methodology 444
Life cycle 14, 383, 408
 assessment (LCA) 71, 420
 costs 444
 design 432
 modeling 43
 processes 432
 product modelling 489

Maintenance 80, 408
Management duties 3
Management of ecological requirements 395
Manufacturing 131
Market model 195
Meta-data 467
Meta-information-services 93
Method 116
Methodical design approach 489
Methodology 349
Methods 475
Modeling 183
Modelling of requirements 281
Modelling tool 586
Multi-agent systems 195
Multicriteral optimisation 281
Multicriteria algorithms 576
Multiple-future 158

NC 515
NC programming using machining and
 measuring objects 356
Near net shape technologies 293
Negotiation 195
Network-thinking 158
New products 146

Object-oriented approach 589

Packaging line 576
Petri net 221
Planning 56
Plastics 383
Polymers 383
Post mass production paradigm 370
Product
 development 31, 158, 325
 design 80, 475
 life cycle 3, 80, 349, 370
 model 467
 modelling 80
 and process development 14
 realization 146
Production control 195, 576
Production planning 131
Profits 146

Keyword index

Quality 183
   control 80
   function deployment 146

Rapid
   prototyping 281, 293, 325
   shape definition 527
   product development 281
Re-engineering 383
Recycle 383
Recycling 3, 233, 444, 489
Requirements modeling 395

Scenario-management 158
Scenario-planning 158
Scheduling 131
Shortening operator 256
Simulation 183
Simultaneous engineering 475, 515
Soft machines 370
Solid freeform manufacturing 293
Solid model 313
Solution patterns 281
Specifications 116
Standardisation 553
STEP 356

Stereolithography 313
Strategic goals 146
Strategic quality deployment 146
Structured methodologies 146
Sustainable
   development 395
   industrial production 31
   manufacturing 349
   product and process design 420
System life cycle 349

Taguchi methods 146
Tessellation 301
Textile industry 131
Tolerances 515
Tooling 325
Tools 56
Total quality management 146

Utilisation problems 444

Value benchmarking 146
Virtual
   clay modelling 527
   organization 566

Available

Advanced CAD/CAM Systems
State-of-the-art and future trends in feature technology
Edited by René Soenen and Gus Olling

Benchmarking – Theory and Practice
Edited by Asbjørn Rolstadås

Computer Applications in Production Engineering
Proceedings of CAPE '95
Edited by Quangnan Sun, Zesheng Tang and Yijun Zhang

Simulation Games and Learning in Production Management
Edited by Jens O. Riis

Balanced Automation Systems
Architectures and design methods
Edited by Luis M. Camarinha-Matos and Hamideh Afsarmanesh

Artificial Intelligence in Reactive Scheduling
Edited by Roger Kerr and Elizabeth Szelke

Virtual Prototyping
Virtual environments and the product design process
Edited by Joachim Rix, Stefan Haas and José Teixeira

Forthcoming

Re-engineering the Enterprise
Edited by Jim Browne and David O'Sullivan

Integrated Manufacturing Systems Engineering
Edited by François B. Vernadat and Pierre Ladet

Life Cycle Modelling for Innovative Products and Processes
Edited by Helmut Jansen and Frank-Lothar Krause

Environmental Software Systems
Edited by Ralf Denzer

Formal Design Methods for CAD
Edited by John S. Gero

Knowledge-Intensive CAD
Volume 1
Edited by Tetsuo Tomiyama, Martti Mantyla and Susan Finger